Sustainable Construction

Sustainable Construction

Green Building Design and Delivery

Third Edition

Charles J. Kibert

WILEY

JOHN WILEY & SONS, INC.

Cover image: Photo courtesy of Nanyang Technological University, Singapore
Cover design: Andrew Liefer

This book is printed on acid-free paper. ♾

Published by John Wiley & Sons, Inc., Hoboken, New Jersey
Published simultaneously in Canada

For general information about our other products and services, please contact our Customer
Care Department within the United States at (800) 762-2974, outside the United States at
(317) 572-3993 or fax (317) 572-4002.

Wiley publishes in a variety of print and electronic formats and by print-on-demand. Some
material included with standard print versions of this book may not be included in e-books or
in print-on-demand. If this book refers to media such as a CD or DVD that is not included in the
version you purchased, you may download this material at http://booksupport.wiley.com. For
more information about Wiley products, visit www.wiley.com.

Library of Congress Cataloging-in-Publication Data:

Kibert, Charles J.
    Sustainable construction : green building design and delivery / Charles J. Kibert.—3rd ed.
      p. cm.
    Includes index.
    ISBN 978-0-470-90445-9 (cloth); ISBN 978-1-118-33013-5 (ebk); ISBN 978-1-118-33203-0 (ebk);
ISBN 978-1-118-33284-9 (ebk);ISBN 978-1-118-38457-2 (ebk); ISBN 978-1-118-38458-9 (ebk)
      1. Sustainable buildings—United States—Design and construction.   2. Green technology—
United States. 3. Sustainable architecture. I. Title. II. Title: Green building design and
delivery.
   TH880.K53 2012
   690.028'6—dc23

                                           2012023633

Printed in the United States of America
10 9 8 7 6 5 4 3 2 1

For Charles, Nicole, and Alina,
and in memory of two friends and sustainability stalwarts,
Ray Anderson and Gisela Bosch

Contents

Chapter 6

Part III

Chapter 7

Chapter 8

Chapter 12

Indoor Environmental Quality   389

Part IV

Green Building Implementation   433

Chapter 13

Construction Operations and Commissioning   435

Chapter 14

Green Building Economics   461

Foreword

The Roman architect, Vitruvius, once defined the purposes of architecture as creating commodity, firmness, and delight—roughly translated as usefulness, stability, and beauty. To that list, we now must add a fourth purpose, harmony, by which I mean the fit between buildings and the built environment broadly with the ecologies of particular places. In contrast to architecture as utilitarian or as form making, place making poses unique challenges. The first rule of place making is to ruin no other place. This requires considerable care, competence, and foresight in managing the upstream and the downstream effects of buildings from materials selection and construction to long-term operations and maintenance.

The challenge of creating commodity, firmness, delight, and harmony will be tougher in a world of 7 billion people predicted to grow to 10 billion by 2100 and facing worsening climate destabilization and its collateral economic, social, and political effects. In other words, ecological and economic constraints in the years ahead will limit what can be built, where, and how. Higher temperatures, larger storms, stronger winds, longer droughts, and rising sea levels will require more planning, better design, and more stringent engineering standards. Financial and climatic constraints could interact to diminish the role that architecture has played historically as a source of delight at a time when we will need a great deal of it. Vitruvius emphasized the importance of careful site selection for buildings and cities in order to maximize the salubrious effects of sun, wind, water, and shade. Those factors will become more important but less predictable in an age of rapid climate change. Moreover, designers can no longer assume that energy will be cheap and reliable. Military planners have said repeatedly that the US electric grid is highly vulnerable to terrorism, operator error, technological accident, and larger storms. Much the same could be said of the systems that provision us with water and food.

We have entered the rapids of human history and will need to respond with a new era of design. How architects, engineers, builders, and building managers respond to the new realities will have a larger impact on the human prospect than we thought even a few years ago. Building construction and operations are responsible for roughly 40 percent of global carbon emissions. If we are to make the necessary transition to climate stability, that number will have to decline dramatically as the number of buildings increases to accommodate a projected 40 percent rise in population. At the same time, the capacity of governments to respond to the climate emergency is being challenged both by those who want less government and by increasingly difficult economic circumstances. The upshot is that a great deal rides on the design and building professions and the private sector.

Against this background, the green building movement and the remarkable rise of the US Green Building Council and its counterparts elsewhere is a great success story, in no small measure due to the work of Charles Kibert and the Powell Center at the University of Florida in Gainesville. From modest beginnings in the 1990s to the present, the art and science of high-performance building is becoming the default for renovation and construction worldwide. It is now well documented that high-performance buildings have lower operating, maintenance, and environmental costs and generate better long-term economic values and higher human satisfaction and productivity.

The next design challenge is to take the logic, methodology, and economics of green building to a community, city, and regional scale with the goal of improving resilience, which is defined as the capacity of the system to "absorb disturbance and to undergo change and still retain essentially the same function, structure, and feedbacks."[1] It is a concept long familiar to engineers, mathematicians, ecologists, designers, and military planners. Resilient systems are characterized by redundancy so that failure of any one component does not cause the entire system to crash. They consist of diverse components that are easily repairable, widely distributed, cheap, locally supplied, durable, and loosely coupled. The goal of resilience raises questions that go beyond the specifics of single buildings to those having to do with how entire communities are provisioned with food, energy, water, materials, and livelihood in a more constrained and less predictable world.

Resilience as a design goal includes much of what is subsumed in the words "sustainable design," but differs in one critical respect. Sustainability is sometimes described as an end state as if it can be achieved once and for all. The goal of resilience, on the other hand, implies the capacity to make ongoing adjustments to changing political, economic, and ecological conditions. In practical terms, resilience is a design strategy that aims to reduce vulnerabilities by shortening supply lines, improving redundancy in critical areas, bolstering local capacity, and solving for a deeper pattern of dependence and disability. The less resilient the country, the more military power is needed to protect its far-flung interests and client states and hence the greater the likelihood of wars fought for oil, water, food, and materials. Resilient societies, on the other hand, do not send their young to fight and die in far-away battlefields because they lack wit, foresight, and design intelligence.

The goal of resilience presumes but does not end with sustainable building practices. It makes little sense to design high-performance buildings that exist as islands in a larger sea of unsustainability and that rest on a scaffolding of supply chains and infrastructure dependent on cheap fossil fuels. Design, accordingly, must be broadened to encompass a full spectrum of issues at a community and regional scale, including water, food, energy, education, economic development, policy and law, and urban planning. The challenge to designers now is to create the methodologies and practical tools to integrate diverse sectors, professions, and interest groups into systems so that each of the parts reinforces the resilience and durability of the whole community. "Full-spectrum" design is a fancy phrase to describe design strategies implicit in the writings of Lewis Mumford and Buckminster Fuller, as well as more recent work of contemporary architects and designers such as Bob Berkebile and William McDonough. It is a strategy rooted in the ancient meaning of the word "religion," which means "bind together." It is manifest in law and policy in the National Environmental Policy Act (1969) and in practical grassroots work such as the Transition Town movement that began in Totnes in the United Kingdom. In every instance, it is predicated on the belief that the whole is more than the sum of its parts and that we should take thought for the morrow.

David W. Orr

---

[1] Brian Walker and David Salt, *Resilience Thinking* (Washington, DC: Island Press, 2006), p. 32. See also the classic treatment of the subject by Amory and Hunter Lovins, *Brittle Power* (Andover, MA: Brick House Press, 1982), especially chapter 13; and Lovins et al., *Small Is Profitable* (Snowmass, CO: Rocky Mountain Institute, 2002).

Preface

Much has changed in the world of green building since the publication of the Second Edition of *Sustainable Construction: Green Building Design and Delivery*, and the Third Edition is an attempt to both capture the shifts in thinking and practice and to improve the content of this book. The enormous threat of climate change demands more attention to reducing energy consumption and understanding the carbon footprint of the built environment, and both of these issues have received significant attention in this edition. There is expanded coverage of the building hydrologic cycle, and stormwater considerations have been included in this discussion. All materials on LEED and Green Globes have been consolidated into chapters dedicated to these assessment systems. The major international building assessment systems such as BREEAM, Green Star, CASBEE, and DGNB have expanded coverage and case studies. The concept of net zero buildings has emerged in the last few years, and the subject of net zero energy and net zero water buildings is addressed in several chapters. Progress is being made toward producing green building standards and codes that, in essence, result in green buildings being standard in jurisdictions that adopt them. This shift in thinking and practice is covered in this edition.

In many of the chapters, a thought piece, or essay, by a top thinker in the field is included to further round out the discussion on a particular topic. I owe a special thanks to all the thought piece authors, who include Bill Reed, Ray Cole, Ravi Srinivasan, Brad Guy, John Chyz, and Kim Sorvig. A significant number of case studies from other countries, particularly Germany, are included to better define the cutting edge of high-performance green buildings. Helmut Meyer of Transsolar Energietechnik GmbH in Stuttgart provided truly excellent information and insights into the design of the very low energy buildings that are becoming commonplace in Germany. Christian Luft was extremely kind and helpful in describing the approach of his company, Drees & Sommer, located in Stuttgart, in designing high-performance green buildings that are certified via DGNB, the German building assessment system. Dr. Christine Lemaitre, CEO of DGNB, was generous with her time and provided current information about the progress of green building certification in the German building market. I also met Martin Haas and David Cook from Behnisch Architekten and was provided with a large number of case studies of high-performance buildings that their firm had designed over the past few years. I also was hosted by Stefan Zimmerman of the Karlsruhe Institute of Physics when I visited a new building on their campus in Karlsruhe designed by Behnisch Architekten. My good friend, Thomas Lützkendorf, who was one of the authors of the German building assessment program and who is in part responsible for the unique structure of DGNB, provided me with invaluable information and insights into the logic of this approach. Thomas was kind enough to spend most of a day with me and gave me the inspiration to pursue case studies in Germany. Additionally, Thomas has been a source of excellent information and input on the valuation of green buildings, a very important issue because the higher value of green construction is giving significant impetus to the major shift toward higher-performance buildings that is currently under way, especially in the United States.

The Third Edition has about double the graphics of the Second Edition, and a large number of organizations and companies were kind enough to permit the publication of their content in this edition. Thanks to all the contributors of these invaluable materials.

Thanks to Paul Drougas at John Wiley & Sons for once again guiding me through the publication process and to Mike New at John Wiley & Sons for keeping me on track. Donna Conte, Senior Production Editor at John Wiley & Sons, was instrumental in implementing the wide range of new material in this edition. The Third Edition would not have been possible without the enormous contributions of Tracy Wyman and Dustin Stephany, graduate assistants, who were extremely dedicated to helping produce a comprehensive, quality outcome. I owe an enormous debt to both of them for their very hard work and dedication.

Charles J. Kibert
Gainesville, Florida

Chapter 1

Introduction and Overview

Two decades after the first significant efforts to apply the *sustainability* paradigm to the built environment, the resulting *sustainable construction* movement has gained significant strength and momentum. In some countries, for example, the United States, there is growing evidence that this responsible and ethical approach is dominating the market for commercial and institutional buildings, including major renovations. Over 32,000 building projects have been registered with the US Green Building Council (USGBC), the major American proponent of built environment sustainability, in effect declaring the project team's intention to achieve the status of an officially recognized or certified green building. Nowhere has this shift been more evident than in American higher education. In August 2011, Harvard University announced its 50th green building certified in accordance with the requirements of the USGBC, including six projects with the highest, or platinum, rating and including more than 1.5 million square feet (484,000 square meters) of labs, dormitories, libraries, classrooms, and offices. An additional 3 million square feet (968,000 square meters) of space is registered and pursuing official recognition as green building projects. The sustainable construction movement is now international in scope, with almost 60 national green building councils establishing ambitious performance goals for the built environment in their countries (see Figure 1.1).[1] In addition to promoting green building, these councils develop and supervise building assessment systems that provide ratings for buildings based on a holistic evaluation of their performance against a wide array of environmental, economic, and social requirements. The outcome of applying sustainable construction approaches to creating a responsible built environment is most commonly referred to as *high-performance green buildings*, or simply, *green buildings.*

The contemporary high-performance green building movement was sparked by finding answers to two important questions: What is a high-performance green building? How do we determine if a building meets the requirements of this definition? The first question is clearly important—having a common understanding of what comprises a green building is essential for coalescing effort around this idea. The answer to the second question is to implement a *building assessment* or *building rating* system that provides detailed criteria and a grading system for these advanced buildings. The breakthrough in thinking and approach first occurred in 1989 in the United Kingdom with the advent of a building assessment system known as *BREEAM* (Building Research Establishment Environmental Assessment Method). BREEAM was an immediate success because it proposed both a standard definition for green building and a means of evaluating its performance against the requirements of the building assessment system. BREEAM represented the first successful effort at evaluating buildings on a wide range of factors that included not only energy performance but also water consumption, indoor environmental quality, location, materials use, environmental impacts, and contribution to ecological system health, to name but a few of the general categories that can be included in an assessment. To say that BREEAM was a success is a

Nation	Label
Australia	Nabers / Green Star
Brazil	AQUA / LEED Brasil
Canada	LEED Canada / Green Globes / Built Green Canada
Czech Rep	SBToolCZ
China	GBAS
Finland	PromisE
France	HQE
Germany	DGNB / CEPHEUS
Hong Kong	HKBEAM
India	Indian Green Building Council (IGBC) / (GRIHA)
Indonesia	Green Building Council Indonesia (GBCI) / Greenship
Italy	LEED / Italy / Protocollo Itaca / GBCouncil Italia
Japan	CASBEE
Jordan	EDAMA
Malaysia	GBI Malaysia

Nation	Label
Mexico	LEED Mexico
Netherlands	BREEAM Netherlands
New Zealand	Green Star NZ
Philippines	BERDE / Philippine Green Building Council
Portugal	Lider A
Taiwan	China Green Building Network
Singapore	Green Mark
South Africa	Green Star SA
South Korea	KGBC
Spain	VERDE
Switzerland	Minergie
United States	LEED / Living Building Challenge / Green Globes
UAE	Estidama
UK	BREEAM

Figure 1.1 Some of the almost 60 countries that either have or are developing green building assessment systems are shown in this diagram.

huge understatement because over 1 million buildings have been registered for certification and about 200,000 have successfully navigated the certification process. Canada and Hong Kong subsequently adopted BREEAM as the platform for their national building assessment systems, thus providing their building industries with an accepted approach to green construction. In the United States, the USGBC developed an American building rating system with the acronym *LEED* (Leadership in Energy and Environmental Design), which, when launched as a fully tested rating system in 2000, rapidly dominated the market for third-party green building certification. Similar systems were developed in other major countries, for example, *CASBEE* (Comprehensive Assessment System for Building Environmental Efficiency) in Japan (2004) and *Green Star* in Australia (2006). In Germany, which has always had a strong tradition of high-performance buildings, the German Green Building Council and German government collaborated in 2009 to develop a building assessment system known as *DGNB* (Deutsche Gesellschaft für Nachhaltiges Bauen), which is perhaps the most advanced evolution of building assessment systems. BREEAM, LEED, CASBEE, Green Star, and DGNB represent the cutting edge of today's high-performance green building assessment systems, both defining the concept of high performance and providing a scoring system to indicate the success of the project in meeting its sustainability objectives.

In the United States, the green building movement is often considered to be the most successful of all the American environmental movements. It serves as a template for engaging and mobilizing a wide variety of stakeholders to accomplish an important sustainability goal, in this case dramatically improving the efficiency, health, and performance of the built environment. The green building movement provides a model for other sectors of economic endeavor about how to create a consensus-based, market-driven approach that has rapid uptake, not to mention broad impact. This movement has become a force of its own and, as a result, is compelling professionals engaged in all phases of building design, construction, operation, financing, insurance, and public policy to fundamentally rethink the nature of the built environment.

The Shifting Landscape for High-Performance Buildings

As we enter the second decade of the 21st century, circumstances have changed significantly since the onset of the sustainable construction movement. In 1990, the global population was 5.2 billion, climate change was just entering the public consciousness, the United States had just become the world's sole superpower, and Americans were paying just $1.12 for a gallon of gasoline. Fast-forwarding almost a quarter century and the world's population is approaching 7.5 billion, the effects of climate change are becoming evident at a pace far more rapid than predicted, the global economic system is floundering with debt crises in the United States and Europe, Japan is recovering from the impacts of a tsunami and nuclear disaster, and energy and food prices are rapidly increasing. In spite of a global recession, gasoline prices in the United States are approaching $4.00 per gallon, and food prices are increasing at a rate of as much as 40 percent annually. According to the World Bank Food Index, food prices rose at an annualized rate of 32 percent in the first quarter of 2012, putting millions more at risk for starvation. The convergence of financial crises, climate change, and higher food and energy prices has produced an air of uncertainty that grips governments and institutions around the world. What is still not commonly recognized is that all of these problems are coupled and that population and consumption remain the twin horns of the dilemma that confronts humanity. Population pressures, increased consumption by wealthier countries, the understandable desire for a good quality of life among the 5 billion impoverished people on the planet, and the depletion of finite, nonrenewable resources are all factors in creating the wide range of environmental, social, and financial crises that are characteristic of contemporary life in the early 21st century (see Figure 1.2).

These changing conditions are affecting the built environment in significant ways. First, there is an increased demand for buildings that are resource-efficient, that use minimal energy and water, and whose material content will have value for future populations. In 2000, the typical office building in the United States consumed over 300 kilowatt-hours per square meter annually ($kWh/m^2/yr$); today's high-performance buildings are approaching 100 $kWh/m^2/yr$.[2] In Germany, the energy profiles of high-performance buildings are even more remarkable, in the range of 50 $kWh/m^2/yr$. It is important to recognize that reduced energy consumption generally causes a proportional reduction in climate change impacts. Reductions in water consumption in high-performance buildings are also noteworthy. A high-performance building in the United States can reduce potable water consumption by 50 percent simply by opting for the most water-efficient fixtures available, including high-efficiency toilets (HETs) and high-efficiency urinals (HEUs). By using alternative sources of water such as rainwater and graywater, potable water consumption can be reduced by another 50 percent, to one-fourth that of a conventionally designed building water system. This is also referred to as a Factor 4 reduction in potable water use. Similarly impressive impact reductions are emerging in materials consumption and waste generation.

Second, it has become clear over time that building location is a key factor in reducing energy consumption because transportation energy can amount to two times the operational energy of the building.[3] Not only does this significant level of energy for commuting have environmental impacts, but it also represents a significant cost for the employees who make the daily commute. It is clear that the lower the energy consumption of the building, the greater is the proportion of energy used in commuting. For example, a building that consumes 300 $kWh/m^2/yr$ of operational energy and 100 $kWh/m^2/yr$ of commuting energy by its occupants

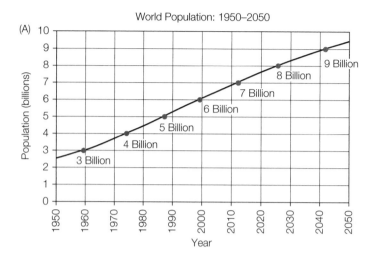

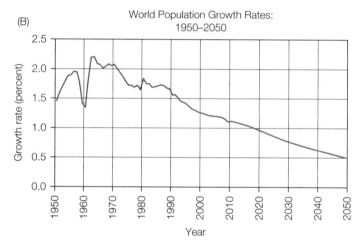

Figure 1.2 World population continues to increase, but the growth rate is declining, from about 1.2 percent in 2012 to a forecasted 0.5 percent in 2050. (*Source:* US Census Bureau, International Database, June 2011)

has 25 percent of its total energy devoted to transportation. A high-performance building in the same location with an energy profile of 100 kWh/m^2/yr and the same commuting energy of 100 kWh/m^2/yr now has 50 percent of its total energy consumed by transportation. Clearly, it makes sense to reduce transportation energy along with building energy consumption to have a significant impact on total energy consumption (see Figure 1.3).

Third, the threat of climate change is enormous (see Figure 1.4) and must be addressed across the entire life cycle of a building, including the energy invested in producing its materials and products and in constructing the building, commonly referred to as *embodied energy*. The energy invested in building materials and construction is significant, amounting to as much as 20 percent of the total life-cycle energy of the facility. Furthermore, significant additional energy is invested by maintenance and renovation activities during the building's life cycle, sometimes exceeding the embodied energy of the construction materials. Perhaps the most noteworthy effort to address the built environment contribution to climate change is the *Architecture 2030 Challenge* whose goal is to achieve a dramatic reduction in the greenhouse gas (GHG) emissions of the built environment by changing the way buildings and developments are planned, designed, and constructed.[4] The goal of the 2030 Challenge is to require that all new buildings and major renovations be climate neutral by 2030; that is, they will not use GHG-emitting fossil fuels for their operation. The *2030 Challenge for Product* addresses the GHG emissions of building materials and products and sets a goal of reducing the maximum carbon-equivalent footprint to 35 percent below the product category average by 2015 and eventually to 50 percent below the product category

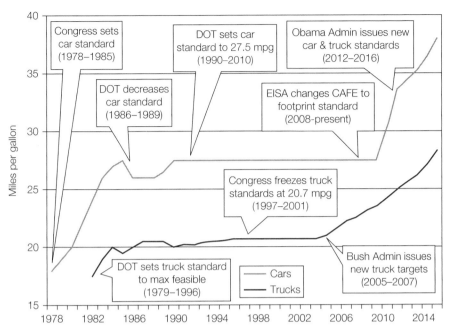

Figure 1.3 The fuel efficiency of US vehicles has languished, with federal standards improving dramatically in the 1970s due to the energy crises of that decade and more recently because of rising energy prices. More recent requirements have dramatically increased the miles per gallon of both automobiles and trucks. (Center for Climate and Energy Solutions)

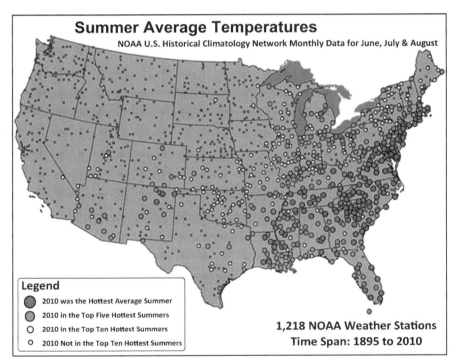

Figure 1.4 Record high temperatures are being experienced at an increasing rate in the United States. This diagram shows the locations where record summer average temperatures were experienced in 2010. (*Source:* National Oceanic and Atmospheric Administration.)

average by 2030. The emerging concept of *net zero energy* (NZE), which, in its simplest form, suggests that buildings generate as much energy from renewables as they consume on an annual basis, also supports the goals of the 2030 Challenge. Every unit of energy generated by renewables that displaces energy generated from fossil fuels results in less climate change impact. An NZE building would, in effect, have no climate change impacts due to its operational energy. It is clear that influencing energy consumption and climate change requires a comprehensive approach that addresses all forms of energy consumption, including operational energy, embodied energy, and commuting energy.

In summary, high-performance building projects are beginning to address three major shifts: (1) the demand for resource-efficient buildings, (2) the location of buildings to minimize transportation energy, and (3) the challenge of climate change. Building assessment systems such as LEED are being affected by these changes as is the very definition of green buildings. As time advances and more is learned about the future and its challenges, the design, construction, and operation of the built environment will adapt to meet this changing future landscape.

Sustainable Development and Sustainable Construction

The main impetus behind the high-performance green building movement is the sustainable development paradigm, which is changing not only physical structures but also the workings of the companies and organizations that populate the built environment, as well as the hearts and minds of the individuals who inhabit it.[5] Fueled by examples of personal and corporate irresponsibility and negative publicity resulting from events such as the collapse of the international finance system that triggered the Great Recession of 2008–2010, there is increased public concern about the behavior of private and public institutions. As a result, accountability and transparency are becoming the watchwords of today's corporate world. Heightened corporate consciousness has embraced comprehensive sustainability reporting as the new standard for corporate transparency. Corporate transparency refers to complete openness of companies about all financial transactions and all decisions that affect their employees and the communities in which they operate. Major companies such as DuPont, the Ford Motor Company, and Hewlett-Packard now employ triple bottom-line reporting,[6] which refers to a corporate refocus from mere financial results to a more comprehensive standard that includes environmental and social impacts. By adopting the cornerstone principles of sustainability in their annual reporting, corporations acknowledge their environmental and social impacts and ensure improvement in all arenas.

Still, other major forces such as climate change and the rapid depletion of the world's oil reserves threaten national economies and the quality of life in developed countries. Both are connected to our dependence on fossil fuels, especially oil. Climate change, caused at least in part by increasing concentrations of human-generated carbon dioxide, methane, and other gases in the earth's atmosphere, is believed by many authoritative scientific institutions and Nobel laureates to profoundly affect our future temperature regimes and weather patterns.[7] Much of today's built environment will still exist during the coming era of rising temperatures and sea levels; however, little consideration has been given to how human activity and building construction should adapt to potentially significant climate alterations. Global temperature increases must now be considered when forming assumptions about passive design, the building envelope, materials selection, and the types of equipment required to cope with higher atmospheric energy levels.

The *oil rollover point* describes the time when peak worldwide production of oil will occur and when approximately 50 percent of the world's oil supply will have been depleted (see Figure 1.5).[8] At the rollover point, the energy value of oil (the amount of energy into which the oil can be converted) will be less than the energy needed to extract it. According to the International Energy Agency (IEA), oil production likely peaked in 2006, and by 2035, the world's oil fields will be producing just 20 million barrels per day, a substantial drop in production.[9] In the early 1970s, the net energy for oil extraction was over 25 to 1;

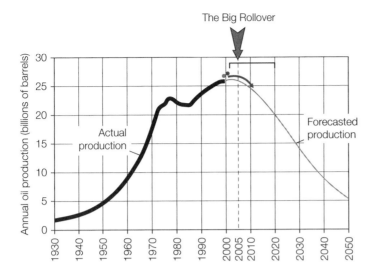

Figure 1.5 The oil rollover point is the year in which the worldwide production of oil will peak. The International Energy Agency reported that this likely occurred when oil production peaked at about 70 million barrels per day in 2006. By 2035, the IEA expects world oil production will have fallen to just 20 million barrels per day. (Illustration courtesy of Bilge Çelik)

that is, 25 units of energy were extracted for each unit of energy invested. Today, the net energy for oil has fallen to 7 to 1, and once it falls to under 3 to 1, it will no longer be economically feasible to extract oil in any significant quantities.

At the precise time that oil production is peaking, the emerging blockbuster economies of the so-called BRIC countries (Brazil, Russia, India, and China) are growing at an enormous pace and demanding their share of the planet's energy supplies. The Chinese economy grew at an official rate of 9.7 percent in 2011 with some estimates that it will continue at above 9 percent over the next few years. China produced about 2 million automobiles in 2000, tripling to about 6 million in 2005, and doubling again to 12 million in 2010. China's burgeoning industries are in heavy competition with the United States and other major economies for oil and other key resources such as steel and cement. The combination of increasingly scarce supplies of oil, the rapid economic growth in China and India, and concerns over the contribution of fossil fuel consumption to climate change will inevitably force the price of gasoline and other fossil fuel–derived energy sources to increase rapidly in the coming decades. At present, there are no foreseeable technological substitutes for the world's rapidly depleting oil supplies. Alternatives such as hydrogen or fuels derived from coal and tar sands threaten to be prohibitively expensive. The expense of operating buildings that are heated and cooled using fuel oil and natural gas will likely increase, along with the cost of fossil fuel–dependent industrial, commercial, and personal transportation. A shift toward hyperefficient buildings and transportation cannot begin soon enough.

The Vocabulary of Sustainable Development and Sustainable Construction

A unique vocabulary is emerging to describe concepts related to sustainability and global environmental changes. Terms such as *Factor 4* and *Factor 10, ecological footprint, ecological rucksack, biomimicry,* the *Natural Step, eco-efficiency, ecological economics, biophilia,* and the *precautionary principle* describe the overarching philosophical and scientific concepts that apply to a paradigm shift toward sustainability. Complementary terms such *as green building, building assessment, ecological design, life-cycle assessment, life-cycle costing, high-performance building,* and *charrette* articulate specific techniques in the assessment and application of principles of sustainability to the built environment.

The sustainable development movement has been evolving worldwide for almost 25 years, causing significant changes in building delivery systems in a relatively short period of time. A subset of sustainable development, sustainable construction, addresses the role of the built environment in contributing to the overarching vision of sustainability. The key vocabulary of this relatively new movement is discussed in the following sections and in Chapter 2. Additionally, a glossary of key terms is included at the end of this book.

SUSTAINABLE CONSTRUCTION

The terms *high performance, green*, and *sustainable construction* are often used interchangeably; however, the term *sustainable construction* most comprehensively addresses the ecological, social, and economic issues of a building in the context of its community. In 1994, the Conseil International du Bâtiment (CIB), an international construction research networking organization, defined sustainable construction as " . . . creating and operating a healthy built environment based on resource efficiency and ecological design."[10] The CIB articulated seven Principles of Sustainable Construction, which would ideally inform decision making during each phase of the design and construction process, continuing throughout the building's entire life cycle (see Table 1.1).[11] These factors also apply when evaluating the components and other resources needed for construction (see Figure 1.6). The Principles of Sustainable Construction apply across the entire life cycle of construction, from planning to disposal (here referred to as *deconstruction* rather than *demolition*). Furthermore, the principles apply to the resources needed to create and operate the built environment during its entire life cycle: land, materials, water, energy, and ecosystems.

GREEN BUILDING

The term *green building* refers to the quality and characteristics of the actual structure created using the principles and methodologies of sustainable construction. Green buildings can be defined as "healthy facilities designed and built in a resource-efficient manner, using ecologically based principles." Similarly, *ecological design, ecologically sustainable design*, and *green design* are terms that describe the application of sustainability principles to building design. Despite the prevalent use of these terms, truly sustainable green commercial buildings with renewable energy systems, closed materials loops, and full integration into the

TABLE 1.1

Principles of Sustainable Construction

1. Reduce resource consumption (reduce).
2. Reuse resources (reuse).
3. Use recyclable resources (recycle).
4. Protect nature (nature).
5. Eliminate toxics (toxics).
6. Apply life-cycle costing (economics).
7. Focus on quality (quality).

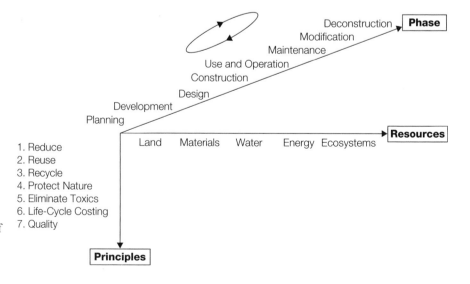

Figure 1.6 Framework for sustainable construction developed in 1994 by CIB Task Group 1 (Sustainable Construction) for the purpose of articulating the potential contribution of the built environment to the attainment of sustainable development. (Illustration courtesy of Bilge Çelik)

landscape are rare to nonexistent. Most existing green buildings feature incremental improvement over, rather than radical departure from, traditional construction methods. Nonetheless, this process of trial and error, along with the gradual incorporation of sustainability principles, continues to advance the industry's evolution toward the ultimate goal of achieving complete sustainability throughout all phases of the built environment's life cycle.

HIGH-PERFORMANCE BUILDINGS, SYSTEMS THINKING, AND WHOLE-BUILDING DESIGN

The term *high-performance building* has recently become popular as a synonym for green building in the United States. According to the Office of Energy Efficiency and Renewable Energy (EERE) of the US Department of Energy, a high-performance commercial building ". . . uses whole-building design to achieve energy, economic, and environmental performance that is substantially better than standard practice."[12] This requires that the design team fully collaborate from the project's inception in a process often referred to as *integrated design*.

Whole-building, or integrated, design considers site, energy, materials, indoor air quality, acoustics, and natural resources, as well as their interrelation with one another. In this process, a collaborative team of architects, engineers, building occupants, owners, and specialists in indoor air quality, materials, and energy and water efficiency uses systems thinking to consider the building structure and systems holistically, examining how they best work together to save energy and reduce the environmental impact. A common example of systems thinking is advanced daylighting strategy, which reduces the use of lighting fixtures during daylight, thereby decreasing daytime peak cooling loads and justifying a reduction in the size of the mechanical cooling system. This, in turn, results in reduced capital outlay and lower energy costs over the building's life cycle.

According to the Rocky Mountain Institute (RMI), a well-respected non-profit organization specializing in energy and building issues, whole-systems thinking is a process through which the interconnections between systems are actively considered and solutions are sought that address multiple problems. Whole-systems thinking is often promoted as a cost-saving technique that allows additional capital to be invested in new building technology or systems. RMI cites developer Michael Corbett, who applied just such a concept in his 240-unit Village Homes subdivision in Davis, California, completed in 1981. Village Homes was one of the first modern-era developments to successfully create an environmentally sensitive, human-scale residential community. The result of designing narrower streets was reduced stormwater runoff. Simple infiltration swales and on-site detention basins handled stormwater without the need for conventional stormwater infrastructure. The resulting $200,000 in savings was used to construct public parks, walkways, gardens, and other amenities that improved the quality of the community. A more recent example of systems thinking is Solaire, a 27-story luxury residential tower in New York City's Battery Park (see Figure 1.7). The façade of Solaire contains photovoltaic cells that convert sunlight directly into electricity, and the building itself uses 35 percent less energy than a comparable residential building. It provides its residents with abundant natural light and excellent indoor air quality. The building collects rainwater in a basement tank for watering roof gardens. Wastewater is processed for reuse in the air-conditioning system's cooling towers or for flushing toilets. The roof gardens not only provide a beautiful urban landscape but also assist in insulating the building to reduce heating and cooling loads. This interconnection of many of the green building measures in Solaire indicates that the project team carefully selected approaches that would have multiple layers of benefit, the core of systems thinking.[13]

Figure 1.7 Solaire, a 27-story residential tower on the Hudson River in New York City, is the first high-rise residential building in the United States specifically designed to be environmentally responsible. (Photograph courtesy of the Albanese Development Corporation)

Sustainable Design, Ecological Design, and Green Design

The issue of resource-conscious design is central to sustainable construction, which ultimately aims to minimize natural resource consumption and the resulting impact on ecological systems. Sustainable construction considers the role and potential interface with ecosystems to provide services in a synergistic fashion. With respect to materials selection, closing materials loops and eliminating solid, liquid, and gaseous emissions are key sustainability objectives. *Closed loop* describes a process of keeping materials in productive use by reuse and recycling rather than disposing of them as waste at the end of the product or building life cycle. Products in closed loops are easily disassembled, and the constituent materials are capable and worthy of recycling. Because recycling is not entirely thermodynamically efficient, dissipation of residue into the biosphere is inevitable. Thus, the recycled materials must be inherently nontoxic to biological systems. Most common construction materials are not completely recyclable, but rather *downcyclable*, for lower-value reuse such as for fill or road subbase. Fortunately, aggregates, concrete, fill dirt, block, brick, mortar, tiles, terrazzo, and similar low-technology materials are composed of inert substances with low ecological toxicity. In the United States, the 160 million tons (145 million metric tons) of construction and demolition waste produced annually make up about one-third of the total solid waste stream, consuming scarce landfill space, threatening water supplies, and driving up the costs of construction. As part of the green building delivery system, manufactured products are evaluated for their life-cycle impacts, to include energy consumption and emissions during resource extraction, transportation, product manufacturing, and installation during construction; operational impacts; and the effects of disposal.

LAND RESOURCES

Sustainable land use is based on the principle that land, particularly undeveloped, natural, or agricultural land (greenfields), is a precious finite resource, and its development should be minimized. Effective planning is essential to creating efficient urban forms and minimizing urban sprawl, which leads to overdependence on automobiles for transportation, excessive fossil fuel consumption, and higher pollution levels. Like other resources, land is recyclable and should be restored to productive use whenever possible. Recycling disturbed land such as former industrial zones (brownfields) and blighted urban areas (grayfields) back to productive use facilitates land conservation and promotes economic and social revitalization in distressed areas.

ENERGY AND ATMOSPHERE

Energy conservation is best addressed through effective building design, which integrates three general approaches: (1) designing a building envelope that is highly resistant to conductive, convective, and radioactive heat transfer; (2) employing renewable energy resources; and (3) fully implementing passive design. Passive design employs the building's geometry, orientation, and mass to condition the structure using natural and climatologic features such as the site's solar *insolation*,[14] thermal chimney effects, prevailing winds, local topography, microclimate, and landscaping. Since 40 percent of domestic primary energy[15] is consumed by buildings in the United States, increased energy efficiency and a shift to renewable energy sources can appreciably reduce carbon dioxide emissions and mitigate climate change.

WATER ISSUES

The availability of potable water is the limiting factor for development and construction in many areas of the world. In the high-growth Sun Belt and western regions of the United States, the demand for water threatens to rapidly outstrip the natural supply, even in normal, drought-free conditions.[16] Climate alterations and erratic weather patterns precipitated by global warming threaten to further limit the availability of this most precious resource. Since only a small portion of the earth's hydrologic cycle yields potable water, protection of existing groundwater and surface water supplies is increasingly critical. Once water is contaminated, it is extremely difficult, if not impossible, to reverse the damage. Water conservation techniques include the use of low-flow plumbing fixtures, water recycling, rainwater harvesting, and xeriscaping, a landscaping method that utilizes drought-resistant plants and resource-conserving techniques.[17] Innovative approaches to wastewater processing and stormwater management are also necessary to address the full scope of the building hydrologic cycle.

ECOSYSTEMS: THE FORGOTTEN RESOURCE

Sustainable construction considers the role and potential interface of ecosystems in providing services in a synergistic fashion. Integration of ecosystems with the built environment can play an important role in resource-conscious design. Such integration can supplant conventional manufactured systems and complex technologies in controlling external building loads, processing waste, absorbing stormwater, growing food, and providing natural beauty, sometimes referred to as *environmental amenity*. For example, the Lewis Center for Environmental Studies at Oberlin College in Oberlin, Ohio, uses a built-in natural system, referred to as a "Living Machine," to break down waste from the building's occupants; the effluent then flows into a reconstructed wetland (see Figure 1.8). The wetland also functions as a stormwater retention system, allowing pulses of stormwater to be

Figure 1.8 The Lewis Center for Environmental Studies at Oberlin College in Oberlin, Ohio, was designed by a team led by William McDonough, a leading green building architect, and including John Todd, developer of the Living Machine. In addition to the superb design of the building's hydrologic strategy, the extensive photovoltaic system makes it an NZE building. (Photograph courtesy of Oberlin College)

stored and thereby reducing the burden on stormwater infrastructure. The restored wetland also provides environmental amenity in the form of native Ohio plants and wildlife.[18]

Rationale for High-Performance Green Buildings

High-performance green buildings marry the best features of conventional construction methods with emerging high-performance approaches. Green buildings are achieving rapid penetration in the US construction market for three primary reasons:

1. Sustainable construction provides an ethical and practical response to issues of environmental impact and resource consumption. Sustainability assumptions encompass the entire life cycle of the building and its constituent components, from resource extraction through disposal at the end of the materials' useful life. Conditions and processes in factories are considered, along with the actual performance of their manufactured products in the completed building. High-performance green building design relies on renewable resources for energy systems; recycling and reuse of water and materials; integration of native and adapted species for landscaping; passive heating, cooling, and ventilation; and other approaches that minimize environmental impact and resource consumption.

2. Green buildings virtually always make economic sense on a life-cycle costing (LCC) basis, although they may be more expensive on a capital, or first-cost, basis. Sophisticated energy-conserving lighting and air-conditioning systems with an exceptional response to interior and exterior climates will cost more than their conventional, code-compliant counterparts. Rainwater harvesting systems that collect and store rainwater for nonpotable uses will require additional piping, pumps, controls, storage tanks, and filtration components. However, most key green building systems will recoup their original investment within a relatively short time. As energy and water prices rise due to increasing demand and diminishing supply, the payback period will decrease. LCC provides a consistent framework for determining the true economic advantage of these alternative systems by evaluating their performance over the course of a building's useful life.[19]

3. Sustainable design acknowledges the potential effect of the building, including its operation, on the health of its human occupants. A 2012 report from the Global Indoor Health Network suggested that, globally, about 50 percent of all illnesses are caused by indoor air pollution.[20] Estimates peg the direct and indirect costs of building-related illnesses, including lost worker productivity, as exceeding $150 billion per year.[21] Conventional construction methods have traditionally paid little attention to sick building syndrome (SBS), building-related illness (BRI), and multiple chemical sensitivity (MCS) until prompted by lawsuits. In contrast, green buildings are designed to promote occupant health, including measures such as protecting ductwork during installation to avoid contamination during construction; specifying finishes with low to zero volatile organic components to prevent potentially hazardous chemical off-gassing; more precise sizing of heating and cooling components to promote dehumidification, thereby reducing mold; and the use of ultraviolet radiation to kill mold and bacteria in ventilation systems.[22]

State and Local Guidelines for High-Performance Construction

At the onset of the green building movement, several state and local governments took the initiative in articulating guidelines aimed at facilitating high-performance construction. The Pennsylvania Governor's Green Government Council (GGGC) uses mixed but very appropriate terminology in its "Guidelines for Creating High-Performance Green Buildings." The lengthy but instructive definition of high-performance green building (see Table 1.2) focuses as much on the collaborative involvement of the stakeholders as it does on the physical specifications of the structure itself.[23]

Similar guidance is provided by the New York City Department of Design and Construction in its "High Performance Building Guidelines," in which the end product, the building, is hardly mentioned and the emphasis is on the strong collaboration of the participants (see Table 1.3).[24]

The "High Performance Guidelines: Triangle Region Public Facilities," published by the Triangle J Council of Governments in North Carolina, focuses on three principles:

- Sustainability, which is a long-term view that balances economics, equity, and environmental impacts
- An integrated approach, which engages a multidisciplinary team at the outset of a project to work collaboratively throughout the process
- Feedback and data collection, which quantifies both the finished facility and the process that created it and serves to generate improvements in future projects

TABLE 1.2

High-Performance Green Building as Defined by the Pennsylvania GGGC

A project created via cooperation among building owners, facility managers, users, designers, and construction professionals through a collaborative team approach.
A project that engages the local and regional communities in all stages of the process, including design, construction, and occupancy.
A project that conceptualizes a number of systems that, when integrated, can bring efficiencies to mechanical operation and human performance.
A project that considers the true costs of a building's impact on the local and regional environment.
A project that considers the life-cycle costs of a product or system. These are costs associated with its manufacture, operation, maintenance, and disposal.
A building that creates opportunities for interaction with the natural environment and defers to contextual issues such as climate, orientation, and other influences.
A building that uses resources efficiently and maximizes use of local building materials.
A project that minimizes demolition and construction wastes and uses products that minimize waste in their production or disposal.
A building that is energy-and resource-efficient.
A building that can be easily reconfigured and reused.
A building with healthy indoor environments.
A project that uses appropriate technologies, including natural and low-tech products and systems, before applying complex or resource-intensive solutions.
A building that includes an environmentally sound operations and maintenance regimen.
A project that educates building occupants and users to the philosophies, strategies, and controls included in the design, construction, and maintenance of the project.

TABLE 1.3

Goals for High-Performance Buildings According to the New York City Department of Design and Construction

Raise expectations for the facility's performance among the various participants.

Ensure that capital budgeting design and construction practices result in investments that make economic and environmental sense.

Mainstream these improved practices through (1) comprehensive pilot high-performance building efforts and (2) incremental use of individual high-performance strategies on projects of limited scope.

Create partnerships in the design and construction process around environmental and economic performance goals.

Save taxpayers money through reduced energy and material expenditures, waste disposal costs, and utility bills.

Improve the comfort, health, and well-being of building occupants and public visitors.

Design buildings with improved performance, which can be operated and maintained within the limits of existing resources.

Stimulate markets for sustainable technologies and products.

Like the other state and local guidelines, North Carolina's "High Performance Guidelines" emphasize collaboration and process, rather than merely the physical characteristics of the completed building. Historically, building owners assumed that they were benefiting from this integrated approach as a matter of course. In practice, however, the actual lack of coordination among design professionals and their consultants often resulted in facilities that were problematic to build. Now the green building movement has begun to emphasize that strong coordination and collaboration is the true foundation of a high-quality building. This philosophy promises to influence the entire building industry and, ultimately, to enhance confidence in the design and construction professions.

Green Building Progress and Obstacles

Until recently considered a fringe movement, in the early 21st century the green building concept has won industry acceptance, and it continues to influence building design, construction, operation, real estate development, and sales markets. Detailed knowledge of the options and procedures involved in "building green" is invaluable for any organization providing or procuring design or construction services. The number of buildings registered with the USGBC for a LEED building assessment grew from just a few in 1999 to more than 6000 registered and certified in late 2006. By 2012, the number of registered buildings had grown to over 32,000 and a total of over 8600 buildings had been certified. The area of LEED certified buildings increased from a few thousand square feet in 1999 to 1.7 billion square feet (177 million square meters) in 2012 for commercial buildings alone. Federal and state governments, many cities, several universities, and a growing number of private-sector construction owners have declared sustainable or green materials and methods as their standard for procurement.

Despite the success of LEED and the US green building movement in general, challenges abound when implementing sustainability principles within the well-entrenched traditional construction industry. Although proponents of green buildings have argued that whole-systems thinking must underlie the design phase of this new class of buildings, conventional building design and procurement processes are very difficult to change on a large scale. Additional impediments may also apply. For example, most jurisdictions do not yet permit the elimination of stormwater infrastructure in favor of using natural systems for stormwater control. Daylighting systems do not eliminate the need for a full lighting system, since buildings generally must operate at night. Special low-emissivity (low-E) window glazing, skylights, light shelves, and other devices increase project cost. Controls that adjust lighting to compensate for varying amounts of available daylight, and occupancy sensors that turn lights on and off depending on occupancy, add additional expense and complexity. Rainwater harvesting systems require dedicated piping, a storage tank or cistern, controls, pumps, and valves, all of which add cost and complexity.

Green building materials often cost substantially more than the materials they replace. Compressed wheatboard, a green substitute for plywood, can cost as much as 4 times more than the plywood it replaces. The additional costs, and those associated with green building compliance and certification, often require owners to add a separate line item to the project budget. The danger is that, during the course of construction management, when costs must be brought under control, the sustainability line item is one of the first to be "value-engineered" out of the project. To avoid this result, it is essential that the project team and the building owner clearly understand that sustainability goals and principles are paramount and that LCC should be the applicable standard when evaluating a system's true cost. Yet, even LCC does not guarantee that certain measures will be cost-effective in the short or long term. Where water is artificially cheap, systems that use rainwater or graywater are difficult to justify financially, even under the most favorable assumptions. Finally, more expensive environmentally friendly materials may never pay for themselves in a LCC sense.

A summary of trends in, and barriers to, green building is presented in Table 1.4. These trends are an outcome of the Green Building Roundtable, a forum held by the USGBC for members of the US Senate Committee on Environment and Public Works in April 2002, and they still apply today.[25]

TABLE 1.4

Trends and Barriers to Green Building in the United States

Trends

1. Rapid penetration of the LEED green building rating system and growth of USGBC membership
2. Strong federal leadership
3. Public and private incentives
4. Expansion of state and local green building programs
5. Industry professionals taking action to educate members and integrate best practices
6. Corporate America capitalizing on green building benefits
7. Advances in green building technology

Barriers

1. Financial disincentives
   a. Lack of LCC analysis and use
   b. Real and perceived higher first costs
   c. Budget separation between capital and operating costs
   d. Security and sustainability perceived as trade-offs
   e. Inadequate funding for public school facilities
2. Insufficient research
   a. Inadequate research funding
   b. Insufficient research on indoor environments, productivity, and health
   c. Multiple research jurisdictions

Book Organization

This book describes the high-performance green building delivery system, a rapidly emerging building delivery system that satisfies the owner while addressing sustainability considerations of economic, environmental, and social impact, from design through the end of the building's life cycle. A building delivery system is the process used by building owners to ensure that a facility meeting their specific needs is designed, built, and handed over for operation in a cost-effective manner. This book will examine the design and construction of state-of-the-art green buildings in the United States, considering the nation's unique design and building traditions, products, services, building codes, and other characteristics. Best practices, technologies, and approaches of other countries will be used to illustrate alternative techniques. Although intended primarily for a US audience, the general approaches described could apply broadly to green building efforts worldwide.

Much more so than in conventional construction delivery systems, the high-performance green building delivery system requires close collaboration among building owners, developers, architects, engineers, constructors, facility managers, building code officials, bankers, and real estate professionals. New certification systems with unique requirements must be considered. This book will focus largely on practical solutions to the regulatory and logistical challenges posed in implementing sustainable construction principles, delving into background and theory as needed. The USGBC's green building certification program will be covered in detail. Other complementary or alternative standards such as the Green Building Initiative's Green Globes building assessment system, the federal government's Energy Star program, and the United Kingdom's BREEAM building certification program will be discussed. Economic analysis and the application of life-cycle costing, which provides a more comprehensive assessment of the economic benefits of green construction, will also be considered.

Following this introduction, the book is organized into four parts, each of which describes an aspect of this emerging building delivery system. Part I, "Green Building Foundations," covers the background and history of green buildings, the basic concepts, ethical principles, and ecological design. Part II, "Assessing High-Performance Green Buildings," addresses the important issue of assessing or rating green buildings, with special emphasis on the two major US rating systems, LEED and Green Globes. Part III, "Green Building Design," more closely examines several important subsystems of green buildings: siting and landscaping, energy and atmosphere, the building hydrologic cycle, materials selection, and indoor environmental quality. In Part IV, "Green Building Implementation," the subjects of construction operations, building commissioning, economic issues, and future directions of sustainable construction are addressed. Additionally, several appendices containing supplemental information on key concepts are provided. To support the readers, a website, www.wiley.com/go/sustainableconstruction, contains hyperlinks to relevant organizations, references, and resources. This website also references supplemental materials, lectures, and other information suitable for use in university courses on sustainable construction.

Trends in High-Performance Green Building

Even though the high-performance green building movement is relatively new, there have already been several shifts in direction as more is learned about the wider impacts of building and the accelerating effects of climate change. Fifteen years ago at the onset of this revolution, the use of the charrette was a relatively new concept as were integrated design, building commissioning, the design-build delivery system, and performance-based fees. All of these are now familiar green building themes, and building industry professionals are familiar with their potential application. In a short span of time, much has changed. Energy prices are rising and likely to increase at an exponential rate as we experience the impacts of peak oil. New technologies such as high-efficiency photovoltaic systems and building information modeling (BIM) are affecting approaches to project design and collaboration. Evidence is mounting that climate change is occurring significantly faster than even the most pessimistic models predicted. Some fundamental thinking about green building assessment has changed, and there is significant impetus toward integrating life-cycle assessment (LCA) far more deeply into project evaluation. The impacts of building location are being taken into account since it has become apparent that the energy and carbon associated with transportation is approaching the levels resulting from construction and operation of the built environment. The following sections address these emerging trends in more detail and provide some insights into how they are affecting high-performance green buildings.

CARBON ACCOUNTING

By virtually all accounts, climate change seems to be accelerating and lining up with the worst-case scenarios hypothesized by scientists. One unexpected event that is rapidly increasing levels of atmospheric CO_2, the primary cause of climate change, is drought, which causes, among other things, the death of rain-forest trees. Researchers calculate that millions of trees died in 2010 in the Amazon due to what has been referred to as a 100-year drought. The result is that the Amazon is soaking up much less CO_2 from the atmosphere, and the dead trees are releasing all the carbon they accumulated over 300 or more years. The widespread 2010 drought followed a similar drought in 2005 (another 100-year drought), which itself will put an additional 5.5 billion tons (5 billion metric tons) of CO_2 into the atmosphere.[26] In comparison, the United States, the world's second largest producer of CO_2 behind China, emitted 6.0 billion tons (5.4 billion metric tons) of CO_2 from fossil fuel use in 2009. The two droughts will end up adding an estimated 14.3 billion tons (13 billion metric tons) to atmospheric carbon and are likely accelerating global warming.

In the last major report by the Intergovernmental Panel on Climate Change (IPCC) in 2007, estimated sea level rises were just 7 to 23 inches (18 to 45 centimeters) by 2100. However, a mere four years later a newer 2011 study presented by the International Arctic Monitoring and Assessment Program (IAMAP) found that feedback loops are already accelerating warming in the Far North, which will rapidly increase the rate of ice melt. As a result, the panel now estimates that sea levels could rise by as much as 5.2 feet (1.7 meters) by the end of the century. The only conclusion that can be reached by observing the many positive feedback loops influencing climate change is that all indicators point to a much higher rate of change than had been predicted.

The result of these alarming changes is that releases of CO_2 into the atmosphere are becoming an increasingly serious issue. Governments around the world are making plans to reduce carbon emissions, which entails tracking or accounting for carbon in order to limit its production. The built environment, with enormous quantities of *embodied energy*[27] and associated operational and transportation energy, is a ripe target for gaining control of global carbon emissions. It is likely that projects that can demonstrate significant reductions in total carbon emissions will be far better received than those with relatively high carbon footprints, which could conceivably be banned. New concepts such as *low-carbon, carbon-neutral*, and *zero-carbon buildings* are emerging in an effort to begin coping with the huge quantities of carbon emissions associated with the built environment. On the order of 40 percent of all carbon emissions are associated with building construction and operation, and it is likely that as much as another 20 percent could be attributable to transportation. Perhaps nowhere in the world has there been more interest and progress in low-carbon building than in the United Kingdom. The Carbon Trust was established by the government as a nonprofit company to take the lead in stimulating low-carbon actions, contributing to UK goals for lower carbon emissions, the development of low-carbon businesses, and increased energy security and associated jobs, with a vision of a low-carbon, competitive economy. We can expect to see control of carbon emissions and other measures to mitigate their impacts becoming an ever more prominent feature of high-performance green buildings.

NET ZERO BUILDINGS

In the early 1990s, William McDonough, the noted American green building architect and thinker, suggested that buildings should, among other things, "live off current solar income." Today, what seemed a rash prediction is becoming reality as the combination of high-performance buildings and high-efficiency, low-cost renewable energy technologies are providing the potential for buildings that, in fact, can live off current solar income. These are commonly now referred to as net zero energy (NZE) buildings. In general, these are grid-connected buildings that export excess energy produced during the day and import energy in the evenings, such that there is an energy balance over the course of the year. As a result, NZE buildings have a zero annual energy bill with the added bonus that they are considered to be carbon neutral with respect to their operational energy.

An excellent example of an NZE building is the Research Support Facility (RSF) designed and built for the National Renewable Energy Laboratory (NREL) in Golden, Colorado. The RSF, completed in 2011, is a 220,000-square-foot (23,000-square-meter), four-story building with a photovoltaic (PV) system on-site. It is interesting to note that a 2007 NREL study concluded that one-story buildings could achieve NZE if the building roof alone were used for the PV system but that it would be extremely difficult for two-story buildings to meet this goal.[28] Clearly, much has been learned in a short time because the RSF has four stories, twice the limit suggested by NREL's own research. The Energy Use Index (EUI) of the RSF is just 32,000 BTU/ft^2/yr (66 kWh/m^2/yr), making it a very low energy building with the potential for producing enough PV energy to meet all its annual energy needs (see Figure 1.9A–D). The relatively narrow building floor plate, just 60 feet (19.4 meters) wide, enables daylighting and natural ventilation for its 800 occupants, and 100 percent of the workstations are daylit. Building orientation and geometry minimize east and west glazing. North and south glazing is optimally sized and shaded to provide daylighting while minimizing unwanted heat losses and gains. The building uses triple-glazed operable windows and window shading to address different orientations and positioning of its glazed openings. The operable windows can be used by the occupants to provide natural ventilation and cooling for the building. Electrochromic windows, which can be darkened using a small amount of electrical current, are used on the west side of the building to control glare and heat gain. The RSF has approximately 42 miles of radiant piping embedded in all floors of the building to provide water for radiant cooling and heating the majority of the work spaces. This radiant system provides thermal conditioning for the building at a fraction of the energy costs of the forced-air systems used in most office buildings. A thermal storage labyrinth under the RSF stores heating and cooling in its concrete structure and is integrated into the building energy recovery system. Outdoor air is heated by a transpired solar collector system located on the façade of the structure. Approximately 1.6 megawatts of on-site photovoltaics are being installed and dedicated to RSF use. Rooftop photovoltaic power will be added through a Power Purchase Agreement (PPA), and photovoltaic power from adjacent parking areas will be purchased through arrangement with a local utility. The RSF was awarded a LEED platinum rating in recognition of the success of its integrated design and the holistic approach of the project team.

The implementation of NZE is now national policy, and the US Department of Energy has programs in place with the objective that all new buildings will be NZE by 2050. In some local jurisdictions such as Austin, Texas, new homes are required to be NZE by 2015. The American Society of Heating, Refrigerating and Air-Conditioning Engineers (ASHRAE) proposed building energy label, known as Energy Quotient, reserves its highest rating for NZE buildings. This is an important new trend that appears to have significant momentum and will also influence the direction of green building evolution.

Figure 1.9 (A) The NREL Research Support Facility in Golden, Colorado, is a four-story NZE building that combines low-energy design with high-efficiency photovoltaics to produce all the energy it requires over the course of a year. (*Source:* National Renewable Energy Laboratory)

Figure 1.9 (B) Ground view of the air intake structure that conducts outside air into the thermal storage labyrinth in the crawl space of the NREL RSF. (*Source:* National Renewable Energy Laboratory)

Figure 1.9   (C) The daylighting system for the NREL RSF was designed using extensive simulation. Shading devices were carefully placed on the exterior and interior to manage both direct and indirect sunlight, distributing it evenly to create a bright, pleasant working environment. (*Source:* National Renewable Energy Laboratory)

Figure 1.9   (D) The fenestration for the NREL RSF was designed to provide excellent daylighting while controlling glare and unwanted solar thermal gain through the use of shading devices, recessed windows, and electrochromic glass. Operable windows allow the occupants to control their thermal comfort and obtain fresh air. (*Source:* National Renewable Energy Laboratory)

BUILDING INFORMATION MODELING

The emergence of building information modeling (BIM) as a design and visualization tool is an important trend for the building industry. Its three-dimensional modeling promises to provide owners with a far better representation of their projects, increase the quality of both design and construction, and increase the speed of construction. BIM makes the handling of complex projects with enormous information requirements far easier. One of the attributes of high-performance green building projects is their reliance on significant additional modeling, additional specification requirements, and the need to track aspects of the construction process such as construction waste management, indoor air quality protection during construction, and erosion and sedimentation control. Additionally, quantities of recycled materials, emissions from materials, and other data must be gathered for green building certification. BIM has the capability of accepting plug-ins that can perform energy modeling and daylighting simulation and provide a platform for the data required by green building certification bodies. BIM software can be used to relatively easily select the optimum site and building orientation to maximize renewable energy generation and daylighting and minimize energy consumption. BIM is an important and potentially powerful tool that can further increase the uptake of green buildings by lowering costs. Although not strictly relevant to green building certification, it makes the process far easier and less costly by providing "one-stop shopping" for information.

LIFE-CYCLE ASSESSMENT

Although a mature concept, life-cycle assessment (LCA) is growing in importance because it allows the quantification of the environmental impacts of design decisions that span the entire life of the project. In the past, LCA has been used to compare products and building assemblies, which provided some indication of how to improve decision making but did not provide information about the long-term effects resulting from building operation. With the emergence of the German DGNB building assessment system, the environmental performance of the whole building—its materials, construction, operation, disposal, and transportation impacts—can be quantified and compared to baselines that have been compiled to allow comparisons. Designers can rapidly consider a wide variety of alternative building systems, materials, and sites and compare them to the norms for the type of building being considered. For example, the global warming and ozone depletion potentials for various alternatives per square meter of building area can be compared to find the least damaging outcome. The Australian Green Star building assessment system considers energy not in energy units but in carbon dioxide equivalents to focus on the impact of climate change. LCA affords the design team the capability of quickly evaluating their energy strategies to find one that improves on the baselines established for carbon or other parameters. In North America, LCA is rewarded to some extent in the Green Globes rating system and is part of the new standard based on the Green Globes rating system and promulgated by the American National Standards Institute (ANSI) and the Green Building Initiative (GBI): ANSI/GBI 01-2010, Green Building Assessment Protocol for Commercial Buildings. LCA is also included as a pilot credit in the LEED system, although a decision has not been made as to whether it will be incorporated fully into the next major revision. The state of California also included LCA as a voluntary measure in its 2010 draft Green Building Standards Code. In the future, as governments struggle to cope with reducing GHG emissions because the effects of climate change are causing economic problems and social dislocations, it is likely that LCA will become a mandatory area of evaluation for building design.

Case Study: Kroon Hall, Yale University, New Haven, Connecticut

Ten years after the onset of the green building revolution in the United States, there are numerous examples of outstanding high-performance buildings, many of which have been awarded the platinum LEED rating, the highest accolade awarded by the USGBC. Although it is difficult to pick the best from among this group of facilities, Kroon Hall at Yale University in New Haven, Connecticut, certainly would be a contender for outstanding high-performance green building. The new $33.5 million building houses the School of Forestry and Environmental Studies at Yale University, the home of such luminaries as Tom Graedel and Stephen Kellert, leaders and provocative thinkers of the contemporary sustainability movement. Kroon Hall is located on the site of a decommissioned power plant, derelict parking lot, and network of service roads that has been transformed into a highly visible place for the study of the environment on Yale's Science Hill Campus. Native plants, shade trees, and walking paths were employed to create a parklike landscape on the site of a former brownfield (see Figure 1.10A–E).

According to Hopkins Architects, the London-based firm that has created a number of other noteworthy high-performance buildings around the world, the similarity to an elegant New England barn was not intentional, but the design certainly fits the character of its New England surroundings. The building has a narrow profile and east–west orientation that contributes to maximizing opportunities for daylighting and renewable energy generation, while at the same time enabling passive heating and cooling. To maximize daylighting, the design team decided to locate the building in the middle of the block in which it sits rather than at the end in order to prevent shading from adjacent structures. Breyer Hill Sandstone is used on the north and south façades, and its vaulted roof is supported by glue-laminated

Figure 1.10   (A) Kroon Hall at Yale University in New Haven, Connecticut, is located on a former industrial area that included a decommissioned power plant, a derelict parking lot, and a network of service roads. A 100-kW photovoltaic array on the roof provides about 25 percent of the building's electricity needs. (Robert Benson Photography)

e was transformed into an attractive
new landscaping around Kroon Hall
d to the research and instructional missions
ntal Studies. [(B) Michael Taylor, Hopkins

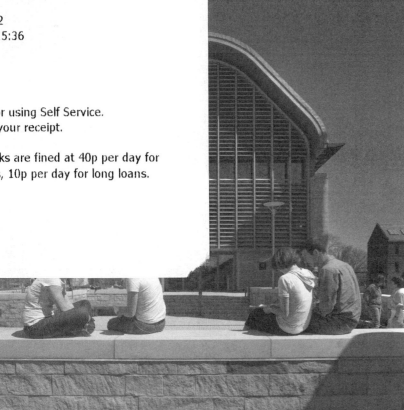

Figure 1.10   (D) The south side of the building is recessed, and window overhangs are integrated into the façade of the building to control glare and thermal loads while maximizing daylighting. (Photograph by Morley Von Sternberg)

Figure 1.10 (E) The daylighting strategy for Kroon Hall produces spectacular results and creates a pleasant connection to the outdoors. (Robert Benson Photography)

beams. The clever use of horizontal shading along the south façade allows solar heat gain in the winter while blocking the heat gain and glare in the summer. Spandrel panels consisting of low-E insulated glass units at the exterior, a 3-inch (8-centimeter) airspace, and a 2.5-inch (6-centimeter) space filled with translucent aerogel insulation were used as part of the building's façade. These remarkable panels transmit 20 percent of visible light, while offering an insulation value at their center of more than R-20. The average insulation value of the curtain wall is about R-8, about four times better than that of a conventional curtain wall.

Kroon Hall is designed to consume 50 percent of the energy of a conventionally designed academic building and to reduce greenhouse gas emissions by 62 percent. The building is conditioned by a displacement ventilation system which introduces air through the floor at low velocity, providing very quiet spaces. Low-velocity fans in the basement circulate air almost imperceptibly and use relatively little energy. A 100-kW photovoltaic array on the rooftop supplies 25 percent of the building's electricity needs; the remainder of the required electricity is purchased from renewable energy sources to help meet the goal of carbon neutrality. Four solar thermal panels are located in the south façade to help provide the building with hot water. Heating and cooling are provided by ground-source heat pumps connected to four 1500 feet (484 meters) deep wells, located near the building. During the fall and spring, the mechanical systems of the building are shut down, and color-coded lights are used to prompt the building occupants to open the windows for cooling and ventilation. Other strategies used to reduce energy consumption include evaporative cooling, operable windows, and exposed concrete slabs that serve as energy sinks to both buffer temperature swings and reduce energy consumption.

A rainwater harvesting system conducts water from the roof and grounds to a courtyard, where aquatic plants filter out sediment and contamination; the water is used for landscape irrigation and for flushing the toilets. The building's hydrologic cycle strategy, which includes the rainwater harvesting system, is predicted to save more than 600,000 gallons (2.3 million liters) of potable water per year. Waterless urinals and low-flow faucets together with toilets flushed by rainwater result in an 81 percent reduction in total potable water consumption for the building. To reduce stormwater runoff, a green roof was installed on one of the galleries, and porous asphalt was used for all the walkways on the site. The green roof also decreases the building's cooling load and limits the urban heat island effect of the project while at the same time providing a pleasant view for occupants and visitors.

Although it was not the primary goal of the project team, Kroon Hall has achieved a platinum LEED rating from the USGBC. The result of an excellent integrated design process was that the team was able to weave green strategies throughout the project in an intelligent and fruitful manner. As a result, Kroon Hall is an exemplar not only for high-performance green building but also for architecture.

Summary and Conclusions

The rapidly evolving and exponentially growing green building movement is arguably the most successful environmental movement in the United States today. In contrast to many other areas of environmentalism that are stagnating, sustainable building has proven to yield substantial beneficial environmental and economic advantages. Despite this progress, however, there remain significant obstacles, caused by the inertia of the building professions and the construction industry and compounded by the difficulty of changing building codes. Industry professionals, in both the design and construction disciplines, are generally slow to change and tend to be risk-averse. Likewise, building codes are inherently difficult to change, and fears of liability and litigation over the performance of new products and systems pose appreciable challenges. Furthermore, the environmental or economic benefit of some green building approaches has not been scientifically quantified, despite their often intuitive and anecdotal benefits. Finally, lack of a collective vision and guidance for future green buildings, including design, components, systems, and materials, may affect the present rapid progress in this arena.

Despite these difficulties, the robust US green building movement continues to gain momentum, and thousands of construction and design professionals have made it the mainstay of their practices. Numerous innovative products and tools are marketed each year, and in general, this movement benefits from an enormous air of energy and creativity. Like other processes, sustainable construction may one day become so common that its unique distinguishing terminology may be unnecessary. At that point, the green building movement will have accomplished its purpose: to transform fundamental human assumptions that create waste and inefficiency into a new paradigm of responsible behavior that supports both present and future generations.

Notes

1. The World Green Building Council (WGBC) is a forum for national green building councils that provides a platform for collaboration, exchange of research, and support for developing building assessment systems. The WGBC website is at www .wgbc.org. The diagram in Figure 1.1 is derived from the approach used by Christian Luft of Drees & Sommer in some of his presentations on green building assessment.
2. The energy consumption figures for buildings in the United States refer to purchased or metered energy.

3. See *Environmental Building News* (2007).

4. The Architecture 2030 Challenge was started by Ed Mazria in 2002. A parallel effort known as the 2030 Challenge for Products was initiated in 2011 to reduce the contributions of building materials to climate change.

5. The origin of the word *sustainability* is an item of controversy. In the United States, sustainability was first defined in 1981 by Lester Brown, a well-known American environmentalist and for many years the head of the Worldwatch Institute. In "Building a Sustainable Society," he defined a sustainable society as ". . . one that is able to satisfy its needs without diminishing the chance of future generations." In 1987, the Brundtland Commission, headed by then prime minister of Norway, Gro Harlem Brundtland, adapted Brown's definition, referring to sustainable development as ". . . meeting the needs of the present without compromising the ability of future generations to meet their needs." Sustainable development, or sustainability, strongly suggests a call for intergenerational justice and the realization that today's population is merely borrowing resources and environmental conditions from future generations. In 1987, the Brundtland Commission's report was published as a book, *Our Common Future*, by the UN World Commission on Environment and Development.

6. The World Business Council for Sustainable Development (WBCSD) promotes sustainable development reporting by its 170-member international companies. The WBCSD is committed to sustainable development via the three pillars of sustainability: economic growth, ecological balance, and social progress. Its website is www.wbcsd.org.

7. In November 1992, more than 1700 of the world's leading scientists, including the majority of the Nobel laureates in the sciences, issued the "World Scientists' Warning to Humanity." The preamble of this warning stated: "Human beings and the world are on a collision course. Human activities inflict harsh and often irreversible damage on the environment and critical resources. If not checked, many of our current practices put at serious risk the future that we wish for human society and the plant and animal kingdoms, and may so alter the living world that it may be unable to sustain life in the manner we know. Fundamental changes are urgent if we are to avoid the collision our present course will bring about." The remainder of this warning addresses specific issues, global warming among them, and calls for dramatic changes, especially on the part of the high-consuming developed countries, particularly the United States.

8. See, for example, Campbell and Laherrere (1998).

9. The information about peak oil production is from the *World Energy Outlook 2010* published by the International Energy Agency (IEA).

10. At the First International Conference on Sustainable Construction held in Tampa, Florida, in November 1994, Task Group 16 (Sustainable Construction) of the CIB formally defined the concept of sustainable construction and articulated six principles of sustainable construction, later amended to seven principles.

11. Sustainable construction and the model are described in Kibert (1994).

12. The *Whole Building Design Guide* can be found at www.wbdg.org.

13. The design approach used in creating Solaire in Battery Park, New York City, plus updates on construction progress, can be found at www.batteryparkcity.org. Another website with detailed information and illustrations is www.thesolaire.com.

14. *Insolation* is an acronym for incoming solar radiation.

15. Primary energy accounts for energy in its raw state. The energy value of the coal or fuel oil being input to a power plant is primary energy. The generated electricity is metered or purchased energy. For a 40 percent efficient power plant, 1 kWh of purchased electricity requires 2.5 kWh of primary energy.

16. A description of the severe water resource problems beginning to emerge even in water-rich Florida can be found in the May/June 2003 issue of *Coastal Services*, an online publication of the National Oceanic and Atmospheric Administration (NOAA) Coastal Services Center, available at www.csc.noaa.gov/magazine/2003/03/florida.html. A similar overview of water problems in the western United States can be found in Young (2004).

17. An overview of xeriscaping and the seven basic principles of xeriscaping can be found at http://aggie-horticulture.tamu.edu/extension/xeriscape/xeriscape.html.

18. The Adam Joseph Lewis Center for Environmental Studies at Oberlin College was designed by a highly respected team of architects, engineers, and consultants and is a cutting-edge example of green buildings in the United States. An informative website, www.oberlin.edu/envs/ajlc, shows real-time performance of the building and its photovoltaic system.

19. "The Cost and Benefits of Green Buildings," a 2003 report to California's Sustainable Buildings Task Force, describes in detail the financial and economic benefits of green buildings. The principal author of this report is Greg Kats of Capital E. Several other reports on this theme by the same author are available online. See the references for more information.

20. As stated in "Indoor air quality in our homes, schools and workplace questioned," at the website of Environmental Protection on April 12, 2012 at http://eponline.com/articles/2012/04/26/new-report-about-indoor-air-quality-in-our-homes-schools-and-workplaces.aspx.

21. The losses are estimated productivity losses as stated by Mary Beth Smuts, a toxicologist with the US Environmental Protection Agency, in Zabarsky (2002).

22. From "Ultra-Violet Radiation Could Reduce Office Sickness" (2004).

23. See "Guidelines for Creating High-Performance Green Buildings" (1999).

24. Excerpted from "High Performance Building Guidelines" (1999).

25. The outcomes of the Green Building Roundtable can be found in *Building Momentum* (2003).

26. The quantities of CO_2 released by Amazonian forests dying from drought is from a study by Lewis (2011) and his colleagues at the University of Leeds.

27. The embodied energy of a product refers to the energy required to extract raw materials, manufacture the product, and install it in the building, and includes the transportation energy needed to move the materials comprising the product from extraction to installation.

28. As suggested in Griffith et al. (2007).

References

Brown, Lester. 1981. *Building a Sustainable Society.* New York: Norton.

Campbell, C. J., and J. H. Laherrere. 1998. "The End of Cheap Oil." *Scientific American* 273(3): 78–83.

City of New York Department of Design and Construction. 1999. "High Performance Building Guidelines." Available at www.nyc.gov/html/ddc/downloads/pdf/guidelines.pdf.

Environmental Building News. 2007. "Driving to Green Buildings: The Transportation Energy Intensity of Buildings."16 (9).

Griffith, B., N. Long, P. Torcellini, and R. Judkoff. 2007. *Assessment of the Technical Potential for Achieving Net-Zero Energy Buildings in the Commercial Sector.* National Renewable Laboratory Technical Report NREL/TP-550–41957. Available at www.nrel.gov/docs/fy08osti/41957.pdf.

"High Performance Guidelines: Triangle Region Public Facilities," Version 2.0.2001. Available at www.tjcog.org/docs/regplan/susenerg/grbuild.pdf.

IEA. 2010. *World Energy Outlook 2010.* Paris: International Energy Agency. The IEA website is www.iea.org.

Kats, Gregory H. 2003. "The Costs and Financial Benefits of Green Buildings." A report developed for California's Sustainable Building Task Force. Available at the Capital E website, www.cap-e.com.

———. 2006. "Greening America's Schools."Available at www.cap-e.com.

Kats, Gregory H., and Jeff Perlman. 2005. "National Review of Green Schools."A report for the Massachusetts Technology Collaborative. Available at www.cap-e.com.

Kibert, Charles J. 1994. "Principles and a Model of Sustainable Construction."Proceedings of the First International Conference on Sustainable Construction, November 6–9, Tampa, FL, 1–9.

Lewis, Simon, Paulo M. Brando, Oliver L. Phillips, Geertje M. F. van der Heijden, and Daniel Nepstad.2011. "The 2010 Amazon Drought." *Science* 331 (6017): 554.

Pennsylvania Governor's Green Government Council (GGGC). 1999. "Guidelines for Creating High-Performance Green Buildings."Available at www.portal.state.pa.us/portal/server.pt/community/green_buildings/13834/high_performance_green_building_guide/588208.

"Ultra-violet Radiation Could Reduce Office Sickness." 2004. Available at the website of the Lung Association of Saskatchewan, www.sk.lung.ca/index.php/about-us-mainmenu-279/newsroom/extras/649-xtra0129.

UN World Commission on Environment and Development. 1987. *Our Common Future.* Oxford: Oxford University Press. *Our Common Future* is also referred to as the Brundtland Report, because GroHarlem Brundtland, prime minister of Norway at the time, chaired the committee writing the report.

US Green Building Council. 2003. *Building Momentum: National Trends and Prospects for High-Performance Green Buildings.* Available at www.usgbc.org/Docs/Resources/043003_hpgb_whitepaper.pdf.

World Health Organization. 1983. "Indoor Air Pollutants: Exposure and Health Effect."EURO Report and Studies, No. 78.

Young, Samantha. 2004. *Las Vegas Review Journal*, March 10. Available at www.reviewjournal.com/lvrj_home/2004/Mar-10-Wed-2004/news/23401764.html.

Zabarsky, Marsha. 2002. "Sick Building Syndrome Gains a Growing Level of National Awareness." *Boston Business Journal*, August 16. Available at www.bizjournals.com/boston/stories/2002/08/19/focus9.html.

Part I

Green Building Foundations

This book is intended to guide construction and design professionals through the process of developing commercial and institutional high-performance green buildings. A green building can be defined as a facility that is designed, built, operated, and disposed of in a resource-efficient manner using ecologically sound approaches and with both human and ecosystem health as goals. The nonprofit US Green Building Council (USGBC) has successfully defined the parameters of a nonresidential green building in the United States.[1] The organization's Leadership in Energy and Environmental Design (LEED) building assessment system provides design guidance for the vast majority of US buildings currently described as green and has been implemented in several other countries.[2] From 1998 to the present, the number of LEED-certified buildings has almost doubled each year in both number and area. In 2011, the value of buildings registered for LEED certification equaled as much as 15 percent of the total value of commercial/institutional buildings in this country and is expected to approach 50 percent by 2015. More recently, an alternative to LEED, known as *Green Globes*, has been competing with LEED as a tool for assessing and certifying high-performance green buildings in the United States.[3]

This book addresses the application of building assessment systems such as LEED and Green Globes in the United States, as well as several noteworthy building assessment systems used in other countries. Part I addresses the background and history of the sustainable construction movement, various green building rating systems, the concept of life-cycle assessment, and green building design strategies. It is intended to provide the working professional with sufficient information to implement the techniques necessary to create high-performance green buildings. This part contains the following chapters:

Chapter 2: Background

Chapter 3: Ecological Design

Chapter 2 describes the emergence of the green building movement, its rapid evolution and growth over the past decade, and current major influences. This chapter also addresses the unusual scale of resource extraction, waste, and energy consumption associated with construction, and it examines the resource and environmental impacts of the built environment. Although this book focuses on the United States, the context, organizations, and approaches of other countries are also mentioned.

General design strategies for green building are covered in Chapter 3. Fundamentally, green design is based on an ecological model or metaphor commonly referred to as *ecological design.* The recent works of Sim Van der Ryn and Stuart Cowan, Ken Yeang, and David Orr, along with earlier works by R. Buckminster Fuller, Frank Lloyd Wright, Ian McHarg, Lewis Mumford, John Lyle, and Richard Neutra, are reviewed in this chapter.

In spite of the impulse to apply the highest ecological ideals to the built environment, a vast majority of contemporary designers lack an adequate understanding of ecology. Claims of a building's "ecological design" are often tenuous in fact, and greater participation by ecologists and industrial ecologists is necessary to reduce the gap between the ideal of ecological design and its expression in reality. To that end, the LEED and Green Globes building assessment systems are probably the first step in a long process of achieving truly ecological design. The products, systems, techniques, and services needed to create buildings in harmony and synergy with nature are rare. Buildings are often assembled from components produced by a variety of manufacturers that have paid little or no attention to the environmental impacts of their activities. Installation is performed by a workforce largely unaware of the impacts of the built environment and often results in enormous waste. Conventional buildings are designed by architects and engineers who often have little or no training in sustainable construction. In spite of these obstacles, certified green buildings are usually superior to conventional projects in terms of energy and water efficiency, materials selection, building health, waste generation, and site utilization. The USGBC has created an ambitious training and publicity program to disseminate LEED concepts. Innovative products for sustainable construction have become more prevalent, greatly easing the process of materials selection. Of equal importance, the green building process has necessitated a deeper integration of the client, the designer, and the general public. New projects are generally initiated via the charrette, which includes construction and design professionals as well as community members, who together brainstorm the project's initial design.

Exceeding the requirements of the contemporary assessment standards such as LEED and Green Globes is the next rung on the ladder of truly sustainable construction. The following are some of the features of future sustainable construction:

- The built environment would fully adopt closed-loop materials practices, and the entire structure, envelope, systems, and interior would be composed of products easily disassembled to permit ready recycling. Waste material throughout the structure's life cycle would be capable of biological (composting) or technological recycling. The building itself would be deconstructable; in other words, it would be possible to disassemble it economically for reuse and recycling. Only materials with future value, either to human or to biological systems, would be incorporated into buildings.

- Buildings would have a synergistic relationship with their natural environment and blend with the surrounding environment. Materials exchanges across the building-nature interface would benefit both sides of the boundary. Building and occupant waste would be processed to provide nutrients to the surrounding biotic systems. Toxic or harmful emissions of air, water, and solid substances would be eliminated.

- The built environment would incorporate natural systems at various scales, ranging from individual buildings to bioregions. The underexplored integration of natural systems with the built environment has staggering potential to produce superior human habitats at lower cost. Landscaping would provide shade, food, amenities, and stormwater uptake for the built infrastructure. Wetlands would process wastewater and stormwater and often eliminate the need for enormous and expensive infrastructure. The integration of nature, which is barely addressed in building assessment systems, is currently considered under the comprehensive category of design innovation. Ideally, the integration of human and natural systems would be standard practice rather than being considered an innovation.

- Energy use by buildings would be reduced by a Factor 10 or more below that of conventional buildings.[4] Rather than the typical 100,000 BTU/ft^2 (292 kWh/m^2) or more consumed by today's commercial and institutional structures, truly green buildings would be relatively deenergized, using no more than 10,000 BTU/ft^2 (29 kWh/m^2). The source of this energy would be the sun or other solar-derived sources such as wind power or biomass. Alternatively, geothermal and tidal power, both nonsolar energy sources, would also be employed as renewable forms of energy derived from natural sources.

In summary, the green building movement has come a long way in a short time. Its exponential growth promises its longevity, and numerous public and private organizations support its agenda. It is exciting to contemplate the possibility of extending the boundaries of ecological design and construction as global environmental problems become exigent and as solutions, if not survival itself, demand a radical departure from conventional thinking. The evolution of products, tools, services, and, ultimately, Factor 10 buildings cannot occur soon enough. Only then may we alter the trajectory of the human quality of life from one of certain disaster to one that finally exists within the carrying capacity of nature. Although humanity is halfway through the race, the ultimate question remains unanswered: Can we change the built environment rapidly enough to save both nature and ourselves?

Notes

1. The USGBC (www.usgbc.org) is now the de facto US leader in promoting commercial and institutional green buildings. The greening of single-family-home residential construction and land development is far more decentralized and varies from state to state. An example of an organization leading change at the state level in the residential and land development sectors is the Florida Green Building Coalition (FGBC) (www.floridagreenbuilding.org). The Florida Green Residential Standard and the Florida Green Development Standard can be downloaded from the FGBC website.
2. Although intended for the greening of US buildings, LEED is being adopted by other countries, such as Canada, Spain, and Korea.
3. The genesis of Green Globes was the Building Research Establishment Environmental Assessment Method (BREEAM), which was developed in the United Kingdom in the early 1990s, brought to Canada in 1996, and eventually developed as an online assessment and rating tool. In 2004, the Green Building Initiative (GBI) acquired the rights to distribute Green Globes in the United States. In 2005, the GBI became the first green building organization to be accredited as a standards developer by the American National Standards Institute (ANSI) and began the process of establishing Green Globes as an official ANSI standard. The GBI ANSI technical committee was formed in early 2006 and the ANSI/GBI 01 standard based on Green Globes was published in 2010.

4. Factor 10, a concept developed by the Wuppertal Institute in Wuppertal, Germany (www.wupperinst.org), suggests that long-term sustainable development can be achieved only by reducing resource consumption (energy, water, and materials) to 10 percent of its present levels. Another concept, Factor 4, suggests that technology presently exists to reduce resource consumption immediately by 75 percent. The book *Factor Four: Doubling Wealth, Halving Resource Use*, by Ernst von Weizsäcker, Amory Lovins, and L. Hunter Lovins (London: Earthscan, 1997), popularized this concept.

Chapter 2
Background

The shift to a high-performance built environment is being propelled by three major forces. First, there is growing evidence of accelerated destruction of planetary ecosystems, alteration of global biogeochemical cycles, and enormous increases in population and consumption. The threat of global warming, depletion of major fisheries, deforestation, and desertification are among likely outcomes that some environmentalists have labeled the *Sixth Extinction*, referring to the human species' massive destruction of life and biodiversity on the planet.[1]

Second, increasing demand for natural resources is pressuring developed and developing countries such as the so-called BRIC (Brazil, Russia, India, and China) countries, resulting in shortages and higher prices for materials and agricultural products. China adds about 11 million people each year to its population of 1.3 billion, and its economy is expanding at a rate of about 9.5 percent annually. China's growing economy and improving standard of living have increased the demand for, and prices of, meat and grain. The negative consequences of rapid urban expansion in China have included water shortages and increasing desertification, leading to the growth of the Gobi Desert by 4000 square miles (10,400 square kilometers) per year.

The growing Chinese economy has a huge appetite for materials, which is contributing to shortages and driving up prices around the world. China produced over 46 percent of the world's steel in 2011 and is increasing production at a prodigious rate, from approximately 12 million tons (11 million metric tons) per month in 2001 to 64 million tons (60 million metric tons) per month in 2011, an annual rate increase of 768 million tons (720 million metric tons) and rising rapidly. Global steel prices increased by over 50 percent in 2011, due in large part to Chinese demand. In comparison, steel production in the United States has been relatively flat in the past decade, totaling 88 million tons (81 million metric tons) in 2011, a small fraction of the Chinese level of production. Chinese demand for fossil fuels is growing at a rate of 30 percent per year, just as world oil production is peaking. Copper prices have increased 10-fold in 10 years. Even the price of relatively abundant Appalachian coal in the United States increased 80 percent in 2010. The manufacturing sector is experiencing higher prices for virtually every commodity used in the production system. Rare earths, which, as their name implies, are not abundant materials but indispensable elements such as lanthanum, neodymium, and europium, are essential for the magnets, motors, and batteries used in electric cars, wind generators, hard-disk drives, mobile phones, and other high-tech products. They are in critically short supply and affecting industries worldwide. Japanese industry had to reduce its imports of rare earths by 17 percent in June 2011 due to extremely high prices, thus setting back an economy already on its heels after the March 2011 tsunami and subsequent nuclear plant accidents at the Fukushima Daiichi nuclear power station.[2] In short, prices for nonrenewable materials and energy resources are on a strong upward trend that shows no sign of abating. The construction industry, a major consumer of these resources, must change in order to remain healthy and solvent.

Third, the green building movement is coinciding with similar transformations in manufacturing, tourism, agriculture, medicine, and the public sector, which have adopted various approaches toward greening their activities. From redesigning entire processes to implementing administrative efforts such as adopting green procurement policies, new concepts and approaches are emerging that deem the environment, ecological systems, and human welfare to be of equal importance to economic performance. For example, the Xerox Corporation has announced the strategic environmental goal of creating "waste-free products and waste-free facilities for waste-free workplaces." A recently introduced Xerox product, the DocuColor iGen4 EXP Press, uses nontoxic dry inks and has a transfer efficiency of almost 100 percent. Up to 97 percent of the machine's parts and 80 percent of its generated waste can be reused or recycled. Furthermore, by reclaiming copy machines at the end of their useful life, recovering components for reuse and recycling, and instituting sophisticated remanufacturing processes, Xerox conserves materials and energy, dramatically reduces waste, and limits its potential liability by eliminating hazardous materials.[3]

In the automotive industry, the European end-of-life vehicles (ELV) directive has been in effect since the year 2000 (see Figure 2.1). This legislation requires manufacturers to accept the return of vehicles at the end of their useful life, with no charge to the consumer. The measure requires extensive recycling of the returned vehicles and minimizes the use of hazardous materials in automobile production. Spurred by European efforts, Ford Motor Company is using European engineering expertise at its research center in Aachen, Germany, to develop recycling technologies that will raise the recovery yield of recycled materials above their current 80 to 85 percent level. Construction is generally seen as a wasteful industry, and efforts to increase the reuse and recycling of building materials are beginning to emerge as part of the high-performance green building movement (see Figure 2.2). The European automobile industry, although a different economic sector, provides ample lessons for reducing waste and closing materials loops in construction.

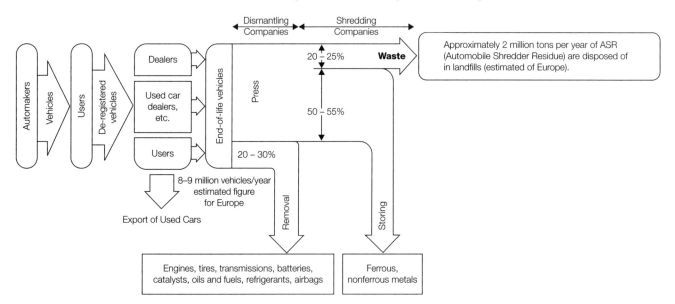

Figure 2.1   The European ELV directive requires manufacturers to accept the return of vehicles at the end of their useful life, with no charge to the consumer. This diagram shows the extensive recycling of returned vehicles and greatly reduced waste generation in automobile production.

Figure 2.2   The structural system for Rinker Hall, a Leadership in Energy and Environmental Design (LEED) certified building at the University of Florida in Gainesville, is steel. Steel is an excellent material due to its high recycled content, almost 100 percent for some building components, and is readily deconstructable and recyclable. Rinker Hall is the only building out of the thousands certified by the US Green Building Council to have been awarded an innovation credit for its deconstructability. Although some would consider metals such as steel to be "green" building materials, their embodied energy, that is, the energy required for resource extraction, manufacturing, and transport, is fairly high and results in the consumption of nonrenewable fossil fuels and the generation of global warming gases and air pollution. Consequently, whether steel can truly be considered a green building material is controversial and depends on the criteria used in the evaluation. Of all the challenges in creating high-performance green buildings, finding or creating truly environmentally friendly building materials and products is the most difficult task facing construction industry professionals.

This chapter describes the effect of these three forces on the green building movement and their influence on defining new directions for design and construction of the built environment. It lays out the ethical arguments supporting sustainability and, by extension, sustainable construction. It explores the relatively new vocabulary associated with various efforts that attempt to reduce human environmental impact, increase resource efficiency, and ethically confront the dilemmas of population growth and resource consumption. Finally, it covers the history of the green building movement in the United States, acknowledging that an understanding of its roots is necessary to appreciate its evolution and current status.

Ethics and Sustainability

In the context of sustainable development and sustainable construction, ethics must be broadened to address a wide range of concerns that are not usually a basis for consideration. Ethics addresses relationships between people by providing rules of conduct that are generally agreed to govern the good behavior of contemporaries. Sustainable development requires a more extensive set of ethical principles to guide behavior because it addresses relationships between generations, calling for what is sometimes referred to as *intergenerational justice*. The classic definition of sustainable development is ". . . meeting the needs of the present without compromising the ability of future generations to meet their needs."[4] It is clear that intertemporal considerations, the responsibility of one generation to future generations as well as the rights of future generations vis-à-vis a contemporary population, are fundamental concepts of sustainable development. The result of intertemporal or intergenerational considerations with respect to morality and justice must be an expanded concept of ethics that extends not only to future generations but also to the nonhuman living world and arguably to the nonliving world because the alteration or destruction of nonhuman living and nonliving systems affects the quality of life of future generations by reducing their choices. The result of destroying biodiversity today, for instance, is the removal of important information for future populations that could have been the basis for biomedicines, not to mention the removal of at least some portion of *environmental amenity*.[5] It is clear that the choices of a given population in time will directly affect the quantity and quality of resources remaining for future inhabitants of earth, impact the environmental quality they will experience, and alter their experience of the physical world. With this in mind, the purpose of this section is to expand on the foundations of classical ethics to provide a robust set of principles that are able to address questions of intergenerational equity.

THE ETHICAL CHALLENGES

Humans are unique among all species with respect to control over their destiny. Gary Peterson (2002), an ecologist, articulated this very well when he stated:[6]

> Humans, individually or in groups, can anticipate and prepare for the future to a much greater degree than ecological systems. People use mental models of varying complexity and completeness to construct views of the future. People have developed elaborate ways of exchanging, influencing, and updating these models. This creates complicated dynamics based upon access to information, ability to organize, and power. In contrast, the organization of ecological systems is a product of the mutual reinforcement of many interacting structures and processes that have emerged over long periods of time. Similarly, the behavior of plants and animals is the product of successful evolutionary experimentation that has occurred in the past. Consequently, the arrangement and behavior of natural systems are based upon what has happened in the past, rather than looking in anticipation toward the future. The difference between forward-thinking human systems and backwards-looking natural systems is fundamental. It means that understanding the role of people in ecological systems requires not only understanding how people have acted in the past, but also how they think about the future.

Following this line of thinking, humans are certain to create materials and develop processes that have not evolved in a natural sense, that have no precedent in nature. The question then becomes: What constraints should society place on the development of new materials, products, and processes? The ongoing debates about genetically modified organisms (GMOs) and cloning are indicative of the uncertainty about the outcomes of human tinkering with the blueprints of life, not to mention the creation of materials that have uncertain long-term impacts. Other major developments such as biotechnology, genetic engineering, nanotechnology, robotics, and nuclear energy, to name but a few, present fundamental challenges to human society. Decisions about implementing technologies with no precedent in nature and with potentially unprecedented negative and irreversible impacts must be carefully considered, especially since once deployed, it is extremely difficult to reverse course if negative consequences are discovered. Decisions about how to move forward must be based on (1) an ethical framework that represents society's general moral attitudes toward life and future generations, (2) an understanding of and willingness to accept risk, and (3) the economic costs of implementation and resulting impacts.[7]

INTERGENERATIONAL JUSTICE AND THE CHAIN OF OBLIGATION

The choices of today's generations will directly affect the quality and quantity of resources remaining for future inhabitants of earth and environmental quality. This concept of obligation that crosses temporal boundaries is referred to as *intergenerational justice.* Furthermore, the concept of intergenerational justice implies a *chain of obligation* between generations that extends from today into the distant future. Richard Howarth expresses this obligation by stating, ". . . unless we ensure conditions favourable to the welfare of future generations, we wrong existing children in the sense that they will be unable to fulfill their obligation to their children while enjoying a favourable way of life themselves."[8] Howarth also suggests that the actions and decisions of the present generation affect not only the welfare but also the composition of future generations. He argues that when we create conditions that change resource availability or that alter the environment, future populations will be compositionally different than if the resource base and environmental conditions had been passed on, from one generation to future generations, unchanged. For instance, one can envision that mutations caused by excessive ultraviolet radiation through an ozone layer depleted by human activities, or by synthetic toxic chemicals used without adequate safeguards, will certainly result in different people and conditions. Consequently, the chain of obligation that underpins the key sustainability concept of intergenerational justice includes parents' responsibility for enabling their offspring to meet their moral obligations to their children and beyond. Clearly, this would include educating the offspring about these obligations and the basis for them.

DISTRIBUTIONAL EQUITY

There is an obligation to ensure the fair distribution of resources among present people so that the life prospects of all people are addressed. This obligation can be referred to as *distributional equity* or *distributive justice* and refers to the right of all people to an equal share of resources, including goods and services, such as materials, land, energy, water, and high environmental quality. Distributional equity is based on principles of justice and the reasonable assumption that all individuals in a given generation are equal and that a uniform distribution of resources must be a consequence of *intragenerational equity.* The principle of distributional equity can be extended to relationships between generations because a given generation has a moral responsibility for providing for their offspring, which is referred to as *intergenerational equity.* Thus, distributional equity also underpins the chain of obligation concept. Distributional equity is a complex concept, and a number of principles underpin and are related to it: (1) the difference principle, (2) resource-based principles, (3) welfare-based principles, (4) desert-based principles, (5) libertarian principles, and (6) feminist principles.

THE PRECAUTIONARY PRINCIPLE

The *precautionary principle* requires the exercise of caution when making decisions that may adversely affect nature, natural ecosystems, and global biogeochemical cycles. According to the Center for Community Action and

Environmental Justice (CCAEJ), the precautionary principle states that "when an activity raises threats of harm to human health or the environment, precautionary measures should be taken even if some cause and effect relationships are not fully established scientifically." Global climate change is an excellent example of the need to act with caution. Notwithstanding debate about the effects of man-made carbon emissions on future planetary temperature regimes, the potentially catastrophic outcome should motivate humankind to behave cautiously and attempt to limit the emission of carbon-containing gases such as methane and carbon dioxide. The CCAEJ lists the four tenets of the precautionary principle:[9]

1. People have a duty to take anticipatory action to prevent harm.
2. The burden of the proof of harmlessness of a new technology, process, activity, or chemical lies with the proponents, not the general public.
3. Before using a new technology, process, or chemical or starting a new activity, people have an obligation to examine a full range of alternatives including the alternative of not doing it.
4. Decisions applying the precautionary principle must be open, informed, and democratic and must include the affected parties.

With respect to the precautionary principle, a hypothetical danger of *nanotechnology* is the creation of so-called *gray goo*. Nanotechnology is an approach to building machines at the submicrometer level, that is, on an atomic scale. K. Eric Drexler suggested that one of the hallmarks of nanotechnology will be the ability of these invisible machines to self-replicate, with enormous potential benefits to humanity, but with the attendant danger that the replication will bring an out-of-control conversion of matter into machines. Drexler warned that "we cannot afford certain kinds of accidents with replicating assemblers," which can be restated as "we cannot afford the irresponsible use of powerful technologies."[10] Thermodynamics and energy requirements will limit the effects of the gray goo conversion process, but significant harm may still be the consequence. Similar concerns exist with regard to genetic engineering and nuclear engineering: that they will put future generations at risk. Clearly, the precautionary principle should be applied to each of these scenarios to eliminate as much as possible risks to future populations, both human and nonhuman, from the consequences of technologies that are not fully understood.

Despite the wisdom of exercising caution when addressing complex issues that may have unknown, far-reaching effects, the precautionary principle is controversial and is sometimes perceived as a threat to progress, since it fails to consider the negative consequences of its application. For example, refusing to use new drugs because society has not fully established their effects on nature and people may foreclose options for advancing human health. Nonetheless, the consequences of not applying the precautionary principle are becoming apparent in several areas. Most notably, the widespread use of estrogen-mimicking chemicals is believed to damage the reproductive systems of animal species and probably that of humans. With these concerns in mind, in 1999 the National Science Foundation (NSF) developed the Biocomplexity in the Environment Priority Area to address the interaction of human activities with the environment and on climate change and biodiversity.[11] At least the debate surrounding application of the precautionary principle has focused greater attention on the environmental impacts of technology and has pressured technologists to acknowledge the potential consequences of their efforts on humans and nature.

THE REVERSIBILITY PRINCIPLE

Making decisions that can be undone by future generations is the foundation of the *reversibility principle.* Renowned science fiction author Arthur C. Clarke suggested a rule that well describes this principle: "Do not commit the irrevocable."[12] At its core, this principle calls for a wider range of options to be considered in decision making. Addressing the issue of energy choices is an excellent example because a rapidly growing global economy is faced with looming energy shortages, exacerbated by depletion of finite oil supplies. In the United States, a shift is under way to reconsider nuclear plants as a major source of energy because they can probably generate electricity at an acceptable cost and also be a source of thermal energy for producing hydrogen from water for use in fuel cells. The reversibility principle would force today's society to confront the issue of whether or not the choice of nuclear energy as an option is reversible by a future society. Two questions would immediately emerge from this consideration. First, is the technology safe enough for widespread use? The nuclear industry suggests that over the past two decades of a national hiatus from building new plants, the technology has advanced to the point where a Chernobyl or Three Mile Island incident has been eliminated. The second question is: How would a future society cope with the nuclear waste from these plants? Converting the waste to harmless materials via a new technology is highly unlikely, and the power plants built today would force future generations to store and be put at risk by the radionuclides in the spent fuel rods. A subset of questions on this same subject would result as a consequence of assuming that if storage of the radioactive waste for periods of time in the 10,000-year range is feasible, what are the storage options? In addressing this question, Gene I. Rochlin suggests that there are two options.[13] One is to deposit the waste deep in a stable rock formation where it could be recovered, for example, if leaks in the storage containers were detected by future generations. A second option is to deposit it in inaccessible locations, for example, by allowing the waste to melt through the polar ice, or to place it deep in the ocean, where sliding continental plates would gradually cover it. The former solution allows future generations access to the waste to take corrective action, while the latter forgoes the option.

The reversibility principle is related to the precautionary principle because it lays out criteria that must be observed prior to the adoption of a new technology. It is less stringent than the precautionary principle in some respects because it suggests reversibility as the primary criterion for making a decision to employ the technology, whereas the precautionary principle requires that a technology not be implemented if the effects of it are not fully understood and the risks are unacceptable.

THE POLLUTER PAYS PRINCIPLE AND PRODUCER RESPONSIBILITY

The fundamental premise of the precautionary and reversibility principles is that those who are responsible for implementing technologies must be prepared to address the consequences of their implementation. The precautionary principle suggests that technologists should demonstrate the efficacy of their products and processes prior to allowing them to impact the biosphere. The reversibility principle permits implementation in the face of some level of risk as long as any negative effects can be undone. The *polluter pays principle* addresses existing technologies that have not been subject to these other principles and places the onus for mitigating damage and consequences on the individuals causing the impacts. The polluter pays principle originated with the Organisation for Economic Co-operation and Development (OECD) in 1973 and is based on the premise that polluters should pay the costs of dealing with pollution for which they are responsible. Historically, the polluter pays principle has focused on retrospective liability for pollution; for example, an industry causing pollution would have to pay for the cleanup costs arising from it.

More recently, the focus of the polluter pays principle has shifted toward avoiding pollution and addressing wider environmental impacts through producer responsibility. Producer responsibility is an example of the extended version of the polluter pays principle, as it applies to waste and resource management, placing responsibility for the environmental impact associated with a product on the producers of that product. Producer responsibility is intended to address the whole life-cycle environmental problems of the production process, from initial minimization of resource use, through extended product life span, to recovery and recycling of products once they have been disposed of as waste. Producer responsibility is increasingly used throughout the world as a means of addressing the environmental impacts of certain products. The European Union (EU) has applied producer responsibility through directives on packaging and packaging waste, waste electronic and electrical equipment, and end-of-life vehicles.

PROTECTING THE VULNERABLE

There are populations, including those of the animal world, that are vulnerable to the actions of portions of the human species, due to the destruction of ecosystems under the guise of development, introduction of technology (including toxic substances, endocrine disruptors, and genetically modified organisms), and general patterns of conduct (war, deforestation, soil erosion, eutrophication, desertification, and acid rain, to name a few). People who are essentially powerless due to governing and economic structures are vulnerable to the decisions of those who are powerful because of their wealth or influence. This asymmetrical power arrangement is governed by moral obligation. Those in power have a special obligation to *protect the vulnerable*, those dependent on them. In a family, children's dependence on their parents gives them rights against their parents. Future generations are also vulnerable because they are subject to the effects of decisions we make today. In a technological society, many portions of the human population and certainly the animal world can be exposed to harm by the actions of individuals or companies performing medical research or because the government that is charged with protecting them fails in its responsibilities when it comes to pollution, the use of toxic substances, and a wide variety of other poorly controlled actions. Breaches of ethics are not uncommon when it comes to vulnerable populations such as prisoners, mentally disabled people, women, and people in developing countries. And, as noted above, today's actions have consequences for future generations that have only recently been considered. Future people are certainly vulnerable to our actions, and both their existence and quality of life are potentially compromised by short-term thinking and decisions based solely on the comfort and wealth of past populations. The ethical principle of protecting the vulnerable places an enormous responsibility on earth's present population, one made even more difficult due to rampant global poverty.

PROTECTING THE RIGHTS OF THE NONHUMAN WORLD

The nonhuman world refers to plants and animals and could be extended to include bacteria, viruses, mold, and other living organisms. The principle of protecting this world is an extension of the principle of protecting the vulnerable, particularly animals but also plants that are in danger of extinction. Animal rights fall under this principle. The nonliving portion of the earth is essential to supporting life, and a set of sustainability principles should address the requirements for protecting this key element of the life support system. Some would argue that ethics should require the character of beautiful places such as the Grand Canyon to be protected in perpetuity. This principle is an important one because humans have become disconnected from both the living and the nonliving human worlds when, in fact, we are utterly dependent on them for our survival. Indeed, the *biophilia hypothesis*, described in a subsequent section of this chapter, states that humans crave a connection with nature and that our health, at least in part, is dependent on being able to connect on a routine basis with nature. Human ingenuity in the form of technology is having quite the opposite effect. As noted by Andrew J. Angyal, ". . . this destructive myth of a technological wonderland in which nature is bent to every human whim is turning the earth into a wasteland and threatening human survival. Western spiritual traditions have not been able to impede these lethal tendencies, but have encouraged them as part of god's plan for human domination of the earth, and these traditions have understood human destiny as primarily involving a heavenly spiritual redemption. With their preoccupation with redemption and their neglect of creation, modern religious traditions are unable to offer a spirituality adequate to experience the divine in ordinary life or in the natural world."[14] Thomas Berry describes 10 precepts based on nature deriving its rights from universal law, and not human law, that provide an ethical framework for the rights of the nonhuman world:[15]

1. Rights originate where existence originates. That which determines existence determines rights.

2. Since it has no further context of existence in the phenomenal order, the universe is self-referent in its being and self-normative in its activities. It is also the primary referent in the being and activities of all derivative modes of being.

3. The universe is a communion of subjects, not a collection of objects. As subjects, the component members of the universe are capable of having rights.

4. The natural world on the planet earth gets its rights from the same source that humans get their rights, from the universe that brought them into being.

5. Every component of the earth community has three rights: the right to be, the right to habitat, and the right to fulfill its role in the ever-renewing processes of the earth community.

6. All rights are species specific and limited. Rivers have river rights. Birds have bird rights. Insects have insect rights. Difference in rights is qualitative, not quantitative. The rights of an insect would be of no value to a tree or a fish.

7. Human rights do not cancel out the rights of other modes of being to exist in their natural state. Human property rights are not absolute. Property rights are simply a special relationship between a particular human "owner" and a particular piece of "property" so that both might fulfill their roles in the great community of existence.

8. Since species exist only in the form of individuals, rights refer to individuals and to their natural groupings of individuals into flocks, herds, packs, not simply in a general way to species.

9. These rights as presented here are based upon the intrinsic relations that the various components of earth have to each other. The planet earth is a single community bound together with interdependent relationships. No living being nourishes itself. Each component of the earth community is immediately dependent on every other member of the community for the nourishment and assistance it needs for its own survival. This mutual nourishment, which includes the predator-prey relationships, is integral with the role that each component of the earth has within the comprehensive community of existence.

10. In a special manner humans have not only a need for but a right of access to the natural world to provide not only the physical need of humans but also the wonder needed by human intelligence, the beauty needed by human imagination, and the intimacy needed by human emotions for fulfillment.

Clearly, putting nature on an equal footing with humans is a difficult leap for many people, but vigorously protecting nature is in the best interests of humanity. Indeed, simply protecting nature does not quite meet the imperatives of this principle. Rather, humans should consider restoring nature in all activities, righting the wrongs of the past, and in the process restoring the badly damaged link between humans and nature.

RESPECT FOR NATURE AND THE LAND ETHIC

Respect for nature follows from acknowledging the rights of the nonhuman world described in the previous sections. An ethics of respect for nature is based on the fundamental concepts that (1) humans are members of the earth's community of life, (2) all species are interconnected in a web of life, (3) each species is a teleological center of life pursuing good in its own way, and (4) human beings are not superior to other species. This last concept is based on the other three and shifts the focus from *anthropocentrism*, or a human-centered viewpoint, to a *biocentric* outlook.[16]

Humans are part of precisely the same evolutionary process as all other species. All other species that exist today faced the same survival challenges as humans. The same biological laws that govern other species—for example, the laws of genetics, natural selection, and adaptation—apply to all living creatures. Earth does not depend on humans for its existence. On the contrary, humans are the only species that has ever threatened the existence of earth itself. As relative latecomers, humans appeared on a planet that had contained life for 600 million years, and not only have to share earth with other species but are totally dependent on them for survival. Human beings threaten the soundness and health of the earth's ecosystems by their behavior. Technology results in the release of toxic chemicals, radioactive materials, and endocrine disruptors. Forestry and agriculture destroy biologically dense and diverse forests. Emissions pollute land, water, and air. Unlike natural extinctions of the past from which the earth recovered, the present human-induced extinction is causing disruption, destruction, and alteration at such a high rate that, even with the self-extinction of the human species, the planet may never recover. An ethics based on biocentrism would result in humans realizing that the integrity of the entire biosphere would benefit all communities of life, including nonhumans. It is debatable whether this concept is merely an ethical one because it is also a biological fact that humans cannot survive without the ecosystems on which they depend. However, human beings have the capability to act and change behavior based on knowledge, in this case being aware of the causal relationship of behavior to the survival of other species. An ethics of respect for nature consists not only of realizing this causal relationship but also of adopting behaviors that respect the rights of nonhuman species to both exist and thrive.

In addition to respecting the rights to survival of other species, as a consequence of careful observation and the application of scientific principles and the scientific method, humans understand the unique qualities and aspects of other organisms. These observations allow us to see these organisms as unique teleological centers of life, each struggling to survive and realize its good in its own way. This does not mean that organisms need to have the characteristic of consciousness, that is, self-awareness, to be "good" because each is oriented toward the same ends: self-preservation and well-being. The ethical concept here is that because each species is a teleological center of life, its universe or world can be viewed from the perspective of its life. Consequently, good (finding food), bad (being injured or killed), and indifferent (swimming in the ocean) events can be said to occur in each species' life, as is the case for the human species. Having respect for nature means that humans can view life events for nonhuman species in much the same fashion as they would for other humans.

Aldo Leopold suggested that there should be an ethical relationship to the land and that this relationship should and must be based on love, respect, and admiration for the land.[17] Furthermore, this ethical relationship, referred to as the *land ethic*, not only should exist because of economic value but also should be based on value in the philosophical sense. The land ethic makes sense because of the close relationship and interdependence of humans with land that provides food and amenities and contributes to good air and water quality. Humans have tended to become disconnected from the land because of technological developments that give apparent but not actual independence from the land. Substitutes for natural material (e.g., polyester instead of cotton) further the notion that land is not essential for survival and that technology can provide suitable substitutes. Farm mechanization has also tended to separate the farmer from the land, the result being less care and attention for a critical resource.

Basic Concepts and Vocabulary

Although probably the greatest success story of the contemporary American environmental movement, sustainable construction is only one part of a larger transformation taking place via a wide range of activities throughout numerous economic sectors. Progressive ideas articulated with new vocabulary serve as the intellectual foundation for this evolution. The most notable and important include the concepts of sustainable development, industrial ecology, construction ecology, biomimicry, design for the environment, ecological economics, carrying capacity, ecological footprint, ecological rucksack, embodied energy, the biophilia hypothesis, eco-efficiency, the Natural Step, life-cycle assessment, life-cycle costing, the precautionary principle, Factor 4, and Factor 10. These concepts are briefly described in the following sections.

SUSTAINABLE DEVELOPMENT

Sustainable development, or *sustainability*, is the foundational principle underlying various efforts to ensure a decent quality of life for future generations. The Brundtland Report, more properly known as *Our Common Future* (1987), defines sustainable development as ". . . meeting the needs of the present without compromising the ability of future generations to meet their needs" (see Figure 2.3). This classic definition implies that the environment and the quality

Figure 2.3 The publication of *Our Common Future* in 1987 is generally accepted as marking the initiation of the contemporary sustainable development movement.

of human life are as important as economic performance and suggests that human, natural, and economic systems are interdependent. It also implies intergenerational justice; highlights the responsibility of the present population for the welfare of millions yet unborn; and implies that we are borrowing the planet, its resources, and its environmental function and quality from future generations. Intergenerational justice raises the question of how far into the future we should consider the impacts of our actions. Although no clear answer to this important question is readily apparent, the Native American philosophy of thinking seven generations, or 200 years, into the future is instructive. If in two centuries few contemporary buildings will be standing, we must ask whether our present stock of materials will provide recyclable resources for future generations or saddle them with enormous and difficult waste disposal problems. It is this question, originating in the philosophy of sustainability, that marks the fork in the road of our current industrial processes. Those on the path of "business as usual" will view the environment as an infinite source of materials and energy and a repository for waste. In contrast, those on the more ethical "road less traveled" will regard the quality of life of our descendants and question whether we are permanently stealing, versus temporarily borrowing, the environmental capital of future generations. At the philosophical core of the green building movement is the decision to embark on the latter path.

INDUSTRIAL ECOLOGY

The science of *industrial ecology*, which emerged in the late 1980s,[18] refers to the study of the physical, chemical, and biological interactions and interrelationships both within and among industrial and ecological systems.[19] Applications of industrial ecology involve identifying and implementing strategies for industrial systems to emulate more closely harmonious and sustainable ecological ecosystems. The first major effort of industrial ecology was to reduce the massive quantities of waste generated by traditional manufacturing processes, from which only an estimated 6 percent of extracted resources end up as final products.[20] The first well-known example of the resulting process, referred to as *industrial symbiosis*, is the industrial complex in Kalundborg, Denmark, where excess heat energy, waste, and water are shared among five major partners: (1) the Asnaes power station, Denmark's largest coal-fired power plant, with a 1500-megawatt capacity; (2) the Statoil refinery, Denmark's largest, with a present capacity of 4.8 million tons per year; (3) Gyproc, a plasterboard factory producing 14 million square meters of gypsum wallboard annually (roughly enough to build all the houses in six towns the size of Kalundborg); (4) Novo Nordisk, an international biotechnological company, with annual sales of over $2 billion, that manufactures industrial enzymes and pharmaceuticals, including 40 percent of the world's supply of insulin; and (5) the city of Kalundborg district heating system, which supplies heating to 20,000 residents and water to its homes and industries. The Kalundborg complex (diagrammed in Figure 2.4) was the world's first eco-industrial park; since its inception, similar waste exchange complexes have been created around the world.[21] Since the early 1990s, the concept of industrial ecology has expanded to encompass issues of design for the environment, product design, closing materials loops, recycling, and other environmentally conscious practices. Industrial ecology can be considered a comprehensive approach to implementing sustainable industrial behavior.

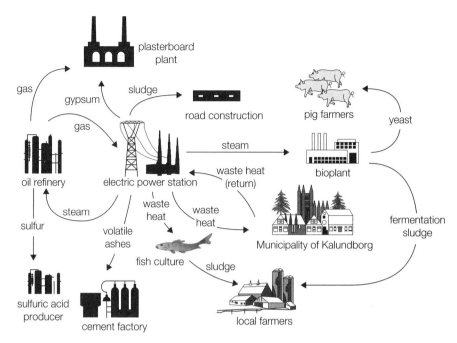

Figure 2.4 The industrial complex in Kalundborg, Denmark, exchanges energy, water, and materials among its member companies and organizations, demonstrating industrial symbiosis, one of the basic concepts of industrial ecology. (*Source: Ecodecision*, Spring 1996: 20)

CONSTRUCTION ECOLOGY

Construction ecology is a subcategory of industrial ecology that applies specifically to the built environment. Construction ecology employs principles of industrial ecology combined with ecological theory that differentiates buildings from other industrial products such as automobiles, refrigerators, and copying machines. Construction ecology also supports the design and construction of a built environment that (1) has a closed-loop materials system integrated with eco-industrial and natural systems, (2) depends solely on renewable energy sources, and (3) fosters the preservation of natural system functions. Application of these principles should result in buildings that (1) are readily deconstructable at the end of their useful lives; (2) have components that are decoupled from the building for easy replacement; (3) are composed of products designed for recycling; (4) are built using recyclable, bulk structural materials; (5) have slow "metabolisms" due to their durability and adaptability; and (6) promote the health of their human occupants.[22]

BIOMIMICRY

The term *biomimicry* was popularized by Janine Benyus in her book, *Biomimicry: Innovation Inspired by Nature*, and has since received widespread attention as a concept that demonstrates the direct application of ecological concepts to the production of industrial objects.[23] According to Benyus, biomimicry is the "conscious imitation of nature's genius," and it suggests that most of what we need to know about energy and materials use has been developed by natural systems over almost 4 billion years of trial and error. Biomimicry advocates the possibility of creating strong, tough, and intelligent materials from naturally occurring materials, at ambient temperatures, with no waste, and using current solar "income" (sunlight) to power the manufacturing process. For example, nature produces strong, elegant, functional, and beautiful ceramic seashells from local materials in seawater at ambient temperatures. At the end of their useful lives as aquatic habitat, they degrade and provide future resources in a waste-free manner. In contrast, production of ceramic clay tiles requires high fire temperatures of at least 2700°F and the extraction and transport of clay and energy resources, and results in emissions and waste. Unlike their natural counterparts, clay tiles do not degrade into useful products and are likely to be disposed of in landfills at the end of their useful lives.

DESIGN FOR THE ENVIRONMENT

Design for the environment (DfE), sometimes referred to as *green design*, is a practice that integrates environmental considerations into product and process engineering procedures and considers the entire product life cycle.[24] The related concept of front-loaded design advocates the investment of greater effort during the design phase to ensure the recovery, reuse, and/or recycling of the product's components. Although DfE typically describes the process of designing products that can be disassembled and recycled, depending on the context, DfE may encompass design for disassembly, design for recycling, design for reuse, design for remanufacturing, and other approaches. DfE's application to building design implies that, to be considered green, significant effort must be made in product design to enable the reuse and recycling of the product's components. A window assembly designed using DfE strategies, for example, would be easy to remove from the building and to disassemble into its basic metal, glass, and plastic components. Furthermore, the materials must possess and maintain value in order to motivate the industrial system to keep them in productive use. As applied to the built environment, DfE implies that entire buildings should be designed to be taken apart, or deconstructed, to recover components for further disassembly, reuse, and recycling.

ECOLOGICAL ECONOMICS

Contemporary, or neoclassical, economics fails to consider or adequately address the problems of resource limitations or the environmental impact of waste and toxic substances on productive ecological systems. In contrast, ecological economics posits that healthy, natural systems and the free goods and services provided by nature are essential to economic success. *Ecological economics* is a fundamental requirement of sustainable development that specifically addresses the relationship between human economies and natural ecosystems. Since the human economy is embedded in the larger natural ecosystem and depends on it for exchanging matter and energy, both systems must coevolve. Ecological economic philosophy counters the human propensity to ignorantly or deliberately degrade ecosystems by extracting useful, high-quality matter and energy, which are ultimately transformed into useless, low-quality waste and heat. Ecological economics values nature's provision of goods, energy, services, and amenities, as well as humanity's cultural and moral contributions.[25] Valuing nature—that is, assigning a monetary worth to its goods and services—although antithetical to some, is essential to appreciate and understand the worth of natural system resources and services in the human economy.

Unfortunately, obstacles exist to replacing the shortsighted approach of contemporary neoclassical economics with the ecological economics consideration of the contributions and limitations of natural systems. Our present limited understanding of complex nonlinear natural systems, as well as the difficulty of accurately representing these systems in relevant economic models, presents challenges. Nonetheless, ecological economics illuminates the dismal science of traditional economics and provides a more comprehensive framework for applying economic principles in the evolving, transformative era of sustainable development.

CARRYING CAPACITY

The term *carrying capacity* attempts to define the limits of a specific land's capability to support people and their activities. According to the Carrying Capacity Network,

> Carrying capacity is the number of people who can be supported in a given area within natural resource limits, and without degrading the natural social, cultural, and economic environment for present and future generations. The carrying capacity for any given area is not fixed. It can be altered by improved technology, but mostly it is changed for the worse by pressures which accompany a population increase. As the environment is degraded, carrying capacity actually shrinks, leaving the environment no longer able to support even the number of people who could formerly have lived in the area on a sustainable basis. No population can live beyond the environment's carrying capacity for very long.[26]

Carrying capacity focuses on the relationship between land area and human population growth and suggests the point at which the system may break down. Much debate surrounds the carrying capacity of the planet in general and the United States in particular. Although the United States may be able to carry 1 billion people with adequate resources, it is doubtful that a population of this magnitude is desirable. The concept of carrying capacity is also linked to the precautionary principle, discussed earlier in this chapter.

ECOLOGICAL FOOTPRINT

Mathis Wackernagel and William Rees suggested that an *ecological footprint*, referring to the land area required to support a certain population or activity, could serve as a surrogate measure for total resource consumption, thus allowing a simple comparison of the resource consumption of various lifestyles.[27] The ecological footprint is the inverse of carrying capacity and represents the amount of land needed to support a given population. An ecological footprint calculation indicates that, for example, the Dutch need a land area 15 times larger than that of the Netherlands to support their population. The population of London requires a land area 125 times greater than its physical footprint. If everyone on earth enjoyed a North American lifestyle, it would take up to five planet earths, owing to the increasingly consumptive US lifestyle and the burgeoning world population, which exceeds 7 billion at the time of this writing. The ultimate problem that must be solved, especially in the context of sustainable development, is how all people can have a decent quality of life without destroying the planetary systems that support life itself. A partial solution requires developed countries to dramatically reduce consumption and to ensure that developing countries receive resources sufficient for more than mere survival. Such resource sharing lies at the heart of the original formulation of sustainable development, which values the goal of moving the developing world from mere survival to the ability to sustain a reasonably good quality of life. As William Rees notes in the preface to the book *Our Ecological Footprint*, coauthored with Mathis Wackernagel, "on a finite planet, at human carrying capacity, a society driven mainly by selfish individualism has all the potential for sustainability of a collection of angry scorpions in a bottle."

TABLE 2.1

Ecological Rucksack* of Some Well-Known Materials

Material	Ecological Rucksack
Rubber	5
Aluminum	85
Recycled aluminum	4
Steel	21
Recycled steel	5
Platinum	350,000
Gold	540,000
Diamond	53,000,000

*The rucksack indicates how many units of mass must be moved to produce one unit mass of the material. For example, 1 kilogram (2.2 pounds) of aluminum from bauxite requires displacing 85 kilograms (187 pounds) of materials, compared to moving only 4 kilograms (9 pounds) to produce 1 kilogram of recycled aluminum.

THE ECOLOGICAL RUCKSACK AND MATERIALS INTENSITY PER UNIT SERVICE

The term *ecological rucksack*, coined by Friedrich Schmidt-Bleek, formerly of the Wuppertal Institute in Wuppertal, Germany, attempts to quantify the mass of materials that must be moved in order to extract a specific resource. The concept of the ecological rucksack was developed to demonstrate that prosperity attributable to certain human activities has been achieved only by the destruction of natural resources through excavation, mining, channeling rivers and lakes, and processing gigatons[28] of materials to extract dilute resources. Schmidt-Bleek suggested that since these activities are responsible for significant environmental damage, extracted materials could be said to carry a "rucksack," or extraction burden. For example, the 10 grams of gold contained in a typical gold wedding band are extracted and concentrated from 300 tons of raw material.

The European Environment Agency (EEA) defines ecological rucksack as the material input of a product or service minus the weight of the product itself.[29] The material input is defined as the life-cycle-wide total quantity (in pounds or kilograms) of natural material physically displaced in order to generate a particular product (see Table 2.1).[30] The environmental stress caused by an activity is proportional to the quantity of materials moved. The greater the mass moved, the higher is the environmental impact. The concept of ecological rucksack focuses on these large displacements of earth and rock rather than on minute quantities of toxic materials. It has been the large land transformations occasioned by increasing material demands, coupled with depleted deposits of rich materials, that have been historically neglected by environmentalists and policy makers.

Materials intensity per unit service (MIPS) is another concept originated by Schmidt-Bleek to assist in understanding the efficiency with which materials are used. MIPS measures how much service a given product delivers. The higher or greater the service, the lower is the MIPS value. MIPS is also an indicator of resource productivity, or eco-efficiency, and products with greater service are said to possess greater eco-efficiency and resource productivity.

THE BIOPHILIA HYPOTHESIS

E.O. Wilson, the eminent Harvard University entomologist, suggested that humans have a need and craving to be connected to nature and living things. He coined the term *biophilia hypothesis* to propose the concept that humans have an affinity for nature and that they "tend to focus on life and lifelike processes." The biophilia hypothesis asserts the existence of a fundamental genetically based human need and propensity to affiliate with life and lifelike processes. Various recent studies have shown that even minimal connection with nature, such as looking outdoors through a window, increases productivity and health in the workplace, promotes healing of patients in hospitals, and reduces the frequency of sickness in prisons. Prison inmates whose cells overlooked farmlands and forests needed fewer health-care services than inmates whose cells overlooked the prison yard.[31]

In their book *The Biophilia Hypothesis*, Wilson and Stephen Kellert, a professor in the School of Forestry and Environmental Studies at Yale University, collected invited papers to both support and refute this hypothesis. Kellert states that for green buildings to eventually become truly successful, they must relate to natural processes and help humans achieve meaning and satisfaction. He suggests that there are nine values of biophilia, which offer a broad design template for sustainable building: (1) the utilitarian value emphasizes the material benefit that humans derive from exploiting nature to satisfy various needs and desires; (2) the aesthetic value emphasizes a primarily emotional response of intense pleasure at the physical beauty of nature; (3) the scientific value emphasizes the systematic study of the biophysical patterns, structures, and functions of nature; (4) the symbolic value emphasizes the tendency for humans to use nature for communication and thought; (5) the naturalistic value emphasizes the many satisfactions people obtain from the direct experience of nature and wildlife; (6) the humanistic value emphasizes the capacity for humans to care for and become intimate with animals; (7) the dominionistic value emphasizes the desire to subdue and control nature; (8) the moralistic value emphasizes right and wrong conduct toward the nonhuman world; and (9) the negativistic value emphasizes feelings of aversion, fear, and dislike that humans have for nature.[32]

Anecdotal evidence emerging about the effects of daylighting and views to the outside indicates that human health, productivity, and well-being are promoted by access to natural light and views of greenery. Hundreds of studies have demonstrated that stress reduction results from connecting humans to nature. Consequently, facilitating the ability of humans to interact with nature, even at a distance, from inside a building, is emerging as an issue for consideration in the creation of high-performance green buildings.

ECO-EFFICIENCY

Originated by the World Business Council for Sustainable Development (WBCSD) in 1992, the concept of *eco-efficiency* includes environmental impacts and costs as a factor in calculating business efficiency. The WBCSD considers the term *eco-efficiency* to describe the delivery of competitively priced goods and services that satisfy human needs and enhance the quality of life while progressively reducing ecological impacts and resource intensity throughout the products' life cycles to a level commensurate with the earth's estimated carrying capacity. The WBCSD has articulated seven elements of eco-efficiency (see Table 2.2).[33]

Furthermore, the WBCSD has identified four aspects of eco-efficiency that render it an indispensable strategic element in the contemporary knowledge-based economy:[34]

- Dematerialization: companies are developing ways of substituting knowledge flows for material flows.
- Closing production loops: the biological designs of nature provide a role model for sustainability.
- Service extension: the world is moving from a supply-driven economy to a demand-driven economy.
- Functional extension: companies are manufacturing smarter products with new and enhanced functionality and are selling services to enhance the products' functional value.

The WBCSD suggests that business can achieve eco-efficiency gains through:

- Optimized processes: moving from costly end-of-pipe solutions to approaches that prevent pollution in the first place
- Waste recycling: using the by-products and wastes of one industry as raw materials and resources for another, thus creating zero waste

TABLE 2.2

Seven Elements of Eco-Efficiency as Defined by the WBCSD

1. Reducing the material requirements of goods and services
2. Reducing the energy intensity of goods and services
3. Reducing toxic dispersion
4. Enhancing materials recyclability
5. Maximizing sustainable use of renewable resources
6. Extending product durability
7. Increasing the service intensity of goods and services

- Eco-innovation: manufacturing "smarter" by using new knowledge to make old products more resource-efficient to produce and use
- New services: for instance, leasing products rather than selling them, which changes companies' perceptions, spurring a shift to product durability and recycling
- Networks and virtual organizations: sharing resources to increase the effective use of physical assets

As a concept, eco-efficiency describes most of the foundational principles underpinning the concept of sustainable development. Its promotion by the WBCSD, essentially an association of major corporations, is a positive sign that the business community is beginning to take sustainability seriously.

THE NATURAL STEP

Developed by Swedish oncologist Karl-Henrik Robèrt in 1989, the *Natural Step* provides a framework for considering the effects of materials selection on human health. Robèrt suggested that many human health problems, particularly those of children, result from materials we use in our daily lives. The extraction of resources such as fossil fuels and metal ores from the planet's crust produces carcinogens and results in heavy metals entering the earth's surface biosphere. The abundance of chemically produced synthetic substances that have no model in nature has similar deleterious effects on health. The Natural Step articulates the four systems conditions, or basic principles, that should be followed to eliminate the effects of materials practices on our health. The four systems conditions are listed here.[35] Their potential application to construction projects is described in greater detail in Chapter 11.

1. In order for a society to be sustainable, nature's functions and diversity are not systematically subject to increasing concentrations of substances extracted from the earth's crust.
2. In order for a society to be sustainable, nature's functions and diversity are not systematically subject to increasing concentrations of substances produced by society.
3. In order for a society to be sustainable, nature's functions and diversity are not systematically impoverished by overharvesting or other forms of ecosystem manipulation.
4. In a sustainable society, resources are used fairly and efficiently in order to meet basic human needs globally.

LIFE-CYCLE ASSESSMENT

Life-cycle assessment (LCA) is a method for determining the environmental and resource impacts of a material, a product, or even a whole building over its entire life. All energy, water, and materials resources, as well as all emissions to air, water, and land, are tabulated over the entity's life cycle. The life cycle, or time period considered in this evaluation, can span the extraction of resources, the manufacturing process, the installation in a building, and the item's ultimate disposal. The assessment also considers the resources needed to transport components from extraction through disposal. LCA is an important, comprehensive approach that examines all impacts of material selection decisions, rather than simply an item's performance in the building. LCA and the tools used to produce an LCA are described in greater detail in Chapter 11.

LIFE-CYCLE COSTING

The ability to model a building's financial performance over its life cycle is necessary to justify measures that may require greater initial capital investment but yield significantly lower operational costs over time. Using *life-cycle costing* (LCC), a cost/benefit analysis is performed for each year of the building's probable life. The present worth of each year's net benefits is determined using an appropriate discount rate. Net benefits for each year are tabulated to calculate the total present worth of a particular feature. For example, the financial return for installation of a photovoltaic system would be determined by amortizing the system's costs over its probable life; the worth of the energy generated each year would then be calculated to determine the net annual benefit. Application of LCC may determine whether the payback for this system meets the owner's economic criteria. LCC analysis can also be combined with LCA results to weigh the combined financial and environmental impact of a particular system. LCC is covered in more detail in Chapter 14.

EMBODIED ENERGY

Embodied energy refers to the total energy consumed in the acquisition and processing of raw materials, including manufacturing, transportation, and final installation. Products with greater embodied energy usually have higher environmental impact due to the emissions and greenhouse gases associated with energy consumption. However, another calculation, which divides the embodied energy by the product's time in use, yields a truer indicator of the environmental impact. More durable products will have a lower embodied energy per time in use. For example, a product with high embodied energy such as aluminum could have a very low embodied energy per time in use because of its extremely high durability. Additionally, certain products have relatively low embodied energy when recycled. Recycled aluminum has just 10 percent of the embodied energy of aluminum made from bauxite ore. Similarly, recycled steel has about 20 percent of the embodied energy of steel made from ores. A list of typical embodied energies for common construction materials is presented in Table 2.3.[36]

FACTOR 4, FACTOR 5, AND FACTOR 10

The concepts of *Factor 4 and Factor 10* provide a set of guidelines for comparing design options and for evaluating the performance of buildings and their component systems. The notion of Factor 4 was first suggested in the book *Factor Four: Doubling Wealth, Halving Resource Use*, written in 1997 by Ernst von Weizsäcker, Amory Lovins, and L. Hunter Lovins (see Figure 2.5).[37] Factor 4 suggests that for humanity to live sustainably today, we must rapidly reduce resource consumption to one-quarter of its current levels. Fortunately, the technology to accomplish Factor 4 reductions in resource consumption already exists and requires only public policy prioritization and implementation. A parallel approach originated by Friedrich Schmidt-Bleek hypothesizes that, in order to achieve long-term sustainability, we must reduce resource consumption by a factor of 10.[38] An example of applying this principle to the built environment is provided by Lee Eng Lock, a Chinese engineer in Singapore. He has challenged many of the fundamental assumptions made by mechanical engineers in their systems design and layout. Rather than oversizing chillers, air handlers, pumps, and other equipment, he ensures that they are precisely the correct size for the job. This commonsense approach achieves the same cooling and comfort while using only 10 percent of the energy of conventional designs, thus accomplishing a Factor 10 reduction in energy.[39] The

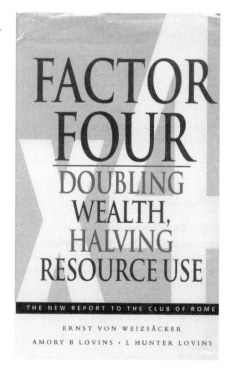

Figure 2.5 The Factor 4 concept originated in the book *Factor Four: Doubling Wealth, Halving Resource Use*, by Ernst von Weizsäcker, Amory Lovins, and L. Hunter Lovins.

TABLE 2.3

Embodied Energy of Common Construction Materials

Material	Embodied Energy MJ/kg*	MJ/m^3†
Aggregate	0.1	150
Concrete (30 MPa)	1.3	3,180
Lumber	2.5	1,380
Brick	2.5	5,170
Cellulose insulation	3.3	112
Mineral wool insulation	14.6	139
Fiberglass insulation	30.3	970
Polystyrene insulation	117.0	3,770
Gypsum wallboard	6.1	5,890
Particleboard	8.0	4,400
Plywood	10.4	5,720
Aluminum	227.0	515,700
Aluminum (recycled)	8.1	21,870
Steel	32.0	251,200
Steel (recycled)	8.9	37,210
Zinc	51.0	371,280
Copper	70.6	631,164
Polyvinyl chloride (PVC)	70.0	93,620
Linoleum	116.0	150,930
Carpet (synthetic)	148.0	84,900
Paint	93.3	117,500
Asphalt shingles	9.0	4,930

*Megajoules per kilogram of material.
†Megajoules per cubic meter of material.

Factor 10 concept has had a significant effect internationally and is now being implemented by the EU. The Factor 4 concept was revisited by Ernst von Weizsäcker and several Australian colleagues in 2009, and they concluded there was a potential Factor 5 of available efficiency improvements for entire sectors of the economy, without losing the quality of service or well-being.[40]

Major Environmental and Resource Concerns

TABLE 2.4

Major Environmental Issues Connected to Built Environment Design and Construction

Climate change
Ozone depletion
Soil erosion
Desertification
Deforestation
Eutrophication
Acidification
Loss of biodiversity
Land, water, and air pollution
Dispersion of toxic substances
Depletion of fisheries

Concerns about environmental degradation, resource shortages, and human health impacts are promoting widespread acceptance of green building, the ultimate goal of which is to mitigate the enormous pressures on planetary ecosystems caused by human activities. The major environmental issues to be addressed by sustainable construction methods are shown in Table 2.4. Some of these are covered in more detail in the following sections.

CLIMATE CHANGE

As defined by the National Oceanic and Atmospheric Administration (NOAA), *climate change* consists of long-term fluctuations in temperature, precipitation, wind, and all other aspects of the earth's climate. The UN Convention on Climate Change describes the phenomenon as a change of climate attributable directly or indirectly to human activity that alters the composition of the global atmosphere and that is, in addition to natural climate variability, observable over comparable time periods.

A strong majority of Nobel Prize–winning scientists believe there is very strong evidence that the average temperature of the planet's surface will increase 4 to 6°C in the 21st century. The likely result will be rapidly rising sea levels, substantially reduced crop yields, drought, and more energetic hurricanes and cyclones, all threatening the very survival of the human species. Increasing concentrations of climate change gases produced by human activities, particularly carbon dioxide, are the primary force pushing up global average temperatures. Carbon dioxide concentrations are now about 387 parts per million (ppm) compared to 280 ppm at the start of the Industrial Revolution 225 years ago in the late 18th century. The geological record indicates that an average carbon dioxide level of 450 ppm defines the boundary between an ice-free planet, when water levels were 220 feet higher than today, and a planet with ice sheets. Climate scientists generally suggest that 350 ppm is the safe upper limit for carbon dioxide, one we have already surpassed. The continuation of business as usual (BAU) will likely cause carbon dioxide levels to increase from the current concentration of 380 ppm to over 450 ppm between 2030 and 2050. Clearly, strong and drastic action is needed to stop and reverse atmospheric carbon dioxide concentrations to avoid the worst outcomes of climate change. It should be noted that carbon dioxide is the most common greenhouse gas by far, but there are more potent greenhouse gases such as methane that have a much higher impact per molecule (see Tables 2.5 and 2.6).

TABLE 2.5

Current and Preindustrial Concentrations of Greenhouse Gases

Gas	Chemical Formula	Preindustrial Level	Current Level*	Increase since 1750
Carbon dioxide	CO_2	280 ppm	389 ppm	109 ppm
Methane	CH_4	700 ppb	1745 ppb	1045 ppb
Nitrous oxide	N_2O	270 ppb	314 ppb	44 ppb
CFC-12	CCl_2F_2	0	533 ppt	533 ppt
HCFC-22	$CHClF_2$	0	206 ppt	206 ppt

*Concentrations are given in ppm (parts per million), ppb (parts per billion), and ppt (parts per trillion).
Source: IPCC Fourth Assessment Report (2007).

TABLE 2.6

Atmospheric Lifetime and Global Warming Potential (GWP) of Greenhouse Gases

Gas	Chemical Formula	Lifetime (years)*	GWP†
Carbon dioxide	CO_2	Variable	1
Methane	CH_4	12	72
Nitrous oxide	N_2O	114	289
CFC-12	CCl_2F_2	100	11,000
HCFC-22	$CHClF_2$	12	5,160

*The lifetime of carbon dioxide is variable because about 50 percent is removed in a century while about 20 percent is resident in the atmosphere for thousands of years.
†The GWP of carbon dioxide is defined as 1. The GWP of 72 for methane means that each molecule is 72 times more potent than a carbon molecule at trapping energy. Although methane is far more potent, its atmospheric concentration is relatively small.
Source: IPCC Fourth Assessment Report (2007).

The Intergovernmental Panel on Climate Change (IPCC) was established by the World Meteorological Organization (WMO) and the United Nations (UN) in 1988 to assess, on a comprehensive, objective, open, and transparent basis, the scientific, technical, and socioeconomic information relevant to understanding the scientific basis of the risk of human-induced climate change, its potential impacts, and options for adaptation and mitigation. The Fourth Assessment Report of the IPCC, published in 2007, is the latest report to the global community.[41] According to this report, the best estimate range of projected temperature increase is 3.1 to 7.2°F (1.8 to 4.0°C) by the end of the century. As noted above, tropical cyclones (hurricanes and typhoons) are likely to become more intense, with higher peak wind speeds and heavier precipitation associated with warmer tropical seas. Extreme heat, heat waves, and heavy precipitation are very likely to continue becoming more frequent. Sea ice is projected to shrink in both the Arctic and the Antarctic under all model simulations. Some projections show that, by the latter part of the century, late-summer Arctic sea ice will disappear almost entirely. The IPCC states that it is very likely that circulation in the Atlantic Ocean will be 25 percent slower on average by 2100 (with a range from 0 to 50 percent). Nevertheless, Atlantic regional temperatures are projected to rise overall due to more significant warming from increases in heat-trapping emissions. Increasing atmospheric carbon dioxide concentrations will lead to increasing acidification of the ocean, with negative repercussions for all shell-forming species and their ecosystems.

The models used by the IPCC project that, by the end of this century, the global average sea level will rise between 7 and 23 inches (0.18 and 0.59 meter) above the 1980–1999 average. Also, if the observed contributions from the Greenland and Antarctic ice sheets between 1992 and 2003, the IPCC states, "were to grow linearly with global average temperature change," the upper ranges of sea-level rise would increase by 3.9 to 7.9 inches (0.1 to 0.2 meter). In other words, in this example, the upper range for sea-level rise would be 31 inches (0.79 meter).

Moreover, there could be changes in climate variability, as well as in the frequency and intensity of some extreme climate phenomena. It is important to note that systems theory shows that the behavior of global systems such as climate is nonlinear. Each increase in carbon dioxide will not necessarily produce a proportional change in global temperature. However, the dynamic, chaotic character of the earth's climate is such that climate can suddenly "flip" from one temperature regime to another in a relatively short time. Indeed, fossil records indicate that previous flips have occurred, with temperatures increasing or decreasing almost 10°F (5.6°C) in about a decade. The potential for climate change has profound implications for every aspect of human activity on the planet. Shifting temperatures, more violent storms, rising sea levels, melting glaciers, and other effects will displace people, affect food supplies, reduce biodiversity, and greatly reduce the average quality of life. The responsibility for creators of the built environment, which is a major energy consumer, is to dramatically reduce energy consumption, particularly reliance upon fossil fuels.

In addition to causing climate change, certain chemicals used in building construction and facility operations have been thinning the ozone layer, the protective sheath of the atmosphere consisting of three-molecule oxygen (O_3), which is located 10 to 25 miles (16 to 40 kilometers) above the earth and serves to attenuate harmful ultraviolet radiation. In 1985, scientists discovered a vast hole the size of the continental United States in the ozone layer over Antarctica. By 1999, the size of the hole had doubled. Ozone depletion is caused by the interaction of halogens—chlorine- and bromine-containing gases such as chlorofluorocarbons (CFCs) used in refrigeration and foam blowing, and halons used for fire suppression. Table 2.7 provides a summary of the main

TABLE 2.7

Gases Used for Typical Building Functions

Halogen Gas*	Lifetime† (years)	Global Emissions (1000s of metric tons/year)	Ozone Depletion Potential (ODP)‡
Chlorine			
CFC-12	100	130–160	1
CFC-113	85	10–25	1
CFC-11	45	70–110	1
HCFCs	1–26	340–370	0.02–0.12
Bromine			
Halon 1301	65	~3	12
Halon 1211	16	~10	6

*The chlorine gases are used in refrigerants, the bromine gases in fire suppression systems.
†Lifetime refers to their duration in the atmosphere, and ODP is their ozone depletion impact.
‡The ODP of CFC-11 is defined as 1. With an ODP of 12, Halon 1301 depletes ozone at a rate 12 times greater than that of CFC-11.

contributors to the destruction of the ozone layer.[42] In one of the few successful examples of international environmental cooperation, the UN Montreal Protocol of 1987 produced an international agreement to eventually halt the production of ozone-depleting chemicals. Assuming that the Montreal Protocol is faithfully adhered to by the international community, the ozone layer is projected to be fully restored by the year 2050.

DEFORESTATION, DESERTIFICATION, AND SOIL EROSION

Natural forests are estimated to contain half of the world's total biological diversity, possessing the greatest level of biodiversity of any type of ecosystem. Sadly, worldwide *deforestation* is occurring at a rapid rate, with 2 acres (0.8 hectare) of rainforest disappearing every second[43] and temperate zone forests losing about 10 million acres (4 million hectares) per year. Although about one-third of the total land area is forested worldwide, about half of the earth's forests have disappeared. In the United States, only 1 to 2 percent of the original forest cover still remains. This pattern of large-scale forest removal, known as *deforestation*, is linked to negative environmental consequences such as biodiversity loss, global warming, soil erosion, and desertification (see Figure 2.6).

Deforestation defeats the capability of forests to "lock up," or sequester, the large quantities of carbon dioxide stored in tree mass; instead, it is released into the atmosphere as gaseous compounds, which contribute to accelerated climate change. Between 1850 and 1990, worldwide deforestation released 134 billion tons (122 billion metric tons) of carbon. Currently, deforestation releases about 1.8 billion tons (1.6 billion metric tons) of carbon per year, compared to the burning of fossil fuels such as oil, coal, and gas, which releases about 6.6 billion tons (6 billion metric tons) per year. And because trees and their root systems are necessary to prevent soil erosion, landslides, and avalanches, their removal contributes to soil loss and changes the rate at which water enters the watershed. Forest-sustained freshwater supplies are an important source of oxygen, which fosters biodiversity, especially in rainforests. Additionally, large-scale deforestation affects the albedo, or reflectivity, of the earth, altering its surface temperature and energy, rate of surface water evaporation, and, ultimately, patterns and quantity of rainfall.

Figure 2.6 Deforestation, such as this clear-cut in northern Florida, destroys animal habitat, causes soil erosion, and affects biodiversity. Green building standards call for the use of wood products from sustainably managed forests. (Photograph courtesy of M. R. Moretti)

Figure 2.7 Desertification in southern Niger is consuming not only land but also local villages.

Deforestation also causes soil erosion, a key factor in land degradation. More than 2 billion tons (1.8 billion metric tons) of topsoil are lost annually due to human agricultural and forestry land development. More than 5 billion acres (2 billion hectares) of land, an area equal to the United States and Mexico combined, is now considered degraded.[44] In arid and semiarid regions, degradation results in *desertification*, or the destruction of natural vegetative cover, which prevents desert formation. The UN Convention to Combat Desertification, formed in 1996 and ratified by 179 countries, reports that over 250 million people are directly affected by desertification.[45] Furthermore, drylands susceptible to desertification cover 40 percent of the earth's surface, putting at risk a further 1.1 billion people in more than 100 countries dependent on these lands for survival. China, with a rapidly growing population and economy, loses about 300,000 acres (121,000 hectares) of land each year to drifting sand dunes (see Figure 2.7).

EUTROPHICATION AND ACIDIFICATION

Two environmental conditions that frequently threaten water supplies are eutrophication and acidification. *Eutrophication* refers to the overenrichment of

Figure 2.8 Agricultural runoff, urban runoff, leaking septic systems, sewage discharges, eroded stream beds, and similar sources can increase the flow of nutrients and organic substances into aquatic systems. The result is an overstimulation of algae growth, causing eutrophication and interfering with the recreational use of lakes and estuaries and adversely affecting the health and diversity of indigenous fish and animal populations. (Photograph courtesy of M. R. Moretti)

water bodies with nutrients from agricultural and landscape fertilizer, urban runoff, sewage discharge, and eroded stream banks (see Figure 2.8). Nutrient oversupply fosters algae growth, or algae blooms, which block sunlight and cause underwater grasses to die. Decomposing algae further utilize dissolved oxygen necessary for the survival of aquatic species such as fish and crabs. Eventually, decomposition in a completely oxygen free, or *anoxic*, water body can release toxic hydrogen sulfide, poisoning organisms and making the lake or seabed lifeless. Eutrophication has led to the degradation of numerous water-ways around the world. For example, in the Baltic Sea, huge algae blooms, now common after unusually warm summers, have decreased water visibility by 10 to 15 feet (3 to 4.6 meters) in depth.

Acidification is the process whereby air pollution in the form of ammonia, sulfur dioxide, and nitrogen oxides, mainly released into the atmosphere by burning fossil fuels, is converted into acids. The resulting acid rain is well known for its damage to forests and lakes. Less obvious is the damage caused by acid rain to freshwater and coastal ecosystems, soils, and even ancient historical monuments. The acidity of polluted rain leaches minerals from soil, causing the release of heavy metals that harm microorganisms and affect the food chain. Many species of animals, fish, and other aquatic animal and plant life are sensitive to water acidity. As a result of European directives that forced the installation of desulfurization systems and discouraged the use of coal as a fossil fuel, Europe experienced a significant decrease in acid rain in the 1990s. Nonetheless, a 1999 survey of forests in Europe found that about 25 percent of all trees had been damaged, largely due to the effects of acidification.[46]

LOSS OF BIODIVERSITY

Biodiversity refers to the variety and variability of living organisms and the ecosystems in which they occur. The concept of biodiversity encompasses the number of different organisms, their relative frequencies, and their orga-nization at many levels, ranging from complete ecosystems to the biochemical structures that form the molecular basis of heredity. Thus, biodiversity expresses the range of life on the planet, considering the relative abundances of ecosystems, species, and genes. Species biodiversity is the level of biodiversity most commonly discussed. An estimated 1.7 million species have been described out of a total estimated 5 to 100 million species. However, deforestation and climate change are causing such a rapid extinction of many species that some biologists are predicting the loss of 20 percent of existing species over the next 20 years.

Deforestation is particularly devastating, especially in rainforests, which comprise just 6 percent of the world's land but contain more than 500,000 of its species. Biodiversity preservation and protection is important to humanity since diverse ecosystems provide numerous services and resources, such as protection and formation of water and soil resources, nutrient storage and cycling, pollution breakdown and absorption, food, medicinal resources, wood products, aquatic habitat, and undoubtedly many undiscovered applications.[47] Once lost, species cannot be replaced by human technology, and potential sources of new foods, medicines, and other technologies may be forever forfeited.

Furthermore, destruction of ecosystems contributes to the emergence and spread of infectious diseases by interfering with natural control of disease vectors. For example, the fragmentation of North American forests has resulted in the elimination of the predators of the white-footed mouse, which is a major carrier of Lyme disease, now the leading vector-borne infectious illness in the United States. Finally, species extinction prevents discovery of potentially useful medicines such as aspirin, morphine, vincristine, taxol, digitalis, and most antibiotics, all of which have been derived from natural models.[48]

TOXIC SUBSTANCES AND ENDOCRINE DISRUPTORS

One dangerous by-product of the human propensity to invent has been the creation of an enormous number of chemical compounds that have no analogue in nature and often affect biological systems toxically. A *toxic substance* is a chemical that can cause death, disease, behavioral abnormalities, cancer, genetic mutations, physiological or reproductive malfunctions, or physical deformities in any organism or its offspring, or that can become poisonous after concentration in the food chain or in combination with any other substances.[49] Toxic substances can be carcinogenic or mutagenic, and they affect developmental, reproductive, neurological, or respiratory systems. Ignitable or corrosive substances are also classified as toxic. As an aside, *toxins* are biological poisons that are the by-products of living organisms. A toxin may be obtained naturally, that is, from secretions of various organisms, or it may be synthesized.

The rate of synthetically produced chemicals in the United States has increased from 1 million tons (0.9 million metric tons) per year in 1940 to over 125 million tons (113 million metric tons) per year in 1987. And in spite of the fact that, in 2012, approximately 67 million commercially available chemicals were listed with the Chemical Abstracts Service (CAS), the National Academy of Sciences stated that adequate information to assess public health hazards existed for only 2 percent of these chemicals. Each year, more than 6000 new chemical compounds are developed; however, industry is required to report the environmental release of only 320 specific substances. Over 3 billion pounds (1.4 billion kilograms) of toxic chemicals enter the environment each year, with official hazardous waste production amounting to 1400 trillion pounds (635 trillion kilograms) per year. Each year, US industry produces about 12 pounds (5.4 kilograms) of toxic waste per capita.[50] Since 1987, industries have been required to report the release of certain chemicals to the government through the Toxics Release Inventory (TRI), but the TRI does not cover all chemicals or all industries, and only the largest facilities are required to report. A report by the US Public Interest Research Group (PIRG) Education Fund found that the following amounts of chemicals were released into the atmosphere in the year 2000:[51]

- Cancer-causing chemicals: 100 million pounds (45.4 million kilograms), with dichloromethane being the most frequent
- Chemicals, such as toluene, linked to developmental problems: 138 million pounds (63 million kilograms)
- Chemicals, such as carbon disulfide, related to reproductive disorders: 50 million pounds (23 million kilograms)
- Respiratory toxicants: 1.7 billion pounds (0.8 billion kilograms), most commonly acid aerosols of hydrochloric acid
- Dioxins: grams (15.4 pounds)
- Persistent toxic substances: lead [275,000 pounds (125,000 kilograms)], lead compounds [1.3 million pounds (0.6 million kilograms), mercury [30,000 pounds (13,600 kilograms)], and mercury compounds [136,000 pounds (62,000 kilograms)]

During the past few decades, it has become apparent that many chemicals damage animal and human hormonal systems. *Endocrine-disrupting chemicals* (EDCs) interfere with the hormones produced by the endocrine system, a complex network of glands and hormones that regulates the development and function of bodily organs, physical growth, development, and maturation. Some commonly known EDCs are dioxin, polychlorinated biphenyls (PCBs), dichlorodiphenyltrichloroethane (DDT), and various pesticides and plasticizers. EDCs have been implicated in the occurrence of abnormally swollen thyroid glands in the eagles, terns, and gulls found in the fish-bird food chain of the Great Lakes. EDCs have contributed to the appearance of alligators with diminished reproductive organs and are blamed for the declining alligator populations in Lake Apopka, Florida. The most notorious example occurring in the human population was the use of diethylstilbestrol (DES), a synthetic estrogen prescribed until 1971 to prevent miscarriages in pregnant women. DES has since been linked to numerous health problems in offspring exposed to DES in the womb, including reproductive complications and infertility in DES daughters.[52] Although a "better life through chemistry," the tagline of American industry of the 1950s, can still be claimed, the unexpectedly high price tag is still being tallied.

DEPLETION OF METAL STOCKS

The depletion of key resources needed to support the energy and materials requirements of today's technological, developed world societies is a threat to the high quality of life enjoyed by North Americans, Europeans, Japanese, and the other countries that make up modern industrialized societies. The subject of oil depletion is covered in Chapter 1 of this book, and evidence to date seems to indicate that we have maximized our ability to extract oil and that we are in an era of probably far higher prices for oil-based products, among them gasoline, diesel fuel, jet fuel, and oil-based polymers. A similar scenario is playing out with other key resources, most notably metals. A study of the supply and usage of copper, zinc, and other metals has determined that supplies of these resources—even if recycled—may fail to meet the needs of the global population.[53] Even the full extraction of metals from the earth's crust and extensive recycling programs may not meet future demand if all countries try to attain the same standard of living enjoyed in developed nations. The researchers, Robert Gordon, Marlen Bertram, and Thomas Graedel, based their study on metal still in the earth, in use by people, and lost in landfills. Using copper stocks in North America as a starting point, they tracked the evolution of copper mining, use, and loss during the 20th century. They then applied their findings and additional data to an estimate of the global demand for copper and other metals if all nations were fully developed and used modern technologies. The study found that all of the copper in ore, plus all of the copper currently in use, would be required to bring the world to the level of the developed nations for power transmission, construction, and other services and products that depend on copper. Globally, the researchers estimate that 26 percent of extractable copper in the earth's crust is now lost in nonrecycled wastes, while lost zinc is estimated at 19 percent. Interestingly, the researchers said that current prices do not reflect those losses because supplies are still large enough to meet the demand, and new methods have helped mines produce material more efficiently. While copper and zinc are not at risk of depletion in the immediate future, the researchers believe that scarce metals, such as platinum, are at risk of depletion in this century because there is no suitable substitute for their use in devices such as catalytic converters and hydrogen fuel cells. And because the rate of use for metals continues to rise, even the more plentiful metals may face similar depletion risks in the not too distant future. The impact on metal prices due to a combination of demand and dwindling stocks has been dramatic. Between 2002 and 2012, copper experienced a 500 percent rise in price, and the price of metals such as nickel, brass, and stainless steel rose by about 250 percent. In spite of the higher prices, the good news is that there is a renewed emphasis on recycling, using only the quantity of metals required and ensuring that all in-plant scrap is recovered during manufacturing.[54]

The Green Building Movement

More than any other human endeavor, the built environment has direct, complex, and long-lasting impacts on the biosphere. In the United States, the production and manufacture of building components, along with the construction process itself, involves the extraction and movement of 6 billion tons of basic materials annually. The construction industry, representing about 8 percent of the US gross domestic product (GDP), consumes 40 percent of extracted materials in the United States. Some estimates suggest that as much as 90 percent of all materials ever extracted reside in today's buildings and infrastructure. Construction waste is generated at a rate of about 0.5 ton (0.45 metric ton) per person each year in the United States, or about 5 to 10 pounds per square foot (24 to 49 kilograms per square meter) of new construction. Waste from renovation occurs at a level of 70 to 100 pounds per square foot (344 to 489 kilograms per square meter). The demolition process results in truly staggering quantities of waste, with little or no reuse or recycling occurring (see Figure 2.9). Of the approximately 145 million tons (132 million metric tons) of construction and demolition waste generated each year in the United States, about 92 percent is demolition waste, with the remainder being waste from construction activities. In addition to the enormous quantities of waste resulting from built environment activities, questionable urban planning and development practices also have far-reaching consequences. Since transportation consumes about 40 percent of primary energy consumption in the United States, the distribution of the built environment and the consequent need to rely on automobiles for movement between work, home, school, and shopping results in disproportionate

Figure 2.9 Annual construction and demolition waste in the United States is estimated to be about 160 million tons (145 million metric tons), or about one-half ton per capita. Buildings are not generally designed to be disassembled, and the result is that only a small percentage of demolition materials can be recycled. The partial demolition of the Levin College of Law library at the University of Florida in Gainesville in mid-2004 illustrates the quantities of waste typically generated in renovation projects, on the order of 70 to 100 pounds per square foot (344 to 489 kilograms per square meter). (Photograph courtesy of M. R. Moretti)

energy consumption, air pollution, and the generation of carbon dioxide, which contributes to global warming.

The green building movement is the response of the construction industry to the environmental and resource impacts of the built environment. As was noted in Chapter 1, the term *green building* refers to the quality and characteristics of the actual structure created using the principles and methodologies of sustainable construction. In the context of green buildings, *resource efficiency* means high levels of energy and water efficiency, appropriate use of land and landscaping, the use of environmentally friendly materials, and minimizing the life-cycle effects of the building's design and operation.

GREEN BUILDING ORGANIZATIONS—UNITED STATES

Key American organizations promoting the implementation of sustainable construction practices include the US Green Building Council, the Green Building Initiative, the US Department of Energy, the US Environmental Protection Agency, the National Association of Home Builders, the US Department of Defense, and other public agencies and nonprofit organizations. The private sector has been led by several manufacturers. Notably, the late Ray Anderson, founder and former chairman of InterfaceFLOR, guided the company's transition from a conventional carpet tile manufacturer to one with a corporate philosophy based on industrial ecology (see Figure 2.10). Anderson's efforts to move Interface toward sustainability prompted competition among other manufacturers to produce "green" carpet tiles, among them Milliken and Collins and Aikman. In the US commercial building arena, the prime green building organization is the US Green Building Council (USGBC), located in Washington, DC. A relatively new organization, the Green Building Initiative (GBI), which is headquartered in Portland, Oregon, acquired the rights to a

Figure 2.10 Ray Anderson, founder and former chairman of InterfaceFLOR, is considered one of the essential leaders of the US green building movement. His abiding belief that sustainability was an ethical imperative was supported by his strong actions to shift a major building products supplier from a business-as-usual mode to being the most sustainable company on earth. (Photograph courtesy of Interface, Inc.; © Lynne Siller)

Canadian building assessment standard known as Green Globes in 2004. The GBI has adapted Green Globes to the US building market and is offering it as an alternative to the USGBC LEED building rating systems.

Homebuilding and residential development are represented by a proliferation of organizations, many of which preceded the USGBC and arose independently in homebuilding organizations and municipalities across the United States. The city of Boulder, Colorado, took an aggressive stance in 1998 with respect to green building by passing an ordinance requiring specific measures. Pennsylvania established the Governor's Green Government Council (GGGC) in part to address the implementation of green building principles in the state. The city of Austin, Texas, is perhaps best known for its efforts in green building and was the recipient of an award at the first UN Conference on Sustainable Development in Rio de Janeiro, Brazil, in 1992. Local residential green building movements have emerged in Denver, Colorado; Kitsap County, Washington; Clark County, Washington; Baltimore, Maryland, with the suburban builders association; and, more recently, in Atlanta, Georgia, with the Earthcraft Houses program.

The National Association of Home Builders now provides guidance to its 800 state and local associations to assist in implementing green building programs. Reliable and independent information and critical analysis is published by BuildingGreen, Inc., in its monthly newsletter, *Environmental Building News*. BuildingGreen also publishes *GreenSpec*, a directory of products addressed to high-performance building needs, and provides the Green Building Advisor, computer software that facilitates green building design.

GREEN BUILDING ORGANIZATIONS—INTERNATIONAL

The international green building movement came of age in the early 1990s owing to the activities of task groups within the Conseil International du Bâtiment (CIB), a construction research networking organization based in Rotterdam, Netherlands, and the International Union of Laboratories and Experts in Construction Materials, Systems, and Structures (RILEM, from the name in French), based in Bagneux, France. In 1992, CIB Task Group 8 on building assessment provided international impetus for the development and implementation of building assessment tools and standards. CIB Task Group 16 on sustainable construction helped consolidate international standards regarding the application of sustainability principles to the built environment. And the relatively new International Initiative for a Sustainable Built Environment (iiSBE)[55] provides a clearinghouse for an extensive range of green building information. The iiSBE also organizes the biannual green building challenge and sustainable building conference and facilitates international sustainable building assessment with its main assessment method, the Sustainable Building Tool (SBTool), which is used at biannual conferences to assess or rate entrant exemplary buildings worldwide.

HISTORY OF THE US GREEN BUILDING MOVEMENT

The green building movement has a long history in the United States, with its philosophical roots traceable to the late 19th century. Subsequently, it developed in tandem with the country's environmental movement, and since the 1990s, it has been enjoying a renaissance. Notable dates include 1970, the year the first Earth Day was celebrated and the US Environmental Protection Agency was created, both events marking a major philosophical shift. Other influential events include the publication of Rachel Carson's landmark book *Silent Spring* in 1962 and the efforts of early environmentalists such as Barry Commoner, Lester Brown, Denis Hayes, and Donella Meadows. Concern over resource availability, particularly reliance on fossil fuels, was magnified by the oil shocks of the early 1970s, which resulted from the Arab-Israeli conflict of the time. This further piqued public interest in energy efficiency, solar technologies, retrofitting homes and commercial buildings with insulation, and energy recovery systems. As a result, the federal government began to provide tax credits for investment in solar energy and funded development and testing of innovative technologies ranging from solar air conditioning to eutectic salt energy storage batteries. By the late 1970s, many new efficiency standards were embodied in the model energy codes adopted by the states. After this burst of activity, however, interest in energy conservation began to wane as energy prices began to decline.

The early 1990s saw a renewed interest in energy and resource conservation as humans began to seriously consider more complex global environmental issues such as ozone depletion, global climate change, and destruction of major fisheries. Three events in the late 1980s and early 1990s helped to focus attention on problems associated with global environmental impacts: the publication in 1987 of *Our Common Future*, commonly referred to as the Brundtland Report; the 1989 meeting of the American Institute of Architects (AIA), at which it established its Committee on the Environment (COTE); and the UN Conference on Sustainable Development in 1992, commonly known as the Rio Conference.

The recent American resurgence in sustainable construction was precipitated in 1993 by a joint meeting of the International Union of Architects (Union Internationale des Architectes; UIA) and the AIA, known as "Architecture at the Crossroads." The UIA/AIA World Congress of Architects promulgated the Declaration of Interdependence for a Sustainable Future, which articulated a code of principles and practices to facilitate sustainable development (see Figure 2.11).

Although many energy-efficient buildings emerged after the oil crises of the 1970s, the first US buildings that considered a wider range of environmental and resource issues did not emerge until the 1980s. The earliest examples of green buildings were the result of major US environmental organizations requiring holistic approaches to the design of their office buildings. In 1985, William McDonough was hired by the Environmental Defense Fund to design its New York offices. The design featured natural materials, daylighting, and excellent indoor air quality, all part of a green solution for then endemic sick building problems. In 1989, the Croxton Collaborative, a design firm founded by Randy

DECLARATION OF INTERDEPENDENCE FOR A SUSTAINABLE FUTURE

UIA/AIA WORLD CONGRESS OF ARCHITECTS

CHICAGO, 18-21 JUNE 1993

RECOGNISING THAT:

A sustainable society restores, preserves, and enhances nature and culture for the benefit of life, present and future; ■ a diverse and healthy environment is intrinsically valuable and essential to a healthy society; ■ today's society is seriously degrading the environment and is not sustainable.

We are ecologically interdependent with the whole natural environment; ■ we are socially, culturally and economically interdependent with all of humanity; ■ sustainability, in the context of this interdependence, requires partnership, equity, and balance among all parties.

Building and the built environment play a major role in the human impact on the natural environment and on the quality of life; ■ a sustainable design integrates consideration of resources and energy efficiency, healthy buildings and materials, ecologically and socially sensitive land-use, and an aesthetic sensitivity that inspires, affirms, and ennobles; ■ a sustainable design can significantly reduce adverse human impacts on the natural environment while simultaneously improving quality of life and economic well-being.

WE COMMIT OURSELVES,

As members of the world's architectural and building-design professions, individually and through our professional organizations, to:

■ place environmental and social sustainability at the core or our practices and professional responsibilities;

■ develop and continually improve practices, procedures, products, curricula, services and standards that will enable the implementation of sustainable design;

■ educate our fellow professionals, the building industry, clients, students and the general public about the critical importance and substantial opportunities of sustainable design;

■ establish policies, regulations and practices in government and business that ensure sustainable design becomes normal practice; and

■ bring all the existing and future elements of the built environment – in their design, production, use, and eventual reuse – up to sustainable design standards.

Olfemi Majekodunmi
President,
International Union of Architects

Susan A. Maxman
President,
American Institute of Architects

Figure 2.11 The joint Declaration of Interdependence for a Sustainable Future, promulgated by the UIA/AIA World Congress of Architects during a joint meeting in Chicago, Illinois, in 1993, was an important event in the history of the high-performance green building movement. (*Source:* International Union of Architects and American Institute of Architects)

Croxton, designed the offices of the Natural Resources Defense Council in the Flatiron District of New York City. In this project, natural lighting and energy-conserving technologies were employed to reduce energy consumption by two-thirds compared to conventional buildings. The 1992 renovation of Audubon House, also in New York City, was a significant early effort in the contemporary green building movement (see Figure 2.12). The organization sought to reflect its values as a leader of the environmental movement and directed architect Randy Croxton to design the building in the most environmentally friendly and energy-efficient manner possible. In the process of achieving that goal, the extensive collaboration required by the many building team members provided a model of cooperation that has now become a hallmark of the contemporary green building process in the United States.[56]

The first highly publicized green building project in the United States, the "Greening of the White House," was initiated in 1993 and included renovation of the Old Executive Office Building, a 600,000-square-foot (55,700-square-meter) structure across from the White House (see Figure 2.13). The participation in this project of a wide array of architects, engineers, government officials, and environmentalists drew national attention and resulted in dramatic energy cost savings (about $300,000 per year), emissions reductions [845 tons (767 metric tons) of carbon per year], and significant reductions in water and solid waste associated costs. The success of the White House project spurred the federal government's sustainability efforts and prompted the US Postal Service, the Pentagon, the US Department of Energy, and the General Services Administration to address sustainability concerns within their organizations. The US National Park Service, too, opened green facilities at several national parks, including the Grand Canyon, Yellowstone, and Denali. The Naval Facilities Engineering Command (NAVFAC), the US Navy's construction arm, began a series of eight pilot projects to address sustainability and energy conservation concerns. The highly visible effort at its 156,000-square-foot (14,500-square-meter), 150-year-old headquarters in the Washington Navy Yard reduced energy consumption by 35 percent and resulted in annual savings of $58,000.[57]

In addition, several important guides to green building or sustainable design appeared in the early to mid-1990s. The *Environmental Building News*, first published in 1992, remains an independent, dispassionate, and authoritative guide to sustainable construction.[58] In 1994, the AIA first published its *Environmental Resource Guide*, followed by a more detailed version in 1996.[59] The "Guiding Principles of Sustainable Design," produced by the National Park Service in 1994, provided one of the first overviews of green building production.[60] Similarly, the *Sustainable Building Technical Manual* was developed and published jointly by the US Department of Energy and Public Technology, Inc., in 1996.[61] The Rocky Mountain Institute's *A Primer on Sustainable Building*, published in 1995, also contributed to the public understanding of sustainable construction.

Other international efforts and organizations interacted with and influenced the US movement during this period. The British green building rating system, the Building Research Establishment Environmental Assessment Method (BREEAM), was developed in 1992. As noted previously, the CIB convened Task Group 8 (Building Assessment) and Task Group 16 (Sustainable Construction) in 1992, which held influential international conferences in 1994 in the United Kingdom and Tampa, Florida. Also, as noted earlier, the USGBC, headquartered in Washington, DC, was formed in 1993 and held its first major meeting in March 1994.[62] Early articulations of the organization's LEED standard appeared at this time, along with green building standards developed by the American Society for Testing and Materials (ASTM). The ASTM standards were eventually set aside in favor of the USGBC's LEED assessment standard.

Development of the USGBC's LEED building rating system took four years and culminated in a 1998 test version known as LEED 1.0. It was enormously successful, and the Federal Energy Management Program sponsored a pilot effort to test its assumptions. Eighteen projects comprising more than 1 million square feet (93,000 square meters) were evaluated in the beta-testing phase. A greatly improved LEED 2.0 was launched in 2000 and provided for a maximum of 69 credits and four levels of building certification: platinum, gold, silver, or bronze. A further refined LEED was published in 2003 and labeled LEED for New Construction version 2.1 (LEED-NC 2.1). The name of the lowest level of certification, "bronze," was also changed to "certified." In 2005, further improvements, such as moving the rating system online, occurred, resulting in the issuance of LEED-NC 2.2. Major revisions to LEED occurred in 2009 (LEED 3.0) and 2012 (LEED 4.0), including reweighting credits and restructuring the rating system. The LEED rating system is covered in far more detail in Chapter 5.

Figure 2.12 (A) Audubon House in New York City was designed by the Croxton Collaborative as the headquarters of the Audubon Society. It is one of the projects marking the start of the contemporary US green building movement. (B) Desk illumination from a skylight in Audubon House. (Photographs courtesy of Croxton Collaborative Architects, P.C.)

Figure 2.13 The "Greening of the White House" project was the first widely publicized federal government green building project. (Illustration courtesy of View by View, Inc.)

New approaches, including the GBI's Green Globes for New Construction and Green Globes for Continual Improvement of Existing Buildings, as well as the National Association of Home Builders Model Green Home Guidelines, are reinforcing the enormous growth in green building by providing a variety of approaches to rating green buildings and creating competition to improve green building rating systems.

The first green building standards, as distinguished from green building rating systems such as LEED, began to emerge in 2010. The American Society of Heating, Refrigerating and Air-Conditioning Engineers (ASHRAE) issued ASHRAE 189.1-2009, Standard for the Design of High-Performance Green Buildings Except Low-Rise Residential Buildings (see Figure 2.14A). If adopted by a code body, for example, the Standard Building Code, it would be code-enforceable, effectively making green buildings standard practice.[63] Another similar standard was issued by the American National Standards Institute (ANSI) and the Green Building Initiative (GBI) in the form of ANSI/GBI 01-2010, Green Building Assessment Protocol for Commercial Buildings, which was derived from the Green Globes environmental design and assessment rating system for new construction (see Figure 2.14B).[64]

The International Green Construction Code (IgCC) was also issued in 2010. According to the International Code Council (ICC), the IgCC addresses site development and land use, including the preservation of natural and material resources, as part of the process. The code is designed to improve indoor air quality and support the use of energy-efficient appliances, renewable energy systems, water resource conservation, rainwater collection and distribution systems, and the recovery of used or graywater. The IgCC emphasizes building performance, including features such as a requirement for building system performance verification along with building owner education, to ensure the best energy-efficient practices are being carried out. The IgCC references ASHRAE 189.1-2009as an alternative jurisdictional compliance option within the IgCC. Governments across the United States and around the globe can adopt the code immediately to reduce energy usage and their jurisdiction's carbon footprint.

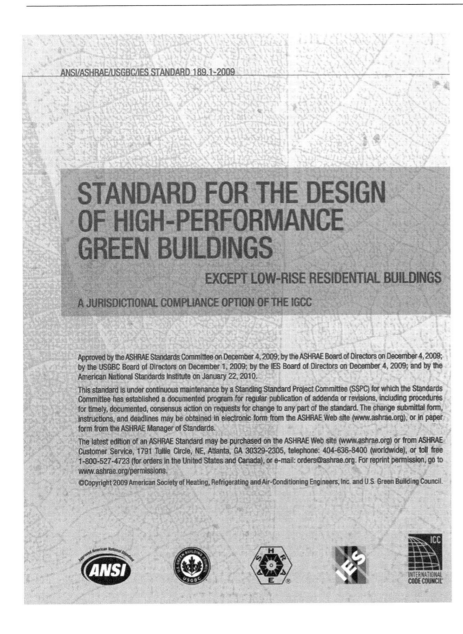

Figure 2.14 (A) ASHRAE 189.1-2009, Standard for the Design of High-Performance Green Buildings Except Low-Rise Residential Buildings, addresses site development and land use, including the preservation of natural and material resources as part of the process.

NEW DIRECTIONS IN HIGH-PERFORMANCE GREEN BUILDING

Perhaps the most interesting recent development in high-performance buildings in the United States is the emergence of two movements with the same basic approach, that is, they both call for a dramatic transformation of the requirements for green buildings. The first of these movements is the Living Building Challenge (LBC), which originated in the Cascadia Green Building Council, originally founded to represent the US Green Building Council in the northwestern United States and Vancouver, Canada. The LBC is exactly what its name implies, a challenging set of 20 prerequisites that a building project must attain in order to achieve certification from the International Living Future Institute as a green building. Unlike other green building rating systems, such as LEED, which bases a building rating on a point system, there are only mandatory requirements. Again, unlike LEED, which has several levels of certification ranging from certified to platinum, the LBC provides either certification or renewal certification. Among the mandatory requirements are that

ANSI/GBI 01-2010

**Green Building Assessment Protocol
for Commercial Buildings**

An American National Standard

April 1, 2010

Figure 2.14 (B) ANSI/GBI 01-2010,
Green Building Assessment Protocol for
Commercial Buildings, was derived from
the Green Globes environmental design
and assessment rating system for new
construction.

the building must be net zero energy (NZE), net zero water, and nontoxic;
provide for habitat restoration on sister sites; and incorporate urban agriculture.
The 20 LBC imperatives, all of which must be addressed, go well beyond the
efficiency standards that are generally used to declare a project "sustainable."
The first two projects to achieve full LBC certification in late 2010 were the
Omega Center for Sustainable Living in Rhinebeck, New York, and the Tyson
Living Learning Center in Eureka, Missouri (see Figures 2.15 and 2.16). The
latter project provides a good example of the choices that often must be made to
meet the LBC mandates. Although it achieved NZE performance by the end of
its first year in operation, producing almost 3800 kWh of electricity more than it
needed, the Tyson Living Learning Center needed some adjustments to achieve
the NZE level because the building was using more electricity than calculated.
When commissioning had been completed and the dynamic behavior of the
building indicated it would not achieve NZE performance, the team had

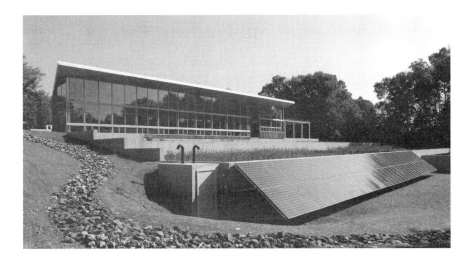

Figure 2.15 The Omega Center for Sustainable Living in Rhinebeck, New York, was one of the first projects to achieve the Living Building Challenge certification in late 2010. (© Omega Institute for Holistic Studies)

Figure 2.16 The Tyson Living Learning Center in Eureka, Missouri. Miscalculations initially resulted in this building falling short of energy performance goals. Due to the ambitious sustainability requirements of the Living Building Challenge, postconstruction adjustments were made by adding insulation in several areas, retrofitting storm windows, and adjusting the HVAC system. These improvements led to the project achieving the desired NZE performance by the end of its first year in operation. (David Kilper, WUSTL)

the choice of adding more photovoltaic (PV) panels to the building or finding another solution. Choosing the latter route, the project team added insulation in several areas; retrofitted storm windows; and adjusted the heating, ventilation, and air conditioning (HVAC) system to improve the energy performance of the building. Approximately 70 other projects were pursuing LBC certification in early 2012. The Living Building Challenge is covered in more detail in Chapter 4.

Similarly, *Architecture 2030* was established in response to the global climate change crisis by architect Edward Mazria in 2002. The mission of this organization is to rapidly transform the built environment in order to achieve enormous reductions in greenhouse gas emissions by changing the way buildings are planned, designed, and constructed. Specifically, the Architecture 2030 Challenge calls for rapid reductions in building energy consumption and associated greenhouse gas emissions such that, by the year 2030, all new buildings would be carbon neutral. Several local jurisdictions in the United States have adopted the targets set by Architecture 2030. In July 2006, Sarasota County, Florida, was the first county to formally adopt the Architecture 2030 Challenge as policy. In February 2007, two bills were introduced in the California legislature that duplicate the Architecture 2030 Challenge targets for energy consumption reductions for new residential and nonresidential buildings.

Case Study: OWP 11, Stuttgart, Germany

The Drees & Sommer Group is a top international engineering firm headquartered in Stuttgart, Germany, with 32 offices and 1125 employees worldwide. For 40 years, Drees & Sommer (DS) has provided a wide range of services in project management, real estate consulting, and engineering for public- and private-sector owners and investors in all aspects of real estate. DS refers to its approach to business as the "blue way" because its approach is to combine the traditional services provided by full-service engineering companies, such as economy, functionality, and process quality, with considerations of ecology, architecture, and human comfort. By virtue of this approach, DS demonstrates a philosophy of ensuring client success by thinking and acting in an integrated and sustainable manner. This comprehensive approach is apparent in the design of its own office building, commonly referred to as OWP 11, which is located in Stuttgart, Germany, and which, for its exemplary design and performance, was awarded a gold certification by the German Sustainable Building Council (Deutsche Gesellschaft für Nachhaltiges Bauen; DGNB) for meeting the criteria of the DGNB building assessment system (see Chapter 4 for more about DGNB).

OWP 11 consists of the renovation and expansion of an existing building located on an awkwardly shaped site. The hallmark of the building's exterior is a metal façade that faces north onto Pascalstrasse, underscoring the high-tech nature of much of the company's business (see Figure 2.17). The foyer is the essential interior element and functions without dominating the ensemble of two buildings that it links (see Figure 2.18). The shared courtyard and grounds are integrated as a key element of the overall architectural design. The management of DS paid special attention to the interaction of its workforce in the new facility. To ensure unexpected, chance meetings of colleagues to stimulate the generation of new ideas, the building was laid out to maximize the potential of these interactions. Additionally, management was aware that the self-esteem of the employees has a

Figure 2.17 The signature exterior of OWP 11 in Stuttgart, Germany, is its metal façade, which is the outer layer of a well-insulated wall system that reduces internal heating and cooling loads to very low levels. (© Dietmar Strauβ, Besigheim, Germany, and © Martin Duckek, Ulm, Germany)

lot to do with the location of their offices within the building. Those who are located close to pedestrian traffic areas feel they are closer to the action and therefore important. The pedestrian traffic and the movement of workers throughout the building act as a sort of "brain wave," which enhances communication and inter-action, increasing opportunities for innovation and new ideas. As a result, the design of the building maximizes placement of offices along the pedestrian corri-dors, while at the same time providing the opportunity for workers to function in quiet areas at the appropriate time (see Figure 2.19).

Step 1 in saving energy was to minimize heating and cooling loads. This required optimal thermal insulation of the building and demand-driven external solar protection in the form of a combination of computer-controlled exterior blinds and manually controlled interior blinds. The building is heavily insulated, with a 6- to 11-inch (16- to 27-cm) layer of mineral fiber insulation in the walls. The aluminum-framed windows have triple-glazed, low-emissivity (low-E) glazing. The U-value, that is, the thermal conductance of the window frames, is particularly low, some 20 percent lower than commercially available frames. During warmer seasons, oper-able windows under the control of the office workers are used for ventilating the work spaces. During the heating season, fresh air is pumped into the building by a mechanical ventilation system with heat recovery.

Step 2 in energy savings was heating and cooling the building with the mini-mum possible temperature difference between the required room temperature and the heating and cooling elements. In OWP 11, as a result of the optimum thermal insulation, heating and cooling loads are so low that energy-efficient, low-temper-ature heating (under 90°F or 32°C) and high-temperature cooling (64°F or 18°C) was able to be used (see Figure 2.20). Because this strategy requires large heat transfer surfaces, normal radiators could not be used. Instead, structural heating and cooling was installed in the office areas in the form of pipes that carry warm or cool water, depending on the season, through the reinforced concrete floors. The floor provides extremely effective and economical heating in winter and cooling in summer. The only downside of this arrangement is the thermal inertia of the system, meaning that because of the high thermal storage capacity of the structure, room temperature cannot be quickly changed. The DS engineers, in collaboration with their Zent-Frenger consultants, found the solution to this potential problem by using a supplemental system that responds rapidly to changing conditions. The structural heating and cooling within the reinforced concrete ceilings covers the base loads and is supplemented by additional heating elements with a fast response time that

Figure 2.18 The interior foyer of the building links the new and old wings and provides a spectacular entryway and circulation corridor for OWP 11. (© Dietmar Strauß, Besigheim, Germany, and © Martin Duckek, Ulm, Germany)

Figure 2.19 The work spaces of OWP 11 balance the desire to be at the center of the action with the need for quiet spaces for productive work. Note that there is no general lighting in the space; all lighting needs are provided by a floor-mounted indirect lighting system. (© Dietmar Strauß, Besigheim, Germany, and © Martin Duckek, Ulm, Germany)

Figure 2.20 (A) The edge strip heating elements placed in the formwork. (B) Technical installation in the reinforced concrete floor prior to pouring. (C) Heating, cooling, and ventilation pipes are integrated into the reinforcement. (© Dietmar Strauβ, Besigheim, Germany, and © Martin Duckek, Ulm, Germany)

allows individual room temperature regulation. The fast-response system is composed of a 4-centimeter-thick, prefabricated slab with a special, highly conductive concrete mix design with thermal insulation on top. These fast-response heating elements have a separate water supply network and are laid parallel to the façade. A simple control allows users to regulate their room temperature for responsive individualized heating or cooling.

Step 3 was to use alternative energy sources made possible by the low temperature differences required for heating and cooling, as described in step 2 above. Geothermal energy is tapped from rock underneath the building using ground-coupled heat exchangers. Eighteen holes, 8 inches (20 centimeters) in diameter, were drilled at least 19 feet (6 meters) apart to a depth of 170 feet (55 meters). At this depth, the ground temperature year-round is 52 to 54°F (11 to 12°C). Plastic pipes were then inserted into the bore holes and a mixture of water and glycol circulated through the system. During the heating season, the glycol-water solution is first heated 6 to 9°F (3 to 5°C) by the heat recovery system and then boosted by an electrically powered heat pump to about 90°F (32°C). The heating load requires primary energy of approximately 21 kWh per square meter per year (kWh/m^2/yr) by comparison to a conventional office building in Germany, which would require 130 kWh/m^2/yr, more than six times as much. During the summer, the glycol-water solution, after passing through the ground-coupled heat exchangers, is pumped at a temperature of approximately 54 to 59°F (12 to 15°C) and raised to a temperature of about 64°F (18°C) by a heat exchanger. The only electrical energy required for this cooling process is for pumping the heat exchange fluid and the cold water in the building cooling circuit. The entire building can be cooled for 1.50 to 2.00 euros per day on an extremely hot summer day. The overall primary energy requirement for climate control—that is, for heating, ventilation, and air conditioning—is about 36 kWh/m^2/yr, less than a fifth the energy demand of a conventional office building with oil heating and compressor-driven cooling.

German buildings are required to display an "energy passport" in a public location that indicates the primary energy consumption of the facility (see Figure 2.21). The passport for OWP 11 tells an interesting story: the building uses just 76.9 kWh/m^2/yr of primary energy. In comparison, the best US buildings consume about 100 kWh/m^2/yr of site or metered energy. Primary energy is the energy consumed at the power plant to produce the metered energy. For an all-electric building, primary energy is about three times the metered energy. As a result, 100 kWh/m^2/yr energy of metered energy is over 300 kWh/m^2/yr of primary energy. Therefore, OWP11 and other similar German buildings consume a small fraction of the energy consumed by the best US buildings.

Summary and Conclusions

Significant global environmental problems increasingly threaten food supplies, water and air quality, and the survival of ecosystems upon which humanity depends for a wide variety of goods and services. Because it uses enormous quantities of resources and replaces natural systems with human artifacts, the built environment sector of the economy has disproportionate environmental impacts on the planet. Consequently, the construction industry has a special obligation to behave proactively and shift rapidly from wasteful, harmful practices to a paradigm under which construction and nature work synergistically rather than antagonistically. This new model of sustainable construction is referred to as *high-performance green building*.

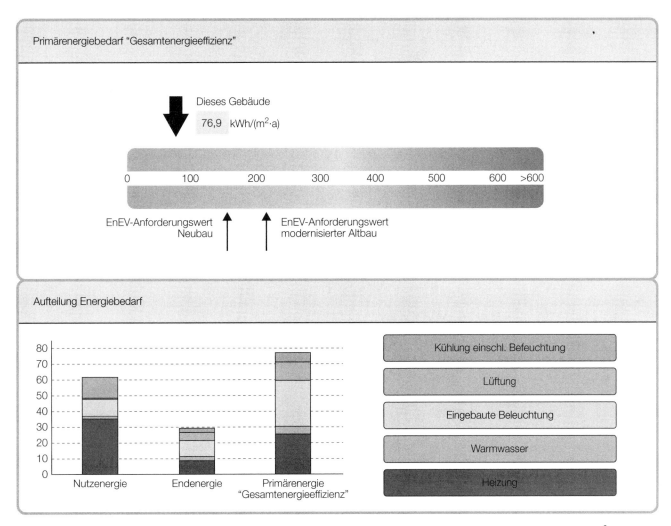

Figure 2.21 The *Energieausweis*, or energy passport, for OWP 11 indicates a primary energy consumption of 76.9 kWh/m²/yr, which is far lower than the German energy code for new construction (*EnEV-Anforderungswert Neubau*) limit of about 160 kWh/m²/yr. (© Dietmar Strauβ, Besigheim, Germany, and © Martin Duckek, Ulm, Germany)

The green building movement is a relatively recent phenomenon, and in the United States, it is presently growing at an exponential rate. The USGBC's LEED building assessment standard has emerged as the definitive guideline. It articulates the parameters for green buildings in the United States and several other countries. Parallel efforts in other economic sectors are occurring simultaneously as manufacturers attempt to design and produce goods with low environmental impact. The concepts of closed materials loops, efficient resource use, and the redesign of products and buildings to emulate natural systems are indispensable to preserve humanity's quality of life, along with the constant acknowledgment that nature is the source of that quality.

Notes

1. The five prior extinctions were the Ordovician (440 million years ago), Devonian (365 million years ago), Permian (245 million years ago), Triassic (210 million years ago), and Cretaceous (66 million years ago). The as yet unnamed sixth extinction is not being caused by major geologic upheavals, as was the case for the previous five, but instead by the activities of just one of the millions of species inhabiting the planet: humans.

2. There are 17 rare earth elements, and 97 percent of global production is in China, giving China a stranglehold over the world's high-tech industries. See IBT (2011).

3. Xerox's activities to redesign its product line and incorporate sustainability into the company's philosophy are described by Maslennikova and Foley (2000).

4. This commonly accepted definition of sustainable development is that proposed in *Our Common Future* (1987).

5. *Environmental amenity* refers to enjoyment that nature provides because of its many positive effects on human beings.

6. From Peterson (2002).

7. For a far more detailed discussion of the ethics underpinning sustainability, see Kibert et al. (2011).

8. From Howarth (1992).

9. As stated on the website of the Center for Community Action and Environmental Justice, www.ccaej.org.

10. From Drexler (1987).

11. A description of the NSF's Biocomplexity in the Environment Priority Area can be found at www.nsf.gov/news/priority_areas/biocomplexity/index.jsp.

12. From Goodin (1983).

13. From Rochlin (1978).

14. From Angyal (2003).

15. From Berry (2002).

16. From Taylor (1981).

17. From Leopold (1949).

18. In Frosch and Gallopoulos (1989), the term *industrial ecology* was used for the first time in the popular scientific press. This marked the beginning of the widespread use of this phrase to describe a wide variety of environmentally responsible approaches to industrial production.

19. This definition of industrial ecology is from Garner and Keoleian (1995).

20. Robert Ayres has written extensively on the subject of industrial materials flows. More detailed information on the problem of enormous waste can be found in Ayres (1989).

21. An excellent summary of industrial ecology in general and the Kalundborg plant specifically can be found at the website of Indigo Development, www.indigodev.com. Several excellent references and handbooks are also available from the website. Indigo Development, founded by Ernie Lowe, is devoted to furthering the development of industrial ecology, which he refers to as ". . . an interdisciplinary framework for designing and operating industrial systems as living systems that are interdependent with natural systems."

22. Construction ecology is defined in the context of industrial ecology and sustainable construction in Kibert, Sendzimir, and Guy (2002).

23. In addition to Janine Benyus's book on biomimicry, a useful website providing an overview of this concept is www.biomimicry.net.

24. This definition of DfE is from Keoleian and Menerey (1994).

25. An excellent short overview of ecological economics by Stephen Farber of the Graduate School of Public and International Affairs, University of Pittsburgh, can be found at www.fs.fed.us/eco/s21pre.htm.

26. The definition of carrying capacity is from the Carrying Capacity Network at www.carryingcapacity.org.

27. A thorough description of the ecological footprint concept can be found in Wackernagel and Rees (1996).

28. A gigaton is 1 billion tons.

29. The EEA has an excellent online glossary of environmental terms at http://glossary.eea.europa.eu/.

30. The ecological rucksack quantities are derived from a number of online and published sources. A good description of the concept, along with a diagram showing relative ecological rucksacks for a variety of materials, can be found in von Weizsäcker, Lovins, and Lovins (1997).

31. From Kahn (1997).

32. From Stephen R. Kellert, "Ecological Challenge, Values of Nature, and Sustainability," cited in Kibert (1999) and Kahn (1997).

33. From World Business Council for Sustainable Development (1996).
34. As stated on the WBCSD website, www.wbcsd.org.
35. The website of the US branch of the Natural Step is www.naturalstep.org.
36. Excerpted from the website of *Canadian Architect*, www.cdnarchitect.com.
37. The book *Factor Four: Doubling Wealth, Halving Resource Use* was written as a report to the Club of Rome as a follow-up to the 1972 book *The Limits to Growth*, written by Donella Meadows, Dennis Meadows, Jorgen Randers, and William Behrens III, which was the original report to the club. *Limits to Growth* stated that exponential growth in population and the world's industrial system would force growth on the planet to be halted within a century, a result of environmental impacts and resource shortages.
38. The Factor 10 concept continues to be fostered by the Factor 10 Club and the Factor 10 Institute, whose publications and activities can be found at www.factor10-institute.org.
39. According to the authors of Factor 4, Lee Eng Lock's supply fans use 0.061 kW/ton of cooling versus 0.60 kW/ton in conventional practice. Similarly, his chilled-water pumps use 0.018 kW/ton versus 0.16 kW/ton, condenser water pumps use 0.018 kW/ton versus 0.14 kW/ton, and cooling towers use 0.012 kW/ton versus 0.10 kW/ton.
40. See von Weizsäcker et al. (2009).
41. The IPCC assessment reports are published every six years. The most current version is the Fourth Assessment Report (AR4) was published in 2007 and the next Fifth Assessment Report (AR5) will be published in 2013. See IPCC (2007) in the References.
42. Excerpted from "Twenty Questions and Answers about the Ozone Layer" (2002) at www.esrl.noaa.gov/csd/assessments/ozone/2010/twentyquestions/.
43. The rate of rainforest destruction, according to the Rainforest Action Alliance, is available at www.rainforest-alliance.org.
44. Data are from the *Global Environmental Outlook 2002 Report* (GEO-3) (2002) at www.unep.org/geo/geo3.
45. The website of the UN Convention to Combat Desertification is www.unccd.int.
46. A group of Swedish nongovernmental organizations maintains a website promoting knowledge about the effects of acid rain, www.acidrain.org.
47. See "Global Environmental Problems: Implications for U.S. Policy" (2003).
48. Excerpted from "The Loss of Biodiversity and Its Negative Effects on Human Health" (2004).
49. The definition of toxic substances is adapted from the definition provided on the Great Lakes website of Environment Canada, www.on.ec.gc.ca/water/raps.
50. Excerpted from reports on the website of the Center for Community Action and Environmental Justice, www.ccaej.org.
51. Excerpted from Dutzik, Bouamann, and Purvis (2003).
52. Information on endocrine disruptors can be found on the website of the National Resources Defense Council (NRDC), www.nrdc.org.
53. From Gordon, Bertram, and Graedel (2006).
54. From "Materials Prices Dictate Creative Engineering" (2006).
55. The iiSBE website is www.iisbe.org.
56. The story of the Audubon House design process is recounted in Croxton Collaborative and the National Audubon Society (1992).
57. An excellent detailed overview of the history of the US green building movement can be found in the "White Paper on Sustainability" (2003). This publication also contains other important background information about the green building movement and suggests an action plan to help improve and ensure the quality and outcomes of green building design and construction.
58. BuildingGreen, Inc., publishes *Environmental Building News* and produces a range of other useful products, including the *GreenSpec* directory. All of its publications are also available by subscription at www.buildinggreen.com.
59. The *Environmental Resource Guide* is a thorough guide to the environmental and resource implications of construction materials. The first version was published by the AIA in 1994; the second, expanded version was published by John Wiley & Sons in 1996.
60. The current National Park Service Sustainable Building Implementation Plan is available at www.nps.gov/sustainability/sustainable/implementation.html.

61. The "Sustainable Building Technical Manual" is available at http://smartcommunities .ncat.org/pdf/sbt.pdf.

62. The USGBC's earliest organizers were David Gottfried and Michael Italiano, and its first president was Rick Fedrizzi, who, at the time, was with Carrier Corporation. The first annual meeting of the USGBC was held in Washington, DC, in March 1994 and featured as its keynote speakers Paul Hawken, who had just completed the groundbreaking book *Ecology of Commerce*, and William McDonough, recognized as one of the major architectural figures in the US green building movement and the author of the Hannover Principles.

63. As described by ASHRAE, ASHRAE 189.1-2009 provides a "total building sustainability package" for those who strive to design, build, and operate green buildings. From site location to energy use to recycling, this standard sets the foundation for green buildings by addressing site sustainability; water use efficiency; energy efficiency; indoor environmental quality; and the building's impact on the atmosphere, materials, and resources. ASHRAE 189.1-2009 serves as a jurisdictional compliance option to the Public Version 2.0 of the International Green Construction Code (IgCC) published by the International Code Council. The IgCC regulates construction of new and remodeled commercial buildings.

64. According to the GBI, the ANSI/GBI 01-2010 standard was developed following ANSI's highly regarded consensus-based guidelines, which are among the world's most respected for the development of consensus standards and ensure a balanced, transparent, and inclusive process. A variety of stakeholders, including sustainability experts, architects, engineers, environmental nongovernment organizations (ENGOs), and industry groups, participated in its development.

References

Angyal, Andrew J. 2003. "Thomas Berry's Earth Spirituality and the 'Great Work." *Ecozoic Reader* 3: 35–44.

Ausubel, J. H., and H. E. Sladovich, eds. 1989. *Technology and the Environment.* Washington, DC: National Academy Press.

Ayres, Robert U. 1989. "Industrial Metabolism." In *Technology and the Environment*, edited by J. H. Ausubel and H. E. Sladovich. Washington, DC: National Academy Press.

Benyus, Janine. 1997. *Biomimicry: Innovation Inspired by Nature.* New York: William Morrow.

Berry, Thomas. 2002. "Rights of the Earth: Earth Democracy." *Resurgence* 214: 28–29.

Croxton Collaborative and the National Audubon Society. 1992. *Audubon House: Building the Environmentally Responsible, Energy Efficient Office.* Hoboken, NJ: John Wiley & Sons.

Demkin, Joseph, ed. 1996. *Environmental Resource Guide.* Hoboken, NJ: John Wiley & Sons.

Drexler, K. Eric. 1987. *Engines of Creation.* New York: Anchor Books.

Dutzik, Tony, Jeremiah Bouamann, and Mehgan Purvis. 2003. "Toxic Releases and Health: A Review of Pollution Data and Current Knowledge on the Health Effects of Toxic Chemicals." Written for the US PIRG Education Fund. Available at www .uspirg.org/reports/toxics03/toxicreleases1_03report.pdf.

Frosch, Robert, and Nicholas Gallopoulos. 1989. "Strategies for Manufacturing." *Scientific American* (September): 144–152.

Garner, Andy, and Gregory Keoleian. 1995. *Industrial Ecology: An Introduction.* Ann Arbor: National Pollution Prevention Center in Higher Education, University of Michigan.

Goodin, Robert E. 1983. "Ethical Principles for Environmental Protection."In *Environmental Philosophy*, edited by R. Elliot and A. Gare. London: Open University Press.

Gordon, R. B., M. Bertram, and T. E. Graedel.2006. "Metal Stocks and Sustainability." *Proceedings of the National Academy of Sciences* 103 (5): 1209–1214.

Howarth, Richard B. 1992. "Intergenerational Justice and the Chain of Obligation." *Environmental Values* 1: 133–140.

IBT. 2011. "Rising Rare Earth Prices Cripple Japan's June Imports." *International Business Times*, June 17, 2011. Available at www.ibtimes.com/articles/189028/20110729/rising-rare-earth-prices-cripple-japan-s-june-imports-us-prices-rise.htm.

IPCC. 2007. *Climate Change 2007: Synthesis Report*. Available at www.ipcc.ch/publications_and_data/publications_ipcc_fourth_assessment_report_synthesis_report .htm.

Kahn, Peter H., Jr. 1997. "Developmental Psychology and the Biophilia Hypothesis: Children's Affiliations with Nature." *Developmental Review* 17: 1–61.

Kellert, Stephen R., and E. O. Wilson, eds. 1993. *The Biophilia Hypothesis*. Washington, DC: Island Press.

Keolelan, Gregory, and D. Menerey. 1994. "Sustainable Development by Design." *Air and Waste* 44: 645–668.

Kibert, Charles J., ed. 1999. *Reshaping the Built Environment: Ecology, Ethics, and Economics*. Washington, DC: Island Press.

Kibert, Charles J., Martha Monroe, Anna Peterson, Richard Plate, and Leslie Thiele. 2011. *Working toward Sustainability: Ethical Decision Making in a Technological World*. Hoboken, NJ: John Wiley & Sons.

Kibert, Charles J., Jan Sendzimir, and G. Bradley Guy, eds. 2002. *Construction Ecology: Nature as the Basis for Green Buildings*. London: Spon Press.

Leopold, Aldo. 1949. *A Sand County Almanac*. New York: Oxford University Press.

Lopez Barnett, Dianna, and William D. Browning. 1995. *A Primer on Sustainable Building*. Snowmass, CO: Rocky Mountain Institute.

Maslennikova, Irina, and David Foley. 2000. "Xerox's Approach to Sustainability." *Interfaces* 30 (3): 226–233.

"Materials Prices Dictate Creative Engineering." 2006. *Engineeringtalk*. Available at www.engineeringtalk.com/news/lag/lag102.html.

Meadows, Donella H., Dennis I. Meadows, Jorgen Randers, and William W. Behrens III. 1972. *The Limits to Growth*. New York: Universe Books.

Peterson, Gary. 2002. "Ecology of Construction." In *Construction Ecology: Ecology as the Basis for Green Buildings*, edited by Charles J. Kibert, Jan Sendzimir, and Bradley Guy. London: Spon Press.

Rochlin, Gene I. 1978. "Nuclear Waste Disposal: Two Social Criteria." *Science* 195: 23–31.

Taylor, Paul W. 1981. "The Ethics of Respect for Nature." *Environmental Ethics* 3: 206–218.

"The Loss of Biodiversity and Its Negative Effects on Human Health." 2004. Available at the website of Students for Environmental Awareness in Medicine, Seamglobal .com/lossofbiodiversity.html.

UN Environmental Programme. 2002. *Global Environmental Outlook 2002 Report* (GEO-3). Available at www.unep.org/geo/geo3.

UN World Commission on Environment and Development. 1987. *Our Common Future*. Oxford: Oxford University Press.

US Department of Energy and the Public Technology, Inc., Initiative. 1996. *Sustainable Building Technical Manual*. Available at http://smartcommunities.ncat.org/pdf/sbt .pdf.

von Weizsäcker, E., K. Hargroves, M. Smith, C. Desha, and P. Stasinopoulos. 2009. *Factor 5: Transforming the Global Economy through 80% Increase in Resource Productivity*. London: Earthscan.

von Weizsäcker, Ernst, Amory Lovins, and L. Hunter Lovins. 1997. *Factor Four: Doubling Wealth, Halving Resource Use*. London: Earthscan.

Wackernagel, Mathis, and William Rees. 1996. *Our Ecological Footprint*. Gabriola Island, BC: New Society Publishers.

Watson Institute for International Studies, Brown University. 2003. "Global Environmental Problems: Implications for U.S. Policy."Available at www.choices.edu.

"White Paper on Sustainability: A Report on the Green Building Movement." 2003. *Building Design and Construction*. Available at www.bdcnetwork.com.

World Business Council for Sustainable Development. 1996. "Eco-Efficient Leadership for Improved Economic and Environmental Performance."Available atwww.wbcsd.org.

Chapter 3
Ecological Design

The key to creating a high-performance green building is the ability of the design team to understand and apply the concept of *ecological design*. Sim Van der Ryn and Stuart Cowan defined ecological design as "any form of design that minimizes environmentally destructive impacts by integrating itself with living processes."[1] Although a design rooted in ecology and nature should be integral to creating a green building, ecological design is in the early stages of evolution, and it will take considerable time and experimentation before a robust version matures. Meanwhile, designers often must use their best judgment when making decisions from among the myriad choices available. The ability to minimize the direct impact of the project on the site due to the construction footprint and construction operations and landscape modifications such as tree removal and alteration of natural habitats requires a fairly high level of understanding of the available options, especially in the context of sustainability. Developing a low-energy scheme demands a significant level of knowledge and experience with selecting approaches that maximize the potential for passive heating, cooling, lighting, and ventilating the building; with understanding the best orientation and massing for storing and releasing energy on a time scale compatible with building operation; and with understanding the myriad energy trade-offs that must be considered, for example, between daylighting and solar heat gain. When considering materials and product selection, the best choices can be far from obvious. In addition to the environmental implications, performance and cost criteria must be addressed in the selection process. These are just but a few of the many decisions a project team must make that are far better informed when the team has knowledge of, and experience with, ecological design as applied to high-performance green buildings.

One of the outcomes of the high-performance green building movement has been the advent of green building rating systems such as Leadership in Energy and Environmental Design (LEED) and Green Globes in the United States and the Building Research Establishment Environmental Assessment Method (BREEAM) in the United Kingdom. These rating systems allow a project team to simply use a checklist of measures derived from one of the building rating systems that, if followed, produces a green building, at least in the eyes of the rating system proponent, without the need for a deeper understanding of ecological design being required. The proposed outcome is that the project team, without ever having studied or pondered the diverse and complex issues of the construction industry's environmental impact, can design and build a high-performance building. Using a standard building rating system as guidance for green building design is certainly an advantage in that this approach has rapidly increased the penetration of green buildings in the marketplace. Yet, simple adherence to a checklist without deeper thinking could ultimately result in building stereotypes that stagnate rather than advance the art of green building. Commitment to a design approach that is rooted in an understanding of natural

Figure 3.1 The Federal Building in San Francisco, California, exemplifies ecological design by employing local natural forces such as the prevailing winds and sunlight to provide cooling and daylighting. Detailed analysis of natural airflows induced by wind and thermal processes was accomplished using sophisticated computational fluid dynamics (CFD) modeling. (Illustration courtesy of Morphosis Architects)

systems and in the behavior of ecosystems, and that is concerned with resource conservation, will undoubtedly produce a high-performance building of higher economic and aesthetic value. The bottom line is that high-performance green buildings that are truly exceptional beyond the points and certifications do require an integration with nature that is not achievable with a mere checklist.

In the brief history of the green building movement, several design approaches have been articulated, including ecological design, *environmental design, green design, sustainable design*, and *ecologically sustainable design*. Fundamentally, each approach seeks to acknowledge, facilitate, and/or preserve the interrelationship of natural system components and buildings. In doing so, particular questions and problems recur, such as:

- What can be learned from nature and ecology that can be applied to buildings?
- Should ecology serve as model or metaphor for green buildings?
- How can natural systems be directly incorporated to improve the functioning of the built environment?
- How can the human-nature interface best be managed for the benefit of both systems?
- When does the natural system metaphor break down and is another approach required?

These profound questions have no easy answers, yet responses to them are critical to the evolution of truly sustainable buildings (see Figure 3.1). Clearly, the progress of green building requires greater understanding and consideration of the environmental and human impact of the built environment, as well as incorporation of nature's lessons into the building process. The striking lack of understanding of ecology among design and construction professionals is less surprising when one considers that the green building movement was not created by ecologists, but rather by building professionals and policy makers with only glancing familiarity with the dynamic discipline of ecology. Yet, without greater understanding of ecology and ecological theory, green buildings may cease to evolve beyond merely fanciful, intuitive structures that are green in name only. With this in mind, this chapter reviews fundamental principles of ecological, or green, design and explores the philosophy and rationale of practitioners and academics whose life's work has centered on these issues. An overview of the history and current efforts to connect ecological thinking to buildings provides a starting point; further study of ecology, industrial ecology, and related fields is recommended.

Design versus Ecological Design

According to Van der Ryn and Cowan, *design*, in its simplest form, can be defined as ". . . the intentional shaping of matter, energy, and process to meet a perceived end or desire."[2] This broad definition means that literally everyone is a designer because we are all using resources to achieve some end; consequently, the responsibility for design does not rest solely with those who might be called the design professionals, the most prominent of whom are architects. The world we design collectively is a rather simple one compared to the design of nature. In our world, we use a limited number of models and templates to produce an impoverished urban and industrial landscape largely devoid of true imagination and creativity. It is clear that this human-designed and -engineered landscape often replaces the natural landscape with unrecyclable and toxic products manufactured by wasteful industrial processes that were implemented with little regard for the consequences for humans or ecological systems. It is often said that the environmental problems we face today, such as climate change and biodiversity loss, reflect a failure of design. The disconnection of human design from nature is precisely the problem that high-performance green building, through the application of ecological design, seeks to redress.

In contrast to their definition of design, Van der Ryn and Cowan define ecological design as that which transforms matter and energy using processes that are compatible and synergistic with nature and that are modeled on natural systems. Thus, unlike design that destroys landscapes and nature, ecological design, in the context of the built environment, seeks solutions that integrate human-created structures with nature in a symbiotic manner; that mimic the behavior of natural systems; and that are harmless to humans and nonhumans in their production, use, and disposal. Some would widen the concept of ecological design to an even broader concept, that of sustainable design, which would address the triple bottom-line effects of creating buildings: environmental impacts, social consequences, and economic performance. Clearly, the larger context and impacts of building design and construction need to be kept in mind by all the players in the process. Ecological design focuses on the human-nature interface and uses nature rather than the machine as its metaphor.

The key problem facing ecological design is a lack of knowledge, experience, and understanding of how to apply ecology to design. Complicating the issue is that there are several major approaches to understanding ecology, even among ecologists. Systems ecology, for example, focuses on energy flows, whereas proponents of adaptive management study processes.[3] Nature functions across scales and time horizons that are virtually unimaginable to human designers, who continue to struggle to apply even relatively simple ecological concepts such as resilience and adaptability to their work. An even deeper flaw is that building professionals have little or no background or education in ecology; hence, any application of so-called ecological or green design is likely to be shallow and perhaps even trivial. Equally problematic is that an enormous legacy of machine-oriented design is in place in the form of buildings and infrastructure, and the industrial products comprising buildings are still being created based on concepts, design approaches, and processes that have their roots in the Industrial Revolution. Thus, contemporary ecological designers are engaged in a struggle on several fronts in their attempt to shift to a form of thinking that would reconnect humans and nature. These "fronts" can be itemized as follows:

1. Understanding ecology and its applicability to the built environment
2. Determining how to use nature as the model and/or metaphor for design
3. Coping with an industrial production system that operates using conventional thinking
4. Reversing at least two centuries of design that used the machine as its model and metaphor

The classic approach to building design has been for the architect to define and lead the design effort, with input from the building owner but with scant input from other entities affected by the project. Contemporary ecological design changes this thinking dramatically by engaging a wide range of stakeholders in the design process from the onset of the effort. The key point of ecological design is to obtain the maximum amount of input from as many parties to the project as possible.

TABLE 3.1

Benefits of Sustainable Design

	Economic	Societal	Environmental
Siting	Reduced costs for site preparation, parking lots, roads	Improved aesthetics, more transportation options for employees	Land preservation, reduced resource use, protection of ecological resources, soil and water conservation, restoration of brownfields, reduced energy use, less air pollution
Water Efficiency	Lower first costs, reduced annual water and wastewater costs	Preservation of water resources for future generations and for agricultural and recreational uses, fewer wastewater treatment plants	Lower potable water use and reduced discharge to waterways, less strain on aquatic ecosystems in water-short areas, preservation of water resources for wildlife and agriculture
Energy Efficiency	Lower first costs, lower fuel and electricity costs, reduced peak power demand, reduced demand for new energy infrastructure	Improved comfort conditions for occupants, fewer new power plants and transmission lines	Lower electricity and fossil fuel use, less air pollution and fewer carbon dioxide emissions, lowered impacts from fossil fuel production and distribution
Materials and Resources	Decreased first costs for reused and recycled materials, lower waste disposal costs, reduced replacement costs for durable materials, reduced need for new landfills	Fewer landfills, greater markets for environmentally preferable products, decreased traffic due to the use of local/regional materials	Reduced strain on landfills, reduced use of virgin resources, better-managed forests, lower transportation, energy and pollution, increase in recycling markets
Indoor Environmental Quality	Higher productivity, lower incidence of absenteeism, reduced staff turnover, lower insurance costs, reduced litigation	Reduced adverse health impacts, improved occupant comfort and satisfaction, better individual productivity	Better indoor air quality, including reduced emissions of volatile organic compounds, carbon dioxide, and carbon monoxide
Commissioning; Operations and Maintenance	Lower energy costs, reduced occupant/owner complaints, longer building and equipment lifetimes	Improved occupant productivity, satisfaction, health, and safety	Lower energy consumption, reduced air pollution and other emissions

BENEFITS OF ECOLOGICAL DESIGN

For green buildings to be successful, the benefits of designing them must be known to those purchasing construction services and facilities. Because sustainability addresses a broad range of economic, environmental, and social issues, the benefits of ecological or sustainable design are potentially enormous. A list of these benefits recently published by the Federal Energy Management Program (FEMP) provides an overview of the promise of a shift to sustainable design (see Table 3.1).[4]

Historical Perspective

Although the green building movement is a relatively recent phenomenon, it has its roots in the work and thinking of several previous generations of architects and designers, dating back at least to the end of the 19th century. In the American context, several key figures laid the foundation for today's ecological or green design, among them R. Buckminster Fuller, Frank Lloyd Wright, Richard Neutra, Lewis Mumford, Ian McHarg, Malcolm Wells, and John Lyle.

A brief introduction to each of these thinkers is presented here. The following section, "Contemporary Ecological Design," covers the synthesis of this foundational thinking about ecological design into an emerging, coherent process for green building design. To articulate today's thinking, the efforts of William McDonough, Ken Yeang, Sim Van der Ryn, Stuart Cowan, and David Orr are described.

R. BUCKMINSTER FULLER

Perhaps more than any other figure, R. Buckminster Fuller (1895–1983) laid the foundation for the green building revolution in the United States (see Figure 3.2). His list of accomplishments is long, among them the design of the aluminum Dymaxion car in 1933; the design of the autonomous Dymaxion House in the 1920s, one of which was built in Wichita, Kansas, in 1946; and, of course, the creation of the geodesic dome in the 1950s (see Figure 3.3). Fuller has been called an inventor, architect, engineer, mathematician, poet, and cosmologist. He was, at heart, an ecologist. His designs emphasized resource conservation: the use of renewable energy in the form of sun and wind; the use of lightweight, ephemeral materials such as bamboo, paper, and wood; and the concept of design for deconstruction. His geodesic dome has been called the lightest, strongest, and most cost-effective structure ever devised.

Fuller is also credited with originating the term *Spaceship Earth* to describe how dependent humans are on the planet and its ecosystems for their survival and how the waste we create ends up in the biosphere, to the peril of everyone. His Dymaxion Map and World Game were designed to allow players to observe world resources and create strategies for solving global problems by matching human needs with the planet's resources. Fuller understood the issue of renewable and nonrenewable resources, and his research showed that all energy needs could be provided by renewables. In the United States, he showed that, at the time, wind energy alone could provide three and a half times the country's

Figure 3.2   The R. Buckminster Fuller postage stamp was issued by the US Postal Service in July 2004 to commemorate the 50th anniversary of Fuller's patents for the geodesic dome, said to be the lightest, strongest, and most cost-effective structure ever devised. (Stamp Designs © 2004 United States Postal Service. Displayed with permission. All rights reserved)

Figure 3.3   R. Buckminster Fuller's Dynamic Maximum Tension, or Dymaxion House, in Wichita, Kansas, was the first serious attempt to create an autonomous house. It was designed for mass production, weighed just 3000 pounds (1364 kilograms) compared to the 150 tons (137 metric tons) of a typical house, featured a built-in wind turbine for generating power, and had a graywater system. (Courtesy, The Estate of R. Buckminster Fuller)

Figure 3.4 Frank Lloyd Wright (1867–1958) laid some of the early foundations for the contemporary high-performance green building movement through his fusion of site, structure, and context. (*Source:* Library of Congress)

total energy needs.[5] His work influenced many of today's green building movement participants, so much so that he is sometimes referred to as the "father of environmental design."

Fuller was also a prolific author; he is credited with writing 28 books, among them *Operating Manual for Spaceship Earth* (1969), in which he imagines humans as the crew of the planet, all bound together by a shared fate on what amounts to a tiny spaceship in an infinite universe. The question he posed to his fellow planetary inhabitants was: How do we contribute to the safe operation of Spaceship Earth? In the book, he describes many of his basic concepts, two of which are synergy and ephemeralization. Another notable book by Fuller was *Critical Path* (1981), in which he explored social issues, marking him as one of the first people to connect the issues of environment, economics, and humans, labeled many years later by Lester Brown as *sustainability*. In *Critical Path*, Fuller analyzes how humanity has found itself at the limits of the planet's resources and facing political, economic, environmental, and ethical crises. Fuller, labeled "the planet's friendly genius," was an extraordinary member of the planet's "crew."

FRANK LLOYD WRIGHT

Frank Lloyd Wright (1867–1958) is well known as an important figure in architecture (see Figure 3.4). Less well known is that his thinking on nature and building laid some of the early foundations for the contemporary high-performance green building movement. His early exposure to nature had a profound effect on both his life and his architecture. Under the tutelage of his mother, who employed Friedrich Froebel's nature-based training, he learned about nature's forms and geometries. His architecture reflects this influence, relying on the underlying structure of nature. Wright's goal was to create buildings that were, as he put it, integral to the site, to the environment, to the life of the inhabitants, and to the nature of the materials. He also introduced the term *organic architecture* into the design vocabulary to reflect, at least in part, how his thinking had evolved from that of his mentor, Louis Sullivan. Sullivan's mantra, "form follows function," was modified by Wright to "form and function are one,"a change inspired by his observations of nature. Wright preferred an approach that emulated rather than imitated nature. Nature is an integrated whole, with seamless design. However, people filter and reinterpret nature's principles and the result is outcomes that are like nature, but not precisely like nature. He advocated a similar outcome for architecture by integrating spaces into a coherent whole and fusing site, structure, and context into one idea (see Figure 3.5). The building's design should be carefully considered to make it an organic whole. Every element of the building should be designed to make it integral to this organic whole: windows, doors, chairs, floors, roof, walls, spatial form, all related to one another, emulating the order in nature. Materials and motifs are repeated throughout the building, geometries are selected for their compatibility with a central theme, again emulating nature. Wright's provocative thinking and writing on organic architecture are important cornerstones of today's greening revolution and the frequent reference to him as "America's first green architect" is certainly well deserved.

RICHARD NEUTRA

Richard Neutra (1892–1970), a pupil of Wright's, recognized how flawed the products of human creation were compared to those of nature (see Figure 3.6). He noted that human artifacts were static and unable to self-regenerate or

Figure 3.5 Taliesin West in Scottsdale, Arizona, designed by Frank Lloyd Wright, illustrates organic architecture. (*Source:* National Register of Historic Places)

Figure 3.6 Richard Neutra (1892–1970) recognized the need of humans to be connected to nature and was one of the originators of the concept of biophilia. (Photograph courtesy of the J. Paul Getty Trust)

self-adjust, unlike nature's creations, which are dynamic and self-replicating. He observed that nature's form and function emerge simultaneously, whereas humans must first create a building's form and then allow it to function. Neutra was one of the first to recognize the concept of *biophilia*, the need or craving of humans to be connected to nature, a concept that has been expounded on more recently by E.O. Wilson and Stephen Kellert.[6]

Neutra advocated the close connection of living spaces to the "green world of the organic." According to Neutra, imitating nature is not simply flattery on the part of humans; it is the copying of systems that function in an extraordinarily successful fashion. He was also one of the first architects to recognize the connection between human health and nature and the need to consider this relationship in building design. In designing what became known as the Health House, a Los Angeles residence for Dr. P.M. Lovell, a naturopath, or integrated medical practitioner, Neutra explored the health relationship between nature and structure (see Figure 3.7). In today's green buildings, health issues are of

Figure 3.7 Neutra explored the health relationship between nature and structure as evidenced in the Health House in Los Angeles, California. (*Source:* National Register of Historic Places)

Figure 3.8 Lewis Mumford (1895–1990) was an architecture critic and advocate for ecological consciousness over technology. (Courtesy of the Estate of Lewis and Sophia Mumford)

paramount importance, and connections between nature and health are again being explored in a wide variety of building experiments.

LEWIS MUMFORD

Lewis Mumford (1895–1990) was renowned for his writings on cities, architecture, technology, literature, and modern life (see Figure 3.8). His long-term connection with the built environment was forged over a 30-year stint as architectural critic for the *New Yorker*. He was also a cofounder of the Regional Planning Association of America, which advocated limited-scale development and the region as significant for city planning. He wrote *The Brown Decades* in 1931 to detail the architectural achievements of Henry Hobson Richardson, Louis Sullivan, and Frank Lloyd Wright. Mumford was particularly critical of technology, and in *The Myth of the Machine*, written in 1967, he argued that the development of machines threatened humanity itself, citing, for example, the design of nuclear weapons. He argued in *Values for Survival*, written in 1946, for the restoration of organic human purpose and for humankind to exert "... primacy over its biological needs and technological pressures" and to "... draw freely on the compost from many previous cultures." Mumford advocated the implementation of *ecotechnics*, technologies that rely on local sources of energy and indigenous materials in which variety and craftsmanship add ecological consciousness, as well as beauty and aesthetics. He drew his conclusions from observations of how cities evolved, from preindustrial cities that respected nature to post–Industrial Revolution metropolises that sprawled and destroyed compact urban forms, caused resources to be wasted, and virtually had no connection to nature.

IAN MCHARG

The disconnect between buildings and nature in the Industrial Age was also noted and articulated by Ian McHarg (1920–2001), particularly the lack of a multidisciplinary effort to produce a built environment that was responsive to nature. He decried the lack of environmental consideration in planning; the lack of interest on the part of scientists in planning; and the absence of consideration of life itself in many of the sciences such as geology, meteorology, hydrology, and soil science (see Figure 3.9). According to McHarg, the compartmentalization and specialization of disciplines have created conditions that at present may make truly ecological design difficult or impossible to achieve.

McHarg's 1969 book, *Design with Nature*, is a modern classic, especially for the discipline of green building. McHarg called for environmental planning on a local level and advocated taking everything in the environment (such as humans, rocks, soils, plants, animals, and ecosystems) into account when planning the built environment. He was also one of the first people to realize that the best way to preserve open space is to sustain urban areas, which contain existing resources (such as sewer systems and streets) to handle human growth. He also noted that it was critical that everyone have an ecological education in order to be able to make the best-informed decisions about growth and development.

Figure 3.9 Ian McHarg (1920–2001) was an advocate of planning for a built environment that is responsive to nature. (*Source:* The Japan Prize Foundation)

MALCOLM WELLS

Malcolm Wells (1926–2009) was generally critical of architects for failing to be aware of or moved by the biological foundations of both life and art. In his 1981 work *Gentle Architecture*, he asked a key question: "Why is it that every

Figure 3.10   Malcolm Wells (1926–2009) significantly influenced today's green building movement through his "tread gently on the earth" approach. (Photograph courtesy of Karen Wells)

Figure 3.11   The underground art gallery of Karen Wells (wife of Malcolm Wells). (Photograph courtesy of Karen Wells)

architect can recognize and appreciate beauty in the natural world yet fail to endow his own work with it?" (see Figure 3.10). His solution was a simple but very effective one: Leave the surface of the planet alone and submerge the built environment underground so that the earth's surface can continue to provide unimpeded services, as shown in the Wells art studio in Figure 3.11. Wells' approach was to tread gently on the earth, minimize the use of asphalt and concrete, and use local natural resources and solar energy as the primary resources for the built environment. He is known as the "father of gentle architecture" or of earth-sheltered architecture, and although he claims that his work has not had the effect he had hoped for, his thinking has significantly influenced today's green building movement. He suggests that buildings should consume their own waste, maintain themselves, provide animal habitat, moderate their own climate, and match nature's pace—all notions that are frequently presented in the increasing number of green building forums throughout the United States.

JOHN LYLE

Landscape is perhaps the most neglected and underrated issue in green design, but one man, John Lyle (1934–1998), pursued the goal of creating regenerative landscapes. His book, *Design for Human Ecosystems*, originally published in 1985, is his classic text. In it, he explores methods of designing landscapes that function in the sustainable ways of natural ecosystems (see Figures 3.12 and 3.13). The book provides a framework for thinking about and understanding ecological design, highlighted by a wealth of real-world examples that bring Lyle's key ideas to life. Lyle traces the historical growth of design approaches involving natural processes and presents an introduction to the principles, methods, and techniques that can be used to shape landscape, land use, and natural resources in an ecologically sensitive and sustainable manner. He

Figure 3.12   John Lyle (1934–1998) promoted the idea of creating regenerative landscapes through ecological design. (Photograph courtesy of the Lyle Center for Regenerative Studies, California State Polytechnic University, Pomona)

Figure 3.13 The Center for Regenerative Studies at the California State Polytechnic University in Pomona. (Photograph courtesy of the Lyle Center for Regenerative Studies, California State Polytechnic University, Pomona)

articulates the problems inherent in imposed and artificial infrastructures, which are part of a linear industrial system in which materials extracted from nature and the earth end up as useless waste.

Unlike its natural counterpart, the urban landscape does not produce food; store, process, or treat stormwater; or provide diverse habitat for wildlife; is not part of an ecological system; and does not contribute to biological diversity. And the artificial landscape is not sustainable because it is highly dependent for its survival on fossil fuel, chemicals, and large quantities of water. In contrast, Lyle's regenerative landscape is characterized by the qualities of locality, fecundity, diversity, and continuity. A regenerative landscape grows out of a particular place (locality) in a manner unique to that place. It is fertile and continually grows and renews itself through reproduction, the heart of regeneration (fecundity). The regenerative landscape is composed of a wide variety of plants and organisms, each occupying a niche in its environment (diversity). And the regenerative landscape is not fragmented; it changes gradually over space and time (continuity).

Contemporary Ecological Design

The influence of these architects, designers, and philosophers on today's green building movement has been profound. In addition to establishing the foundations for ecological design, they influenced a large number of today's practitioners. Even though ecological design is still in development, the green building movement is driving efforts to refine its meaning and to explore in detail the connection between ecology and the built environment. Today's green building movement builds on the thoughts and work of figures like Fuller, Wright, Neutra, Mumford, Lyle, and McHarg. To a few voices on the subject of ecological design prior to 1990 are now added the intellectual capital and professional output of thousands of individuals, organizations, and companies.

The process of discovery and implementation will be a long but exciting journey as design, practices, materials, methods, and technologies adapt to a world that is truly in need of a refined approach to the built environment.

Perhaps the first step in describing where ecological design is today is to sort through the terminology being used in association with this concept. Christopher Theis, professor of architecture at Louisiana State University, in a paper published in 2002 on the website of the Society of Building Science Educators, suggests that we first have to deal with several differing sets of nomenclature floating around in the building community.[7] A variety of terms, including those already introduced in this book, are being used to describe the approach to delivering high-performance buildings: sustainable design, green design, ecological design, and ecologically sustainable design. Theis advocates the use of *ecological* to describe the design strategy needed to produce a high-performance green building. Although using the word *sustainable* to describe this design strategy may be more comprehensive, doing so leads to levels of complexity that are not resolvable in designing a building, because it is necessary to consider the three major aspects of sustainability: social, economic, and environmental. This is a nearly impossible task for the building team because their task is to take projects awarded to them by an owner or client and meet the requirements spelled out in their contract. This is not to say that the building team should be unaware of sustainability issues, and as much as possible, they should consider the ramifications of all their decisions with respect to sustainability. In fact, the building team can exert a powerful influence on owners by educating them about these broad issues, both directly, through an articulation of their philosophy, and indirectly, by their approach to building design.

As for ecological design itself, Peter Wheelwright, chair of the Department of Architecture at the Parsons School of Design, described two often contradictory and conflicting approaches to ecological design currently proffered in schools of architecture: the organic one, which combines an activist social agenda with a "Wrightian" design ethic, and the technological one, which is "futurist in orientation and scientific in method." In fact, they coexist, with designers seeking to create solutions rooted in nature, yet applying technology as appropriate.

Figure 3.14 William McDonough developed the principles of sustainable design commonly known as the *Hannover Principles* in 2000. (*Source:* Boise State University)

Key Green Building Publications: Early 1990s

The early 1990s marked the start of the green building movement in the United States. Three publications of this era provided an early articulation of green building design: *The Hannover Principles* in 1992, *The Local Government Sustainable Buildings Guidebook* in 1993, and *The Sustainable Building Technical Manual* in 1996. In addition, in 1992, *Environmental Building News*, the first and still the most authoritative publication on green building issues, was launched and featured a checklist for green design. Each of these key publications is briefly reviewed below.

THE HANNOVER PRINCIPLES

In 1992, the city manager of Hannover, Germany, Jobst Fiedler, commissioned William McDonough, one of the early major figures in the emergence of green buildings, to work with the city to develop a set of principles for sustainable design for the 2000 Hannover World's Fair (see Figures 3.14 and 3.15). The principles were not intended to serve as a how-to for ecological design but as a foundation for ecological design. One of the contributions that emerged from

Figure 3.15 Holland Pavilion at the Hannover Expo 2000. (Hans Werlemann)

TABLE 3.2

The Hannover Principles

1. Insist on the rights of humanity and nature to coexist.
2. Recognize interdependence.
3. Respect relationships between spirit and matter.
4. Accept responsibility for the consequences of design.
5. Create safe objects of long-term value.
6. Eliminate the concept of waste.
7. Rely on natural energy flows.
8. Understand the limitations of design.
9. Seek constant improvement by the sharing of knowledge.

this relatively early attempt to articulate principles for the green building movement was a definition of sustainable design as the "conception and realization of ecologically, economically, and ethically responsible expression as part of the evolving matrix of nature." These principles, commonly known as the *Hannover Principles*, are listed in Table 3.2.[8]

THE LOCAL GOVERNMENT SUSTAINABLE BUILDINGS GUIDEBOOK AND THE SUSTAINABLE BUILDING TECHNICAL MANUAL

In the 1990s, several publications attempted to provide an orientation to the current era of ecological design, especially as driven by the emergence of the LEED building assessment system. Two of the first publications on the subject of designing a green building were produced by Public Technology, Inc. (PTI): *The Local Government Sustainable Buildings Guidebook*, in 1993, and *The Sustainable Building Technical Manual*, in 1996. At the time of their publication, the USGBC was a very new organization, and the first drafts of the LEED standard were just beginning to emerge from its committees.

The Local Government Sustainable Buildings Guidebook reveals some of the very first thoughts on the direction of the US green building movement. A number of the guiding principles noted in the guidebook are shown in Table 3.3.[9]

In contrast to the guidebook, *The Sustainable Building Technical Manual* was in essence a stopgap measure to serve the rapidly growing interest in green building. The manual provides a list of areas that should be considered in

TABLE 3.3

Design Considerations and Practices for Sustainable Building

Resources should be used only at the speed at which they naturally regenerate, and should be discarded only at the speed at which local ecosystems can absorb them.

Material and energy resources must be understood as a part of a balanced human/natural cycle. Waste occurs only to the extent that it is incorporated back into that cycle and used for the generation of more resources.

Site planning should incorporate resources naturally available on the site, such as solar and wind energy, natural shading, and drainage.

Resource-efficient materials should be used in construction of the building and in furnishings to lessen local and global impact.

Energy and materials waste should be minimized throughout the building's life cycle from design through reuse or demolition.

The building shell should be designed for energy efficiency.

Material and design strategies should strive to produce excellent total indoor environmental quality, of which indoor air quality is a major component.

The design should maximize occupant health and productivity.

Operation and maintenance systems should support waste reduction and recycling.

Location and systems should optimize employee commuting and customer transportation options and minimize the use of single-occupancy vehicles. These include using alternative work modes such as telecommuting and teleconferencing.

Water should be managed as a limited resource.

designing a green building. These are summarized in Table 3.4.[10] The manual emphasizes the need for an integrated, holistic approach to design, with the building being considered a system rather than an assemblage of parts. This marked one of the first public statements of this key aspect of green building. As noted previously, the notion of a systems approach has emerged as one of the dominant themes of green building, even though in practice it is difficult to achieve due to the large quantities of information being processed, the many actors involved, and the same difficulties in communication that occur in conventional design.

ENVIRONMENTAL BUILDING NEWS

The most prominent US publication on green building is *Environmental Building News* (*EBN*), a monthly newsletter/journal dedicated to the subject of high-performance buildings. Periodically, it has featured checklists on various subjects related to green building, among them one for environmentally responsible design. Although not considered a philosophical approach, it does provide an overview of the major issues that should be considered in designing green buildings. Table 3.5 presents this checklist.[11]

TABLE 3.4

Overview of Building Design Issues as Stated in *The Sustainable Building Technical Manual*

Passive Solar Design
Daylighting
Building envelope
Renewable energy
Building Systems and Indoor Environmental Quality
HVAC, electrical, and plumbing systems
Indoor air quality
Acoustics
Building commissioning
Materials and Specifications
Materials
Specifications

TABLE 3.5

EBN Checklist for Environmentally Responsible Design

Smaller is better. Optimize use of interior space through careful design so that the overall building size—and the resources used in constructing and operating it—are kept to a minimum.

Design an energy-efficient building. Use high levels of insulation, high-performance windows, and tight construction. In southern climates, choose glazings with low solar heat gain.

Design buildings to use renewable energy. Passive solar heating, daylighting, and natural cooling can be incorporated cost-effectively into most buildings. Also consider solar water heating and photographvoltaics—or design buildings for future solar installations.

Optimize material use. Minimize waste by designing for standard ceiling heights and building dimensions. Avoid waste from structural overdesign (use optimum-value engineering/advanced framing). Simplify building geometry.

Design water-efficient, low-maintenance landscaping. Conventional lawns have a high impact because of water use, pesticide use, and pollution generated from mowing. Landscape with drought-resistant native plants and perennial groundcovers.

Make it easy for occupants to recycle waste. Make provisions for storage and processing of recyclables—recycling bins near the kitchen, undersink compost receptacles, and the like.

Look into the feasibility of graywater. Water from sinks, showers, or clothes washers (graywater) can be recycled for irrigation in some areas. If current codes prevent graywater recycling, consider designing the plumbing for easy future adaptation.

Design for durability. To spread the environmental impacts of building over as long a period as possible, the structure must be durable. A building with a durable style ("timeless architecture") will be more likely to realize a long life.

Avoid potential health hazards—radon, mold, pesticides. Follow recommended practices to minimize radon entry into the building and provide for future mitigation if necessary. Provide detailing to avoid moisture problems, which could cause mold and mildew growth. Design insect-resistant detailing to make minimizing pesticide use a high priority.

Figure 3.16 Ken Yeang developed principles for applying ecology directly to architecture. (Photograph courtesy of Ken Yeang)

Key Contemporary Publications about Ecological Design

In addition to the publications just described, in the mid-1990s, two landmark books on the subject of contemporary ecological design were published: *Designing with Nature*, written in 1995 by Ken Yeang, a Malaysian architect, and *Ecological Design*, authored by Sim Van der Ryn and Stuart Cowan, in 1996. Although there are several other volumes on the subject of designing buildings in a manner that employs either the metaphor or model of nature, these two are particularly noteworthy for their deeper thinking on the subject of ecological design.

DESIGNING WITH NATURE: KEN YEANG (1995)

Designing with Nature was perhaps the first publication to attempt to tackle the tremendous challenge of how to apply ecology directly to architecture. Ken Yeang (see Figures 3.16 and 3.17) uses the terms *green architecture* and *sustainable architecture* interchangeably, defining them as "designing with nature and designing with nature in an environmentally responsible way." He approaches this problem by making several important assumptions:[12]

- The environment must be kept biologically viable for people.
- Environmental degradation by people is unacceptable.
- Destruction of ecosystems by humans must be minimized.
- Natural resources are limited.
- People are part of a larger closed system.
- Natural system processes must be considered in planning and design.
- Human and natural systems are interrelated and essentially one system.
- Changing anything in the system affects everything else.

Figure 3.17 The National Library of Singapore designed by Ken Yeang. (Photograph courtesy of Ken Yeang)

TABLE 3.6

Bases for Ecological Design as Suggested by Ken Yeang

1. Design must be integrated not only with the environment, but also with the ecosystems that are present.
2. Because earth is essentially a closed system, matter, energy, and ecosystems must be conserved and the biosphere's waste assimilation capacity considered.
3. The context of the ecosystem, that is, its relationship with other ecosystems, must be considered.
4. Designers must analyze and use each site for its physical and natural structures to optimize the design.
5. The impact of the design must be considered over its entire life cycle.
6. Buildings displace ecosystems, and the matter-energy impacts must be considered.
7. Due to the complex impacts of built environments on nature, design must be approached holistically rather than in a fragmented manner.
8. The limited assimilative capacity of ecosystems for human-induced waste must be factored into design.
9. Design should be responsive and anticipatory, and as much as possible result in beneficial effects for natural systems.

Yeang also suggests several premises or bases for ecological design (see Table 3.6).[13]

In the actual implementation of ecological design, Yeang suggests that there are three major steps:

- Define the building program as an ecological impact statement (analysis).
- Produce a design solution that comes to grips with the probable environmental interactions (synthesis).
- Establish the performance of the design solution by measuring inputs and outputs throughout the life cycle (appraisal).

Yeang continues his efforts to develop his concept of ecological design, and he is particularly well known for his work on tall greening buildings. He has written several other books on the subject of ecological design and in 2008 published an updated work on the general subject of ecological design called *EcoDesign: A Manual for Ecological Design.* He has also written extensively on the greening of skyscrapers in *The Green Skyscraper: The Basis for Designing Sustainable Intensive Buildings* (1999), *Eco Skyscrapers* (2007), and *Eco Skyscrapers*, Volume 2 (2011).

ECOLOGICAL DESIGN: SIM VAN DER RYN AND STUART COWAN (1996)

Sim Van der Ryn and Stuart Cowan also delved deeply into the subject of ecological design in their book by the same name. *Ecological Design* was written to provide a context for green design rather than specific details. The main feature of the book is the articulation of five ecological design principles:

1. *Solutions grow from place.* Each location has its own character and resources; hence, design solutions are likely to differ accordingly. Solutions should also take advantage of local style, whether it be the adobe architecture of New Mexico or the cracker architecture of Florida. Sustainability has to be embedded in the process so that choices can be

made about how a project can interact with local ecosystems and, ideally, improve on the conditions that presently exist—for example, to clean up contaminated industrial sites or brownfields for productive uses.

2. *Ecological accounting informs design.* For true ecological design to take place, the impact of all decisions must be taken into account. These include the effects of energy and water consumption; solid, liquid, and gaseous wastes; and toxic materials use and waste. Moreover, materials selection should support the design of facilities that minimize resource consumption and environmental effects. In regard to materials selection, life-cycle assessment (LCA) is appropriate to determine the total resource consumption and emissions over the entire life of the building and to find the solution with the minimum total impact.

3. *Design with nature.* Ecological design should foster collaboration with natural systems, and the result should be buildings that coevolve with nature. Buildings should mimic nature, where, for example, there is essentially no waste because, in nature, waste equals food. Buildings are one stage in a complex industrial system that has to be redesigned with this strategy in mind to ensure that waste is minimized and closed-loop behavior rather than large-scale waste is the result. A synergistic relationship with nature is desirable, one in which matter-energy flows across the human-nature interface and is beneficial to both subsystems, human and natural. The heating and cooling systems in buildings can be assisted by landscaping; waste can be processed by wetlands; trees can take up vast quantities of stormwater; and waste generated by a building's occupants can provide nutrients for the landscape.

4. *Everyone is a designer.* The participatory process is emerging as a key ingredient of ecological design; that is, including a wide array of people affected by a building provides more creative and interesting results. Schools of architecture need to be reinvigorated, reoriented to teach about building holistically, and to include ecological design as a foundation for the curriculum. A new ecological design discipline should be created to address not only issues that may be connected to the built environment but also issues such as industrial product design and the materials supply chain.

5. *Make nature visible.* Having lost their connection with nature, humans have forgotten details as simple as where their water and food originate and how they are processed and moved to humans for consumption. Ecological design should reveal nature and its workings as much as possible, celebrate place, and reverse the trend from denatured cities to urban spaces with life and vitality. Drainage systems, normally hidden, might be exposed. The disposal areas for waste, sewage systems, wastewater treatment plants, and landfills should be located closer to the human waste generators to expose them to the consequences of wasteful behavior. By the same token, the elegant and complex behavior of natural systems in the form of natural wetlands that treat effluent can serve to educate people about integration with nature. As part of the design and construction process, the regenerative approaches advocated by John Lyle can be employed to restore areas once damaged by human activities to their natural state.

Van der Ryn and Cowan provide a framework for designers—that is, everyone—for creating a nature- and ecology-based process that is flexible, adaptable, and useful for the building project and the place. Again, their framework does not give details on how to accomplish this process, because the details would be immense in scope and volume. Rather, it provides a strong philosophical underpinning for high-performance green building design that, if faithfully followed, will produce human-made structures that cooperate rather than compete with nature.

THE NATURE OF DESIGN: ECOLOGY, CULTURE, AND HUMAN INTENTION: DAVID ORR (2002)

In 2002, David Orr addressed ecological design in his book, *The Nature of Design: Ecology, Culture, and Human Intention.* Orr takes a much broader view, addressing the full array of human interaction with nature, to include how we acquire and use food, energy, and materials and what we do for a living (see Figure 3.18). Although he is not a professional in a built environment

Figure 3.18 In his book *The Nature of Design: Ecology, Culture, and Human Intention*, David Orr addresses the full array of human interaction with nature, including how we acquire and use food, energy, and materials and what we do for a living.

Figure 3.19 The Lewis Center for Environmental Studies at Oberlin College in Oberlin, Ohio, was built in the late 1990s and designed by an elite team of architects and other professionals, among them William McDonough, one of the leading green building architects, and John Todd, creator of the Living Machine, a waste treatment system that uses natural processes to break down the components of the building's wastewater stream. (Courtesy of Oberlin College)

Figure 3.20 David Orr broadens our thinking about ecological design by comparing it to the Enlightenment of the 18th century, with its connections to politics and ethics. He describes ecological design as an emerging field that seeks to recalibrate human behavior to, in effect, synchronize it to nature and connect people, places, ecologies, and future generations in ways that are fair, resilient, secure, and beautiful. (Photograph courtesy of David Orr)

discipline, Orr has made a significant impact on today's green building movement by virtue of his ability to clearly elucidate a vision of ecological design. Orr broadens our thinking about ecological design by comparing it to the Enlightenment of the 18th century, with its connections to politics and ethics. He describes ecological design as an emerging field that seeks to recalibrate human behavior to, in effect, synchronize it to nature and connect people, places, ecologies, and future generations in ways that are fair, resilient, secure, and beautiful. According to Orr, changing the behavior of both the public and private sectors is badly needed to transform our production and consumption patterns.

In addition to his work as an author and as a proponent of environmental literacy, Orr successfully raised funds for what is perhaps the most important green building project of the late 1990s: the Lewis Center for Environmental Studies at Oberlin College in Oberlin, Ohio (see Figures 3.19 and 3.20). The Lewis Center was designed by an elite team of architects and other professionals, among them William McDonough, one of the leading green building architects, and John Todd, creator of the Living Machine, a waste treatment system that uses natural processes to break down the components of the building's wastewater stream. Orr sees buildings as contributing to a pedagogy for environmental literacy and cites numerous examples of how designers can create structures that teach as well as function. For example, buildings can teach us how to conserve energy, recycle materials, integrate with nature, and contribute rather than detract from their surroundings. The landscape around the Lewis Center, by virtue of its design, helps teach ecological competence in horticulture, gardening, natural systems agriculture, forestry, and aquaculture, as well as techniques to preserve biodiversity and ecological restoration. As Orr notes, we need a national effort to engage students of every discipline in ecological design because our current system of production and consumption is poorly designed. This is perhaps the key challenge facing us: understanding how nature can inform design of all types, including that of buildings.

Future Ecological Design

At present, sustainable construction is constrained by an inability to come to grips with a more precise notion of what ecological design is and what can and cannot be achieved through its application. A wide variety of hypotheses about ecological design have been presented in addition to those mentioned in the previous discussion of the history of ecological design. Some of the major hypotheses suggested by designers, industrial ecologists, and others are as follows:

1. General management rules for sustainability (Barbier 1989; Daly 1990)
2. Design principles for industrial ecology (Kay 2002)
3. The golden rules for ecodesign (Bringezu 2002)
4. Adaptive management (Peterson 2002)
5. Biomimicry (Benyus 1997; briefly described in Chapter 2)
6. Factor 4 and Factor 10 (von Weizsäcker, Lovins, and Lovins 1997; briefly described in Chapter 2)
7. Cradle to cradle (McDonough and Braungart 2002)
8. The Natural Step (Robèrt 1989; described in Chapter 2 and Chapter 11)
9. Natural capitalism (Hawken, Lovins, and Lovins 1999)

In the following sections, these major contributions to ecological design are presented as the basis for a future more robust and more refined version of ecological design that can serve as both a philosophical and technical basis for sustainable construction.

GENERAL MANAGEMENT RULES FOR SUSTAINABILITY

Proponents of ecological economics have formulated several pragmatic rules for "managing" sustainability.[14] According to the first rule, the use of renewable resources should not exceed the regeneration rate. In order to operationalize this demand, one has to consider that the use of either naturally or technically renewable materials always requires some inputs of nonrenewables (e.g., mineral fertilizer for the loss of nutrients due to leaching in agriculture, and the requirements for materials and energy for recycling processes). As a consequence, the total life cycle of products has to be checked for the use of renewables and nonrenewables. The former will have to be distinguished according to criteria on sustainable modes of production in agriculture, forestry, and fishery. An example in the construction sector would be the origin of timber products from sustainable cultivation.

The second rule states that nonrenewable resources may be used only if physical or functional substitutes are provided—for example, investments in solar energy systems from gains from fossil fuels. Here the basic assumption is that man-made capital may be substituted for natural capital (weak sustainability). The central requirement from an economic perspective is that the sum of natural and man-made capital is not reduced. However, from a natural systems perspective, it may be argued that there are minimum requirements of nature that may not be depleted without risk for life-support functions. Therefore, man-made capital should not be substituted (permanently) for natural capital (strong sustainability). Under this assumption, the second rule would require minimization of the use of nonrenewables.

The third rule states that the release of waste matter should not exceed the absorption capacity of nature. This can be operationalized by comparing critical loads of water, soil, and air compartments with actual levels of emission rates. After measures have been successfully applied to reduce pollution problems, the after-end-of-pipe approach to limit critical loads is also important. The implementation of the third rule is usually based on substance-specific analyses. This approach has some limitations. Generally, we must acknowledge that we are aware of only the tip of the iceberg with respect to the potential future impacts of all materials and substances released to the environment. Many natural functions react in a nonlinear manner. The complex interactions of natural substances like carbon dioxide, not to mention thousands of synthetic chemicals, cannot be foreseen in total.

From experience, we know that the effects of certain emissions become obvious after release and after the change of the environment takes place. There is a huge time lag between the scientific finding, public perception, and political reaction. Thus, the chances for comprehensive and precautionary materials management are extremely limited. A long-term effective implementation of the third rule should begin before the end of pipe and should aim to minimize the environmental impact potential of anthropogenic material flows. This impact potential is generally determined by the volume of the flow times the specific impacts per unit of flow. The second term is unknown for most materials released to the environment. The first term, the volume or weight used or released in a certain time period, can be made available for nearly every material handled. It may be used to indicate a generic environmental impact potential. As long as detailed information on specific impacts is lacking, it may be assumed that the impact potential is growing with the volume of the material flow. The overall volume of outputs from the *anthroposphere*, the portion of the earth affected by human activities, can only be reduced when the inputs to this system are diminished.[15] This is especially important for construction material flows with large scale and significant retention time within the anthroposphere. Starting from a situation in which the assimilation capacity of nature is overloaded by a variety of known substances, the long-term implementation of the third rule requires a reduction of the resource inputs of the anthroposphere in order to lower the throughput and ultimate output to the environment.

Another rule that has not yet attracted sufficient attention may be derived from the relation of inputs and outputs of the anthroposphere. Currently, the input of resources exceeds the output of wastes and emissions in industrialized as well as developing countries. As a consequence, the economies of these countries are growing physically (in terms of new buildings and infrastructure). The stock of materials in the anthroposphere is therefore increasing. In Germany, for example, the rate of net addition to stock was about 10 tons per capita annually in the mid-1990s. Associated with this accumulation of stock is an increase in built-up land area and a consequent reduction in reproductive and ecologically buffering land. Keeping in mind the limited space on our planet, this development cannot continue infinitely. Thus, a flow equilibrium between input and output must be expected. However, a question naturally arises: When will the economy stop growing physically and to what physical level?

DESIGN PRINCIPLES FOR INDUSTRIAL ECOLOGY

James Kay, the late ecologist from the University of Waterloo, proposed a set of principles that would govern the production-consumption system.[16] They are based on the premise that all man-made systems should contribute to the survival of natural systems.

1. *Interfacing.* The interface between societal systems and natural ecosystems reflects the limited ability of natural ecosystems to provide energy and absorb waste before their survival potential is significantly altered, and the fact that the survival potential of natural ecosystems must be maintained.

2. *Bionics.* The behavior of large-scale societal systems should be as similar as possible to that exhibited by natural systems.

3. *Appropriate biotechnology.* Whenever feasible, the function of a societal system should be carried out by a subsystem of a natural biosphere.

4. *Nonrenewable resources.* Nonrenewable resources are used only as capital expenditures to bring renewable resources online.

The interfacing and appropriate biotechnology principles are related to intermediate ecological design in that they call for natural systems to interface with human systems in a synergistic manner to the benefit of both systems. Natural systems could provide services that would otherwise be performed by expensive engineered systems, such as stormwater control and waste processing. The bionics principle is closely related to strong ecological design but notably for large-scale functions. The nonrenewable resources principle has its roots in ecological economics, where investing limited nonrenewables in transitioning to renewable resources is a key tenet. In effect, Kay's design principles are a mix of various levels of several types of ecological design, and he does not state that one version is most preferable.

THE GOLDEN RULES FOR ECODESIGN

To assist engineers, architects, and planners in the production of an environmentally benign built environment, Stefan Bringezu of the Wuppertal Institute suggested five "golden rules of ecological design":[17]

1. Potential impacts to the environment should be considered on a life-cycle basis (from cradle to cradle).

2. The intensity of use of processes, products, and services should be maximized.

3. The intensity of resource use (material, energy, and land) should be minimized.

4. Hazardous materials should be eliminated.

5. Resource input should be shifted toward renewables.

The first golden rule aims to avoid shifting problems between different processes and actors. For instance, if the energy requirements for heating or cooling during the use phase of buildings were not considered in the planning phase, the options with the highest potential for energy efficiency would be neglected. And if one considers only the direct material inputs for construction, the environmental burden associated with the upstream flows will be hidden.

The second golden rule reflects the fact that most building products are not used much of the time. For a considerable part of each day and each week, homes, offices, and public buildings are essentially unoccupied. Nevertheless, economic, environmental, and probably also social costs have to be paid for maintenance. Multifunctionality and more flexible models of use may reduce the demand for additional construction and contribute to lower costs for the users. The model of car sharing may also be applied for construction. Part-time employees already share the same office. And there is even potential for more efficient building use beyond normal working hours. The third golden rule may be specified with the Factor 4 to 10 target for material requirements, including energy carriers, and should be applied to average products and services. In order to reach these goals, it seems essential to invest more intellectual power in the search for alternative options to provide the services and functions demanded by users. The fourth golden rule calls for the elimination of hazardous substances, at face value a very sensible rule, but very difficult to implement from the perspective of today's economy. The use of nuclear energy violates this rule, and self-replicating nanomachines or genetically modified organisms may also be considered hazardous according to some criteria. The fifth and final golden rule is a restatement of a key concept of ecological economics, namely, that supplies of nonrenewables will clearly diminish over time as they are consumed. For example, recent studies of copper consumption in the United States indicate that only one-third of the original dowry of copper ore exists today. The logic is that as these resources disappear, a shift to renewable resources must occur, and that, in fact, the consumption of nonrenewables should support the development of renewable resources. In the case of copper, a substitute renewable material may not be easy to develop.

ADAPTIVE MANAGEMENT

Ecology, like other fields, has several distinct schools. One of them is adaptive management, as articulated by Gary Peterson, who described it as an approach to ecosystem management that argues that ecosystem functioning can never be totally understood.[18] As Peterson notes, ecosystems are continually changing due to internal and external forces. Internally, ecosystems change due to the growth and death of individual organisms, as well as fluctuations in population size, local extinction, and the evolution of species traits. Ecosystems are also changed by external events such as the immigration of species, alterations in disturbance frequency, and shifts in the diversity and amount of nutrients entering the ecosystems. To cope with these changes, management must continually adapt. Management becomes adaptive when it persistently identifies uncertainties in human-ecological understanding and then uses management intervention as a tool to strategically test the alternative hypotheses implicit in these uncertainties. Consequently, basing the design of human systems on ecosystem function means creating materials, products, and processes using models that are not very well understood. Clearly, this means that it is probably impossible to implement strong ecological design in other than one-dimensional, virtually trivial applications.

Adherents to this line of thinking are also responsible for posing the fundamental and crucial question: "Why are systems of people and nature not just ecosystems?"[19] As noted in the Chapter 2 discussion of ethics and sustainability, the qualities of humankind that make them the only forward-looking and thinking species on this planet can result in humans thinking of themselves as "apart" from nature rather than "a part" of nature. Coupled with the ability to infer the laws of nature and physics and the ability to create materials and products that have no precedent in nature, the challenge is how to address the results of human inventiveness.

BIOMIMICRY

Janine Benyus described biomimicry as the conscious emulation of life's genius[20] (see Figure 3.21). In her popular book on the subject, she states that "'Doing it nature's way' has the potential to change the way we grow food,

Figure 3.21 Janine Benyus describes biomimicry as the conscious emulation of life's genius and outlines 10 lessons for corporations, based on the emulation of nature, as the model for human-designed systems. (Mark Bryant Photography, 2011)

make materials, harness energy, heal ourselves, store information, and conduct business." She goes on to say, "In a biomimetic world, we would manufacture the way animals and plants do, using sun and simple compounds to produce totally biodegradable fibers, ceramics, plastics, and chemicals." Farms would be modeled on prairies, new drugs would be based on plant and animal chemistry, and even computers would use carbon-based rather than silicon-based structures (see Figure 3.22). Proponents of biomimicry point to the 3.8 billion years of research and development that nature has invested in evolving a wide range of materials and processes that could benefit humans. Benyus also laid out 10 lessons for corporations that are based on the emulation of nature as the model for human-designed systems:

1. Use waste as a resource.
2. Diversify and cooperate to fully use the habitat.
3. Gather and use energy efficiently.
4. Optimize rather than maximize.
5. Use materials sparingly.
6. Don't foul the nest.
7. Don't draw down resources.
8. Remain in balance with the biosphere.
9. Run on information.
10. Shop locally.

Benyus also suggests "four steps to a biomimetic future":

1. *Quieting.* Immerse ourselves in nature.
2. *Listening.* Interview the flora and fauna of our own planet.
3. *Echoing.* Encourage biologists and engineers to collaborate, using nature as a model and measure.
4. *Stewarding.* Preserve life's diversity and genius.

With respect to step 3, echoing, she provides 10 questions for testing innovation or technology for its acceptability, and all 10, according to Benyus, should be answered affirmatively.

1. Does it run on sunlight?
2. Does it use only the energy it needs?
3. Does it fit form to function?
4. Does it recycle everything?
5. Does it reward cooperation?
6. Does it bank on diversity?
7. Does it utilize local expertise?
8. Does it curb excess from within?
9. Does it tap the power of limits?
10. Is it beautiful?

In the area of materials, Benyus states that nature has four approaches:

1. Life-friendly manufacturing processes
2. An ordered hierarchy of structures

Figure 3.22 The "Stickybot" (A) is a biomimicry design developed at Stanford University with adhesive "feet" that mimic the setae on a gecko's feet (B), enabling it to climb vertical surfaces. (Photographs courtesy of (A) Mark Cutkosky, Stanford University, and (B) Ali Dhinojwala, University of Akron)

3. Self-assembly

4. Templating of crystals with proteins

As she points out, nature does produce a wide range of complex and functional materials. Abalone (twice as tough as high-tech ceramics), silk (five times stronger than steel), mussel adhesive (works underwater), and many other natural materials are remarkable in their performance. Each is created out of the local environment and biodegrades back to the environment in a harmless manner at the end of its useful life.

Biomimicry has many drawbacks when it is applied to the design of products and materials in the human sphere. Nature manufactures its products at a built-in evolved rate that is a function of information and local resources. In contrast, humans have learned to make products at an astoundingly rapid pace and, over time, to dematerialize and deenergize their production systems. Humans can and do observe nature and natural phenomena and apply their observations to create all manner of products, not all of them beneficial. The strength of biomimicry is that it provides us with a deeper appreciation for the elegant designs of nature and instructs us about how to design systems that are materials and energy conserving, that largely close materials loops, that use renewable energy, and that are niche players in complex ecosystems. The value of biomimicry as a teacher is probably far greater than as a provider of specific information about the chemical composition and structure of materials, and in this regard it should be part of the toolbox of ecological design.

CRADLE-TO-CRADLE DESIGN

The concept of cradle-to-cradle design describes approaches that contrast to designs that employ a cradle-to-grave approach or mentality. More recently, this concept has been popularized in *Cradle to Cradle: Remaking the Way We Make Things*, by William McDonough and Michael Braungart.[21] In laying the foundation for the cradle-to-cradle concept, they suggest that people and industry should set out to create the following:

- Buildings that, like trees, produce more energy than they consume and purify their own wastewater
- Factories that produce effluents that can be used as drinking water
- Products that, when their useful life is over, do not become useless waste but can be tossed on the ground to decompose and become food for plants and animals and nutrients for soil; or, alternatively, that can return to industrial cycles to supply high-quality raw materials for new products
- Billions, even trillions, of dollars' worth of materials accrued for human and natural purposes each year
- A world of abundance, not one of limits, pollution, and waste

McDonough and Braungart suggest that the solution is to follow nature's model of eco-effectiveness. This entails separating the materials we use in human activity into biological substances (which can be returned to the natural ecosystem, where they can benefit other creatures as nutrients) and technical substances (which can, with proper design, be 100 percent recollected and recycled or even upcycled, producing, in second use, products of greater value than their original use, with zero waste). Carpets and shoes, for example, could be made of two layers—a biological outer layer that abrades over time, whose fibers could serve as nutrients in the soil or compost, and a much more durable technical inner layer that would be 100 percent recyclable, after its long life, into another identical product. A biological nutrient is a material or product that is designed to return to the biological cycle. McDonough and Braungart state that packaging, for example, can be designed as biological nutrients, so that, at the end of its use, it can be, as they put it, thrown on the ground or compost heap. A technical nutrient is a material or product that is designed to be returned to the technical cycle, the industrial metabolism from which it comes. The authors also define a class of materials they refer to as the *unmarketables*, which are neither technical nor biological nutrients.

The cradle-to-cradle approach has a number of shortcomings that make it difficult to implement. Biological nutrients, for example, are not easily defined. Is a biopolymer, produced from corn or cellulose and biodegradable, a biological nutrient? Is a biodegradable synthetic material a biological nutrient or a technical nutrient? The fact is that biomaterials such as biopolymers use natural materials as feedstock but result in alterations to the basic feedstock and produce materials that have no precedent in nature. Furthermore, the consequences of their biodegradation are not well known. Whether or not biodegradation results in nutrients or waste has not been firmly established.

McDonough and Braungart suggest implementing changes to products and systems based on five steps to eco-effectiveness:

Step 1. *Get rid of known culprits.* These include X substances, that is, materials that are bioaccumulative: mercury, cadmium, lead, and polyvinyl chloride (PVC), to name a few.

Step 2. *Follow informed personal preferences.* Prefer ecological intelligence, being sure that a product or substance does not contain or support substances or practices that are blatantly harmful to human and environmental health. This also includes admonitions to prefer respect and prefer delight, celebration, and fun.

Step 3. *Create a "passive positive" list concerning harm in manufacture or in use.* This is the X list, involving the X substances in step 1. It includes substances that are carcinogens or problematic as defined by the International Agency for Research on Cancer (IARC) and Germany's Maximum Workplace Concentration (MAK) list. MAK defines two lists of substances, the gray list and the P list. The gray list includes problematic substances not urgently in need of phaseout. The P list consists of benign substances.

Step 4. *Activate the positive list.* Redesign products focusing on the P list substances.

Step 5. *Reinvent.* Totally reinvent products such as the automobile to be "nutri-vehicles."

Dave Pollard describes this process more elegantly in his blog:[22]

1. Free ourselves from the need to use harmful substances (e.g., PVC, lead, cadmium, and mercury).

2. Begin making informed design choices (materials and processes that are ecologically intelligent, respectful of all stakeholders, and which provide pleasure or delight).

3. Introduce substance triage: (a) phase out known and suspected toxins, (b) search for alternatives to problematic substances, and (c) substitute for them "known positive" substances.

Figure 3.23 The Herman Miller Mirra chair, which was certified by the Cradle to Cradle Products Innovation Institute, is made with recycled content, and 96 percent of its components break down for easy recycling. (Photograph courtesy of Herman Miller)

4. Begin comprehensive redesigns to use only "known positives," separate materials into biological and technical, and ensure zero waste in all processes and products.

5. Reinvent entire processes and industries to produce "net positives"—activities and products that actually improve the environment.

Cradle-to-cradle design provides an interesting framework for designing materials and products and focuses attention on waste and on the proliferation of toxic substances used in the production system. Clearly, these are important issues that deserve significant attention when selecting building systems and products for a high-performance built environment. A Cradle-to-Cradle certification process has been developed by McDonough and Braungart, and several products, such as the Herman Miller Mirra chair, have been successfully assessed under this scheme (see Figure 3.23).

Thermodynamics: Limits on Recycling and the Dissipation of Materials

One of the notions repeatedly suggested by McDonough is that human designs should behave like natural systems. One of his oft-stated principles is, "There is no waste in nature," with the implication that human systems should be designed to eliminate the concept of waste. In fact, zero-waste systems are not possible due to the laws of physics, more specifically the laws of thermodynamics. Nicholas Georgescu-Roegen dealt with the implications of the entropy law and the second law of thermodynamics for economic analysis.[23] He described the important difference between primary factors of production (energy and materials) and the agents (capital and labor) that transform those materials into goods and services. The agents are produced and sustained by a flow of energy and materials that enter the production process as high-quality, low-entropy inputs and ultimately exit as low-quality, high-entropy wastes. This restricts the degree to which the agents of production (capital and labor) can

Figure 3.24 A Mercedes Benz can be quickly disassembled for end-of-vehicle life recycling. (Photograph courtesy of Mercedes Benz GmbH, Stuttgart, Germany)

substitute for depleted or lower-quality stocks and flows of energy and material inputs from the environment. Thermodynamics can inform us about ultimate limits. There are irreducible thermodynamic minimum amounts of energy and materials required to produce a unit of output that technical change cannot alter. In sectors that are largely concerned with processing and/or fabricating materials, technical change is subject to diminishing returns as it approaches these thermodynamic minimums. Matthias Ruth uses equilibrium and non-equilibrium thermodynamics to describe the materials-energy-information relationship in the biosphere and in economic systems.[24] In addition to illuminating the boundaries for material and energy conversions in economic systems, thermodynamic assessments of material and energy flows, particularly in the case of effluents, can provide information about depletion and degradation that are not reflected in market prices.

What are the implications of thermodynamics and the entropy law for materials recycling? Georgescu-Roegen argued that materials are dissipated in use, just as energy is, so complete recycling is impossible. He elevated this observation to a fourth law of thermodynamics—or law of matter entropy—describing the degradation of the organizational state of matter. The bottom line for Georgescu-Roegen is that due to material dissipation and the generally declining quality of resource utilization, materials in the end may become more crucial than energy. However, Georgescu-Roegen's fourth law has been criticized by a number of analysts in both economics and the physical sciences.

A paper by Reuter et al. addresses the dissipation of materials in recycling by examining the technical feasibility of an EU mandate for 95 percent end-of-life vehicle (ELV) recycling by 2015 (see Figure 3.24), with an intermediate goal of 85 percent by 2006.[25] One of the conclusions is that while the 85 percent target is achievable, the basic constraints of thermodynamics make it virtually impossible to reach the 95 percent goal. Consequently, at least 5 percent of the automobile mass dissipates into the biosphere. This is true of all recycling activities; the materials being recycled are dissipating to background concentrations, as dictated by the second and perhaps the fourth (according to Georgescu-Roegen) laws of thermodynamics. Indeed, the dissipation of materials in the recycling process begs a number of questions; among them is: What are the health and ecological impacts of recycling as practiced and as envisioned for a sustainable future?

A 1998 US Geological Survey report by Michael Fenton indicated some of the practical problems with so-called cradle-to-cradle strategies.[26] Steel and iron scrap, for which there is high demand, is not recycled at a very impressive rate. Fenton's report stated that, in 1998, an estimated 75 million metric tons of steel and iron scrap was generated. The recycling efficiency was 52 percent, and the recycling rate was 41 percent.

In short, materials will be lost in recycling processes and, due to entropy, will naturally seek to return to background concentrations for naturally occurring substances and to very low concentrations for synthetic materials. Cradle-to-cradle and other approaches do not address this potentially difficult issue when suggesting that recycling of technical nutrients is desirable. Again, recycling, like most other issues involved in improving materials cycles, is a matter of ethics, risk, and economics.

NATURAL CAPITALISM

The concept of natural capitalism was articulated by Hawken, Lovins, and Lovins in its most recent form in a book with the same name.[27] Implementing natural capitalism entails four basic shifts in business practice:

> **Shift 1: Radical resource productivity.** Dramatically increase the productivity of natural resources.
>
> **Shift 2: Ecological redesign.** Shift to biologically inspired models.
>
> **Shift 3: Service and flow economy.** Move to solutions-based business models.
>
> **Shift 4: Investment in natural capital.** Reinvest in natural capitalism.

Each of these shifts is echoed in the other previously mentioned sets of principles and approaches. Relative to Shift 1, the productivity of natural resources can certainly be increased. However, natural renewable resources have little role in the creation of buildings, the vast bulk of which are made of human-designed materials. The authors claim that the industrial manufacturing system converts 94 percent of extracted materials into waste, with just 6 percent becoming product. It is unclear how accurate these numbers are or if they reflect the actual situation. The ultimate goal is to reduce resource extraction, which can be accomplished in several ways:

1. Dematerialization of products
2. Increasing the recycling rate of products at the end of their life cycle
3. Increasing the durability of products

If the industrial system were to double each of these factors, a Factor 8 increase in resource productivity would occur. And each of these is achievable over the short term.

Shift 2, to biologically inspired models, is also echoed time and again and focuses on developing systems with closed-loop behavior. However, as pointed out by Reuter et al., the laws of thermodynamics and separation efficiency dictate that closed loops are not closed loops at all; that some fraction of the materials being recycled will dissipate into the environment; and that ultimately, after many recycling loops, materials will, for all practical purposes, be totally dissipated.

Shift 3, to a service and flow economy, is a proposal that has been made numerous times over the past decade and has received little serious attention. Having manufacturers retain ownership of building components and maintain responsibility for reusing or recycling them makes good sense on paper. However, maintaining the link between manufacturer and product, even after decades of use, would be extremely difficult, and the logistics system that would be required to dismantle buildings and return materials to their originators would be enormously complicated.

Shift 4, reinvesting in natural capital, is an important point, and its implementation in the built environment context can be strongly reinforced. It is indeed possible to restore damaged sites and to ensure that the net ecological value of many sites is greater than it was prior to the alterations caused by building.

BIOLOGICAL MATERIALS, BIOMATERIALS, AND OTHER NATURE-BASED MATERIALS

One of the shifts advocated by many of the approaches described above is a shift from nonrenewable to renewable resources. Natural capitalism, the Natural Step, and cradle-to-cradle design, for example, suggest that this shift is fundamental for sustainability in general. A shift to renewable resources implies a shift in the materials sector to biological materials, biomaterials, and other natural or nature-based materials. Biological materials and biomaterials are two distinct classes of materials. *Biological materials* are natural systems products such as wood, hemp, and bamboo, while *biomaterials* are materials with novel chemical, physical, mechanical, or "intelligent" properties, produced through processes that employ or mimic biological phenomena.[28] Biomaterials include several emerging classes of biopolymers such as polylactic acid (PLA) and polyhydroxyalkanoate (PHA). Long-chain molecules synthesized by living organisms, such as proteins, cellulose, and starch, are natural biopolymers. Synthetic biopolymers are generated from renewable natural sources, are often biodegradable, and are not toxic to produce. Synthetic biopolymers can be produced by biological systems (i.e., microorganisms, plants, and animals) or chemically synthesized from biological starting materials (e.g., sugars, starch, and natural fats or oils). Biopolymers are an alternative to petroleum-based polymers (traditional plastics). (Bio)polyesters have properties similar to those of traditional polyesters. Starch-based polymers are often a blend of starch and other plastics [e.g., polyethylene (PE)], which allows for enhanced environmental properties.

Biological materials, such as wood pulp and cotton, can pose environmental problems. Unsound agricultural or silvicultural practices can quickly turn a fertile tract into a disaster area. Because biological resources are renewable, there is a tendency to think of them as unlimited. Nothing could be further from the truth. If cultivated carefully, crops can be planted in perpetuity. But if the land is pushed past its carrying capacity or otherwise abused, permanent damage can be done.[29]

A widespread shift to biological materials for both energy and materials has other implications because large quantities of land may be required to provide ethanol, biological materials, and the feedstock for biomaterials such as biopolymers. An ethical debate is shaping up over taking excess land from food production and shifting it to these other applications, causing increases in food prices and impacting the poor and hungry of the world.

The fact that these materials are biodegradable and compostable means that they are recyclable via a biological route. However, there is a great deal of uncertainty about the quality and utility of the degraded materials and the logistics for effectively using these nutrients of unknown quality in agriculture or the support of natural systems.

Finally, there is little evidence that biologically based materials can replace the synthetic materials that have become common in construction, especially structural materials such as steel and concrete, not to mention copper and aluminum wiring, glass, and the wide variety of polymers used in myriad applications.

SYNTHESIS

After the range of principles and approaches that describe how to create an environmentally sound and sustainable built environment have been examined, and taking into account the orientation of the human species toward the future, the development and deployment of new materials and products will likely be based on ethics, risk, and economics. Clearly, many lessons have been learned about the introduction of toxins and estrogen mimickers into the environment, the impacts of emissions on human and natural systems, the effects of extraction on the environment and human communities, the impacts of waste, and all the other well-known negatives of the production system. Changing the decision system, screening all substances for a broad range of impacts, is badly needed to ensure that the risks to nature and humans are minimized. Certainly, nature's materials and processes provide inspiration for human-designed materials and products, and the behavior of natural systems can inform human systems. But many novel materials and products will continue to be produced, and a systematic approach to examining the extraction, production, use, recycling, and disposal of these resources is needed. This would include LCA, but with application of toxicology and other screens to produce a fuller understanding of the risks associated with the entire life cycle of materials. Beyond the question of materials is responsibility for products and ensuring their potential for disassembly. In the context of the built environment, one other level of disassembly, that of the whole building, must be considered for closing materials loops. Economics, underpinned by policy in the form of taxes that penalize negative behavior in the production and consumption system, will also help dictate the future. In the final analysis, ethics will have to govern the decision system. It must also address how humans use knowledge of potential negative impacts and, ideally, require detailed screening of all new chemicals and processes to ensure that their effects are well understood. Knowing this would allow risk assessment and the ultimate decision as to whether the benefits outweigh the costs.

THOUGHT PIECE: REGENERATIVE DESIGN

Bill Reed, an internationally known architect and thinker, suggests that we are at the beginning of a shift in thinking about the design of human systems that ultimately needs to be restorative and regenerative, that we are faced with the necessity of actually having to help revive nature after the enormous damage done by human activities over centuries. Bill's work with Regenesis Group is to lift building and community planning into full integration and coevolution with living systems—through an integrative, whole-, and living-systems design process. The purpose of this work is to improve the quality of the physical, social, and spiritual life of our living places.

Regenerative Development and Design: Working with the Whole
Bill Reed, AIA, LEED, Hon. FIGP, Integrative Design Collaborative

Regeneration is both a practice philosophy and a process. Success in regeneration means to evolve and continually develop new potential. Its dictionary definition addresses both the action and the source of this new potential (1) to create anew and (2) to be born of a new spirit.

In practical terms, regeneration means to contribute to the value-generating processes of the living systems of which we are part. Without adding value —*with a conscious awareness of the ongoing, cocreative, and emergent processes of life*— life shifts to a degenerating state. The imperative in any design process is to intentionally develop the understanding required to participate in improving the resiliency of living relationships such as ecosystems, human social systems, businesses, families, and so on. Without a process of continually adding value to living systems, sustainability is not possible.

In order to understand regeneration in the context of the sustainability movement, it is necessary to understand that the practice of targeting of conservation, zero, or neutral conditions—while worthy and necessary aims—will not address what is required for a sustainable condition (even if it is possible to reach this level of perfection). Zero damage is not the same as understanding how we interact with the complexities of life and how to avoid the inevitable, unintended consequences of our actions. Nor does zero damage address how to continually participate in the dance of evolution—the entry-level condition to join the game of life.

There are a few reasons behind why we approach sustainability from this zero-based perspective: these aims are seen primarily from a technical perspective; we perceive life as a mechanical process of interactive components rather than understanding that living wholes are greater than the sum of their parts; humans are seen as the doers, not participants; and the environment is seen as something other than us.

There is a distinction between environmental and ecological thinking. By definition, an environment is the context within which something exists. Environment contains an "us" and a "not us" in its meaning. Ecology, by contrast, sees all aspects as part of a working dynamic whole—it's all us.

There is a need to fill a significant gap in our culture's work toward achieving a sustainable condition. The gap: the development of a state of consciousness that has the ability to hold life, all life, as a living entity that works as a whole, integrated, and evolving living system. The whole, from a living-systems perspective, includes everything, every process, and every dimension of consciousness and existence—whether we can perceive these things or not.

It is difficult for a reductionist culture to understand that working with the complexity of a living system is possible in the first place, and second, how it can be addressed without reducing it to manageable parts. This is where working with pattern understanding comes into play. For practitioners familiar with working with patterns, it is actually easier to assess living patterns and reach definitive conclusions from these distinct patterns than it is to try to make sense of thousands of pieces.

We are quite good at this when it comes to assessing a whole person: we intuitively know that we will not be able to understand the distinct nature (or essence) of a friend if there are only a few organs and bones available for inspection. Even if all his or her component parts were available, all the genetic sequencing, etc., it is obvious that the nature of the person can only be described mechanically, if at all. Yet, with observation, we are able to describe the uniqueness of individuals. We do this by looking at the patterns of how they, as a whole entity, are in relation to other entities—friends, colleagues, family members, their community, a dog in the street, and so on. It is how they are in relationship, what value they add to the relationship, the role they serve and provide that begin to triangulate "who" they are, not just "what" they are.

Often, practitioners mistake the "flows" of a system as the indicator of relationship. Flows of water, energy, habitat, and sun are certainly important, yet, continuing to use human relationship as an analogue, we would not describe our relationship to a friend only in terms of flows. The aspects of relationship are energetic, often invisible, and full of extremely complex and nuanced exchanges.

A living system—or place, or watershed, or community—is a "being" or "organism." It is necessary to be in relationship with it; if we are not, then abuse, neglect, or misunderstood interventions are the result. This nature of relationship is the big leap for the design and building industry. The land is not simply dirt that we build upon. Various aboriginal peoples had this understanding; everything in space and time, including the consciousness of "who" they were, was inextricably part of the whole.

The Navajo term for *mountain* refers to a "whole set of relationships and the ongoing movement inherent in those relationships. These relationships include the life cycles of the animals and plants which grow at different elevations, the weather patterns affected by the mountain, as well as the human's experience of being with the mountain. All of these processes form the dynamic interrelationship and kinetic processes that regenerate and transform life." Since this motion of the mountain is not separate from the entire cosmic process, one can only really come to know the mountain by learning about "the kinetic dynamics of the whole."

All this is not to say that working in pieces and parts with quantitative measurement is wrong. It is just the wrong place to start. As Wendell Berry observes, "A good solution is good because it is in harmony with those larger patterns, solves more than one problem, and doesn't create new ones." He goes on to explain that health is to be valued above any cure, and coherence of pattern above almost any solution produced piecemeal or in isolation. Adopting one or two green or regenerative technologies into a green building practice without understanding the underlying principles that make the approach wholly regenerative is not as effective and, at worst, produces unintended counterproductive consequences.

Western and Eastern medicine practices may be a useful comparison. Neither is right nor wrong in itself. Green design, as it is practiced in a mechanical manner, can be compared to working on the heart or intestinal system as a specialist might—curing the particular issue but not addressing the overall systemic nature of the cause, whether it is diet, environment, stress, or genetics. Integrative design, an organized process to find synergies among building and living systems, has an analogy in integrative medicine—many specialists getting together to diagnose and address relatively complex cause and effects. Regeneration might be compared to naturopathic and Eastern medicine—cranial sacral therapy, acupuncture, and so on—these practices start with the energetic patterns of the whole body. In practice, all these practices should come into play. Yet, it is always better to start with the nature of the larger environmental influences and interrelationships before solving for the symptom and cutting the body open.

From the perspective of architecture and planning, our responsibility is not to design "things" but to positively support human and natural processes in order to achieve long-term quality of life—that is, evolution with the necessary corollary of positive potential for all life.

- This means that the act of creating a building is not a conclusion but a beginning and catalyst for positive change.
- This sets the building within and connects it to a larger system and is concerned with an overall systems approach to design.
- This considers "place"—an expression of integrated ecologies of climate, resources, and culture—critical to the shaping of building, human, and natural development.

There are current designs and policy practices that approach this nature of interrelationship with the places we inhabit. Ecosystems have been seen to recover their health and demonstrate even greater levels of potential than imagined—deserts being turned into food-producing gardens with minimal water use; water being brought back to the desert by appropriate planting and techniques of slowing down water flows; damaged, low-diversity, and desertified ecosystems brought back in to full flower along with increased animal and plant habitat by replicating preindustrial animal habitat patterns; urban areas brought back to civility and high quality of life through paying attention to the nature of human and natural patterns in each unique place. Examples include Jane Jacobs' work in New York City neighborhoods as noted in *The Death and Life of Great American Cities*, in which she uses the term *regeneration* for her work. Alan Savory's work in creating new health in damaged ecosystems is another example. Regenesis in Santa Fe, New Mexico, looks at the socioecological whole and unites these "sectors" as a whole system of healthy evolution.

There are many positive stories, evidenced around the world, about regenerative design. We have frequently seen the first glimmer of new health and wholeness in nature and human habitat appear within a span of 18 months—the qualifier is *if* we understand that every place (neighborhood, city, region) has a pattern of life and that these places are both unique and nested within each other; that the smallest unit of place-sourced design is the watershed (water activates soil health and therefore life); that humans are nature and not separate from it; and that becoming conscious of the need to be in caring relationship with all life is the foundation of a positive and thriving coexistence—and thus moves us into the realm of true coevolution.

Summary and Conclusions

Clearly, a shift is afoot in the design of buildings in the United States. In 2005, just 2 percent of nonresidential new construction was green buildings; by 2008, this had increased to 12 percent, and by 2010 had grown to between 28 and 35 percent. By 2015, an estimated 40 to 48 percent of new nonresidential construction by value will be green, equating to a $120 to $145 billion market. It is estimated that for the four-year period from 2009 to 2013, green building will support 7.9 million US jobs and pump $554 million into the US economy.[30] At a minimum, it can be said that high-performance green building has been a tremendous success. Whether green building rating systems such as LEED can be said to be driving building design to be environmentally friendly and resource-efficient is another question. Certainly, considering the rapid deterioration of our planet's fragile health, any measures that help reduce the destruction of its ecological systems, minimize waste, and use resources more effectively are helpful.

At the very least, high-performance green building can be said to be making deep inroads in addressing the disproportionate impact of the built environment on the earth. In particular, this movement makes design and construction professionals more aware of their ethical responsibilities for providing high-quality, healthy buildings that have the potential for complementing and working synergistically with nature.

Notes

1. As defined by Van der Ryn and Cowan (1996).
2. As defined by Van der Ryn and Cowan (1996).
3. Coincidentally, the founder of systems ecology, Howard T. Odum (1924–2002), and the founder of adaptive management, Crawford (Buzz) Holling (1930–), worked together at the University of Florida for an extended period of time in the last two decades of the 20th century. Odum's book, *Systems Ecology* (1983), and Holling's book, *Adaptive Environmental Assessment and Management* (1978), are landmark works that redefined how scientists think about ecological systems.
4. Excerpted from *The Business Case for Sustainable Design*, published by the Federal Energy Management Program (FEMP) in 2003.
5. Excerpted from the Buckminster Fuller Institute website, www.bfi.org.
6. Wilson and Kellert (1993) addressed the strong and fundamental connection between humans and nature.
7. Theis's 2002 paper can be found at the website of the Society of Building Science Educators, www.sbse.org.
8. *The Hannover Principles: Design for Sustainability* is available at the McDonough and Partners website, www.mcdonough.com/principles.pdf.

9. Excerpted from *The Local Government Sustainable Buildings Guidebook* (1993).
10. Excerpted from *The Sustainable Building Technical Manual* (1996).
11. The checklist is also available at the BuildingGreen, Inc., website, www.buildinggreen .com, as part of a paid subscription service. The website also provides access to all of the back issues of *Environmental Building News* (*EBN*), the oldest and most prominent source of information in the United States on green building issues.
12. These assumptions are paraphrased from Yeang (1995), chapter 1.
13. The bases for ecological design are paraphrased from Yeang (1995), chapter 1.
14. From Daly (1990) and Barbier (1989).
15. The anthroposphere, also referred to as the technosphere, is part of the biosphere, the part of the planet where life exists.
16. Kay (2002).
17. Bringezu (2002).
18. Peterson (2002).
19. Westley et al. (2002).
20. Benyus (1997).
21. McDonough and Braungart (2002).
22. http://howtosavetheworld.ca/2006/02/page/2/.
23. Georgescu-Roegen (1971, 1979).
24. Ruth (1995).
25. Reuter et al. (2005).
26. Fenton (1998).
27. Hawken, Lovins, and Lovins (1999).
28. As described at the US Department of Agriculture website, http://agclass.nal.usda .gov/glossary.shtml.
29. Hayes (1978).
30. From *Green Outlook 2011: Green Trends Driving Growth*, McGraw-Hill Construction (2010).

References

Barbier, E. B. 1989. *Economics, Natural Resource Scarcity and Development: Conventional and Alternative Views.* London: Earthscan.

Benyus, J. 1997. *Biomimicry: Innovation Inspired by Nature.* New York: William Morrow.

Bringezu, Stefan. 2002. "Construction Ecology and Metabolism."In *Construction Ecology: Nature as the Basis for Green Building*, edited by C. J. Kibert, J. Sendzimir, and G. B. Guy. London: Spon Press.

Daly, H. E. 1990. "Towards Some Operational Principles of Sustainable Development." *Ecological Economics* 2: 1–6.

Fenton, M. D. 1998. *Iron and Steel Recycling in the United States in 1998.* US Geological Survey Open File Report 01-224.

Fuller, R. Buckminster. 1969. *Operating Manual for Spaceship Earth.* Carbondale: Southern Illinois University Press.

———. 1981. *Critical Path.* New York: St. Martin's Press.

Georgescu-Roegen, N. 1971. *The Entropy Law and the Economic Process.* Cambridge, MA: Harvard University Press.

———. 1979. "Energy Analysis and Economic Valuation." *Southern Economic Journal* 45: 1023–1058.

Hawken, P., A. Lovins, and H. Lovins. 1999. *Natural Capitalism.* New York: Little, Brown.

Hayes, D. 1978. "Repair, Reuse, Recycling: First Steps Toward a Sustainable Society." Worldwatch Paper 23, Worldwatch Institute, Washington, DC.

Holling, Crawford S. 1978. *Adaptive Environmental Assessment and Management.* London: John Wiley & Sons.

Jacobs, Jane. 1961. *The Death and Life of Great American Cities.* New York: Random House.

Kay, James. 2002. "Complexity Theory, Exergy, and Industrial Ecology." In *Construction Ecology: Nature as the Basis for Green Building*, edited by C. J. Kibert, J. Sendzimir, and G. B. Guy. London: Spon Press.

Lyle, John Tillman. 1985. *Design for Human Ecosystems: Landscape, Land Use, and Natural Resources*. New York: Van Nostrand Reinhold. (Republished by Island Press, 1999.)

———. 1994. *Regenerative Design for Sustainable Development*. Hoboken, NJ: John Wiley & Sons.

McDonough, William. 1992. *The Hannover Principles: Design for Sustainability*. Charlottesville, VA: William McDonough and Partners. Available at www.mcdonough .com/principles.pdf.

McDonough, William, and Michael Braungart. 2002. *Cradle to Cradle: Remaking the Way We Make Things*. New York: North Point Press.

McGraw-Hill Construction. 2010. *Green Outlook 2011: Green Trends Driving Growth*. New York: McGraw-Hill Construction. Available at http://construction.com/ market_research/.

McHarg, Ian. 1969. *Design with Nature*. Garden City, NY: Natural History Press.

Mumford, Lewis. 1946. *Values for Survival: Essays, Addresses, and Letters on Politics and Education*. New York: Harcourt, Brace.

———. 1955. *The Brown Decades: A Study of the Arts in America*. New York: Dover.

———. 1967. *The Myth of the Machine*. New York: Harcourt, Brace & World.

Odum, Howard T. 1983. *Systems Ecology*. Hoboken, NJ: John Wiley & Sons.

Office of Energy Efficiency and Renewable Energy (EERE). 2003. *The Business Case for Sustainable Design in Federal Facilities*. Washington, DC: Federal Energy Management Program (FEMP), US Department of Energy. Available at www1.eere .energy.gov/femp/pdfs/bcsddoc.pdf.

Orr, David W. 2002. *The Nature of Design: Ecology, Culture, and Human Intention*. New York: Oxford University Press.

———. 2009. *Climate Collapse*. New York: Oxford University Press.

Pearce, David W., and R. Kerry Turner. 1990. *Economics of Natural Resources and the Environment*. Hemel Hempstead, UK: Harvester Wheatsheaf.

Peterson, Gary. 2002. "Using Ecological Dynamics to Move Toward an Adaptive Architecture." In *Construction Ecology: Nature as the Basis for Green Building*, edited by C. J. Kibert, J. Sendzimir, and G. B. Guy. London: Spon Press.

Public Technology. 1993. *The Local Government Sustainable Buildings Guidebook*. Washington, DC: Public Technology.

———. 1996. *The Sustainable Building Technical Manual: Green Design, Construction and Operations*. Washington, DC: Public Technology. Available from the USGBC website, www.usgbc.org.

Reuter, M. A., A. van Schaik, O. Ignatenko, and G. J. de Haan. 2005. "Fundamental Limits for the Recycling of End of Life Vehicles." *Minerals Engineering* 19 (5): 433–449.

Ruth, M. 1995. "Thermodynamic Implications for Natural Resource Extraction and Technical Change in U.S. Copper Mining." *Environmental and Resource Economics* 6: 187–206.

Theis, Christopher C. 2002. "Prospects for Ecological Design Education." Society of Building Science Educators. Available at www.sbse.org.

Van der Ryn, Sim, and Stuart Cowan. 1996. *Ecological Design*. Washington, DC: Island Press.

von Weizsäcker, Ernst, Amory Lovins, and L. Hunter Lovins. 1997. *Factor Four: Doubling Wealth, Halving Resource Use*. London: Earthscan.

Wells, Malcolm. 1981. *Gentle Architecture*. New York: McGraw-Hill.

Westley, F., S. Carpenter, W. Brock, C. S. Hollins, and L. H. Gunderson. 2002. "Why Systems of People and Nature Are Not Just Social and Ecological Systems," In *Panarchy*, edited by L. H. Gunderson and C. S. Holling. Covelo, CA: Island Press.

Wheelwright, Peter M. 2000. "Environment, Technology and Form: Reaction." Paper presented to the Architecture League of New York in response to the symposium Environment, Technology, and Form.

———. 2000. "Texts and Lumps: Thoughts on Science and Sustainability." *ACSA NEWS*, 5–6.

Wilson, E. O., and Stephen Kellert, eds. 1993. *The Biophilia Hypothesis*. Washington, DC: Island Press.

Yeang, Ken. 1995. *Designing with Nature: The Ecological Basis for Architectural Design*. New York: McGraw-Hill.

———. 1996. *The Skyscraper Bioclimatically Considered: A Design Primer*. London: Academy Editors.

———. 1999. *The Green Skyscraper: The Basis for Designing Sustainable Intensive Buildings*. Munich: Prestel.

———.2007. *Eco Skyscrapers*. Mulgrave, Australia: Images Publishing.

———.2008. *EcoDesign: A Manual for Ecological Design*. London: John Wiley & Sons.

———.2009. *Ecomasterplanning*. London: John Wiley & Sons.

———.2011. *Eco Skyscrapers*, vol. 2. Mulgrave, Australia: Images Publishing.

Yeang, Ken, and Arthur Spector, eds. 2011. *Green Design: From Theory to Practice*. London: Black Dog Publishing.

Part II

Assessing High-Performance Green Buildings

At present, high-performance green buildings are defined by the assessment systems that rate and certify them. Building assessment systems simply score a building project on how well it lines up with the general philosophical approach developed by the designers of the assessment system. As a result, a building assessment system provides a standard definition for green building for the country employing it. In the United States, for example, the Leadership in Energy and Environmental Design (LEED) building assessment system, its categories, and the allocation of points for various green attributes define green building for the American marketplace, both public and private. One advantage of relying on building assessment systems for this purpose is that it standardizes the boundaries of what constitutes a high-performance green building, what its important attributes are, and how the performance of the project across a wide variety of categories is measured. A significant disadvantage of these assessment systems is that each is simply one organization's vision of a green building, and often, because of time and financial constraints, assessment systems leave much to be desired. For example, many assessment systems rely on energy modeling to forecast energy consumption rather than using actual energy data as the arbiter of success. The result has been the occasional embarrassing report stating that the actual energy consumption is much higher than was originally forecast by the energy model. The rationale for not using real energy consumption data is that gathering the data takes time, generally a minimum of a year, and the cost of this effort is not insignificant. Another problem associated with overdependence on building assessment systems is that project teams risk losing the ability to think creatively and instead develop "LEED-brain." As clever and useful as LEED and other assessment systems may be, they leave a lot to be desired with respect to a wide range of important issues.

The types of ratings vary widely among assessment systems. LEED, for example, has four ratings based mostly on precious metals: platinum, gold, silver, and certified, from highest rating to lowest. In a similar fashion, Green Globes awards one to four green globes, and the Green Star system used in

Australia, New Zealand, and South Africa provides one to six green stars. In Green Star, however, only the four- to six-star projects are really meaningful. In Japan, the Comprehensive Assessment System for Building Environmental Efficiency (CASBEE) calculates a ratio of environmental benefits to environmental loadings such that a ratio of 3.0 or higher earns the project the top award. Most assessment systems have standards categories such as energy, water, and indoor environmental quality for rating the building. Some have a management category for rating the conduct of the building project while others neglect this aspect as being worthy of scoring in the rating. Clearly, there is no one single approach to building assessment although there seems to be more recent convergence on the need to provide a life-cycle assessment (LCA) of the building's materials and operating impacts.

This part of the book addresses the general subject of building assessment, the major international building assessment systems, and the major building assessment systems used in the United States. The chapters covered in this part of the book are as follows:

Chapter 4: Green Building Assessment

Chapter 5: The US Green Building Council LEED Building Rating System

Chapter 6: The Green Globes Building Assessment System

Chapter 4 is a general overview of green building assessment and covers the main issues relevant to building assessment. The two major building assessment systems used in the United States, LEED and Green Globes, are briefly described. The Living Building Challenge, a truly challenging building assessment system used in the United States and Canada, with stringent requirements for a wide range of green attributes, is covered in some detail because it provides some insights into what may be the shape of future rating systems. Additionally, this chapter provides an overview of each of the building assessment systems used in the United Kingdom, Japan, Australia, and Germany.

The subject of Chapter 5 is the LEED building assessment system. It covers the history of the development of LEED, the structure of the LEED suite of rating systems, the structure of the major LEED rating systems, and a description of all the credits that are available for projects seeking certification under one of the rating systems in the category of LEED Building Design and Construction (LEED-BD&C), which would include LEED for New Construction (LEED-NC), LEED for Schools (LEED-SCH), and LEED for Core and Shell (LEED-CS) projects. The importance of having people specially trained and experienced in the application of LEED to projects is discussed in this chapter. For the LEED building assessment system, the credential LEED Accredited Professional (LEED AP) is awarded to individuals who have had appropriate training on green building and the LEED rating system and who have passed an examination on these topics. The Green Associate (GA) is another credential offered by the US Green Building Council, and it is a prerequisite for becoming a LEED AP. The GA credential identifies individuals who have taken and passed an examination on green building fundamentals.

Chapter 6 addresses a relatively new entrant into the United States, the Green Globes building assessment system, which is supported by the Green Building Initiative located in Portland, Oregon. The history, structure, and credits that are available in Green Globes are covered in this chapter. Similar to LEED, Green Globes offers a structured program for providing projects with individuals who have been accredited as being proficient in the application of Green Globes to building projects. This program includes two levels of accreditation: the Green Globes Professional (GGP) and the Green Globes Assessor (GGA). Similar to the LEED AP, the GGP is qualified to assist project teams navigate the Green Globes building assessment and certification system. The role of a GGA, on the other hand, has no equivalent in LEED, because the GGA provides third-party verification that the project has met the requirements of the GBI for Green Globes certification.

Building assessment systems are evolving over time and the various platforms such as LEED, CASBEE, and Green Star, as examples, learn from each other and adopt the practices that have emerged as being the most useful and well received by the international community.

Chapter 4

Green Building Assessment

D uring the pre-1998 era of sustainable construction in the United States, environmentally friendly buildings were conceptualized by teams of architects and engineers who relied on their collective interpretation of what constituted green building. Beyond the understanding that green buildings should be resource efficient and environmentally friendly, no specific criteria existed to evaluate and compare the merits of green building design. In 1998, however, the US Green Building Council (USGBC) dramatically changed this process with the launch of its *Leadership in Energy and Environmental Design* (LEED) building assessment system for new construction, which identified criteria that specified not only whether or not a building was green but what specific shade of green it was. Versions of LEED are available for various building types and situations. For example, there is a LEED for New Construction (LEED-NC) and a LEED for Schools (LEED-SCH), to name but a few. LEED employs a point system to award a platinum, gold, silver, or certified rating based on how many specific predetermined criteria in several categories the building successfully addresses.

The generic term for LEED and similar systems used in other countries is *building assessment system* or *building rating system*. As mentioned in Chapter 2, the primary building assessment system used in the United Kingdom is the *Building Research Establishment Environmental Assessment Method* (BREEAM), which also was the first widely adopted rating system in the world.[1] The Japanese building assessment system, the *Comprehensive Assessment System for Building Environmental Efficiency* (CASBEE), was developed by the Japan Sustainable Building Consortium.[2] In Australia, *Green Star* is the building assessment system advocated by the Green Building Council of Australia and is fully implemented for a number of building types.[3] The newest major building assessment system is the *Deutsche Gesellschaft für Nachhaltiges Bauen/Bewertungssystem Nachhaltiges Bauen für Bundesgebäude* (DGNB/BNB), which was developed by the German Sustainable Building Council.

Building assessment systems score or rate the effects of a building's design, construction, and operation, among them environmental impacts, resource consumption, and occupant health. This can be a complicated determination, as each aspect has different units of measurement and applies at different physical scales. Environmental effects can be evaluated at local, regional, national, and global scales. Resource impacts are measured in terms of mass, energy, volume, parts per million (ppm), density, and area. Building health can be inferred by the presence or absence of chemical and biological substances within circulating air, as well as the relative health and well-being of the occupants. Comparing arrays of data for various building features presents further complications.

Why consider a building assessment standard or rating at all? In general, building assessment systems are created for the purpose of promoting high-performance buildings, and some, like LEED, are specifically designed to increase market demand for sustainable construction. Building assessment systems generally offer a label or plaque indicating a building's rating, and a plaque on the building awarded as a result of achieving green building certification is a public statement of the building's performance. A superior building assessment rating should create higher market value due to the building's lower operating costs and healthy indoor environment. Competition among owners and developers to achieve high building assessment ratings hopefully results in the development of a high-quality, high-performance building stock. Parallel effects of successful building assessment systems could also help facilitate otherwise difficult political goals, for example, national requirements related to the Kyoto Protocol on climate change, which, in effect, calls on the United States to significantly reduce fossil fuel consumption.[4]

Developers are faced with two major choices when designing a building assessment system: either to use a single number to describe the building's overall performance or to provide an array of numbers for the same purpose. A single number representing a score for the building has the virtue of being easy to understand. But if a single number is used to assess or rate a building, the system must somehow convert the many different units describing the building's resource and environmental impacts (energy usage, water consumption, land area footprint, materials, and waste quantities) and conditions resulting from the building design (building health, built-in recycling systems, deconstruct ability, and percentage of products coming from within the local area) into a series of numbers that can be added together to produce a single overall score. This is a difficult and arbitrary method at best. Paradoxically, however, both the advantage and the disadvantage of the single-number assessment is its simplicity. The LEED standard provides a single number that determines the building's assessment or rating based on an accumulation of points in various impact categories, which are then totaled to obtain a final score.

Alternatively, a building assessment system can utilize an array of numbers or graphs that depict the building's performance in major areas, such as environmental loadings or energy and water consumption, compared to conventional construction. Although this approach yields more detailed information, its complexity makes it difficult to compare buildings, depending on the range of factors considered. The Sustainable Building Tool (SBTool), a system used in the Green Building Challenge conferences to compare building performance in several countries, is an example of an assessment methodology that uses a relatively large quantity of information to assess the merits of a building's design.[5]

This chapter briefly describes the two major US building assessment standards, LEED and Green Globes, and one emerging system, the Living Building Challenge. LEED is described in detail in Chapter 5, and Green Globes is covered in Chapter 6. This chapter also provides information about other major building assessment standards or systems used around the world, including BREEAM (United Kingdom), CASBEE (Japan), Green Star (Australia, New Zealand, and South Africa), and DGNB/BNB (Germany). Several families of rating systems are emerging, some using LEED and others using BREEAM as their platform. Indeed, much of LEED is very similar to BREEAM (1990), which predates LEED (1998) by at least eight years. Green Star, which is similar to LEED, is emerging as a platform of its own. It is clear that the lineage of the major building assessment systems started with BREEAM, which provided the foundation for LEED and Green Star. Each of these three systems is the basis for several other assessment systems. CASBEE at this point in time is used only in Japan and has little similarity to LEED. The German DGNB/BNB rating system is also unique and is emerging as a new potential platform for building assessment systems in other countries, for example, Denmark.

Major Green Building Assessment Systems Used in the United States

There are two major green building assessment systems used in the United States, LEED and Green Globes, and one emerging system, the Living Building Challenge. LEED and Green Globes are described in far more detail in Chapters 5 and 6, respectively, while the Living Building Challenge is addressed in this chapter. While LEED is by far the predominant US building assessment system, Green Globes is gaining traction as an alternative and is being adopted on a large scale by, for example, the US Department of Veterans Affairs, to assess the performance of its existing hospitals undergoing renovation. The Living Building Challenge is, as its name implies, a significant certification challenge because it raises the bar for high-performance buildings to a new high level that for the first time requires building projects to demonstrate they can generate all their energy needs from on-site renewable resources. This represents just one of a number of difficult to achieve goals that the Living Building Challenge poses for project teams and owners.

THE LIVING BUILDING CHALLENGE

The most demanding of all the North American building assessment systems is the *Living Building Challenge*, which originated in 2005 as an outgrowth of programs by the Cascadia Green Building Council in the northwestern United States and western Canada. Its intent was to push the envelope of high-performance building much further than it was likely to be pressed by LEED and other building assessment systems. The Cascadia Green Building Council is unique among green building councils in North America because it represents both the USGBC and the Canada Green Building Council. There are now over 70 buildings registered and pursuing certification under the Living Building Challenge, and 6 buildings have completed the certification process and are awaiting certification.

The Living Building Challenge is based on a few simple but very powerful concepts, among them that a building should produce as much energy as it consumes, provide all the required water, and process all its sewage. One of the more recent projects undergoing Living Building Challenge certification is the Bullitt Center, a new office building located in the Capitol Hill neighborhood of Seattle, Washington, that intends to meet all the Living Building Challenge imperatives (see Figures 4.1 and 4.2). The Bullitt Center will be the home of the Seattle-based Cascadia Green Building Council and will serve as an exemplar of the cutting edge of high-performance buildings in the Pacific Northwest. The Bullitt Center will use less than one-third the energy of a comparable building, and parking will be provided for bikes but not for cars. It has an on-site solar power system that will meet all the requirements of the building yet have a payback of 8 to 10 years. Building materials such as polyvinyl chloride (PVC) plastics, mercury, cadmium, and 360 other substances considered to be hazardous will not be used. The wooden timbers for the six-story frame will originate in forests certified as sustainable by the Forest Stewardship Council (FSC). All steel, concrete, wood, and other heavy materials will come from within a 300-mile radius to help reduce the building's carbon footprint. The stringent requirements of the Living Building Challenge resulted in a collaboration between the building's architect, Miller Hull Partnership, and the University of Washington's Integrated Design Lab, a unit of the Department of Architecture. Three years of brainstorming resulted in an innovative design that

Figure 4.1 The Bullitt Center in Seattle, Washington, will be the home of the Cascadia Green Building Council and is being designed for certification to the Living Building Challenge. Among its other purposes, it will be an exemplar for future commercial buildings for low-impact and responsible construction. (Courtesy of the Miller Hull Partnership)

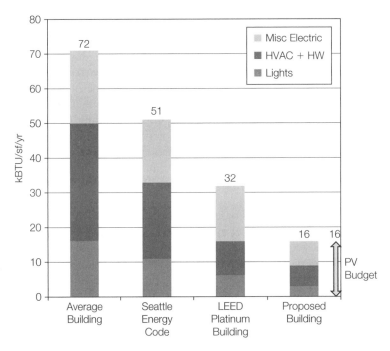

Figure 4.2 The path to net zero energy for the Bullitt Center meant a thorough rethinking of building energy consumption. Energy was reduced from about 72,000 to 16,000 BTU/ft^2/yr (176 to 39 Kwh/m^2/yr), more than a Factor 4 reduction, to be within the energy budget provided by the sun and the capabilities of the photovoltaic system. Even the most ambitious LEED platinum building would likely have used twice the energy of the proposed design.

not only was able to meet the demands of the Living Building Challenge but was also cost-effective in spite of its higher initial costs.

The Living Building Challenge is based on seven performance areas, or *petals*, consisting of Site, Water, Energy, Health, Materials, Equity, and Beauty. Within the seven petals, there are 20 *imperatives* that provide specific guidance for achieving certification (see Table 4.1). Unlike other building assessment systems, the Living Building Challenge requires that the project meet all of the imperatives, not just a sufficient number to gain adequate points for certification. Living Building Challenge certification is awarded only after 12 months of continuous operation that are meant to demonstrate that the building has achieved its performance goals. The imperatives of the Living Building Challenge are all very demanding. For example, Imperative 06 of the Water Petal, Ecological Water Flow, requires that 100 percent of stormwater and building water discharge be

Figure 4.3   Among the innovative ideas emerging as a result of the Living Building Challenge is an off-grid building submitted by Mithun to Canada's Living Building Challenge. As required by the Living Building Challenge, the building is designed to be completely energy- and water-sufficient. It includes greenhouses, rooftop gardens, a chicken farm, and fields for growing produce, providing an integrated system of urban agriculture, another imperative of the Living Building Challenge. (Rendering by Mithun)

TABLE 4.1

The Living Building Challenge has seven categories (or petals) for consideration, and there are a total of 20 imperatives distributed among the petals. All of the imperatives are mandatory, and there are not different levels of certification, just a single certification that the project has met the high standards of the challenge.

Petal	Imperatives
Site	Limits to Growth
	Urban Agriculture
	Habitat Exchange
	Car Free Living
Water	Net Zero Water
	Ecological Water Flow
Energy	Net Zero Energy
Health	Civilized Environment
	Healthy Air
	Biophilia
Materials	Red List
	Embodied Carbon Footprint
	Responsible Industry
	Appropriate Sourcing
	Conservation + Reuse
Equity	Human Scale + Human Places
	Democracy + Social Justice
	Rights to Nature
Beauty	Beauty + Spirit
	Inspiration + Education

managed on-site to meet the project's internal water demands or released onto adjacent sites for management through acceptable time-scale surface flow, groundwater recharge, agricultural use, or adjacent building reuse. The imperatives under the Equity Petal are unique and also demanding, requiring various measures that promote social justice and equity while also preserving the rights of nature. The Beauty Petal is also unique in requiring that the project produce an object of beauty and inspire and educate the community.

One of the outcomes of the Living Building Challenge has been the stimulation of interest in truly advanced, high-performance buildings (Figure 4.3). Having the bar dramatically raised by the Living Building Challenge is creating a renewed sense of interest in the high-performance green building movement to radically change the built environment to make it more responsible for both humans and natural systems.

International Building Assessment Systems

There are several significant building assessment systems that are used in other countries and that provide other perspectives on how to approach the problem of determining how environmentally friendly a given building design may be.

In the following sections, four building assessment systems are described: BREEAM (United Kingdom), CASBEE (Japan), Green Star (Australia), and DGNB/BNB (Germany). SBTool, a building assessment method that is used by countries participating in the Green Building Challenge series of conferences to compare buildings using a uniform approach, is also described.

BREEAM (UNITED KINGDOM)

BREEAM is an acronym for the *Building Research Establishment Environmental Assessment Method* for buildings and is an assessment system designed to describe a building's environmental performance. Launched in 1989 in the United Kingdom, it is the oldest building assessment system and serves as the foundation for many other rating systems, including LEED. Currently, there are over 200,000 BREEAM certified buildings, 20 times the number of LEED certified buildings, and 1 million buildings have been registered for BREEAM certification. BREEAM sets the standard for best practice for sustainable building performance in the United Kingdom. It can be used to rate any type of building, and there are several building-specific BREEAM building assessment systems, each designed for a defined type of building. BREEAM is also used in a country-specific format, for example, the Netherlands, Norway, Sweden, and Spain.

BREEAM for New Construction, one of the building-specific BREEAM rating systems, consists of 49 individual assessment issues spanning 9 environmental categories, plus a 10th category titled Innovation. Each issue addresses a specific building-related environmental impact or issue and has a number of credits assigned to it. BREEAM credits are awarded when a building demonstrates that it meets the best-practice performance levels defined by that issue; for example, it has mitigated an impact or, in the case of health and well-being, addresses a specific building occupant-related issue, for example, good thermal comfort, daylight, or acoustics. Table 4.2 shows the ratings provided by BREEAM for a given building project. Table 4.3 indicates BREEAM's environmental section weightings and a sample calculation for a hypothetical building. The percentage benchmark or threshold for each level of rating is an example of a BREEAM score and rating calculation, and it indicates the BREEAM sections and the credits available for each of the sections. A BREEAM Assessor must determine the BREEAM rating using the appropriate assessment tools and calculators. Table 4.4 provides a breakdown of the issues covered by each of the environmental sections in BREEAM.

TABLE 4.2

BREEAM Rating Benchmarks

BREEAM Rating	Percentage Score	Performance
Outstanding	85	Less than 1% of new UK nondomestic buildings (innovator)
Excellent	70	Top 10% of UK nondomestic buildings (best practice)
Very Good	55	Top 25% of UK nondomestic buildings (advanced best practice)
Good	45	Top 50% of UK nondomestic buildings (intermediate best practice)
Pass	30	Top 75% of UK nondomestic buildings (standard best practice)
Unclassified	<30	

TABLE 4.3

BREEAM Environmental Section Weighting and Sample Rating Calculation

BREEAM Section	Percentage of Weighting	Credits Achieved	Credits Available	Percentage of Credits Achieved	Section Score
Management	12.0%	10	22	45.45%	5.45%
Health and Well-Being	15.0%	8	10	80.00%	12.00%
Energy	19.0%	16	30	53.33%	10.13%
Transport	8.0%	5	9	55.56%	4.44%
Water	6.0%	5	9	55.56%	3.33%
Materials	12.5%	6	12	50.00%	6.25%
Waste	7.5%	3	7	42.86%	3.21%
Land Use and Ecology	10.0%	5	10	50.00%	5.00%
Pollution	10.0%	5	13	38.50%	3.85%
Innovation	10.0%	2	10	20.00%	2.00%
		Final BREEAM Score:			55.66%
		BREEAM Rating:			VERY GOOD

TABLE 4.4

BREEAM Issues

1 Management
Man 01 Sustainable procurement
Man 02 Responsible construction practices
Man 03 Construction site impacts
Man 04 Stakeholder participation
Man 05 Life-cycle cost and service life planning

2 Health and Well-Being
Hea 01 Visual comfort
Hea 02 Indoor air quality
Hea 03 Thermal comfort
Hea 04 Water quality
Hea 05 Safety and security

3 Energy
Ene 01 Reduction of CO_2 emissions
Ene 02 Energy monitoring
Ene 03 External lighting
Ene 04 Low and zero carbon technologies
Ene 05 Energy-efficient cold storage
Ene 06 Energy-efficient transportation system
Ene 07 Energy-efficient laboratory systems
Ene 08 Energy-efficient equipment
Ene 09 Drying space

4 Transport
Tra 01 Public transport accessibility
Tra 02 Proximity to amenities
Tra 03 Cyclist facilities
Tra 04 Maximum car parking capacity
Tra 05 Travel plan

5 Water
Wat 01 Water consumption
Wat 02 Water monitoring

6 Materials
Mat 01 Life-cycle impacts

TABLE 4.4 *(Continued)*

BREEAM Issues

1 Management
   Man 01 Sustainable
   procurement
   Man 02 Responsible
   construction practices
   Man 03 Construction site
   impacts
   Man 04 Stakeholder
   participation
   Man 05 Life-cycle cost and
   service life planning

2 Health and Well-Being
   Hea 01 Visual comfort
   Hea 02 Indoor air quality
   Hea 03 Thermal comfort
   Hea 04 Water quality
   Hea 05 Safety and security

3 Energy
   Ene 01 Reduction of CO_2
   emissions
   Ene 02 Energy monitoring
   Ene 03 External lighting
   Ene 04 Low and zero carbon
   technologies
   Ene 05 Energy-efficient cold
   storage
   Ene 06 Energy-efficient
   transportation system
   Ene 07 Energy-efficient
   laboratory systems
   Ene 08 Energy-efficient
   equipment
   Ene 09 Drying space

4 Transport
   Tra 01 Public transport accessibility
   Tra 02 Proximity to amenities
   Tra 03 Cyclist facilities
   Tra 04 Maximum car parking capacity
   Tra 05 Travel plan

5 Water
   Wat 01 Water consumption
   Wat 02 Water monitoring
   Wat 03 Water leak detection and
   prevention
   Wat 04 Water efficient
   equipment

6 Materials
   Mat 01 Life-cycle impacts
   Mat 02 Hard landscaping and boundary
   protection
   Mat 03 Responsible sourcing of materials
   Mat 04 Insulation
   Mat 05 Designing for robustness

7 Waste
   Wst 01 Construction waste
   management
   Wst 02 Recycled aggregates
   Wst 03 Operational waste
   Wst 04 Speculative floor and
   ceiling finishes

8 Land Use and Ecology
   LE 01 Site selection
   LE 02 Ecological value of site and protection
   of ecological features
   LE 03 Mitigating ecological impact
   LE 04 Enhancing site ecology
   LE 05 Long term impact on biodiversity

9 Pollution
   Pol 01 Impact of refrigerants
   Pol 02 NO_x emissions
   Pol 03 Surface water run off
   Pol 04 Reduction of night time
   light pollution
   Pol 05 Noise attenuation

10 Innovation
   Inn 01 Innovation

TABLE 4.5

For a Building to Be BREEAM Certified, a Rating of "Very Good" Must Be Achieved for Each of the Minimum Standards

Issue	Minimum Standard for BREEAM "Very Good"Rating	Achieved (Y/N)
Man 01	Sustainable procurement	Y
Hea 01	Visual comfort	Y
Hea 04	Water quality	Y
Ene 02	Energy monitoring	Y
Wat 01	Water consumption	Y
Wat 02	Water monitoring	Y
Mat 03	Responsible sourcing of materials	Y
LE 03	Mitigating ecological impact	Y

BREEAM for New Construction also has a number of *minimum standards* that must be met, as indicated in Table 4.5. The USGBC LEED building assessment system also has several so-called prerequisites that are the equivalent of the BREEAM minimum standards. In the case of BREEAM, a rating of at least "Very Good" must be achieved for each of the minimum standards for a project to be certifiable.

BREEAM Case Study: AHVLA Stores Building, Weybridge, United Kingdom

A new two-story stores building was developed to replace an existing building on the Animal Health and Veterinary Laboratories Agency (AHVLA) campus near Weybridge, United Kingdom (see Figure 4.4). For reasons of structural loading and accessibility, the primary storage, receiving, and shipping areas are located on the ground floor, and the office, lockers, and restrooms are located on the first floor.

The building was designed for compact and economical space use and circulation flow in a minimum rectangular envelope. This achieves both a reduced volume of heated space in the building (and so of energy demand) and a reduced external surface area from which heat energy can be lost. The AHVLA stores building was commissioned by the Department for Environment, Food and Rural Affairs (Defra), as part of a wider redevelopment of the campus. The project was BREEAM assessed in accordance with Defra's policy of achieving the highest environmental targets for developments on its estate.

KEY FACTS

- BREEAM rating: Excellent
- Score: 83.76%
- Size: 1500 m^2
- BREEAM version: Industrial 2006

OVERVIEW OF ENVIRONMENTAL FEATURES

- Vertical-axis wind turbines mounted on the roof.
- Biofuel boiler.
- Compact building envelope with good thermal insulation.
- Solar shading.

Figure 4.4 The Animal Health and Veterinary Laboratories Agency (AHVLA) headquarters, located in Weybridge, United Kingdom, achieved a BREEAM score of excellent. (*Source:* Animal Health and Veterinary Laboratories Agency)

- Surface water runoff from roof via "weir" cascade (instead of traditional downpipes) into underground storage and attenuation tank (due to local high water table).
- Rainwater harvested and used for toilet flushing.
- Good thermal insulation and airtightness. The site's relatively exposed and noisy location next to the M25 allowed for noise reduction through the building fabric to be combined with highly insulated external walls and roof. The nature of the building requires a largely windowless external envelope, which also presented opportunities for achieving a good thermal and airtight envelope.

THE BREEAM ASSESSMENT

The AHVLA stores building performed very well across all categories with the top scoring categories being:

- Water and Management: 100% of available credits
- Pollution: 92.31%
- Health and Well-Being: 85.71%
- Energy: 83.33%

BUILDING SERVICES

- Biofuel boiler—running on pure rapeseed oil, which has a low CO_2 emission factor
- 4×6 kW vertical-axis wind turbine units—feeding back into the site's electricity network when the building's use is less than the electricity generated
- Sun pipes—supplementing passive infrared (PIR) controlled lighting to internal areas
- Solar thermal heating to supplement the low-temperature hot water (LTHW) system

GREEN STRATEGY

The client set out the objectives for this project from the very first briefing meetings and was emphatic in aiming for the highest achievable green strategy. As part of earlier initiatives for Defra, the design team had reviewed more than 30 possible options for environmentally sustainable improvements that could be used on the AHVLA campus redevelopment. This allowed them to quickly assess and incorporate the most appropriate elements into the new stores building during the briefing and design stages, so these were fully integrated into the design and not considered as later "add-ons." This approach also enabled the maximum synergy between mutually contributing elements (e.g., water storage, stormwater attenuation, reduction of above- and belowground drainage, and optimization of the site area), giving added value to the BREEAM elements.

CASBEE (JAPAN)

The *Comprehensive Assessment System for Building Environmental Efficiency* (CASBEE) is the Japanese building assessment system. CASBEE was developed by the *Japan Sustainable Building Consortium*, which is composed of academic, industrial, and government entities, specifically for Japanese cultural, social, and political conditions. The key concept in CASBEE is *Building Environmental Efficiency* (BEE), which is a description of the *ecological efficiency*, or *eco-efficiency*, of the built environment. The World Business Council for Sustainable Development (WBCSD) defines eco-efficiency as maximizing economic value while minimizing environmental impacts. Similarly, CASBEE defines BEE as maximizing the ratio of building quality to environmental loadings.

Building quality (Q) is described by CASBEE as the amenities provided for building users and consists of several quantities:

Q1: Indoor environment

Q2: Quality of service

Q3: Outdoor environment on-site

Q: Total quality

Q = Q1 + Q2 + Q3

Similarly, there are several categories of *environmental loadings* (L) in CASBEE:

L1: Energy

L2: Resources and materials

L3: Off-site environment

L: Total loading

L = L1 + L2 + L3

As noted earlier, BEE is simply the ratio of building quality to building environmental loadings. The BEE rating calculation produces a number, generally in the range of 0.5 to 3, that corresponds to a building class, from class S (highest for a BEE rating of 3.0 or higher) to classes A (BEE of 1.5 to 3.0), B$^+$ (BEE of 1.0 to 1.5), B$^-$ (BEE of 0.5 to 1.0), and C (BEE less than 0.5). The relationship of quality (Q) to loading (L) in CASBEE and the resulting BEE letter scores are diagrammed in Figure 4.5. Clearly, it is desirable to have as high a BEE rating as possible. Although simple in concept, extensive data gathering and calculations using CASBEE tools are required to make a determination of the BEE rating. For the example shown in Figure 4.5, the BEE value is 1.4 based on a building quality (Q) value of 59 and a building environmental loading (L) value of 41.

CASBEE has been refined since its inception in 2004 and is now a suite of rating systems, as shown in Figure 4.6 and Table 4.6. Each CASBEE building assessment system provides a numeric score and a one- to five-star rating for the building, depending on its BEE rating, as indicated in Table 4.7.

One interesting use of the BEE approach embedded in CASBEE is for renovation projects. Using the CASBEE-Renovation (CASBEE-RN) building assessment system, the BEE value can be calculated before and after renovation. In the example shown in Figure 4.7, the building had a BEE of 0.6 prior to renovation and a BEE of 1.4 after the renovation, increasing its rank from B$^-$ to B$^+$.

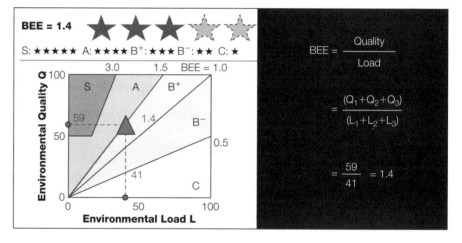

Figure 4.5 The BEE rating is determined by finding the intersection of Q (building quality) and L (building loadings). High ratings (S and A) are achieved by buildings with high environmental quality and performance and low environmental loadings. Higher resource consumption and lower environmental quality produce below standard ratings (B$^-$ or C). (*Source:* Japan Sustainable Building Consortium)

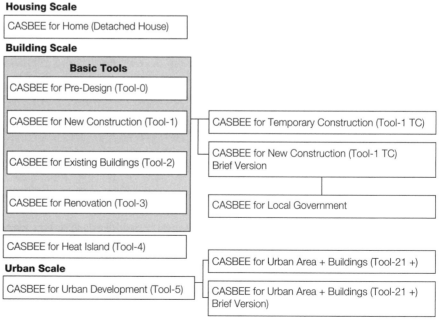

Figure 4.6 The CASBEE family of tools indicating the scaling and building stage assessments built into the system. This assessment system can be applied to individual homes and buildings up to urban scale. It can also be applied not only to new construction but also to existing buildings and renovations of existing buildings.

TABLE 4.6

CASBEE Building Assessment Systems

CASBEE Building Assessment Systems	Short Name	Latest Version
New Construction	CASBEE-NC	2008
Temporary Construction	CASBEE-NC/TC	2005
Existing Buildings	CASBEE-EB	2010
Renovation	CASBEE-RN	2010
Heat Island	CASBEE-HI	2010
Detached Home	CASBEE-DH	2010
Urban Development	CASBEE-UD	2006
Site	CASBEE-S	2010
New Construction—Tenant	CASBEE-NC/T	2010

TABLE 4.7

CASBEE Grading System Based on BEE Value

CASBEE Grade	Stars	BEE Value
S (Excellent)	★ ★ ★ ★ ★	Over 3.0
A (Very Good)	★ ★ ★ ★	Under 3.0, over 1.5
B$^+$ (Good)	★ ★ ★	Under 1.5, over 1.0
B$^-$ (Rather Poor)	★ ★	Under 1.0, over 0.5
C (Poor)	★	Under 0.5

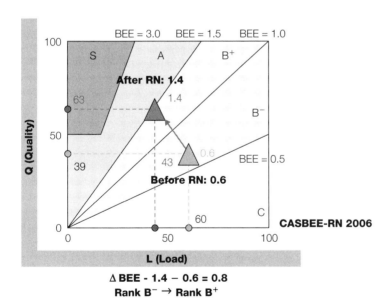

Figure 4.7   The BEE rating can be used to set goals for performance improvement for existing buildings. The building represented in this graphic increased its rating from 0.6 to 1.4. (*Source:* Japan Sustainable Building Consortium)

GREEN STAR (AUSTRALIA)

Green Star is the major Australian green building assessment scheme and is similar in many respects to BREEAM and LEED in its approach and structure. The Green Building Council of Australia (GBCA) developed the Green Star

TABLE 4.8

Green Star Rating Tools

Green Star Rating Tool	Status
Education v1	Active
Healthcare v2	Active
Industrial v3	Active
Multi-Unit Residential v1	Active
Office v3	Active
Office Interiors v1	Active
Retail Centre v1	Active
Office Design v2	Active
Office as Built v2	Active
Public Building	Pilot
Convention Centre Design	Pilot
Custom	Pilot

TABLE 4.9

Categories and Points in Green Star Office v3

Green Star Category	Points	Percentage of Total
Management	12	8.1%
Indoor Environment	27	18.2%
Energy	29	19.6%
Transport	11	7.4%
Water	12	8.1%
Materials	25	17.0%
Land Use and Ecology	8	5.4%
Emissions	19	12.8%
Innovation	5	3.4%
Total	**148**	**100.0%**

Office rating tool in 2002, and now there are a variety of additional tools to cover a wide range of building types (see Table 4.8). Green Star has been adopted by New Zealand and South Africa as the platform for their national building assessment systems.

The Green Star building assessment system awards from one to six green stars, but only those buildings with ratings of four to six green stars have significance with respect to being considered high-performance buildings. The GBCA describes the three highest levels of achievement as follows:

4 Star Green Star Certified Rating (score 45–59) signifies "Best Practice" in environmentally sustainable design and/or construction.

5 Star Green Star Certified Rating (score 60–74) signifies "Australian Excellence" in environmentally sustainable design and/or construction.

6 Star Green Star Certified Rating (score 75–100) signifies "World Leadership" in environmentally sustainable design and/or construction.

Nine categories are assessed in a Green Star rating, and the number of green stars awarded to the project is based on the percentage of the total points available. Table 4.9 shows the points and percentages of total points for each of the nine categories for Green Star Office v3.

Each category has several issues associated with it, as indicated in Table 4.10 for Green Star Office v3. There are several *conditional requirements* in Green Star, one in the Energy category and one in Land Use and Ecology. These are similar to the prerequisites found in the USGBC LEED building assessment system.

TABLE 4.10

Issues in Green Star Office v3

1. Management	6. Materials
Man-1 Green Star Accredited Professional	Mat-1 Recycling Waste Storage
Man-2 Commissioning Clauses	Mat-2 Building Reuse
Man-3 Building Tuning	Mat-3 Reused Materials
Man-4 Independent Commissioning Agent	Mat-4 Shell and Core or Integrated Fit-Out
Man-5 Building Users' Guide	Mat-5 Concrete
Man-6 Environmental Management	Mat-6 Steel
Man-7 Waste Management	Mat-7 PVC Minimization or PVC
	Mat-8 Sustainable Timber or Timber
	Mat-9 Design for Disassembly
	Mat-10 Dematerialization
2. Indoor Environment	7. Land Use and Ecology
IEQ-1 Ventilation Rates	Eco-Conditional Requirement
IEQ-2 Indoor Air Quality	Eco-1 Topsoil
IEQ-3 CO_2 Monitoring and Control	Eco-2 Reuse of Land
IEQ-4 Daylight	Eco-3 Reclaimed Contaminated Land
IEQ-5 Daylight Glare Control	Eco-4 Change of Ecological Value
IEQ-6 High Frequency Ballasts	
IEQ-7 Electric Lighting Levels	
IEQ-8 External Views	
IEQ-9 Thermal Comfort	
IEQ-10 Individual Comfort	
IEQ-11 Hazardous Materials	
IEQ-12 Internal Noise Levels	
IEQ-13 Volatile Organic Compounds	

TABLE 4.10 *(Continued)*

Issues in Green Star Office v3

IEQ-14 Formaldehyde Minimization
IEQ-15 Mold Prevention
IEQ-16 Tenant Exhaust Riser

3. Energy
Ene-Conditional Requirement
Ene-1 Greenhouse Gas Emissions
Ene-2 Energy Submetering
Ene-3 Lighting Power Density
Ene-4 Lighting Zoning
Ene-5 Peak Energy Demand Reduction

8. Emissions
Emi-1 Refrigerant ODP
Emi-2 Refrigerant GWP
Emi-3 Refrigerant Leaks
Emi-4 Insulant ODP
Emi-5 Watercourse Pollution or Stormwater
Emi-6 Discharge to Sewer
Emi-7 Light Pollution
Emi-8 Legionella

4. Transport
Tra-1 Provision of Car Parking
Tra-2 Fuel-Efficient Transportation
Tra-3 Cyclist Facilities
Tra-4 Commuting Mass Transit

9. Innovation
Inn-1 Innovative Strategies and
Technologies
Inn-2 Exceeding Green Star Benchmarks

5. Water
Wat-1 Occupant Amenity Water
Wat-2 Water Meters
Wat-3 Landscape Irrigation
Wat-4 Heat Rejection Water
Wat-5 Fire Water System Consumption

Like all third-party certification systems, Green Star certification is a formal process that involves a project using a Green Star rating tool, for example, Green Star Office v3, to guide the design or construction process during which documentation is gathered for use in the two assessment phases of the project. The Green Building Council of Australia commissions a panel of third-party certified assessors to validate that the documentation for all claimed credits follows the compliance requirements for each rating tool. Project teams are notified of their score based on the recommendation of the assessment panel and, where applicable, of any innovation credits that have been awarded by the GBCA. If a certified rating is awarded, the project receives a framed certificate, award letter, and relevant Green Star logos.

Green Star Case Study

One of the most recent recipients of the highest, six-star rating from Australia's Green Star building assessment system is 1 Bligh Street in Sydney, Australia, co-owned by DEXUS Property Group, DWPF, and Cbus Property (see Figure 4.8). The 28-story building is Australia's first high-rise with a double-skin façade, and it also has a full-building-height, naturally ventilated atrium that helps maximize daylighting at each office floor level. The double-skin façade has internal blinds and external louvers that are automatically adjusted depending on their orientation to the sun (see Figure 4.9). This system conserves energy, eliminates sky glare, and optimizes user comfort. The unique full-height atrium and elliptical-shaped floor plates enable 74 percent of the building to be within 8 meters of either the façade or the atrium,

Figure 4.8 (A) The property at 1 Bligh Street in Sydney, Australia, is one of the most advanced buildings in the world, with a double-skin façade, solar-powered air-conditioning system, and a blackwater recycling system. (B) The naturally ventilated 28-story atrium assists in providing spectacular daylighting for all floors. [Images supplied courtesy of ingenhoven + architectus (Sydney)]

providing large amounts of natural light into the building and spectacular views in all directions. Its energy performance is outstanding, with a 42 percent CO_2 reduction when compared to a similar-sized conventional office tower. On top of the building, 500 square meters of roof mounted solar panels capture solar energy to directly power an absorption chiller to drive the cooling systems, an advanced hybrid of variable air volume (VAV) and chilled-beam air-conditioning technology.

Water is a crucial resource everywhere, but nowhere is it more precious than in Australia, which is in the grip of a decade-long severe drought. New projects, such as 1 Bligh Street, provide an opportunity to demonstrate how to truly minimize potable water consumption. It has the first blackwater recycling system in a highrise office building in Australia, and it will save 100,000 liters of drinking water a day, equivalent to filling an Olympic-size swimming pool every two weeks. Wastewater is mined from the building and nearby sewers, processed, and then distributed around the building for nondrinking purposes, with 75,000 liters used for cooling towers and 25,000 liters used for flushing toilets. The system provides 100 percent recycled water for toilet flushing, as well as 90 percent of cooling tower makeup water. Sydney's goal is to have recycled water provide at least 15 percent of its water supply by 2015, and 1 Bligh Street is an important example because it employs new blackwater recycling technology

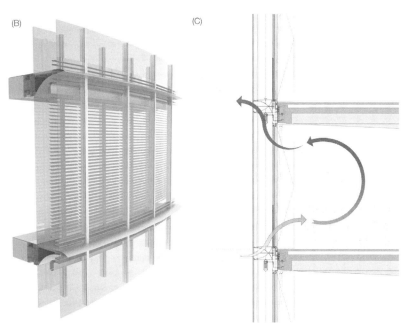

Figure 4.9   (A–B) The double-skin façade of 1 Bligh Street has a system of internal blinds that automatically deploy or adjust to optimize the combination of daylighting and energy transmission while protecting the occupants from glare. (C) Detail of air movement through the façade. (Images supplied courtesy of ingenhoven + architectus (Sydney))

> The use of specially formulated high-strength concrete reduces the number of columns and therefore minimizes the amount of concrete used. Timber and plywood used in the structure is recycled or from FSC accredited sources. The steel used in the project comprises more than 50 percent recycled content. Over 80 percent of all PVC-type products have been replaced with non-PVC materials. Over 37,000 metric tons, amounting to 94 percent of all construction waste produced on the project, was recycled.

DGNB/BNB (GERMANY)

Germany has a long history of designing high-performance buildings but only recently has there been an effort to develop a green building certification program. The first steps in developing a green building certification program and building assessment system started in 2001 with the production of the German Sustainable Building Technical Manual. This served as the genesis of an effort that culminated in the formation of the German Sustainable Building Council (DGNB for Deutsche Gesellschaft für Nachhaltiges Bauen) in 2007 and the emergence of a formal certification system.

There are actually two green building assessment systems in Germany. The first of these is the DGNB, which is directed at nonresidential, commercial buildings. The other building assessment system is the BNB (Bewertungssystem Nachhaltiges Bauen für Bundesgebäude, or Assessment System for Sustainable Construction for Government Buildings), which is used only to assess government buildings. The administration of the two systems is carried out by different organizations that cooperate to ensure uniformity in the application of the rating systems.

Sustainable Building		
Ecology	Economy	Socio-culture
PRO-TECTIVE GOODS — Protection of Natural Resources Global and Local Environment	Capital/ Values	Health User-Satisfaction Functionality Cultural Value
PRO-TECTIVE TARGETS — • Protection of Natural Resources • Protection of the Ecosystem	• Minimization of Life-Cycle Costs • Improvement of Economic Viability • Conservation of Capital/Value	• Health Protection, Safety, and Well-Being • Verification of Functionality • Verification of Design and Urban Quality

Figure 4.10 The developers of the DGNB/BNB assessment system used a top-down approach in its design with the three legs of sustainability as the major points of evaluation (Ecology, Economy, and Socioculture). The process of developing the assessment system included considering what needed to be protected and the specific targets for protection.

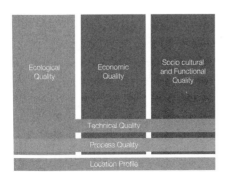

Figure 4.11 The three major sustainability areas of evaluation (Ecological Quality, Economic Quality, and Sociocultural-Functional Quality) each carry 22.5 percent of the possible points in the DGNB/BNB. Technical Quality carries 22.5 percent and 10 percent of the points are allocated to Process Quality. The statement of the Location Profile is an administrative requirement that must be accomplished for certification.

DGNB/BNB is the newest major building assessment system, and it differs substantially from systems employed by other countries. For most other building assessment systems, including LEED, BREEAM, and Green Star, the categories that the developers feel are important are the starting point for creation of the building assessment system. For example in LEED, six categories were deemed important: Sustainable Sites, Water Efficiency, Energy and Atmosphere, Materials and Resources, Indoor Environmental Quality, and Innovation in Design. These were further subdivided into issues. For example, for the Energy and Atmosphere category of LEED, energy consumption, renewable energy, building commissioning, and impacts of refrigerants are the issues that can be awarded points toward certification. This approach is sometimes referred to as a *bottom-up strategy*. In contrast, DGNB/BNB was developed using a *top-down strategy*; that is, the authors of DGNB/BNB based the allocation of points on the three major areas of consideration for sustainability: Ecology, Economy, and Socioculture (see Figure 4.10). The questions asked prior to producing the building assessment and certification system were: What are the sustainability issues relevant to construction? What needs to be protected? How does this protection occur? As shown in Figure 4.11, the three major areas of sustainability were used as the basis for organizing the DGNB/BNB building assessment system, and three other issues were also considered: Technical Quality, Process Quality, and the project's Location Profile. The three major areas of sustainability were determined to be equally important along with Technical Quality, and each of these was allocated 22.5 percent of the available points, a total of 90 percent. The final 10 percent of the total points was allocated to Process Quality, issues such as Integrated Design, Commissioning, and Quality Assurance.

The final outcome of the evaluation of the project is the award of a certificate. As indicated in Figure 4.12, there are three certification levels: gold, corresponding to a minimum 80 percent score; silver, which requires at least 65 percent of the available points; and bronze, requiring a minimum of 50 percent of the total points available. The result of this approach is a very logical and comprehensive rating system that addresses a wide range of factors, while also providing a balanced approach to assessing a building's performance. In the environmental impact section of the evaluation, there is extensive use of life-cycle assessment (LCA), more so than in any other contemporary rating system, and benchmarks have been established for impacts per square meter of building in order to determine the number of points to be awarded for each of the various factors being evaluated. Typical German high-performance office buildings consume on the order of 100 kWh/m^2/yr of primary energy, a number that is not achievable without significant reductions in conventional mechanical cooling system energy consumption. It should be noted that, for the sake of comparison, this number includes all major building systems but does not include plug loads. This very low level of energy consumption is impressive and can be accomplished only with a well-integrated design process and with significant latitude being given to the architects and engineers to use creative approaches to develop this type of advanced building. One other significant factor that differs from the design of buildings in, for example, the United States, is that the occupants in German buildings are willing to accept a significantly larger comfort zone than would be the case in the United States. Due to the nature of naturally ventilated buildings, temperatures are difficult to maintain in a very narrow band; however, the German designers have demonstrated that they can keep temperatures within a reasonable comfort zone with very few annual cases where temperature drifts outside of this zone.

An example of a DGNB certified building, Theaterhaus in Stuttgart, Germany, is covered in detail in Chapter 7.

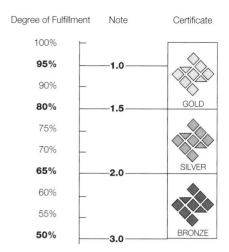

Figure 4.12 The DGNB provides for three levels of certification (gold, silver, and bronze) and a grade (labeled as "Note" in the figure) based on the percentage of points achieved.

SBTOOL

SBTool is a very comprehensive and sophisticated building assessment tool that was developed for the biannual international Green Building Challenge, which was held in 1998 (Paris, France), 2000 (Maastricht, Netherlands), 2002 (Oslo, Norway), 2005 (Tokyo, Japan), 2008 (Melbourne, Australia), and 2011 (Helsinki, Finland). SBTool provides a standard basis of comparison for the wide range of buildings being evaluated in the Green Building Challenge. It requires a comprehensive set of information not only on the building being assessed but also on a benchmark building for use in comparing how well the green building performs compared to the norm. SBTool requires the group using it to establish benchmark values and weights for the various impacts. The tool is implemented in the form of a sophisticated Excel spreadsheet that can be downloaded from the website of the International Initiative for a Sustainable Built Environment (iiSBE). The output from SBTool provides an assessment of the building in seven different categories: Resource Consumption, Environmental Loadings, Indoor Environmental Quality, Service Quality, Economics, Management, and Commuting Transport.

THOUGHT PIECE: BUILDING ASSESSMENT

Ray Cole is perhaps the leading international thinker on the subject of building assessment and has researched and written papers on the subject for about 20 years. In this thought piece, he discusses the role of performance assessment methods and advocates their use in helping to address global societal needs. He also points to regenerative design and its emphasis on "place" as an important subject for assessment because it ties together many of the key important aspects of sustainable construction such as systems thinking, community engagement, and respect for place.

Shifting Emphasis in Green Building Performance Assessment

Raymond J. Cole, School of Architecture and Landscape Architecture, University of British Columbia, Canada

The term *green building* has been used fairly consistently over the past two decades to describe those buildings that have a higher environmental performance compared to that of typical buildings, and the term *green building performance assessment methods* has been used to describe approaches that provide an objective measure of their environmental strengths. The emphasis on green design has been primarily directed at creating buildings that "do less harm" or, more generally, play a key role in reducing the degenerative consequences of human activity on the health and integrity of ecological systems.[6]

Performance assessment methods have unquestionably been instrumental in mainstreaming green building practice and have profoundly influenced the range of considerations deemed important in design. They are now embedded within the parlance of building procurement, design and construction, and operation and, given that the major systems are now global "brands" with considerable organizational support, will continue to play a dominant role for the foreseeable future. While the maintenance of a brand can constrain the type and extent of changes that can be made to their structure and content, they clearly must evolve in terms of scope and emphasis. Indeed, while green building performance assessment tools were initially conceived to engage industry, encourage widespread adoption of green practices, and "transform the market," their scope and application has increasingly expanded. While initial versions were directed at the construction of new buildings—often office buildings—this was followed by an expansion into versions for other building types (hotels, factories, homes, etc.) and conditions (commercial interiors, existing buildings, renovations, etc.). The Japanese CASBEE building assessment system has versions specifically addressing property appraisal that map the performance criteria against increased revenue, reduced costs, reduced risks, and improved image. The focus of these developments was always the performance of *individual* buildings. With their maturation, the scope of assessment systems has been further extended in scale to embrace communities, urban design, and infrastructure planning, for example, LEED for Neighborhood Development, BREEAM Communities, and CASBEE Urban Design.[7] This shift in scale is perhaps the most significant development over the past decade and may be indicative of the increasing need to redefine the spatial and temporal boundaries of consideration, and link the environmental performance of buildings more explicitly with their ecological and physical/infrastructure context. While the performance of the individual building clearly remains important, the scale and emphasis appears to be shifting toward what is considered meaningful, comprehensible, and manageable for society to collectively engage in affecting positive change.

Current building environmental assessment methods have a number of distinct characteristics, including that criteria are technically framed and based on metrics that are quantifiable, measurable, and comparable and which, in aggregate, are assumed to offer an accurate measure and understanding of overall green building performance; and that the overall success of a building is measured through the simple addition of the weighted (either implicitly or explicitly) scores attained for the individual performance issues. Moreover, the need for clear, unambiguous assessment and avoiding "double counting" has required the performance criteria to be kept discrete. The resulting simple listing of performance requirements and scoring inhibits the ability to see how they function as part of an integrated system, both internally and with the context in which they sit. Reed (2007) characterizes this attribute of green design and the associated assessment tools as indicative of the legacy of reductive and fragmented thinking.

In North America, the Living Building Challenge, launched in August 2006, is emerging as a recognized demanding and complementary performance aspiration to the LEED green building rating system.[8] All of its 20 "imperatives" must be met before the designation of "Living Building" is granted. This stands in contrast to LEED where, particularly for the certified, silver, and gold levels, it is possible to select (or "cherry-pick") the credits to attain the necessary overall performance level. Although, similar to LEED, the structure is simply a list of required performance requirements set within seven broad categories. The demanding performance requirements of the Living Building Challenge criteria are, however, challenging many norms and conventions and driving toward greater synergistic design. In a similar way that LEED and other major assessment methods have expanded from individual buildings to communities, the Living Building Challenge evolved to permit "scale jumping" in recognition that different performance issues are more easily or appropriately addressed at different scales, from individual buildings to an entire region.

The term *sustainable building* is often used synonymously with *green building* although the former carries the expectation of extending the range of considerations to include broader social and economic issues. And, with this, "sustainability" assessment methods such as Arup's Sustainable Project Assessment Routine (SPeAR),[9] iiSBE's Sustainable Building Tool (SBTool),[10] the South African Sustainable Building Assessment Tool (SBAT),[11] and the German Sustainable Building Council's Certificate Program[12] have been introduced that explicitly acknowledge this expanded range of performance issues. As with green building assessment methods, sustainability tools have recognizable frameworks that convey their scope, structure, and organization, but these are typically presented graphically. Whereas SPeAR, SBTool, and SBAT frame performance issues within a circle segmented into the key performance areas, the German Certificate Program distinguishes the three "sustainability" quality categories from the technical and process criteria that cut across them, and then presents the output in a circular format. The representation of the criteria within a circular framework as distinct from the list format common to many green building assessment tools is, presumably, seen as evoking potential links and synergies between the various performance criteria. However, as with current green building assessment tools, they still remain discrete, and weighted scores are again simply aggregated.

Expanding the framework to include social, cultural, ecological, and economic considerations moves the assessment into areas where there is greater difficulty and less consensus regarding performance metrics. Perhaps more significantly, buildings, in and of themselves, cannot be sustainable, but can only be designed to support sustainable patterns of living.[13] Such a responsibility clearly shifts the focus on building performance to the larger context in which they are situated. Rather than striving solely for an understanding of an individual building's performance, the potential contribution a building makes to the social, ecological, and economic health of the place within which it functions will perhaps become of equal, if not more, significance.

A number of historical threads that have either been latent or running parallel to green building discourse and practice over the past 40 years are now converging under the umbrella of *regenerative* design and development and, with it, the reframing of approaches to discuss and assess performance. While many of its core tenets—systems thinking, community engagement, respect for place—have long individual histories in architectural design, regenerative design begins to tie them together in a cogent manner. Regenerative design relates to approaches that support the coevolution of human and natural systems in a partnered relationship. Within regenerative development, built projects, stakeholder processes, and inhabitation are collectively focused on enhancing life in all its manifestations—human, other species, ecological systems—through an enduring responsibility of stewardship.[14] Regeneration, in contrast to the emphasis on "doing less harm," which has dominated past green building practice—and the emphasis of most environmental assessment methods—carries the positive message of considering the act of building as one that can give back more than it receives and thereby over time building social and natural capital. Such an approach requires design to acknowledge and respond to the unique attributes of "place" and secures sustained stakeholder engagement to ensure a project's future success.

The structure and emphasis of current green building assessment tools offer little instruction regarding understanding and engaging local ecosystems and their processes or, more generally, of the systems thinking emphasized in regenerative design. Regenerative design requires a fundamental reconceptualization of the act of building design primarily in terms of imagining, formulating, and enabling its role within a larger context.[15] It would therefore seem appropriate that the representation of regenerative design in support tools should reflect this interplay. Indeed, as the notion of regenerative design and development gains increased momentum, it is anticipated that there will be a commensurate demand for support tools to assist those practitioners wishing to engage with it. This could be considered a necessary step to both sharpen regenerative design's theoretical underpinnings and to further a broader discussion and practice.[16]

Capra (1996) illustrates how the reductive approaches to scientific inquiry dominant for over the past few centuries are gradually succumbing to the holistic nature of the disciplines of biology and ecology and how the machine metaphor is being replaced by one of networks. Such a whole-systems approach will invariably guide future building-related initiatives and strategies across all scales, and will clearly have consequences for the scope and emphasis of current assessment methods or the development of complementing approaches to describe and evaluate what constitutes successful performance.

A set of tools and frameworks are emerging directed at representing the priorities and emphasis of regenerative design. For example, the conceptual regenerative design framework—REGEN—proposed by the US architecture firm Berkebile Nelson Immenschuh McDowell (BNIM) for the USGBC;[17] LENSES, created by Colorado State University's Institute for the Built Environment, to help communities and project teams create places where natural, social, and economic systems can mutually thrive and prosper;[18] and the framework developed by Perkins + Will set the resource-related design strategies within cycles— from nature and back to nature—differentiating between those approaches that are primarily executed within the physical bounds of the site and largely within the purview of the design team and owner and those that extend beyond the bounds of the site and must be negotiated with other parties for the implementation and success (Cole et al., 2012). Such approaches can provide a necessary and complementary role to current green building assessment methods. Green building assessment systems were conceived to provide a measure of performance but are also used to guide design by communicating what are deemed priority environmental issues. Plaut et al. (2012) argue that these "offer little guidance in the way of guiding people through the creation, implementation, and operation of projects" and by focusing on "measuring the performance of an end result or product" and can be described as "product based." By contrast, LENSES and the other regenerative frameworks can better be described as what Plaut et al. call "process-based" and are primarily directed at guiding design. Moreover, whereas the product-based tools keep individual environmental performance requirements discrete, the graphic organization of the emerging regenerative design tools expands the issues to include social, cultural, economic, and ecological systems and processes but also emphasizes the relationship between them. In short, they accept the built environment as a complex socioecological system and attempt to offer guidance to designers and other stakeholders in situating projects within it.

At this point, in addition to building new capabilities, other potential implications emerge from shifting from green to regenerative design and the development of associated assessment tools. First: reestablishing regional design practices. The architectural diversity and richness evidenced in the way that indigenous and vernacular practices offered regionally specific solutions is largely absent in current mainstream architectural practice. The central emphasis on "place" within regenerative design provides the necessary frame by which this collective knowledge can be rediscovered and reinterpreted within a contemporary context. Second: establishing common ground between the diverse stakeholders associated with the production and use of a building, something that has often eluded other design approaches. While the integrative design process has been an enormously valuable complement to green design, the more expansive dialogue central to regenerative design and development has the potential to engage and maintain stakeholder commitment. Third: change responsibilities and skills for designers. While green design has required design team members to gain familiarity with a host of environmental strategies and blur professional boundaries, regenerative design will drive designers toward positioning these within a whole-systems setting. In addition to having the potential of reframing what constitutes the nature of design and the role of designers, these and other shifts identified earlier (reductive/holistic, product/process, building/context) have profound consequences for what constitutes "performance" and what constitutes "assessment."

ACKNOWLEDGMENT

The issues and ideas presented in this thought piece are drawn from a *Building Research and Information Special Issue: Regenerative Design—Theory and Practice*, published in 2012 in *Building Research and Information*, Guest Editor: Raymond J. Cole.

Summary and Conclusions

The high-performance building movement worldwide is being propelled by the success of building assessment methods, in particular, LEED in the United States and BREEAM in the United Kingdom. Both methods take complex arrays of numerical and nonnumerical data and provide a score that indicates the performance of a building according to the scoring and weighting system built into the method. Newcomers to the marketplace, such as GBI's Green Globes and the Living Building Challenge, can help to bring the movement and these collective green building design concepts and strategies even further into the mainstream. Internationally, there are a host of green building assessment systems and methods, such as CASBEE in Japan, Green Star in Australia, and DGNB in Germany. Around the world, there are over 40 green building councils that promote green building and building assessment tools, many of which, like Green Mark in Singapore, are local products and not strictly based on other major assessment tools. In addition to creating a competitive atmosphere of promoting high-performance green building, these assessment systems also bring standard definitions of green building to their countries and a common vocabulary, which is essential for increasing the penetration of green buildings around the world.

Notes

1. The Building Research Establishment (BRE) is the national building research organization for the United Kingdom and the developer of BREEAM, which is described in detail at www.breeam.org.
2. The Japan Sustainable Building Consortium developed CASBEE. A detailed description can be found at the consortium's website, www.ibec.or.jp/CASBEE/english/overviewE.htm.
3. At present, Green Star provides a series of assessment tools directed at new offices, existing offices, and office interiors. The Green Building Council of Australia website is www.gbcaus.org.
4. From December 1 through December 11, 1997, more than 160 nations met in Kyoto, Japan, to negotiate binding limitations on greenhouse gases for the developed nations, pursuant to the objectives of the Framework Convention on Climate Change of 1992. The outcome of the meeting was the Kyoto Protocol, in which the developed nations agreed to limit their greenhouse gas emissions relative to the levels emitted in 1990. The United States agreed to reduce its emissions from 1990 levels by 7 percent during the period 2008 to 2012.
5. SBTool, developed by Natural Resources Canada in collaboration with a wide range of academics and practitioners worldwide, has been used by the Green Building Challenge to determine how well buildings compare to base or typical buildings in each category, for example, schools. The tool consists of an Excel spreadsheet. The most recent version is available for research and academic purposes at www.iisbe.org/sbmethod.
6. See McDonough and Braungart (2002) and Reed (2007).
7. See Japan Sustainable Building Consortium (2010).
8. See International Living Building Institute (2010).
9. See www.arup.com/environment/feature.cfm?pageid=1685.
10. See www.iisbe.org.
11. See www.csir.co.za/Built_environment/Architectural_sciences/sbat.html.
12. See http://www.dgnb.de/_en/certification-system/DGNB_Certificate/DGNB_Certificate.php.
13. See Gibberd (2005).
14. See Pedersenand Jenkin (2008), Mang and Reed (2012), and du Plessis (2012).
15. See Mang and Reed (2012).

16. See Cole et al. (2012).
17. See Svec, Berkebile, and Todd (2012).
18. See Plaut et al. (2012).

References

Capra, F.1996. *The Web of Life: A New Scientific Understanding of Living Systems*. New York: Anchor Books.

Cole, R. J., P. Busby, R. Guenther, L. Briney, A. Blaviesciunaite, and T. Alencar. 2012. "A Regenerative Design Framework: Setting New Aspirations and Initiating New Discussions." *Building Research and Information* 40 (1): 95−111.

du Plessis, C. 2012. "Toward a Regenerative Design Paradigm for the Built Environment." *Building Research and Information* 40 (1): 9−22.

Gibberd, J. T. 2005. "Assessing Sustainable Buildings in Developing Countries—The Sustainable Building Assessment Tool (SBAT) and the Sustainable Building Lifecycle (SBL)."Proceedings of the 2005 World Sustainable Building Conference, September 27−29, Tokyo.

International Living Future Institute. 2010. Living Building Challenge Version 2.0.http://ilbi.org/.

Japan Sustainable Building Consortium. 2010. *Comprehensive Assessment System for Built Environment Efficiency (CASBEE) Property Appraisal Manual*.

Mang, P., and W. Reed. 2012. "Designing from Place: Applying the Ecological Worldview and Regenerative Paradigm." *Building Research and Information* 40 (1):23−38.

McDonough, W., and M. Braungart. 2002. *Cradle to Cradle: Remaking the Way We Make*. New York: North Point Press.

Pedersen Zari, M. and S. Jenkin. 2008. *Value Case for a Sustainable Built Environment— Towards Regenerative Development*. Wellington, New Zealand: Ministry for the Environment.

Plaut, J. M., B. Dunbar, A. Wackerman, and S. Hodgin. 2012. "LENSES: A Visionary Framework for Building Dialogue, Guiding Process, and Redefining Success." *Building Research and Information* 40 (1):112−122.

Reed, W. 2007. "Shifting from 'Sustainability' to Regeneration." *Building Research and Information* 35 (6): 674−680.

Svec, P., R. Berkebile, and J. A. Todd. 2012. "REGEN: Toward a Tool for Regenerative Thinking." *Building Research and Information* 40 (1): 81−94.

Chapter 5

The US Green Building Council LEED Building Rating System

The US Green Building Council's (USGBC) Leadership in Energy and Environmental Design (LEED) is the most frequently used building assessment system in the United States. The success of LEED is the result of a long, careful development process that occurred between 1995 and 1998. The earliest attempts at formulating an assessment system, dating from 1993, were conducted under the aegis of the standards structure of the American Society for Testing and Materials (ASTM). This first effort at developing a US rating system was handed over to the then newly formed USGBC in 1995. A pilot version of LEED was issued for beta testing in 1998, and the first operational market version was published in 2000. Perhaps the most important decision of the USGBC members developing LEED that ensured its success was that green building demand should be market-driven rather than being required by regulation, meaning that the building owners would be the ultimate arbiters of the program's success. For commercial green buildings, this meant that they would have to distinguish themselves in the market by having higher resale value than comparable buildings.

A second significant decision in the development of LEED was to create a broad consensus-based process during its formulation. Building assessment systems are typically produced by national building research organizations such as the Building Research Establishment (BRE) in the United Kingdom. The standard is then "sold" to the respective building development market as a tool developed by a reputable institution that will help meet the public demand for more environmentally responsible behavior on the part of the building industry. In contrast, the USGBC was, and remains, a nongovernmental organization comprising a wide range of collaborators from industry, academia, and government. LEED was produced by a cross section of the USGBC's membership during a long, slow, and laborious three-year process that sought to produce a green building rating system that would meet the needs of the wide range of participants in the building industry. The engagement of so many collaborators ensured acceptance when the rating system was completed. In addition, the US Department of Energy (DOE) offered critical funding in the form of grants to support LEED's development. The USGBC was, and continues to be, a nonprofit, nongovernmental organization whose membership is drawn from diverse public and private stakeholders. LEED building assessment products continue to enjoy a high degree of success, largely as a result of the collaborative, consensus-based approach that marks both the products and the contemporary US green building delivery system.

Brief History of LEED

As noted previously, LEED was developed by the USGBC during a three-year process from 1995 to 1998. The first version, known as LEED 1.0, was issued in 1998 as a beta version. Twenty buildings were certified using LEED 1.0 to attain a rating that originally was platinum, gold, silver, or bronze. LEED 2.0 was issued in 2000 as a dramatically changed version of LEED 1.0 and offered to the wider commercial and institutional building market as a final, operational building assessment system. LEED-NC 2.1, the next edition of LEED, issued in 2002, started the process of issuing rating products for specific building types. For example, in the case of the version for new construction, the descriptor NC was appended to the title. LEED-NC 2.1 was virtually identical to LEED 2.0, except that it had greatly simplified documentation requirements. LEED-NC 2.2, issued in 2005, did away with manual documentation submissions and shifted to an Internet portal for this purpose, USGBC LEED-Online. LEED 3.0 was released in 2009 with several major changes to its structure and was an across-the-board change for all LEED building assessment products. Additional points were awarded to projects that focused on regional issues established by local USGBC chapters. A whole new version of LEED-Online was released to facilitate easier communication between the project teams and the certifying bodies. The website interface allows the team to better manage project details and upload supporting files in order to submit data for each of the credits they are seeking. A new version of LEED, called LEED v4. scheduled for release in the near future, will accommodate a larger array of building types, including data centers, hospitality, and warehouse and distribution centers. This new version will adopt the latest American Society of Heating, Refrigerating and Air-Conditioning Engineers (ASHRAE) and other standards and reshape the overall structure of the LEED rating system.

The popularity of LEED certification has continued to grow since its beginnings in 2000. As of 2010, a cumulative total of over a billion square feet of commercial construction projects have been LEED certified, while a year later in 2011 the total had grown to 1.7 billion square feet (see Figure 5.1). The majority of LEED-certified projects are from new construction activity that includes both public and private owners. Figure 5.2 represents the cumulative growth of commercial LEED-certified projects since the beta testing of LEED

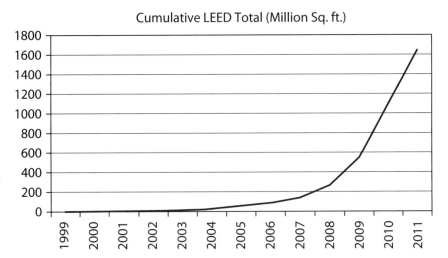

Figure 5.1 Cumulative square footage of LEED-certified projects through 2011. In November 2010, the USGBC announced that the total LEED-certified floor area exceeded 1 billion square feet.

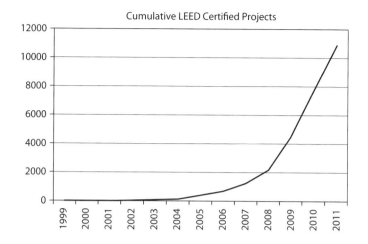

Figure 5.2 The total number of LEED-certified projects in the United States through 2011.

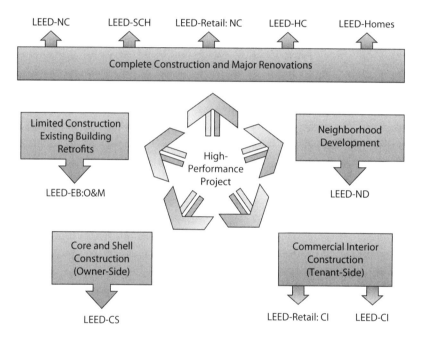

Figure 5.3 The various LEED building rating products address a wide variety of building types. In order to determine which rating product best suits the high-performance project, the project team should identify the construction type and space usage.

1.0 in 1998. Including LEED-certified residential projects increases the overall total to about 26,000 projects as of 2012.

Structure of the LEED Suite of Building Assessment Systems

Although referred to in the singular, LEED is not a single rating system but a *suite* of building rating systems, as shown in Figure 5.3. The current most popular LEED product is LEED for New Construction (LEED-NC), which focuses on building types such as offices, hotels, government buildings, manufacturing plants, institutional facilities such as libraries and churches, and residential buildings with four or more habitable stories. Only one LEED rating can be awarded to a building as a whole and the rating system employed depends on the use of the majority of the building. For example, to apply LEED-NC to a project, the owner must occupy more than 50 percent of the building. If tenant spaces occupy more than 50 percent of the building, then the LEED for Core and Shell (LEED-CS) rating system should be used for owners and developers to provide tenants with a building shell that integrates sustainable design. LEED-CS covers the core building elements, including the structure; building envelope; and heating, ventilation, and air conditioning (HVAC) systems. If tenants are interested in interior space improvement projects, they are to use the LEED for Commercial Interiors (LEED-CI) rating system for assisting in greening their spaces. LEED-CS and LEED-CI are designed to complement one another, with LEED-CS addressing the building and LEED-CI the tenant spaces. When both rating systems are applied in a project, the overall high-performance building would be equivalent to a LEED-NC project. LEED for Retail (LEED-Retail) is used to guide and distinguish high-performance retail projects, including banks, restaurants, and grocery, apparel, electronics, and big-box stores. LEED-Retail has two separate rating systems: (1) LEED-Retail: New Construction (LEED-Retail: NC), which addresses new construction and major renovations, and (2) LEED-Retail: Commercial Interiors (LEED-Retail: CI), which addresses retail interiors. LEED for Schools (LEED-SCH) is similar to LEED-NC but focuses on K−12 schools by addressing issues such as classroom acoustics, master planning, mold prevention, and environmental site assessment. LEED for Healthcare (LEED-HC) can be applied to inpatient, outpatient, licensed long-term-care facilities, medical offices, assisted-living facilities, and medical education and research centers. LEED for Homes focuses on single-family, low-rise homes (less than four stories), affordable housing, and manufactured and modular homes. LEED for Neighborhood Development (LEED-ND) is used for development projects such as neighborhoods, subdivisions, and larger mixed-use developments. LEED for Existing Buildings: Operations and Maintenance (LEED-EB:O&M) is applicable to buildings with commercial occupancies that involve building operation, minor space changes, system and process, upgrades, small additions, and facility alterations.

All building projects now fall under one of the rating systems identified above. The exception is when building tenants want to certify their individual spaces within a building which could be LEED rated. As noted above, a project can be a core and shell project to which LEED-CS applies and have tenant spaces that are being certified using LEED-CI.

LEED Credentials

The USGBC has developed a system of credentials to identify individuals who are knowledgeable in green building practices that support market transformation. Figure 5.4 illustrates these various credentials, which can be achieved through a combination of experience and examination. These include the LEED Green Associate (GA), LEED Accredited Professional (AP) with specialty, and LEED AP Fellow. The LEED GA credential is the fundamental credential and can be pursued by anyone who is employed in a building or environmental field. Green Associates must have basic knowledge of the LEED rating systems, LEED documentation process, sustainable design principles, standard terminology, and LEED resources that are available for identifying green strategies. No "real-world" experience is required to apply for the GA exam. The LEED rating systems provide a point for a project if a LEED AP is a member of the team. No point is awarded if a LEED GA is a team member.

For professionals in the industry, the option of taking the LEED AP with specialty exam is available if the individual has worked on a LEED project within the past three years. The specialty referred to in this context is the specific

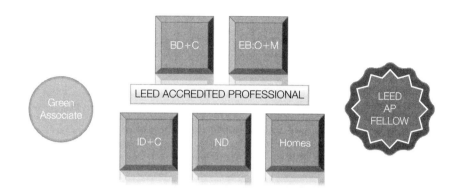

Figure 5.4   The available LEED credentials include the entry-level Green Associate, the five LEED Accredited Professional specialty designations, and the LEED AP Fellow.

rating system that the individual is qualified to manage (see Figure 5.4). There are five specialties that a LEED AP can opt for in terms of taking the LEED AP examination. A LEED AP with a Building Design and Construction (BD&C) specialty can manage projects that are using the New Construction and Major Renovations, Core and Shell, Healthcare, Retail: New Construction, or Schools green building rating systems. A LEED AP with a specialty of Interior Design and Construction (ID&C) may manage both Commercial Interiors and Retail: Commercial Interiors building rating systems. Other LEED AP specialties include Existing Buildings: Operations and Maintenance (EB:O&M), LEED for Homes, and LEED for Neighborhood Development (ND).

LEED APs are required to have in-depth of knowledge of green building practices and must specialize in a particular type of building rating system as described above, for example, LEED BD&C. The LEED AP test is divided into two sections: (1) general knowledge of green building and the LEED rating systems and (2) in-depth knowledge of the building rating systems for the specialty being tested. Both holders of the LEED GA and LEED AP credentials must maintain their status by paying a $50 maintenance fee every two years as well as participating in the Credential Maintenance Program (CMP). The CMP is a structured system used to expand the knowledge and experience base for LEED Professionals. LEED APs are required to have at least 30 hours of coursework every two years. Six of those hours must be LEED-specific. LEED GAs are required to have 15 hours of coursework every two years, including 3 LEED-specific hours.

The most prestigious professional designation currently awarded to an individual is the LEED AP Fellow. The LEED Fellow Program was developed to honor and recognize distinguished LEED APs who have made a significant contribution to the field of green building and sustainability at a regional, national, or international level. In order to become a LEED Fellow, the individual must be nominated by another LEED AP who has a specialty and at least 10 years' experience in the green building field. The nominee must also be a LEED AP with a specialty and have at least 10 years' experience in the green building field. In addition, the nominee must have held the LEED AP credential for at least 8 cumulative years. The nominee is evaluated in four of five mastery elements: technical proficiency, education and mentoring, leadership, commitment and service, and advocacy.

The LEED Process

As noted above, LEED-NC 3.0 is the most recent USGBC rating system for new commercial/institutional buildings and major renovations. It is structured with seven Minimum Program Requirements (MPRs), eight prerequisites, and

TABLE 5.1

LEED Category Allocation for Both LEED-NC 3.0 and LEED-NC v4

LEED-NC 3.0 Categories	Max Points	LEED-NC v4 Categories	Max Points
1. Sustainable Sites (SS)	26	**1.** Integrative Process (IP)	3
2. Water Efficiency (WE)	10	**2.** Location and Transportation (LT)	16
3. Energy and Atmosphere (EA)	35	**3.** Sustainable Sites (SS)	13
4. Materials and Resources (MR)	14	**4.** Water Efficiency (WE)	11
5. Indoor Environmental Quality (IEQ)	15	**5.** Energy and Atmosphere (EA)	26
6. Innovation in Design (ID)	6	**6.** Materials and Resources (MR)	10
7. Regional Priority (RP)	4	**7.** Indoor Environmental Quality (IEQ)	14
		8. Performance (PF)	7
		9. Innovation (IN)	6
		10. Regional Priority (RP)	4
Total Possible Points	**110**	**Total Possible Points**	**110**

a maximum of 110 points divided into seven major categories. As shown in Table 5.1, the category structure and point allocation differ substantially between the LEED-NC 3.0 and proposed LEED-NC v4. In order for a building to be considered for LEED certification, the requirements for all MPRs and all prerequisites must have been met. Further information about MPRs and prerequisites is provided later in this chapter. Lists of these requirements are provided in Tables 5.2 and 5.3, respectively.

The number of points available in each category was established by the developers of LEED-NC 3.0 to indicate the weight they placed on the various major issues addressed by this rating system. As a result, the allocation of points to each category is arbitrary, based solely on the judgment and expert opinion of the developers. Clearly, it is arguable, for example, that Energy and Atmosphere (35 points) is more important than Sustainable Sites (26 points) and more than twice as important as Materials and Resources (14 points). This situation indicates some of the pitfalls inherent in a building assessment system that attempts to reduce complex factors to a single number. Still, it does provide a logical and rational, albeit arbitrary, approach to producing numerical scores in

TABLE 5.2

LEED-NC 3.0 Minimum Program Requirements

1. Comply with environmental laws.
2. Be a complete, permanent building or space.
3. Use a reasonable site boundary.
4. Comply with minimum full-time equivalent (FTE) and floor area requirements.
5. Comply with minimum occupancy rates.
6. Commit to sharing whole-building energy and water usage data.
7. Comply with a minimum building area to site area ratio.

TABLE 5.3

All Prerequisites Listed in LEED-NC 3.0

	Prerequisite	Name of Prerequisite
1.	SSp1	Construction Activity Pollution Prevention
2.	EAp1	Fundamental Commissioning of Building Energy Systems
3.	EAp2	Minimum Energy Performance
4.	EAp3	Fundamental Refrigerant Management
5.	WEp1	Water Use Reduction: 20%
6.	MRp1	Storage and Collection of Recyclables
7.	IEQp1	Minimum Indoor Air Quality Performance
8.	IEQp2	Environmental Tobacco Smoke (ETS) Control

each category. It is important to keep in mind that LEED was developed using an extensive collaborative process; hence, the outcome of this group thought process is probably on target with respect to weighting the points and categories. Thus, in spite of its relative simplicity, it does a good job overall of taking complex information and converting it into a single score and rating level.

The total score from LEED-NC 3.0, computed by adding up the points earned in each category, results in a building rating (see Table 5.4). The platinum and gold ratings are fairly difficult to achieve, and a silver rating is actually a very good assessment and a noteworthy accomplishment.

GREEN BUILDING CERTIFICATION INSTITUTE (GBCI) RELATIONSHIP TO THE USGBC AND LEED

Up until 2008, the USGBC administered building certifications and professional designations in-house. In 2008, a nonprofit organization, the Green Building Certification Institute (GBCI), was founded to provide a balanced third-party certification in order to be recognized by the American National Standards Institute (ANSI). The GBCI is responsible for managing all aspects of LEED professional credentialing, including exam development, registration, delivery, and maintenance, to ensure ongoing excellence and that LEED professionals are proficient in the field. In addition, the GBCI is responsible for managing the LEED project certification program by conducting technical reviews and analysis of submissions to verify and evaluate projects based on how well they have met the requirements of the various LEED rating systems. Project document is submitted through the LEED-Online Internet portal, which is discussed later in this chapter.

The USGBC retains responsibility for creating and implementing new versions of the LEED building rating system by integrating new green building technologies, systems, and strategies into the latest requirements. This includes establishing new reference guides and educational resources and outlining certification and accreditation requirements, among other things. The GBCI is the arbiter of both building certification and LEED GA and AP accreditation, both of which are based on USGBC-generated rules. The relationship between the USGBC and the GBCI is visually depicted in Figure 5.5.

THE LEED CERTIFICATION PROCESS

Prior to certification, the building is referred to as a *LEED-registered project*. Achieving a LEED certification level requires significant dedication from professionals of the project team. This dedication must be maintained from design to the end of construction in order to successfully complete all steps in the certification process. These steps include (1) ensuring that the building is eligible for certification; (2) registering the project with the GBCI; (3) ensuring and documenting that the project meets the MPRs and prerequisites and can attain at least the minimum number of points to achieve the LEED-certified level; (4) submitting the required documentation via LEED-Online; (5) if necessary, appealing points denied by the GBCI; and (6) receiving final notification from the GBCI that the project has achieved LEED certification.

LEED-ONLINE

Over time, the LEED building rating system has shifted from requiring hardcopy documentation for certification to an Internet-based system known as *LEED-Online*. Project teams can submit all of their documentation online in

TABLE 5.4

Points Required for LEED-NC 3.0 Ratings

Rating	Points Required
Platinum	80–110
Gold	60–79
Silver	50–59
Certified	40–49
No rating	39 or less

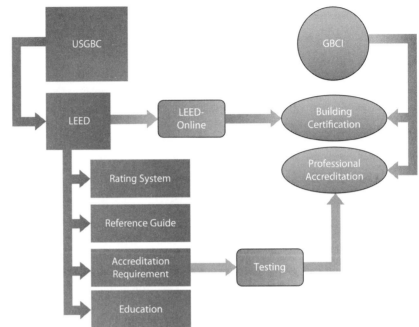

Figure 5.5 The relationship between the USGBC and the GBCI and their respective roles in the LEED green building rating system. The USGBC develops the requirements for LEED certification, and the GBCI ensures the requirements have been met. The GBCI also is responsible for the testing and continuing education of LEED Accredited Professionals.

an easy-to-use format. LEED-Online stores all LEED information, resources, and support in one centralized location. It enables team members to upload credit templates, track Credit Interpretation Rulings (CIRs), view documented responses to questions posed by previous project teams, manage key project details, contact the customer service department, and communicate with reviewers throughout the design and construction review process.

REGISTRATION

The first step in LEED certification is project registration. Projects are registered by visiting the LEED Registration page of the USGBC website, where information about the project is input and a registration fee is paid. Early registration is encouraged because starting the process as early as possible maximizes the potential for achieving certification. Registration establishes contact with the USGBC and provides access to essential information, software tools, and communications. Appointment of a Project Team Administrator occurs when the project is first registered at LEED-Online. The Project Team Administrator invites members of the project team to register with the project and then assigns roles to the individual project team members. Typical roles include architect, landscape architect, civil engineer, owner, and developer, to name just a few. The system also allows the Project Team Administrator to create new roles that are unique to the project, if needed. The Project Team Administrator develops a project description, assigns responsibility for LEED credits to the project team members, and then monitors the submission of documentation to support the LEED credits. The Project Team Administrator should be a LEED AP and is the project team member assigned to steer the project through the certification process. Once a project is registered and responsibilities are assigned, the project team begins to prepare documentation to satisfy the prerequisites and credit submittal requirements.

CREDIT INTERPRETATION RULINGS

If a project team encounters difficulties applying a LEED prerequisite or credit to a specific project, the USGBC encourages the team to sort out the issue themselves first and contact the GBCI only as a last resort. The Credit Interpretation Ruling (CIR) system ensures that rulings are consistent and available to other projects. If there is a gray area for which the project team requires clarification, the team can submit its query through LEED-Online and receive a ruling from the USGBC on the official interpretation of the situation. This latter response by the USGBC is the so-called CIR. All CIRs are contained in a database accessible from LEED-Online, and they can be used as precedents for addressing situations encountered by the project team.

DOCUMENTATION AND CERTIFICATION

To earn LEED certification, the project must meet all the MPRs, satisfy all of the prerequisites, and earn the minimum number of points to at least attain the LEED-certified rating level. For LEED-NC, LEED-SCH, LEED-CS, LEED HC, LEED-Retail, and LEED-CI projects, the project team has the option to divide their review process into two separate reviews, design review and construction review, also known as a split review. The benefit of a split review is that it helps the project team gauge whether or not the project is on track for achieving the anticipated LEED certification level. LEED-Online specifies which points are either design phase credits or construction phase credits, and the project team must submit complete documentation of design phase credits for the design review. The GBCI will then respond with the Preliminary Design Review, indicating which credits are "Anticipated," "Pending," or "Denied" and if any documentation was "Approved" or "Not Approved." The project team then has the option to accept the results of the Preliminary Design Review as final or submit a response. If the project team determines that a response is necessary, they must submit a clear response to the Preliminary Design Review as well as appropriate documentation within 25 business days. The GBCI will then review the submitted documents along with the team's response and within 25 business days reply with the Final Design Review. The Final Design Review will address each credit as "Anticipated," "Pending," or "Denied."During this period, no points are awarded to the project. Closer to project completion, the project team can submit the remaining credits for the Preliminary Construction Review. The structure of this review is the same as the Preliminary Design Review. If the project team feels that they should be awarded a specific credit, whether from the Final Design Review or Final Construction Review, they have the option to appeal.

AWARD OF CERTIFICATION

Upon notification of LEED certification, the project team has 30 days to accept or appeal the awarded certification level of platinum, gold, bronze, or certified. Upon the project's acceptance, or if the project team has not appealed the rating within 30 days, the LEED certification is final. GBCI will refer to the project as a LEED-certified building and the project team will receive an award letter and certificate specifying the LEED certification level. Although in the past LEED-certified projects were awarded plaques by the USGBC, currently a plaque can only be purchased online from an exclusive vendor (see Figure 5.6)[1]. The plaque exhibits the LEED certification level and year of achievement and is typically featured on either the exterior or interior of the high-performance building.

Figure 5.6 The certification plaque from the USGBC is made of recycled glass. (Photograph courtesy of Torii Mor Winery)

APPEALS

If the project team feels that sufficient grounds exist to appeal a credit denied in the Final LEED Review, it has the option to appeal. The appeal fee is $500 per credit appealed. A review of the documentation for the appealed credits occurs within 25 business days, at which time an Appeal LEED Review will be provided to the applicant. All appeals are submitted via LEED-Online. If an appeal is pursued, a different review team will assess the appeal documentation.

LEED REGISTRATION AND CERTIFICATION FEES

A registration fee must be paid for all LEED-NC, LEED-CI, LEED-CS, and LEED-EB projects as part of the registration process. When the project team is prepared to initiate the review process, a certification fee must first be paid prior to action by the GBCI. As noted above, the project team has the option of submitting all documentation either at the conclusion of the construction process or in two phases: (1) a design review, in which all credits that have been completely addressed by the design team are put forward for review; and (2) a construction review, in which all the remaining credits are reviewed. The advantage of the two-phase review process is that it speeds the certification process and allows the project team to decide on and act on appeals far earlier in the process.

LEED Categories

The LEED building rating system is structured to provide points in seven categories: Sustainable Sites (SS), Water Efficiency (WE), Energy and Atmosphere (EA), Materials and Resources (MR), Indoor Environmental Quality (IEQ), Innovation in Design (ID), and Regional Priority (RP). The point allocation varies among the various LEED products for specific building types. Appendix A shows the point allocation for the LEED rating systems. In this chapter, we focus on the details of LEED for New Construction and Major Renovations (LEED-NC). Information on the points and categories for other LEED building rating products can be found online at the USGBC website.

THE SUSTAINABLE SITES (SS) CATEGORY

The structure of the Sustainable Sites category of the USGBC LEED building assessment system is addressed in this section, including an overview of the credits and requirements. For detailed information on how to document points properly for LEED certification, consult the LEED-NC 3.0 Reference Manual. (Note that each rating system product for each version of LEED has its own specific reference manual.) What follows is a detailed explanation of each credit, the required LEED Letter Template, and other miscellaneous documentation.

The Sustainable Sites (SS) category of LEED has a single prerequisite and a maximum of 26 points that can be achieved by employing measures that make the siting of the building as environmentally responsible as possible. LEED requires that all prerequisites be met before a building becomes eligible for LEED certification. Table 5.5 lists the SS credits and points available under LEED-NC 3.0.

TABLE 5.5

Sustainable Sites (SS) Credits and Points under LEED-NC 3.0

Prerequisite/ Credit	Name of Prerequisite/Credit	Maximum Points
SS Prerequisite 1	Construction Activity Pollution Prevention	NA
SS Credit 1	Site Selection	1
SS Credit 2	Development Density and Community Connectivity	5
SS Credit 3	Brownfield Redevelopment	1
SS Credit 4.1	Alternative Transportation—Public Transportation Access	6
SS Credit 4.2	Alternative Transportation—Bicycle Storage and Changing Rooms	1
SS Credit 4.3	Alternative Transportation—Low-Emitting and Fuel-Efficient Vehicles	3
SS Credit 4.4	Alternative Transportation—Parking Capacity	2
SS Credit 5.1	Site Development—Protect or Restore Habitat	1
SS Credit 5.2	Site Development—Maximize Open Space	1
SS Credit 6.1	Stormwater Design—Quantity Control	1
SS Credit 6.2	Stormwater Design—Quality Control	1
SS Credit 7.1	Heat Island Effect—Nonroof	1
SS Credit 7.2	Heat Island Effect—Roof	1
SS Credit 8	Light Pollution Reduction	1
	Total SS Points Available	**26**

SS Prerequisite 1 (SSp1): Construction Activity Pollution Prevention

This prerequisite requires the design and implementation of an erosion and sedimentation control (ESC) plan that prevents soil loss via water or wind and sedimentation of stormwater infrastructure and receiving bodies of water.

SS Credit 1 (SSc1): Site Selection (1 Point Maximum)

The selection of a site with minimal environmental or ecological system impact is a very important feature of a high-performance green building. This credit requires that buildings, roads, or parking areas on portions of sites must not be built on prime farmland; previously undeveloped land whose elevation is lower than 5 feet above the elevation of the 100-year flood; land that is specifically identified as habitat for any species on federal or state threatened or endangered lists; land within 100 feet (30 meters) of any wetlands, or according to state or local regulations if they require greater setback distances from wetlands; previously undeveloped land within 50 feet (15 meters) of a body of water (seas, lakes, rivers, streams, and tributaries that do or could support fish, recreation, or industrial use); and land that, prior to acquisition for the project, was public parkland, unless land of equal or greater value as parkland is accepted in trade by the public landowner.

SS Credit 2 (SSc2): Development Density and Community Connectivity (5 Points Maximum)

Along with reusing disturbed land in preference to greenfields, it makes sense to increase the density of existing development consistent with maintaining or increasing the quality of life of the area. There are two options for earning this credit. Option 1 is to build on a previously developed site that is located within an existing minimum development density of 60,000 square feet per acre (13,800 square meters per hectare) (two-story downtown development). Option 2 is for the project to be on a previously developed site that is within half a mile of a residential zone or neighborhood with an average density of at least 10 units per acre (25 units per hectare). It must also have pedestrian access to (be within half a mile of) 10 so-called basic services (bank, cleaners, day-care facility, pharmacy, post office, and fitness center, to name a few).

SS Credit 3 (SSc3): Brownfield Redevelopment (1 Point Maximum)
Land that has already been impacted by human activities is preferable for a building project to land that is a greenfield. Although brownfields are generally urban sites with access to excellent infrastructure, there are numerous issues with respect to remediating or cleaning up these properties. This is a complex and potentially costly process; hence, it applies to only a very small number of building projects. A site can be designated as a brownfield via an environmental site assessment, a local voluntary cleanup program, or by a federal, state, or local government agency.

SS Credit 4 (SSc4): Alternative Transportation (12 Points Maximum)
The overall purpose of this credit is to reduce dependence on conventional fossil fuel–powered automobiles. SS Credit 4 actually consists of four different subcredits, each with a maximum of 1 point. The requirements for each of these subcredits are described below

SS Credit 4.1 (SSc4.1): Alternative Transportation—Public Transportation Access (6 Points Maximum)
For a building to be truly green, it should be in a location where there is ready access to mass transportation. For the purpose of LEED-NC, this credit requires the building project to be within one-half of a mile (0.8 kilometer) of a commuter rail, light rail, or subway station, or within one-fourth of a mile (0.4 kilometer) of two or more public or campus bus lines usable by building occupants.

SS Credit 4.2 (SSc4.2): Alternative Transportation—Bicycle Storage and Changing Rooms (1 Point Maximum)
Another aspect of green buildings that helps reduce the impacts of their operations is to facilitate the use of bicycles by the occupants. To earn this credit, the building should provide secure bicycle storage with convenient changing/shower facilities [within 200 yards (183 meters) of the building] for 5 percent or more of regular building occupants. For residential buildings, in lieu of changing/shower facilities, covered storage facilities for securing bicycles for 15 percent or more of building occupants must be provided.

SS Credit 4.3 (SSc4.3): Alternative Transportation—Low-Emitting and Fuel-Efficient Vehicles (3 Points Maximum)
Another approach to reducing the impacts associated with occupants having to travel to and from the building is to facilitate the use of alternative-fuel vehicles. This credit can be achieved by providing alternative-fuel vehicles for 3 percent of building occupants and preferred parking for these vehicles, or by providing preferred parking for low-emitting and fuel-efficient vehicles for 5 percent of the total vehicle parking capacity of the site.

SS Credit 4.4 (SSc4.4): Alternative Transportation—Parking Capacity (2 Points Maximum)

This credit emphasizes the reduction of parking capacity for automobiles to the bare minimum needed to meet local zoning requirements. To earn the points associated with this credit, for nonresidential projects the parking capacity must be sized to meet, but not exceed, minimum local zoning requirements and provide preferred parking for carpools or vanpools capable of serving 5 percent of the building occupants; or add no new parking for renovation projects and provide preferred parking for carpools or vanpools capable of serving 5 percent of the building occupants. For residential projects, the parking capacity must be sized to meet, but not exceed, minimum local zoning requirements and facilitate shared vehicle usage through carpool drop-off areas, designated parking for vanpools, car-share services, ride boards, and shuttle service to mass transit.

SS Credit 5 (SSc5): Site Development (2 Points Maximum)

Site clearing, earthwork, compaction, temporary roads and structures, and other operations involving earth movement and construction can have significant environmental impact. This credit has two subcredits, each offering 1 point, for measures that reduce disturbance to the site during construction: SS Credit 5.1 (SSc5.1): Site Development—Protect or Restore Habitat and SS Credit 5.2 (SSc5.2): Site Development—Maximize Open Space. The requirements for each of these subcredits are described below.

SS Credit 5.1 (SSc5.1): Site Development—Protect or Restore Habitat (1 Point Maximum)

The idea behind this credit is to minimize the impacts of the construction process on natural systems by requiring minimal site disturbance during construction. For Case 1, which addresses greenfield sites, site disturbance, including earthwork and clearing of vegetation, must be limited to 40 feet beyond the building's perimeter; 10 feet beyond surface walkways, patios, surface parking, and utilities less than 12 inches in diameter; 15 feet beyond primary roadway curbs, walkways, and main utility branch trenches; and 25 feet beyond constructed areas with permeable surfaces (such as pervious paving areas, stormwater detention facilities, and playing fields) that require additional staging areas in order to limit compaction in the constructed area; or, on previously developed sites, a minimum of 50 percent of the site area (excluding the building footprint) must be restored by replacing impervious surfaces with native or adaptive vegetation. For Case 2, which addresses previously developed sites, restore or protect 50 percent of the site (excluding the building footprint) or 20 percent of the total site area (including the building footprint), whichever is greater, with native or adapted vegetation.

SS Credit 5.2 (SSc5.2): Site Development—Maximize Open Space (1 Point Maximum)

This credit emphasizes the inclusion of requirements for conserving open space and restoring damaged areas into productive ecosystems. Option 1: The point associated with this credit can be achieved by reducing the development footprint (defined as entire building footprint, access roads, and parking) to exceed the local zoning's open-space requirement for the site by 25 percent. Option 2: For areas with no local zoning requirements (e.g., some university campuses and military bases), open-space area adjacent to the building that is equal to the development footprint must be designated. Option 3: If a zoning ordinance exists but there is no requirement for open space, vegetated open space must equal at least 20 percent of the project's site area.

SS Credit 6 (SSc6): Stormwater Design (2 Points Maximum)

Stormwater management is required largely because of the significant reduction in pervious surfaces caused by buildings and their associated parking and paving. This credit comprises two subcredits, both of which are described below.

SS Credit 6.1 (SSc6.1): Stormwater Design—Quantity Control (1 Point Maximum)

The goal of this credit is to minimize the imperviousness of the building site. In cases where there is significant imperviousness, it should be decreased. This can be accomplished by increasing the area of pervious pavement, by using vegetative roofs or eco-roofs, and by other measures that increase the infiltration of water back into the soil. Another approach is to capture stormwater and use it for nonpotable water purposes such as flushing of sanitary fixtures and landscape irrigation. This credit requires that if existing imperviousness is less than or equal to 50 percent, a stormwater management plan must be implemented that prevents the postdevelopment 1.5-year, 24-hour peak discharge rate from exceeding the predevelopment 1.5-year, 24-hour peak discharge rate. Or, if existing imperviousness is greater than 50 percent, a stormwater management plan must be implemented that results in a 25 percent decrease in the rate and quantity of stormwater runoff.

SS Credit 6.2 (SSc6.2): Stormwater Design—Quality Control (1 Point Maximum)

Treating stormwater by using simple approaches and natural systems reduces infrastructure and energy for moving and treating large volumes of water. Mechanical or natural treatment systems such as constructed wetlands, vegetated filter strips, grass swales, bioswales, detention ponds, and filtration basins can be designed to collect and treat the site's stormwater. This credit requires these systems be designed to remove 80 percent of the average annual postdevelopment total suspended solids (TSS) and 40 percent of the average annual postdevelopment total phosphorus (TP) based on the average annual loadings from all storms less than or equal to the 2-year, 24-hour storm.

SS Credit 7 (SSc7): Reducing Heat Island Effects (2 Points Maximum)

The air temperature in urban areas can be 2 to 10°F (1 to 6°C) higher than in the surrounding countryside, a consequence of solar energy absorption and radiation by components of the built environment, particularly dark, nonreflective surfaces used for paving and roofing. This increase in air temperature means that significantly more energy is needed for cooling and even that distinct microclimates are created in the affected areas. Reducing the heat island effect can markedly reduce summertime energy use. This credit comprises the two subcredits, described below.

SS Credit 7.1 (SSc7.1): Heat Island Effect—Nonroof (1 Point Maximum)

The heat island effects of nonroof surfaces can be reduced by providing shade or using light-colored (high-albedo) materials for parking and paving. Using open-grid pavement is another appropriate option for reducing thermal energy buildup, as is locating parking structures underground. Two other heat island reduction strategies are using trees and other vegetation to shade structures and using architectural shading devices where planting vegetation is not feasible. To earn the point associated with this credit, at least 50 percent of the site hardscape must be shaded within five years of occupancy, paving materials must have a Solar Reflectance Index (SRI) of at least 29, or an open-grid pavement system can be used. Optionally, 50 percent of parking spaces can be put undercover (e.g., in underground parking), but the roof must have an SRI of at least 29.

SS Credit 7.2 (SSc7.2): Heat Island Effect—Roof (1 Point Maximum)

Using eco-roofs, vegetative roofs, or light-colored (high-albedo), highly reflective, Energy Star–compliant roofs can greatly reduce the heat island effect associated with this building component. Roofing materials must have an SRI equal to or greater than 78 (for a roof slope of 2:12 or lower) or 29 (for a roof slope greater than 2:12). Optionally, a vegetated roof covering at least 50 percent of the roof area can satisfy the requirements for this credit. A combination of high-albedo roof and vegetated roof can also meet the requirements for this credit.

SS Credit 8 (SSc8): Light Pollution Reduction (1 Point Maximum)

Light pollution is a complex problem that can be caused by both exterior and interior lighting. It can be addressed by adopting site lighting criteria to maintain safe light levels while avoiding off-site lighting and night-sky pollution. Site lighting should be minimized where possible and should be designed using a computer model. Technologies to reduce light pollution include full cutoff luminaires, low-reflectance surfaces, and low-angle spotlights. Exterior illumination should not exceed 80 percent of the lighting power densities for exterior areas and 50 percent for building façades and landscape features as defined in ASHRAE/IESNA 90.1-2004. Interior lighting must be designed such that the angle of maximum candela from interior luminaires does not exit the windows.

THE WATER EFFICIENCY (WE) CATEGORY

The LEED-NC category covering water and wastewater issues is Water Efficiency (WE). A maximum of 10 total points are available in the WE category, as summarized in Table 5.6.

WE Prerequisite 1 (WEp1): Water Use Reduction: 20 Percent

Achieving this prerequisite requires that the design of the building employ a strategy to reduce by 20 percent the water use established in the baseline calculation. The baseline calculation must meet the requirements established by the Energy Policy Act of 2005 as well as the 2006 edition of the Uniform Plumbing Code. The baseline calculation is based on occupant usage and should only incorporate flow and flush fixtures, such as water closets, urinals, lavatory faucets, showers, kitchen sink faucets, and prerinse spray valves. More information on how to calculate a baseline water model can be found in Chapter 10.

TABLE 5.6

Water Efficiency (WE) Credits and Points under LEED-NC 3.0

Prerequisite/Credit	Name of Prerequisite/Credit	Maximum Points
WE Prerequisite 1	Water Use Reduction: 20%	NA
WE Credit 1	Water-Efficient Landscaping	4
WE Credit 2	Innovative Wastewater Technologies	2
WE Credit 3	Water Use Reduction	4
	Total WE Points Available	**10**

WE Credit 1 (WEc1): Water-Efficient Landscaping (4 Points Maximum)

In order to improve the stormwater management system and limit or eliminate the use of potable water for landscape irrigation, two options are available.

Option 1. Reduce potable water use for landscaping by 50 percent. (2 Points Maximum)

Water reduction is based on calculated midsummer baselines. Reducing the use of potable water can be achieved through methods such as specifying plant species that do not require much watering; applying a density factor that reduces the number of plants; improving irrigation efficiency; and using captured rainwater, recycled wastewater, or reclaimed water for irrigation purposes.

Option 2. Further reduce potable water consumption. (2 Points Maximum)

In addition to achieving Option 1, the design must require a 100 percent reduction of potable water irrigation, which can be achieved through measures such as those identified in Option 1.

WE Credit 2 (WEc2): Innovative Wastewater Technologies (2 Points Maximum)

Although the title of this credit indicates that it concerns wastewater, the actual issues addressed are somewhat broader and allow two strategies: (1) reduce potable water used for sewage conveyance by at least 50 percent or (2) treat at least 50 percent of the water to tertiary standards on-site and then use the water on-site or infiltrate it back into the ground.

WE Credit 3 (WEc3): Water Use Reduction (4 Points Maximum)

The reduction of potable water use in the building by 30 percent results in 2 points; increasing the reduction to 35 percent provides an additional point, and a final point is earned by reducing potable water consumption by 40 percent. In calculating the reduction, potable water use is compared to that of a baseline building meeting the fixture requirements of the Energy Policy Act of 1992 (EPAct of 1992). To earn points, the mechanical, electrical, and plumbing (MEP) engineer must provide calculations that demonstrate the reduction compared to the baseline building. As mentioned previously, ways to calculate a water baseline are identified in Chapter 10.

THE ENERGY AND ATMOSPHERE (EA) CATEGORY

The LEED-NC category Energy and Atmosphere (EA) addresses the issues of energy for high-performance buildings; it also covers several issues that connect building systems to environmental impacts on air and the atmosphere—for example, the elimination of hydrochlorofluorocarbons (HCFCs), which, due to their presence in chillers and other mechanical equipment, are implicated in ozone depletion. The following sections discuss LEED-NC 3.0 credits and reporting requirements in the EA category. The numbers associated with these credits correspond to the numbering system used in the LEED-NC standard. Table 5.7 is a summary list of these credits and the points associated with them. Note that the three prerequisites do not carry points and must all be met for a building to be considered for certification.

TABLE 5.7

Energy and Atmosphere (EA) Credits and Points under LEED-NC 3.0

Prerequisite/Credit	Name of Prerequisite/Credit	Maximum Points
EA Prerequisite 1	Fundamental Commissioning of Building Energy Systems	NA
EA Prerequisite 2	Minimum Energy Performance	NA
EA Prerequisite 3	Fundamental Refrigerant Management	NA
EA Credit 1	Optimize Energy Performance	19
EA Credit 2	On-Site Renewable Energy	7
EA Credit 3	Enhanced Commissioning	2
EA Credit 4	Enhanced Refrigerant Management	2
EA Credit 5	Measurement and Verification	3
EA Credit 6	Green Power	2
	Total EA Points Available	**35**

EA Prerequisite 1 (EAp1): Fundamental Commissioning of Building Energy Systems

The purpose of fundamental building commissioning is to ensure that the building operates as intended by the design team. For this to be possible, however, the design team must have adequately carried out its design tasks such that the building's systems have the capability to function as indicated on the plans and in the specifications. EAp1 requires a qualified Commissioning Authority (CxA), which can oversee the process, develop a commissioning plan, and carry out detailed checks of the building energy systems.

EA Prerequisite 2 (EAp2): Minimum Energy Performance

The building must be designed, at a minimum, to meet the mandatory provisions (Sections 5.4, 6.4, 7.4, 8.4, 9.4, and 10.4) of ASHRAE 90.1-2007 and the prescriptive requirements (Sections 5.5, 6.5, 7.5, and 9.5) or performance requirements (Section 11) of ASHRAE 90.1-2007. The building must comply with the mandatory provisions and either the prescriptive or Energy Cost Budget Method performance requirements of the standard.

EA Prerequisite 3 (EAp3): Fundamental Refrigerant Management

Chlorofluorocarbons (CFCs), as previously noted, are ozone-depleting substances with a long history of use in building air-conditioning equipment. This prerequisite has the intent of eliminating CFCs from buildings, thereby protecting the ozone layer. It requires zero use of CFCs in new building HVAC&R systems. For a reuse project with existing equipment, a plan to phase out the CFCs must be submitted prior to project completion.

EA Credit 1 (EAc1): Optimize Energy Performance (19 Points Maximum)

Designing and building an energy-efficient building is important for sustainability reasons as well as for earning a LEED rating. By virtue of having more than half of the EA credits assigned to it, EA Credit 1: Optimize Energy Performance is by far the most important credit in the EA category—in fact, in the entire LEED rating system. ASHRAE 90.1-2007 is the basis for determining how well the high-performance building performs compared to the base case, that is, the building that just meets the standard's minimum requirements. To be successful in obtaining a LEED rating for the building, the design team should ensure that the requirements of ASHRAE 90.1-2007 are exceeded by a substantial margin. The energy-related components of a typical building are its envelope (walls, roof, floor, windows, doors); HVAC equipment; power distribution system; lighting system; and equipment such as pumps, appliances, refrigeration equipment, and elevators.

EAc1 has three options for earning points toward LEED certification, as described below. Buildings must earn at least 2 points under EAc1 to be certified.

Option 1. Whole-Building Energy Simulation (19 Points Maximum)

The project can earn up to 19 points by running a whole-building energy simulation per ASHRAE 90.1-2007 using the Performance Rating Method described in Appendix G. A new building with a 12 percent improvement receives 1 point, with 1 additional point for each 2 percent increase, up to a maximum of 19 points for a 48 percent improvement over the base case. For an existing building, the sliding scale starts at 1 point for an 8 percent improvement over the base case, with 1 additional point for each 2 percent improvement, up to a maximum of 19 points for a 44 percent improvement over the base case.

The building must be designed, at a minimum, to meet the mandatory provisions (Sections 5.4, 6.4, 7.4, 8.4, 9.4, and 10.4) of ASHRAE 90.1-2004. The proposed design must include all energy costs of the proposed design, and the design must be compared to a baseline building as described in Appendix G of the standard. On-site renewable energy generation is included in the modeling to show a reduction in energy demand from external sources.

Option 2. Prescriptive Compliance Path: ASHRAE Advanced Energy Design Guide (1 Point Maximum)

A small office building, that is, one under 20,000 square feet (1860 square meters) in size, can earn 1 point for complying with the prescriptive measures of the ASHRAE Advanced Design Guide appropriate to the project scope.

Option 3. Prescriptive Compliance Path: Advanced Building[TM] Core Performance[TM] Guide (3 Points Maximum)

A building that is not classified as a warehouse, laboratory, or health-care project and is less than 100,000 square feet (9290 square meters) in size can earn up to 3 points by complying with the Advanced Building[TM] Core Performance[TM] Guide developed by the New Buildings Institute.

EA Credit 2 (EAc2): On-Site Renewable Energy (7 Points Maximum)

LEED encourages the consumption of renewable rather than nonrenewable energy for buildings and provides points for on-site or site-recovered renewable energy systems. Eligible renewable energy systems include photovoltaic (PV) or solar thermal systems; active systems; biofuel-based electrical systems; geothermal heating/electrical systems; low-impact hydro, wave/tidal power systems; and wind-based electrical production systems. In order to receive points, projects using renewable systems must calculate project performance by expressing the energy produced as a percentage of the building annual energy cost. The building annual energy costs are calculated either from EAc1 or by the DOE Commercial Buildings Energy Consumption Survey (CBECS) database to estimate electricity use. Up to 7 points are available, from 1 point for providing 1 percent of the total building energy requirements to 7 points if 13 percent is provided.

EA Credit 3 (EAc3): Enhanced Commissioning (2 Points Maximum)

Enhanced commissioning adds several additional requirements to the Fundamental Building Commissioning category. These requirements include reviewing the energy systems design, reviewing contractor submittals, creating a systems manual for building operators, verifying training of operators, and rechecking the building operation within 10 months of occupancy to verify performance.

EA Credit 4 (EAc4): Enhanced Refrigerant Management (2 Points Maximum)

EA Prerequisite 3 calls for zero use of CFC refrigerants in new building HVAC&R equipment. EAc4 provides for measures that further reduce the use of ozone-depleting refrigerants in buildings and addresses the climate change impacts of these substances. EAc4 has two options:

Option 1. Do not use refrigerants.

Option 2. Select refrigerants that minimize contributions to the life-cycle ozone depletion potential (LCODP) and the life-cycle direct global warming potential (LCGWP). In order to account for the potential damage of various refrigerants, a formula is used to quantify the combined impact in the building project.

EA Credit 5 (EAc5): Measurement and Verification (3 Points Maximum)

In addition to motivating significant energy savings, the LEED-NC 3.0 rating system provides an incentive to measure the savings by providing 3 points that calls for a definitive system of sensors that can provide feedback on building operation. The methodology used to measure these savings is the International Performance Measurement and Verification Protocol (IPMVP), Volume III (April 2003), Option B Energy Conservation Measure Isolation, or Option D, Whole Building Calibrated Simulation.

EA Credit 6 (EAc6): Green Power (2 Points Maximum)

Another approach to using renewable energy in a building is to contract for power from a utility that generates energy from renewable sources. This credit requires that the building owner engage in a two-year contract with a source that meets the Center for Resource Solutions (CRS) Green-e products certification process and provides at least 35 percent of the building's electricity. There are three options for purchasing green power. The owner may purchase from a Green-e certified power marketer; from a Green-e accredited utility program, through Green-e accredited Tradeable Renewable Certificates; or from a supply that meets the Green-e renewable power definition.

THE MATERIALS AND RESOURCES (MR) CATEGORY

The materials provisions of LEED-NC with respect to credits and points in the Materials and Resources (MR) category are listed in Table 5.8.

TABLE 5.8

Materials and Resources (MR) Credits and Points under LEED-NC 3.0

Prerequisite/Credit	Name of Prerequisite/Credit	Maximum Points
MR Prerequisite 1	Storage and Collection of Recyclables	NA
MR Credit 1.1	Building Reuse—Maintain Existing Walls, Floors, and Roof	3
MR Credit 1.2	Building Reuse—Maintain Existing Interior Nonstructural Elements	1
MR Credit 2	Construction Waste Management	2
MR Credit 3	Materials Reuse	2
MR Credit 4	Recycled Content	2
MR Credit 5	Regional Materials	2
MR Credit 6	Rapidly Renewable Materials	1
MR Credit 7	Certified Wood	1
	Total MR Points Available	**14**

MR Prerequisite 1 (MRp1): Storage and Collection of Recyclables

A very interesting and progressive LEED requirement is that an easily accessible area must be set aside for the separation, collection, and storage of materials, which, at a minimum, include paper, corrugated cardboard, glass, plastics, and metals.

MR Credit 1 (MRc1): Building Reuse (4 Points Maximum)

Reusing a building to the maximum extent possible is an excellent strategy for reducing the materials impacts of construction and LEED offers a maximum of 4 points for building reuse. MR Credit 1.1 provides one point for reusing 55 percent of the existing walls, floors, and roof, two points for 75 percent, and a third point if 75 percent of these elements are reused. MR Credit 1.2 provides yet another point if 50 percent of the interior nonstructural elements are reused. Interior nonstructural elements refer to such building components as interior walls, doors, floor coverings, and ceiling systems.

MR Credit 1.1 (MRc1.1): Building Reuse—Maintain Existing Walls, Floors, and Roof (3 Points Maximum)

The requirement of this credit is to maintain existing building structural elements, such as walls, floors, and the roof. If 55 percent (by area) of these elements are retained, one point is awarded; an additional point for 75 percent, and a third point if 95 percent are retained. If there is a case when a project includes an addition to an existing building, the square footage of the addition must be less than twice the square footage of the existing building.

MR Credit 1.2 (MRc1.2): Building Reuse—Maintain Existing Interior Nonstructural Elements (1 Point Maximum)

To extend the life cycle of existing building stock and reduce landfill waste, at least 50 percent (by area) of the existing nonstructural elements must be reused. These elements include interior walls, doors, floor coverings, and ceiling systems. Reused elements must perform the same function and, if they differ in functionality, the quantity of this type of reused material would be allocated to MR Credit 3.

MR Credit 2 (MRc2): Construction Waste Management (2 Points Maximum)

Reducing construction waste is a critical part of construction operations for the production of green buildings. In order to achieve points, it is first important to establish a construction waste management plan to identify the materials to be diverted from disposal whether the materials will be sorted on-site or commingled. Quantities of material can be measured either by weight or by volume, but they must be consistent throughout the project. The first point will be allocated based on a diversion rate of 50 percent. The second point can be achieved by diverting 75 percent of the construction and demolition waste from landfills. More information about the construction waste management plan is outlined in Chapter 13.

MR Credit 3 (MRc3): Materials Reuse (2 Points Maximum)

Beyond reducing materials use in the building, reusing components of existing buildings has the greatest benefit in lowering overall materials impacts. Reusing building materials and products can result in the project's receiving up to 2 points. One point is achievable if 5 percent of the total materials cost is used for purchasing salvaged materials; a second point can be earned if the level of reuse is at least 10 percent. If significant MEP system components are reused, their value should be included in the computation; otherwise, the calculation would exclude the MEP systems.

MR Credit 4 (MRc4): Recycled Content (2 Points Maximum)

The use of recycled-content building materials provides up to 2 points in the LEED building assessment process. One point is achieved if the total recycled-content value of the building materials (calculated as the percentage of postconsumer recycled content plus half of the percentage of preconsumer recycled content) is at least 10 percent. A second point is achieved if the total recycled-content value is at least 20 percent. The value of the materials is used in these calculations, not their weight. Mechanical and electrical systems are excluded, but plumbing systems may be included.

MR Credit 5 (MRc5): Regional Materials (2 Points Maximum)

Placing an emphasis on local or regional materials reduces the transportation impacts associated with the life-cycle impacts of the materials. Two points are achievable for this credit. If at least 10 percent of the total value of the materials and products in the project was extracted, harvested, and manufactured within 500 miles of the project site, 1 point is awardable. If 20 percent of the value of these materials was extracted, harvested, and manufactured within the same distance, a second point can be achieved.

MR Credit 6 (MRc6): Rapidly Renewable Materials (1 Point Maximum)

For the purposes of LEED, rapidly renewable materials are defined as those that are derived from plants with a total cycle of growth and harvesting that is 10 years or less. To obtain the point associated with this credit, 2.5 percent of the total materials value used in the project must be from rapidly renewable materials. Materials best considered for this credit include, but are not limited to, bamboo, wool, cotton insulation, agrifiber, linoleum, wheatboard, strawboard, and cork.

MR Credit 7 (MRc7): Certified Wood (1 Point Maximum)

The use of certified wood has been established as one of the key criteria for green buildings. This credit is awarded if a minimum of 50 percent of the wood-based materials and products in the project are certified and have the seal of a Forest Stewardship Council (FSC) certifier. In the context of LEED-NC, wood-based products include structural and general framing; flooring; finishes; furnishings; and temporary structures for formwork, bracing, and pedestrian barriers that are not rented. Initially, for wood products, it was anticipated that only FSC would satisfy the top-tier certification requirements for this credit at this stage. However, as other certification systems for wood improve and as certification systems for bamboo, cork, or agricultural products emerge that meet the USGBC criteria, in the future they may be included under MRc7.

THE INDOOR ENVIRONMENTAL QUALITY (IEQ) CATEGORY

The LEED-NC category of Indoor Environmental Quality (IEQ) has two prerequisites and eight credits that provide a maximum of 15 points. Table 5.9 gives an overview of the EQ structure of LEED-NC.

IEQ Prerequisite 1 (IEQp1): Minimum Indoor Air Quality Performance

This prerequisite establishes that a minimal level of indoor air quality (IAQ) performance must be demonstrable by meeting the requirements of ASHRAE 62.1-2007, Ventilation for Acceptable Indoor Air Quality, and mechanical ventilation systems must be designed using either the Ventilation Rate Procedure or local codes, whichever is more stringent. Naturally ventilated spaces shall comply with ASHRAE 62.1-2007, paragraph 5.1.

TABLE 5.9

Indoor Environmental Quality (IEQ) Credits and Points under LEED-NC 3.0

Prerequisite/Credit	Name of Prerequisite/Credit	Maximum Points
IEQ Prerequisite 1	Minimum Indoor Air Quality Performance	NA
IEQ Prerequisite 2	Environmental Tobacco Smoke (ETS) Control	NA
IEQ Credit 1	Outdoor Air Delivery Monitoring	1
IEQ Credit 2	Increased Ventilation	1
IEQ Credit 3.1	Construction Indoor Air Quality Management Plan—During Construction	1
IEQ Credit 3.2	Construction Indoor Air Quality Management Plan—Before Occupancy	1
IEQ Credit 4.1	Low-Emitting Materials—Adhesives and Sealants	1
IEQ Credit 4.2	Low-Emitting Materials—Paints and Coatings	1
IEQ Credit 4.3	Low-Emitting Materials—Flooring Systems	1
IEQ Credit 4.4	Low-Emitting Materials—Composite Wood and Agrifiber Products	1
IEQ Credit 5	Indoor Chemical and Pollutant Source Control	1
IEQ Credit 6.1	Controllability of Systems—Lighting	1
IEQ Credit 6.2	Controllability of Systems—Thermal Comfort	1
IEQ Credit 7.1	Thermal Comfort—Design	1
IEQ Credit 7.2	Thermal Comfort—Verification	1
IEQ Credit 8.1	Daylight and Views—Daylight	1
IEQ Credit 8.2	Daylight and Views—Views	1
	Total IEQ Points Available	**15**

IEQ Prerequisite 2 (IEQp2): Environmental Tobacco Smoke (ETS) Control

This prerequisite provides three options to prevent or minimize exposure of building occupants, indoor surfaces, and ventilation air distribution systems to environmental tobacco smoke (ETS). The first option is to prohibit smoking in the building and locate designated external smoking areas at least 25 feet (7.6 meters) from entries, operable windows, and air intakes. The second option is the same as the first except, if there is a designated smoking room, the design must ensure no ETS infiltration into the rest of the building. The final option is also the same as the first option, but it applies to residential buildings. Smoking is to be prohibited in common areas, no transmission of ETS between units must be ensured, and all doors to interior corridors must be weatherstripped to minimize air leakage into corridors.

IEQ Credit 1 (IEQc1): Outdoor Air Delivery Monitoring (1 Point Maximum)

Ventilation rates are positively correlated with good IAQ, and this credit addresses the issue of monitoring the ventilation rate to ensure occupant comfort and well-being. For the point associated with this credit to be earned, permanent monitoring equipment is required to ensure that the design ventilation rate is maintained. For mechanically ventilated spaces, carbon dioxide concentrations for normally occupied spaces must be measured at a level 3 to 6 feet (0.9 to 1.8 meters) above the floor, and each mechanical ventilation system serving non-densely occupied spaces [less than 25 people per 1000 square feet (93 square meters)] must be equipped with an airflow measuring device to ensure a minimum outdoor airflow rate with an accuracy of ± 15 percent of the design minimum outdoor air rate as stated in the requirements of ASHRAE 62.1-2007. For naturally ventilated spaces, monitor carbon dioxide concentrations at locations within the room 3 to 6 feet above the floor. Configure all monitoring systems to generate an alarm when the conditions vary by 10 percent or more from a set point, via either a building automation system alarm to the building operator or a visual/audible alert to occupants.

IEQ Credit 2 (IEQc2): Increased Ventilation (1 Point Maximum)

Increasing the ventilation rate can improve IAQ. Increasing the ventilation rate to at least 30 percent above ASHRAE 62.1-2007 for mechanically ventilated spaces is required to earn the point associated with this credit. Natural ventilation systems can also be used to earn this credit, provided that they meet the recommendations in the Carbon Trust Good Practice Guide 237.

IEQ Credit 3 (IEQc3): Construction Indoor Air Quality Management Plan (2 Points Maximum)

This credit has two subcredits, which are described below.

EQ Credit 3.1 (IEQc3.1): Construction Indoor Air Quality Management Plan—During Construction (1 Point Maximum)

This credit addresses measures employed during the construction process that enhance air quality for the eventual building occupants. The requirements are that absorptive materials must be protected from moisture damage; filters with a Minimum Efficiency Reporting Value (MERV) of 8 must be used on return air grilles if the HVAC system is used during construction; all filters must be replaced prior to occupancy; and, during installation of the HVAC system, the approaches recommended in the SMACNA "IAQ Guidelines for Occupied Buildings under Construction" (2nd Edition, 2007) must be followed.

IEQ Credit 3.2 (IEQc3.2): Construction Indoor Air Quality Management Plan—Before Occupancy (1 Point Maximum)

This credit addresses the period of time between the end of construction and building occupancy and has two options. The first option is a building flush-out prior to occupancy but with all interior finishing installed. The flush-out consists of 14,000 cubic feet (396 cubic meters) of outside air per square foot of floor area. The internal building temperature must be at least 60°F (16°C) with a relative humidity no higher than 60 percent. Alternatively, if occupancy is desired prior to completion of flush-out, the building must be flushed out with 3500 cubic feet (99 cubic meters) of outdoor air per square foot prior to occupancy and then ventilated at the rate of 0.30 cubic foot (0.01 cubic meter) per minute of outside air, or the design minimum from EAp1, whichever is greater. The second option is to conduct an air quality test after construction ends and prior to occupancy using the US Environmental Protection Agency (EPA) Compendium of Methods for the Determination of Air Pollutants in Indoor Air. As long the air quality test shows that contaminant levels do not surpass their maximum concentration, then the point will be allocated. The concentrations of formaldehyde, total volatile organic compounds (TVOC), particulates (PM10), 4-phenylcyclohexene (4-PCH), and carbon monoxide (CO), must be measured when using the US EPA procedure.

IEQ Credit 4 (IEQc4): Low-Emitting Materials (4 Points Maximum)

Reducing the volatile organic compounds (VOCs) emitted by building materials is addressed by this credit, which is subdivided into the four subcredits described below.

IEQ Credit 4.1 (IEQc4.1): Low-Emitting Materials—Adhesives and Sealants (1 Point Maximum)

Adhesives, sealants, and sealant primers must comply with the VOC content limits of the South Coast Air Quality Management District (SCAQMD) Rule 1168. Aerosol adhesives must comply with the Green Seal Standard for Commercial Adhesives GS-36 requirements.

IEQ Credit 4.2 (IEQc4.2): Low-Emitting Materials—Paints and Coatings (1 Point Maximum)

Architectural paints, coatings, and primers used in the interior of the building must show they meet the requirements of Green Seal GS-11 with flat paints having no more than 50 grams per liter of VOCs and nonflat paints not exceeding 150 grams per liter. Anticorrosive and antirust paints applied to interior metal surfaces should not exceed 250 grams per liter of total VOC content as stated in Green Seal GC-03, Anti-Corrosive Paints. Similarly, clear wood finishes should not exceed 350 grams per liter for varnish and 550 grams per liter for lacquer; floor coatings must not exceed 100 grams per liter; stains less than 250 grams per liter; and sealers less than 275 grams per liter.

IEQ Credit 4.3 (IEQc4.3): Low-Emitting Materials—Flooring Systems (1 Point Maximum)

All carpet systems must meet the requirements of the Carpet and Rug Institute's Green Label Plus Indoor Air Quality Program, and carpet adhesives must contain no more than 50 grams per liter of total VOCs. All hard-surface flooring must be compliant with the FloorScore standard. Flooring products covered by FloorScore include vinyl, linoleum, laminate flooring, wood flooring, ceramic flooring, rubber flooring, wall base, and associated sundries.

IEQ Credit 4.4 (IEQc4.4): Low-Emitting Materials—Composite Wood and Agrifiber Products (1 Point Maximum)

Composite wood and agrifiber products must contain no urea formaldehyde (UF) resins. This also applies to the laminating adhesives used to fabricate composite wood and agrifiber assemblies.

IEQ Credit 5 (IEQc5): Indoor Chemical and Pollutant Source Control (1 Point Maximum)

Designing the building to minimize the entry of pollutants into occupied spaces is the purpose of this credit. First, it calls for the employment of devices such as 10-foot (3-meter) grilles and grates at building entrances to prevent the entrance of dirt, pesticides, and other materials into the building. Rollout mats may be used with a weekly cleaning contract and as appropriate for the climate. Second, when hazardous gases and chemicals may be present or used, a segregated area with a dedicated exhaust system must be provided. The segregated area must be negatively pressurized and provide self-closing doors and deck-to-deck partitions. Third, in mechanically ventilated buildings, MERV 13 filters or better should be installed prior to occupancy to process both return and outside air.

IEQ Credit 6 (IEQc6): Controllability of Systems (2 Points Maximum)

The ability of the building occupants to control their lighting conditions and thermal comfort has emerged as important issues in providing a high-quality indoor environment. This credit comprises two subcredits, each carrying 1 point maximum: IEQc6.1 for lighting and IEQc6.2 for thermal comfort.

IEQ Credit 6.1 (IEQc6.1): Controllability of Systems—Lighting (1 Point Maximum)

The design must allow at least 90 percent of the occupants to adjust the lighting for their tasks and preferences. For multioccupant spaces, provide lighting system controllability to suit individual tasks and needs.

IEQ Credit 6.2 (IEQc6.2): Controllability of Systems—Thermal Comfort (1 Point Maximum)

The design must allow at least 50 percent of the occupants to adjust the temperature to suit their needs. Operable windows provided within 20 feet (6.1 meters) of occupants or 10 feet (3 meters) to either side can be used in lieu of controls. The areas of operable window must meet the requirements of ASHRAE 62.1-2007. For shared multioccupant spaces, controls are to be provided that allow conditions to be adjusted for the group's needs. ASHRAE 55-2004, Thermal Environmental Conditions for Human Occupancy, defines conditions for thermal comfort, including air temperature, air speed, humidity, and radiant temperature.

IEQ Credit 7 (IEQc7): Thermal Comfort (2 Points Maximum)

Thermal comfort is an important component of indoor environmental quality (IEQ), and the project can acquire 2 points for demonstrating that the criteria for this measure have been met. IEQ Credit 7 has two subcredits as follows, each carrying 1 possible point.

IEQ Credit 7.1 (IEQc7.1): Thermal Comfort—Design (1 Point Maximum)

The project HVAC systems and building envelope must be designed to meet the requirements of ASHRAE 55-2004, Thermal Environmental Conditions for Human Occupancy.

IEQ Credit 7.2 (IEQc7.2): Thermal Comfort—Verification (1 Point Maximum)

The owner must agree to perform a thermal comfort survey of the occupants 6 to 18 months after occupancy. A plan for corrective action must be provided and implemented if the survey indicates that more than 20 percent of the occupants are dissatisfied with the building's thermal comfort.

IEQ Credit 8 (IEQc8): Daylight and Views (2 Points Maximum)

The importance of daylighting and views to the outside is acknowledged by this credit, which is divided into two subcredits, each carrying 1 possible point.

IEQ Credit 8.1 (IEQc8.1): Daylight and Views—Daylight (1 Point Maximum)

There are three options for this credit. The first option requires a computer simulation capable of identifying the various lighting levels within the building. In order to receive a point, 75 percent of all spaces regularly occupied for critical visual tasks must achieve a minimum daylight illumination of 25 foot-candles (fc) and a maximum of 500 fc under clear sky conditions on September 21 at both 9:00 a.m. and 3:00 p.m. The second option is to implement a prescriptive method that uses a combination of site lighting and/or top lighting to achieve a total daylighting zone that is at least 75 percent of all the regularly occupied spaces. The final option is to use physical indoor light measurements that are taken on a 10-foot grid to demonstrate that the daylighting levels of at least 10 foot candles have been achieved. A combination of these options may also be used to provide documentation that the minimum daylight illumination is available for at least 75 percent of occupied spaces.

IEQ Credit 8.2 (IEQc8.2): Daylight and Views—Views (1 Point Maximum)

This credit requires a direct line of sight for occupants to Vision glazing in 90 percent of all regularly occupied spaces. Vision glazing is glazing that is between 3 and 90 inches above the floor. For private offices, the entire area of the office can be counted if 75 percent or more of the area has a direct line of sight. For multioccupant spaces, the actual area with a direct line of sight to perimeter vision glazing is counted.

THE INNOVATION IN DESIGN (ID) CATEGORY

In order to open the door to creative solutions that are not addressed in the LEED rating system, a category of points called *Innovation in Design* was established to provide credit for new thinking and ideas. Additionally, credit for the participation of an individual with LEED AP credentials is also included in this section.

ID Credit 1 (IDc1): Innovation in Design (5 Points Maximum)

These points can be achieved by following one or a combination of two separate options. The first option can be awarded points based on novel approaches to green building. In order to qualify as an innovation, the following must be in writing:

- The intent of the proposed innovation credit
- The proposed requirement for compliance
- The proposed submittals to demonstrate compliance
- The design approach (strategies) used to meet the requirements

The second option can receive up to 3 points through exemplary performance. Points are available by achieving double the credit requirements and/or achieving the next incremental percentage threshold of an existing credit in LEED. For example, an exemplary performance point is available if a project achieves a recycled content of 30 percent of the total value of materials in the project, which is a 10 percent increment over the maximum of 20 percent addressed in the MRc4 credit..

ID Credit 2 (IDc2): LEED Accredited Professional (1 Point Maximum)

At least one principal participant of the project team shall be a LEED Accredited Professional (AP).

Case Study: The Heavener Football Complex, University of Florida, Gainesville

The Heavener Football Complex is one of 27 buildings on the University of Florida campus that have been LEED certified and is the first building on the campus certified to the platinum level. It is also the first athletic facility in the United States certified to the platinum level. The project consists of 62,000 square feet of both renovation and new construction, at a cost of approximately $22 million. The expansion includes offices, meeting rooms, a weight room, an interactive exhibition/reception area, the Gator Room, and support space for the University of Florida's football program (see Figure 5.7).

Direct solar exposure and radiant heat gain are important factors considered in the design of buildings for Florida's climate. Highly reflective paving and roofing materials are installed on the walkway and roof of the facility to reduce the heat island effect. High-efficiency glazing allows daylighting while preventing solar heat gain within the Gator Room. Water consumption and stormwater management are other important factors that were considered in the selection of low-flow plumbing fixtures, a green roof, native plant landscaping, and an efficient drip irrigation system. The irrigation system uses 100 percent reclaimed water from the campus wastewater treatment plant. Integration of all these water-related technologies resulted in an approximately 40 percent reduction in water consumption.

The energy performance of the Heavener Football Complex is projected to be 25 percent lower than the requirements of the Florida Energy Code. A building energy management system (EMS) is used to control the schedules of both HVAC equipment and lighting systems. Demand-controlled ventilation (DCV) was installed along with variable-frequency drives (VFDs) for various pump and fan systems. Lighting efficiency was achieved through a detailed analysis of lighting power density requirements and the installation of occupancy sensors and individual lighting controls to suit individual task needs and preferences. Over 70 percent of the electrical energy requirements were offset through a two-year contract with a local utility company that provides green power.

The project was able to maintain 75 percent of the shell of an existing building. A total of 78 percent of construction and demolition waste was diverted from being landfilled. Over 40 percent of the materials, such as the carpeting and weight room flooring, have recycled and recyclable content (see Figure 5.8). Enhanced materials efficiency was achieved through the use of locally sourced material.

Good indoor environmental quality (IEQ) has been achieved through the use of low-emitting materials, paints, adhesives, sealants, carpet, and composite wood during the construction process. A detailed construction materials-handling program

Figure 5.7 (A) The main entrance to the Heavener Football Complex at the University of Florida in Gainesville. (B) The main entrance to the football complex includes windows with high-efficiency glazing, and reflective sidewalks, which reduce the heat island effect. (C) The Gator Room is used for sports recruiting. Energy-efficient lighting, HVAC, and other high-efficiency equipment have been installed to reduce the electrical load. [(A) D. Stephany. (B–C) Photographs courtesy of Kun Zhang, Dimension Images]

was enforced to ensure all equipment and materials were protected from dust and moisture. When the construction had reached substantial completion, the indoor air was tested to ensure good air quality. Walk-off door mats at all main entrances help decrease pollutants from entering the building. Carbon dioxide monitoring as well as integrated thermal and lighting controls helps provide individual comfort levels year-round.

The scorecard indicating the LEED points achieved by the project team for the Heavener Football Complex is shown in Figure 5.9.

Figure 5.8 Carpeting and other flooring materials in the complex contain materials that are recycled and highly recyclable. (Photograph courtesy of Kun Zhang, Dimension Images)

12 Points	Sustainable Sites	
Y	Prereq 1	**Construction Activity Pollution Prevention**
1	Credit 1	**Site Selection**
1	Credit 2	**Development Density and Community Connectivity**
1	Credit 3	**Brownfield Redevelopment**
1	Credit 4.1	**Alternative Transportation**, Public Transportation Access
1	Credit 4.2	**Alternative Transportation**, Bicycle Storage and Changing Rooms
1	Credit 4.3	**Alternative Transportation**, Low-Emitting and Fuel-Efficient Vehicles
1	Credit 4.4	**Alternative Transportation**, Parking Capacity
1	Credit 5.1	**Site Development**, Protect or Restore Habitat
	Credit 5.2	**Site Development**, Maximize Open Space
1	Credit 6.1	**Stormwater Design**, Quantity Control
1	Credit 6.2	**Stormwater Design**, Quality Control
1	Credit 7.1	**Heat Island Effect**, Nonroof
1	Credit 7.2	**Heat Island Effect**, Roof
0	Credit 8	**Light Pollution Reduction**

5 Points		Water Efficiency	
1	Credit 1.1	**Water Efficient Landscaping**, Reduce by 50%
1	Credit 1.2	**Water Efficient Landscaping**, No Potable Use or No Irrigation
1	Credit 2	**Innovative Wastewater Technologies**
1	Credit 3.1	**Water Use Reduction**, 20% Reduction
1	Credit 3.2	**Water Use Reduction**, 30% Reduction

12 Points		Energy and Atmosphere	
Y	Prereq 1	**Fundamental Commissioning of Building Energy Systems**
Y	Prereq 2	**Minimum Energy Performance**
Y	Prereq 3	**Fundamental Refrigerant Management**
8	Credit 1	**Optimize Energy Performance**
0	Credit 2	**On-Site Renewable Energy**
1	Credit 3	**Enhanced Commissioning**
1	Credit 4	**Enhanced Refrigerant Management**
1	Credit 5	**Measurement and Verification**
1	Credit 6	**Green Power**

7 Points		Materials and Resources	
Y	Prereq 1	**Storage and Collection of Recyclables**
1	Credit 1.1	**Building Reuse**, Maintain Existing Walls, Floors, and Roof
0	Credit 1.2	**Building Reuse**, Maintain Existing Interior Nonstructural Elements
2	Credit 2	**Construction Waste Management**
0	Credit 3	**Materials Reuse**
2	Credit 4	**Recycled Content**
2	Credit 5	**Regional Materials**
0	Credit 6	**Rapidly Renewable Materials**
0	Credit 7	**Certified Wood**

11 Points		Indoor Environmental Quality	
Y	Prereq 1	**Minimum Indoor Air Quality Performance**
Y	Prereq 2	**Environmental Tobacco Smoke (ETS) Control**
1	Credit 1	**Outdoor Air Delivery Monitoring**
0	Credit 2	**Increased Ventilation**
1	Credit 3.1	**Construction Indoor Air Quality Management Plan**, During Construction
1	Credit 3.2	**Construction Indoor Air Quality Management Plan**, Before Occupancy
1	Credit 4.1	**Low-Emitting Materials**, Adhesives and Sealants
1	Credit 4.2	**Low-Emitting Materials**, Paints and Coatings
1	Credit 4.3	**Low-Emitting Materials**, Flooring Systems
1	Credit 4.4	**Low-Emitting Materials**, Composite Wood and Agrifiber Products
1	Credit 5	**Indoor Chemical and Pollutant Source Control**
1	Credit 6.1	**Controllability of Systems**, Lighting
1	Credit 6.2	**Controllability of Systems**, Thermal Comfort

1	Credit 7.1	**Thermal Comfort**, Design
0	Credit 7.2	**Thermal Comfort**, Verification
0	Credit 8.1	**Daylight & Views**, Daylight
0	Credit 8.2	**Daylight & Views**, Views

5 Points	**Innovation in Design**	
4	Credit 1	**Innovation in Design**
1	Credit 2	**LEED Accredited Professional**

52 Total Points LEED-NC 2.2 Ratings

Certified 26-32 points   **Silver** 33-38 points   **Gold** 39-51 points
Platinum 52-69 points

Figure 5.9   The LEED-NC 2.2 scorecard for the Heavener Football Complex indicates the project achieved 52 points, the minimum required for a LEED platinum rating.

THE REGIONAL PRIORITY (RP) CATEGORY

The project can earn 1 to 4 points of the six Regional Priority credits identified by the USGBC regional councils and chapters as having environmental importance for a project's region.[2] A Regional Priority credit is not a new credit as such; it is simply a credit in one of the existing LEED categories, for example, Water Efficiency, which can achieve an additional point. A project in Gainesville, Florida, would be given an extra point for achieving a water use reduction of 40 percent for WEc3, because the local USGBC chapter selected this as one of the six priority areas for this region. The Regional Priority credits are available on a spreadsheet that is downloadable from the USGBC website.

Summary and Conclusions

The LEED building assessment system is a suite of rating products that can be used to guide the life-cycle design, construction, and operation of a high-performance green building. The current version of the LEED assessment system is version 3.0, with the next version, LEED v4, under development. The various LEED rating products provide a wide variety of options from which the project team can select the one most suitable for the situation. A successful certification process results in one of four levels of certification: platinum, gold, silver, and certified. Construction industry professionals who are engaged in green building design and construction can earn a variety of LEED credentials, ranging from LEED Green Associate, to LEED Accredited Professional with specialty, to LEED Fellow. The number of building projects certified by the GBCI as being LEED certified is growing at an exponential rate, and it is expected that by 2015 almost half of all new, nonresidential buildings will be green buildings with the vast majority undergoing LEED certification.

Notes

1. The exclusive website for purchasing LEED plaques is www.greenplaque.com.
2. A database of Regional Priority credits and their geographic applicability is available on the USGBC website at www.usgbc.org.

Chapter 6

The Green Globes Building Assessment System

Green Globes is a building assessment system with roots in Canada that is making inroads in the United States as an alternative to the US Green Building Council's (USGBC) Leadership in Energy and Environmental Design (LEED).[1] It provides a rating of one to four green globes for building projects, depending on the percentage of the maximum points that the project actually achieves (see Figure 6.1). The Green Building Initiative (GBI), the US proponent of Green Globes, describes the Green Globes building assessment system as ". . . a revolutionary green management tool that includes an assessment protocol, a rating system and a guide for integrating environmentally friendly design into commercial buildings." When the assessment protocol has been completed, it also facilitates recognition of the project through third-party verification. It is designed to be an interactive, flexible, and affordable approach to environmental design and building assessment.

The Green Globes building rating system represents more than 15 years of research and refinement by a wide range of prominent international organizations and experts. The genesis of the system was the Building Research Establishment Environmental Assessment Method (BREEAM) in the United Kingdom, which was imported into Canada in 1996. In the same year, the Canadian Standards Association published BREEAM Canada for Existing Buildings. In 2004, the GBI acquired the rights to distribute Green Globes in the United States. The GBI mission is to accelerate the adoption of building practices that result in energy-efficient, healthier, and environmentally sustainable buildings by promoting credible and practical green building approaches. The GBI is committed to continually refining the system to ensure that it reflects changing opinions and ongoing advances in research and technology, as well as involving multiple stakeholders in an open and transparent process.

In 2005, the GBI became the first green building organization to be accredited as a standards developer by the American National Standards Institute (ANSI) and began the process of establishing Green Globes as an

85–100%		Reserved for select building designs which serve as national or world leaders in energy and environmental performance. The project introduces design practices that can be adopted and implemented by others.
70–84%		Demonstrates leadership in energy and environmental design practices and a commitment to continuous improvement and industry leadership.
55–69%		Demonstrates excellent progress in achieving eco-efficiency results through current best practices in energy and environmental design.
35–54%		Demonstrates movement beyond awareness and commitment to sound energy and environmental design practices by demonstrating good progress in reducing environmental impacts.

Figure 6.1 The Green Globes rating levels are based on the percentage of points achieved compared to the maximum available. Achieving a minimum of 35 percent of the available points would provide a certification level of one green globe. (Diagram courtesy of the Green Building Initiative, Inc.)

official ANSI standard. The GBI ANSI technical committee was formed in early 2006 to develop an environmental design and assessment protocol for commercial building under new construction. This protocol, approved in March 2010, is ANSI/GBI 01-2010, Green Building Assessment Protocol for Commercial Buildings.

The Green Globes Process

Green Globes offers building assessment and certification for new buildings, major renovations, additions, and existing buildings. A variety of facilities, including government, education, health-care, industrial, multiunit residential, retail, office, and corporate facilities, have used either the Green Globes New Construction (NC) or the Green Globes Continual Improvement of Existing Buildings (CIEB) rating system. Once the project has been registered for Green Globes certification, a Green Design Facilitator (GDF) must be appointed to begin an internal Green Globes assessment by the project team. The role of the GDF is to outline the overall green design framework for the project by answering a logical sequence of questions that guides the project team in integrating important elements of sustainability (see Figure 6.2). The professional best suited to filling the GD Frole would be a Green Globes Professional (GGP). GGPs are individuals who receive training at the Green Globes user level and are qualified to offer project management and technical support to clients undergoing the building assessment and certification process. GGPs also can assist the project team in developing measurable green design performance requirements to satisfy the overall objectives of the project.

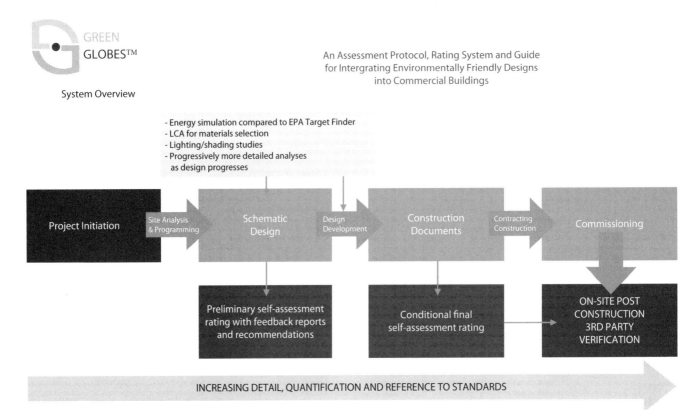

Figure 6.2 Overview of the Green Globes assessment protocol showing the assessment activities at each of the project stages. (Diagram courtesy of the Green Building Initiative, Inc.)

The Green Globes Construction Documents Questionnaire is the foundation of the rating process. However, to benefit fully from the value-added design assistance features of the system—and to obtain a preliminary self-assessment of a building—the project should be registered and the preliminary and subsequent questionnaires should be completed. However, a building cannot be promoted as Green Globes certified until the final assessment site visit and verification report have been completed.

Green Globes Verification and Certification

A Green Globes Assessor (GGA) conducts an extensive third-party assessment of the project by thoroughly reviewing construction documents and conducting a site visit to the project. The site visit consists of interviewing various project team members, reviewing any outstanding documentation, and a walk-through of the project to review the sustainability features and measures selected by the project team. The GGA is an experienced building design and construction industry professional who has been trained on the Green Globes protocol and who has been monitored and mentored by other GGAs prior to assessing a building on her or his own. In the USGBC LEED process, the project team completes documentation online and submits it via LEED-Online, to be reviewed by a team that is not at any time in direct contact with the project team. Unlike LEED, the Green Globes system requires a GGA to actually visit the project, interact directly with the project team, and physically examine the project. At the end of this stage, the GGA verifies point allocation and sends his or her recommendation to the GBI concerning the appropriate certification level. This information is then communicated to the project leaders in the final certification report.

Each environmental assessment area of Green Globes may have criteria that a design and delivery team may deem to be inapplicable to the building. This is an important feature to enable the standard protocol to apply to a wide range of building types and geographic climate zones, and for buildings where codes or regulations may prohibit implementation of specific building enhancement items. In this type of situation, the points do not count, and the total number of achievable points is adjusted accordingly, still requiring certification levels to fall under identified percentages. This approach is different from that of LEED in that it focuses on the work of the project team instead of addressing issues that are outside of the project scope.

STRUCTURE OF GREEN GLOBES

The structure and point allocation of the Green Globes New Construction rating system is shown in Table 6.1. The table also includes the intent for each of the subcategories. Note that in Green Globes, although 1000 points are achievable, points can be indicated as "Not Applicable (NA)" and the total achievable points are reduced by the points designated as NA. In the case study that follows, the project had 58 points indicated as "NA" and the result was that the basis for comparison was reduced from 1000 to 942 points. In contrast, the LEED building rating systems do not permit points being indicated as "NA."

Structure of the ANSI/GBI 01-2010 Standard

ANSI/GBI 01-2010, Green Building Assessment Protocol for Commercial Buildings, is an adaptation of the Green Globes New Construction building

TABLE 6.1

Structure of the Green Globes New Construction Rating System (Total Available Points = 1000)

	Points	Description
A.	**50**	**Project Management—Policies and Practice (5%)**
A.1	20	Integrated Design Process
		To meet the environmental and functional priorities and goals of the project in an effective and cost-efficient manner
A.2	10	Environmental Purchasing
		To select materials, products, and equipment that have minimal impact on the environment in terms of resource use, production of waste, and energy use
A.3	15	Commissioning Plan—Documentation
		To design, construct, and calibrate building systems so they operate as intended
A.4	5	Emergency Response Plan
		To minimize the risk of injury and the environmental impact of emergency incidents
B.	**115**	**Site (11.5%)**
B.1	30	Development Area
		To protect important land uses, lower demands on municipal infrastructure services, and reduce the impact on the site's biodiversity
B.2	30	Minimize Ecological Impacts
		EROSION CONTROL—To avoid the negative effect of erosion on air and water quality and to maintain the ecological integrity of the site
		REDUCED HEAT ISLAND EFFECT—To minimize impact on the microclimate and habitat
		MINIMAL LIGHT POLLUTION—To reduce the impact on the nocturnal environment of fauna and flora
B.3	15	Enhancement of Watershed Features
		To reduce the quantity of stormwater runoff entering storm sewers and increase ground infiltration of stormwater without negatively affecting the building or on-site vegetation
B.4	40	Enhancement of Site Ecology
		To increase the natural biodiversity of the site
C.	**380**	**Energy (38%)**
C.1	100	Building Energy Performance
		To minimize the energy consumption for building operations
C.2	114	Energy Demand Minimization
		SPACE OPTIMIZATION—To achieve efficient utilization of space, minimize the amount of space that will need to be heated or cooled, and provide flexibility for future occupant growth
		RESPONSE TO MICROCLIMATE AND TOPOGRAPHY—To take advantage of site and microclimate opportunities to reduce energy requirements for heating, cooling, and ventilation
		INTEGRATION OF DAYLIGHTING—To reduce the need for electrical lighting
		BUILDING ENVELOPE—To minimize the energy that is gained or lost through the envelope, prevent condensation, and avoid water damage
		INTEGRATION OF ENERGY SUBMETERING—To encourage energy efficiency by monitoring energy consumption
C.3	66	Energy-Efficient Systems
		To reduce energy needed for building systems and equipment
C.4	20	Renewable Sources of Energy
		To minimize the consumption of nonrenewable energy resources and to minimize greenhouse gas emissions
C.5	80	Energy-Efficient Transportation
		To reduce fossil fuel consumption for commuting
D.	**85**	**Water (8.5%)**
D.1	30	Water Performance
		To maximize water efficiency and reduce the burden on municipal supply and treatment systems

D.2   45   Water-Conserving Features
SUBMETERING—To encourage water conservation by measure and monitoring water consumption
INTEGRATION OF WATER-EFFICIENT EQUIPMENT—To minimize the burden on municipal water supply and wastewater treatment systems
MINIMAL USE OF IRRIGATION WATER—To eliminate the use of potable water required for landscape irrigation

D.3   10   Minimization of Off-Site Treatment of Water
To reduce the burden on municipal water supply and wastewater systems

E.   100   Resources, Building Materials, and Solid Waste (10%)
E.1   35   Systems and Materials with Low Environmental Impact
To select materials with the lowest life-cycle environmental burden and embodied energy

E.2   16   Materials That Minimize Consumption of Resources
To conserve resources and minimize the energy and environmental impact of extracting and processing nonrenewable materials

E.3   20   Reuse of Existing Structures
To conserve resources and minimize the energy and environmental impact of extracting and processing nonrenewable materials

E.4   14   Building Durability, Adaptability, and Disassembly
To extend the life of a building and its components and to conserve resources by minimizing the need to replace materials and assemblies

E.5   5   Reuse and Recycling of Construction/Demolition Waste
To divert demolition waste from the landfill

E.6   10   Facilities for Recycling and Composting
To minimize landfill waste generated by occupants

F.   70   Emissions and Effluents (7%)
F.1   15   Minimization of Air Emissions
To minimize air emissions

F.2   25   Minimization of Ozone Depletion
To minimize the emission of ozone-depleting substances

F.3   5   Avoid Contamination of Sewers or Waterways
To avoid contamination of waterways and reduce the burden on municipal wastewater treatment facilities

F.4   25   Pollution Minimization
To minimize risk to occupants' health and impacts on the local environment

G.   200   Indoor Environment (20%)
G.1   55   Ventilation
To provide effective ventilation, thereby helping to ensure occupant well-being and comfort

G.2   50   Source Control of Indoor Pollutants
To minimize contaminants in the indoor air, thereby helping to ensure occupant well-being and comfort

G.3   45   Lighting
DAYLIGHTING—To provide occupants with exposure to natural light, thereby helping to ensure their well-being and comfort
LIGHTING DESIGN—To reduce the energy needed for electrical lighting

G.4   20   Thermal Comfort
To provide a thermally comfortable environment, thereby helping to ensure the well-being and comfort of occupants

G.5   30   Acoustic Comfort
To provide a good acoustic environment, thereby helping to ensure the well-being and comfort of occupants

assessment system into a standard that will eventually become the new version of the Green Globes assessment system. Although its goal and scope remain the same, a more rigorous definition of requirements has been created to provide clear indicators to both the design team and the GBI of the requirements for Green Globes certification. The major differences between the standard and the current rating system include a restructuring of the point system as well as a

TABLE 6.2

Minimum Allowable Percentages That Must Be Fulfilled According to the ANSI/GBI 01-2010 Standard

Environmental Assessment Subject	Total Points Available	Minimum Percentage of Points
Project Management	100	50%
Site	120	24%
Energy Performance Path (A)	300	50%
Prescriptive Path (B)	250	33%
Water	130	26%
Resources/Materials	145	29%
Emissions	45	9%
Indoor Environment	160	32%

required minimum percentage of total possible points for each environmental subject. Minimal percentages are assigned for each environmental assessment subject, as shown in Table 6.2. In addition, the energy section has been split to allow either a performance or a prescriptive path that the project team can follow. The energy section also introduces carbon equivalency measures that are used in combination with energy performance goals.

Currently, the ANSI/GBI 01-2010 standard is not supported by Green Globes online tools. GBI is currently offering a limited number of pilot program buildings to be assessed under the new standard. Once the pilot program is completed, the standard will become available online as a Web-based product and be used as the new standard assessed by GBI for new construction and major building renovation.

The Green Globes Professional and Green Globes Assessor

The Green Globes building assessment system offers a credential similar to the LEED AP described in Chapter 5. The *Green Globes Professional* (GGP) is awarded based on education and industry experience, and its purpose is to train building industry professionals in the Green Globes assessment system so that they can guide the project team through the multistage building certification process. Green Globes also has a second credential, the *Green Globes Assessor* (GGA), which is offered based on education, professional licensure, and relevant industry experience and with the purpose of providing people trained and experienced to the degree that they are able to provide an independent, third-party assessment of a project team's performance in their application for certification. These two credentials and the requirements for their award are shown in Table 6.3. Note that there are two different qualification paths for the GGA, depending on whether the assessor is evaluating new construction (NC) or existing building (CIEB) projects.

Award of the GGP or GGA credential requires that the applicant pass an examination regarding his or her knowledge of the NC or CIEB Green Globes building assessment system. The GGA, in addition to assessing the NC or CIEB project, provides support for the project team and addresses any gray areas that would be covered by the Credit Interpretation Request (CIR) process in LEED.

TABLE 6.3

Qualifications for Application for the Green Globes Professional (GGP) and Green Globes Assessor (GGA) Credential

Qualifications Requirements	Green Globes Professional	Green Globes Assessor CIEB	Green Globes Assessor NC
Industry Experience	5 years	10 years or more total years of applicable industry experience directly pertaining to commercial buildings 5 out of the 10 years of experience must fall within the specific functional areas listed for each assessor designation (CIEB/NC)	
Applicable Functional Categories for Total Industry Experience	Facilities/Operations Management/Maintenance, Architecture/Design/Engineering/Construction, Inspections/Auditing/Appraisals Building Materials/Components/Manufacturing Energy Analysis, Commissioning		
Applicable Categories for Specific Functional Experience	N/A	Facilities Management Operations/Maintenance Architecture/Design Engineering Construction Inspections/Auditing Commissioning	Architecture/Design Engineering Construction Inspections/Auditing Commissioning
Professional Licensure	N/A	N/A	Licensed Architect or Licensed Professional Engineer
Education	N/A	Associates Degree or higher in: Architecture Engineering Facilities/Operations Management Other relevant technical or building science program	Bachelor's Degree or higher in: Architecture Engineering Other relevant technology, science, or environmental program
Building Sustainability in Practice	N/A	Involved in 3 or more projects where building sustainability principles were applied in the areas of energy, water, site, resources/materials, emissions, indoor environment, management	

Case Study: Health Sciences Building, St. Johns River State College, St. Augustine, Florida

The growing campus of St. Johns River State College in St. Augustine, Florida, is phased for construction over the next few decades. The college planning goals require any new construction to be designed in an environmentally sensible manner, which led to the newly constructed Health Sciences Building receiving a three Green Globe rating from GBI's Green Globes building rating system (see Figure 6.3).

An integrated project management team was organized early in the design process to help identify the basis of design, prioritize goals, and create effective policies that allowed design and construction professionals to get on board in developing a high-performance building. Green products were specified for the

Figure 6.3 The Health Sciences Building at St. Johns River State College in St. Augustine, Florida, received a rating of three green globes from the GBI's Green Globes building rating system. (Photograph courtesy of Glen Roberts, St. Johns River State College)

Figure 6.4 Bioswales surrounding the parking lot reduce stormwater runoff into nearby wetland areas. (D. Stephany)

project, as was a commissioning plan to verify the integrity of the building's operations before and during occupancy. The gross area is 32,000 square feet, and the project cost was $8.5 million.

Impact on the site and surrounding areas was reduced by avoiding the disturbance of nearby wetlands and through the implementation of an erosion control plan. The heat island in the parking lot was reduced by planting trees, which will shade the parking areas within five years, and through the use of light-colored roofing materials, including a highly reflective ethylene propylene diene monomer (EPDM) membrane. Stormwater runoff was minimized through the design of bioswales, pervious pavers, and a retention pond (see Figure 6.4). Native plants were used in the landscape to minimize irrigation and maintenance requirements.

Establishing an energy consumption target was the first step toward improving building energy efficiency. Once the target was established, a number of energy-efficient approaches were chosen to meet the energy performance goal. These include optimization of the building orientation, incorporation of high-performance glazing, increased thermal resistance in both walls and windows, incorporation of natural

Figure 6.5   A high-efficiency Smardt centrifugal chiller was installed to replace the outdated unit, providing cooling for the entire St. Johns River State College. The chiller is designed to meet future needs as the campus continues to expand. (Photograph courtesy of Glen Roberts, St. Johns River State College)

ventilation, and installation of daylight sensors in perimeter spaces to optimize natural lighting levels in spaces when occupied. High-efficiency heating, ventilation, and air conditioning (HVAC) systems such as a variable air volume (VAV) system, cooling towers secondary chilled-water pumps, and a high-efficiency water-cooled chiller are controlled by variable-frequency drives for optimum energy performance (see Figure 6.5). The water-cooled chiller contains only one moving part, which is frictionless because it has no bearings but instead rotates in the chilled water itself, minimizing friction and resulting in an energy performance of 0.576 kW/ton. The building automation system (BAS) controls the HVAC system, including the zoned VAV boxes, to continually optimize system operation. For zones with low cooling loads, the BAS adjusts the VAV system to reduce conditioned airflow, resulting in a reduction in pump and fan electrical energy consumption. Energy recovery ventilators (ERVs) are used to pretreat the incoming outside air by air-to-air heat and humidity exchange with the building's exhaust air. Solar tubes are used in some of the interior spaces on the second floor to aid in reducing energy consumption where direct sunlight is not available.

Targets were set for water efficiency and achieved by incorporating waterless urinals, low-flow toilets and faucets, and irrigation systems that use reclaimed water from roof and parking surfaces, coupled with drought-tolerant plants. Additionally, pulsed electromagnetic technology was used to clean cooling tower water, resulting in significant water savings (see Figure 6.6).

The project team reduced the impact of the building on the environment by specifying materials that are durable, capable of extending the life of the building, and sourced locally. For example, polished, stained concrete floors were selected instead of carpeting or other floor coverings (see Figure 6.7). Reclaimed materials or materials with recycled content were used wherever possible. In addition, space for recycling was integrated into the design.

The building reduces its air emissions through the elimination of ozone-depleting chemicals. In addition, the project addresses human health by reducing

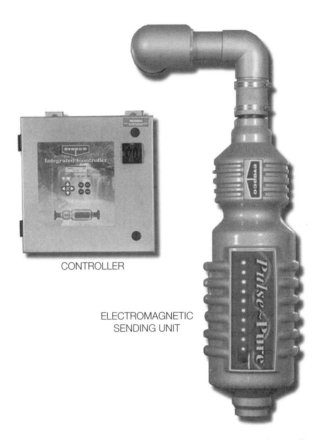

CONTROLLER

ELECTROMAGNETIC
SENDING UNIT

Figure 6.6 Pulsed electromagnetic technology is used to treat the cooling tower water. This technology helps reduce scaling, biological growth, and corrosion, all without the use of chemicals. It also helps reduce water consumption and makes water easier to treat compared to chemically treated systems.

Figure 6.7   Polished, stained concrete floors were specified instead of far less durable floor coverings. These provide a long-lasting, very attractive option and reduce the quantity of materials needed for construction. (Photograph courtesy of Glen Roberts, St. Johns River State College)

harmful chemicals and storing them in a manner that includes adequate ventilation. A vermin prevention plan is integrated into the design, along with a pollution control plan that includes proper gas and chemical prevention measures. Approximately 80 percent of the spaces in the Health Sciences Building are exposed to daylighting (see Figure 6.8). Daylight and occupancy sensors are incorporated to control daylight harvesting and solar exposure. High ventilation rates and zoned thermal controls aid in addressing potential human comfort issues. Acoustics are addressed by specifying an appropriate Sound Transmission Class (STC) for various spaces.

The Health Sciences Building was awarded three Green Globes, similar to a LEED gold rating. The scorecard for this project is shown in Table 6.4.

Figure 6.8   A storage room on the second floor of the Health Sciences Building uses natural light that is projected into the room through solar tubes installed on the roof. (D. Stephany)

TABLE 6.4

The Green Globes assessment of the Health Sciences Building resulted in 71 percent of the available points being achieved, and the project was awarded three Green Globes.

	Description	Points
A.	**Project Management—Policies and Practice (50 Points Applicable)**	**45**
A.1	Integrated Design Process	20
A.2	Environmental Purchasing	10
A.3	Commissioning Plan—Documentation	15
A.4	Emergency Response Plan	0

(Continued)

TABLE 6.4 *(Continued)*

The Green Globes assessment of the Health Sciences Building resulted in 71 percent of the available points being achieved, and the project was awarded three Green Globes.

	Description	Points
B.	**Site (115 Points Applicable)**	**95**
B.1	Development Area	30
B.2	Minimize Ecological Impacts	30
B.3	Enhancement of Watershed Features	15
B.4	Enhancement of Site Ecology	20
C.	**Energy (373 Points Applicable)**	**211**
C.1	Building Energy Performance	30
C.2	Energy Demand Minimization	95
C.3	Energy-Efficient Systems	66
C.4	Renewable Sources of Energy	0
C.5	Energy-Efficient Transportation	20
D.	**Water (81 Points Applicable)**	**67**
D.1	Water Performance	30
D.2	Water-Conserving Features	37
D.3	Minimization of Off-Site Treatment of Water	0
E.	**Resources, Building Materials, and Solid Waste (80 Points Applicable)**	**36**
E.1	Systems and Materials with Low Environmental Impact	0
E.2	Materials That Minimize Consumption of Resources	12
E.3	Reuse of Existing Structures	0
E.4	Building Durability, Adaptability, and Disassembly	9
E.5	Reuse and Recycling of Construction/Demolition Waste	5
E.6	Facilities for Recycling and Composting	10
F.	**Emissions and Effluents (68 Points Applicable)**	**66**
F.1	Minimization of Air Emissions	15
F.2	Minimization of Ozone Depletion	25
F.3	Avoid Contamination of Sewers or Waterways	3
F.4	Pollution Minimization	23
G.	**Indoor Environment (175 Points Applicable)**	**149**
G.1	Ventilation	42
G.2	Source Control of Indoor Pollutants	32
G.3	Lighting	30
G.4	Thermal Comfort	20
G.5	Acoustic Comfort	25

Summary

Project Management—Policies and Practice (45/50 Points)	90%
Site (95/115 Points)	83%
Energy (211/373 Points)	57%
Water (67/81 Points)	83%
Resources, Building Materials, and Solid Waste (36/80 Points)	45%
Emissions and Effluents (66/68 Points)	97%
Indoor Environment (149/175 Points)	85%
Total points (669/942 points)	71%

Summary and Conclusions

Green Globes is an alternative building assessment system for use in the United States for both new and existing buildings. It differs from LEED in several important ways. First, it has a rating category for project management that gives credit for integrated design and environmental purchasing. It also provides credit for conducting life-cycle assessments of building assemblies during the design process. It has a starting base of 1000 points compared to the 110 points offered by LEED. Importantly, in cases where the situation does not apply, the starting base points can be reduced by the number of points that are not applicable. This is an important feature of Green Globes that is not available in LEED projects. The assessment of the project is conducted by third-party Green Globes Assessors who review the construction documents at the end of the design stage and make recommendations to the project team regarding the green attributes of the project. The Green Globes Assessor also visits the project at the end of construction to review the team's self-assessment and documentation for all the credits claimed by the project team. The Assessor also makes a physical inspection of the project to ensure that the as-built project is in compliance with the self-assessment. The Assessor also serves as a resource for the project team and assists the team in resolving gray areas that are not directly covered by the Green Globes questionnaire and support systems. Finally, the Green Globes building assessment system is being revised so that the ANSI/GBI 01-2010 standard will serve as the template for the Green Globes assessment process.

Note

1. Additional information about Green Globes can be found at the Green Building Initiative website www.thegbi.com.

Part III
Green Building Design

This part of the book addresses the major categories of issues covered by most building assessment systems, including Leadership in Energy and Environmental Design (LEED) and Green Globes. These categories include site and landscaping, energy systems, materials and products, the building hydrologic cycle, and indoor environmental quality. Part III contains the following chapters:

Chapter 7: The Green Building Design Process

Chapter 8: The Sustainable Site and Landscape

Chapter 9: Energy and Carbon Footprint Reduction

Chapter 10: Built Environment Hydrologic Cycle

Chapter 11: Closing Materials Loops

Chapter 12: Indoor Environmental Quality

Chapter 7 addresses the high-performance green building delivery system as a distinctly identifiable construction delivery system, analogous to individually recognized design-build systems. A hallmark of the high-performance green building delivery system is the high level of coordination and integration required of the design and construction team members. Additional measures, such as building commissioning and the charrette, are necessary to fully implement this new delivery system. Performance-based design contracts provide financial incentives to implement certain sustainable design features, such as relying on nature for some building services, thus enabling a downsizing of mechanical and electrical systems to reduce energy consumption and cost. Documenting the green building process and gathering system performance data are necessary to demonstrate that the building has met all certification requirements.

Chapter 8 parallels the building assessment categories that include issues such as locating the building near mass transit, siting the building on a brownfield instead of a greenfield, minimizing the ecological footprint of the construction process, and other measures designed to ensure that the building is sited to have the lowest possible environmental impact. This category also covers the potential for enhancing ecosystems as a component of developing green buildings. Stormwater management and alternatives to conventional practices are addressed. The problem of urban heat islands and measures to reduce temperature buildup in urban areas are considered. Light pollution—a health, safety, and environmental problem—is covered, and techniques for preventing excessive light from affecting the surrounding areas are presented.

Chapter 9 covers a range of energy issues, including passive design, design of the thermal envelope, equipment selection, renewable energy systems, green power, and emerging technologies, all of which can help achieve a very low building energy consumption profile. Energy-efficient lighting systems and lighting controls that sense occupancy and can be tied into the lighting system are covered. Innovative practices such as radiant cooling and ground coupling are used as examples of cutting-edge methods for addressing energy issues in green buildings. Smart buildings and building energy management systems are described to show how technology can be used in an effective manner to reduce a building's energy profile.

Chapter 10 focuses on minimizing potable water use, water recycling and reuse, and provisions for minimizing off-site stormwater flows. A strategy for the design of an effective building hydrologic cycle is provided. Technologies are described that can help provide alternative sources of water when potable water is not absolutely necessary. A wastewater strategy for high-performance buildings is also provided to minimize the need to move wastewater off-site for processing. Water-efficient landscaping is described, and its role in a green building hydrologic strategy is covered.

The selection of environmentally friendly construction materials is addressed in Chapter 11, which covers the use of recycled-content materials, used components, embodied energy due to transportation of material, and the minimization of construction waste. Defining green building materials remains the most difficult problem for designers of contemporary green buildings. For example, recycled-content materials are, in principle, green building materials, but many contain industrial and agricultural waste, so it is not clear that recycling these by-products into the built environment is the best solution. Consequently, one objective of this chapter is to promote an understanding of the broad range of issues and problems connected to building materials and products. The chapter also covers the topic of life-cycle assessment (LCA), a method for analyzing the resources, waste, and health effects associated with the entire life of a product or material, from its extraction as raw material to its ultimate disposal.

Indoor environmental quality (IEQ) is covered in Chapter 12. The various types of health-related building problems are described. Selecting low-emission materials; protecting heating, ventilation, and air conditioning (HVAC) systems during construction; monitoring indoor air quality (IAQ); and issues surrounding the health of the construction workforce and future building occupants are explored. Lighting quality, access to daylight and views, and noise as IEQ issues are covered. Best practices for providing building IAQ are also addressed. Additional best practices and checklists that provide assistance in achieving building assessment system points or that address issues not covered in these systems are also provided.

In short, Part III addresses the core issues of the technical side of sustainable construction and discusses approaches that can be employed to limit resource depletion, negative environmental consequences, and impacts on human health that are too often the result of the creation, operation, and disposal of the built environment. Future buildings should contribute to the restoration and regeneration of ecological capacity, recycle water and discharge potable water, generate the energy needed for their operation, contribute to the health of their human occupants, and serve as materials resources for future generations rather than as a disposal headache.

Chapter 7

The Green Building Design Process

The high-performance building movement is changing both the nature of the built environment and the delivery systems used to design and construct the facility according to a client's needs. The result has been the emergence of the *high-performance green building delivery system*, introduced in Chapter 1. This delivery system is distinguishable from conventional practice by the selection of project team members based on their green building expertise, increased collaboration among the project team members, more focus on integrated building performance than on building systems, heavy emphasis on environmental protection during the construction process, careful consideration of occupant and worker health throughout all phases, scrutiny of all decisions for their resource and life-cycle implications, the added requirement of building commissioning, and the emphasis placed on reducing construction and demolition waste. Some of these differences are driven by certification requirements, while others are part of the evolving culture of green building.

This chapter more fully describes the differences between standard practice and the green building process, paying particular attention to the highly collaborative *charrette* process, probably one of the most distinguishing hallmarks of contemporary green building. New tools such as building information modeling (BIM) that produce three-dimensional representations of the model that are linked to energy modeling, daylighting, and life-cycle assessment (LCA) software are increasing the quality of the collaboration and lowering the costs of green building. Plug-ins for BIM that create documentation for green building certification based on building assessment systems are also emerging to further reduce the challenges of creating a high-performance built environment.

Conventional versus Green Building Delivery Systems

Contemporary construction delivery systems in the United States fall into four major categories: *design-bid-build, construction management-at-risk, design-build, and integrated project delivery*. In the following sections, these four systems are briefly described and then compared and contrasted with the emerging high-performance green building delivery system.

DESIGN-BID-BUILD (HARD BID)

The primary objective of a *design-bid-build*, or *hard-bid*, delivery system is low-cost delivery of the project to the owner by the general contractor. The design team is selected by the owner and works on the owner's behalf to produce construction documents that define the location, appearance, materials, and methods to be used in the creation of the building and its infrastructure. General contractors bid on the project, with the lowest qualified bidder receiving the job.

Similarly, the general contractor selects subcontractors based on competitive bidding and awards the specific work—for example, steel erection or masonry—to the lowest qualified bidder. Although the project is theoretically delivered at the lowest cost to the owner, conflicts among the parties to the contract (owner, design team, general contractor, subcontractors, materials suppliers) are frequent, and emotional tension and miscommunication generally permeate the process, often resulting in higher costs from change orders, repairs, and lawsuits. Although there are cases in which high-performance green buildings have been built using the hard-bid construction process, the degree of potential conflict and lack of a collaborative working atmosphere make it the least desirable construction delivery system for this purpose.

CONSTRUCTION MANAGEMENT-AT-RISK (NEGOTIATED WORK)

In the construction management-at-risk delivery system, the owner contracts separately with the design team and the contractor, or *construction manager*, who will work on the owner's behalf. This system is also referred to as *negotiated work* since the construction manager negotiates a fee for management services with the owner. Early in the design process, the construction manager is usually required to guarantee that the total construction cost will not exceed a maximum price, referred to as the *guaranteed maximum price* (GMP). Ideally, both the construction manager and the design team are selected at the start of the project. The construction manager can then provide preconstruction services such as cost analysis, constructability analysis, value engineering, and project scheduling to facilitate an efficient and effective design process followed by a conflict-free construction phase.

Working together, the parties produce construction documents that meet the owner's requirements, schedule, and budget, and prevent physical conflicts among systems, missing information, and other products of miscommunication often found in the construction documents produced for hard-bid projects. Using a bidding process, the construction manager selects subcontractors based on their capabilities and the quality of their work, not merely the lowest bid. Accordingly, the level of conflict in negotiated work is much lower because of the closer working relationships among the parties to the contract. Additionally, construction management firms undertaking negotiated work understand that the primary source of future projects will be current and past clients. Consequently, client satisfaction becomes a primary objective.

DESIGN-BUILD

Although negotiated work reduces the frequency and intensity of conflicts present in a hard-bid construction delivery system, the classic tension between the design team and the construction manager still exists, albeit to a lesser degree. *Design-build* is a method of project delivery in which one entity (the designer-builder) forges a single contract with the owner to provide for architectural/engineering design services and construction services.[1] Design-build is also known as *design-construct* and provides the owner with single-source responsibility. In the typical design-bid-build, or hard-bid, project, the owner commissions an architect or engineer to prepare drawings and specifications under a design contract and subsequently selects a construction contractor by competitive bidding to build the facility. In contrast, the design-build delivery system provides the owner with a single contractual relationship with an entity that combines both design and construction services. This entity may be a firm that possesses in-house design and construction capabilities or a partnership between a design firm and a construction firm. Thus, the design-build delivery system is more likely to reduce typical design-construction conflicts, provide a lower price for the owner, improve quality, speed the project to completion, and facilitate improved communication among the project team members. The design-build delivery system is very compatible with the green building concept, and due to its emphasis on a high degree of collaboration between the design and construction phases, it is very consistent with the design approach required to produce high-performance buildings.

INTEGRATED PROJECT DELIVERY

Integrated project delivery (IPD) is a relatively new construction delivery system that originated in the mid-1990s when a group of construction companies in Orlando, Florida, were attempting to increase productivity and the speed of project delivery without the typical conflicts and stress of construction projects. In May 2007, the IPD Definition Task Group, composed of a variety of stakeholders such as owners, architects, contractors, engineers, and lawyers, collaborated to define the IPD system. This emerging construction delivery system takes advantage of several other relatively new ideas such as integrated process, lean construction, building information modeling (BIM), and other technologies that provide the potential for good collaboration on construction projects. Charles Thomsen, former chairman of 3D International, defines IPD as:[2]

> an approach to agreements and processes for design and construction, conceived to accommodate the intense intellectual collaboration that 21st century buildings require. The inspiring vision of IPD is that of a seamless project team, not portioned by economic self-interest or contractual silos of responsibility, but a collection of companies with a mutual responsibility to help one another meet an owner's goals. To support that vision, architect/engineers, construction managers, and lawyers are crafting management processes and contract terms intended to align the interests of the key project team with the project mission, increase efficiency, reduce waste, and make better buildings.

Thomsen suggests that the following are the main ingredients of IPD:

- A legal relationship
- A management committee
- An incentive pool
- A no-fault working environment
- Design assistance
- Collaborative software
- Green construction
- Integrated leadership

The IPD process is designed to produce shorter delivery times than other construction delivery systems such as design-bid-build. The emergence of collaborative software provides opportunities to improve the flow of documentation, communications, and work to ensure that all parties engaged in the project are working on the same set of documents and collaborating to the same end. *Relational contracts* are the key document in IPD because they define the relationships between all parties to the project. A relational contract is a single agreement signed by the owner, architect/engineer, and the contractor. Although commonly used in the United Kingdom and Australia, these types of contracts are relatively new to the US construction industry. It is also possible for other contractors, for example, subcontractors, to be parties to the contract in what is called an *integrated form of agreement* or *triparty collaborative agreement.* Relational contracts in the IPD process have incentive clauses such that any potential savings are shared among the IPD team members and with the owner. An incentive pool for this purpose is created and put at risk depending on the collaboration of the team. Relative to high-performance green building, IPD is an untested idea because it is relatively new. However, many of the key aspects of IPD dovetail with the leading edge of high-performance green building such as an integrated collaborative process and the application of technology to support the development of truly high-performance buildings. Additionally, IPD has many of the traits of construction delivery systems that are far more compatible with green building certification systems such as Leadership in Energy and Environmental Design (LEED) and Green Globes. Both lean construction and BIM are being merged into the cutting edge of green building, just as they are in IPD. One of the long-standing criticisms of green building has been its higher first cost, and IPD, together with these tools, provides the potential to deliver high-performance buildings at the same or lower cost compared to conventional, code-compliant facilities.

THE HIGH-PERFORMANCE GREEN BUILDING DELIVERY SYSTEM

The evolving high-performance green building delivery system is similar to other construction delivery systems such as negotiated work and design-build, but with additional responsibilities for the project team. Most notably, it requires much greater communication among the project team members. Consequently, initial team building, which engages the widest possible range of stakeholders, ensures that everyone understands the project's goals and the unique specifications. This delivery system also demands special qualifications from its participants, especially an understanding of, and commitment to, the concept of green building and, in the case of projects to be certified using LEED, Green Globes, or the Living Building Challenge, a strong familiarity with the assessment systems and their requirements. The team members should also have experience with the charrette process and be especially willing to engage a wide range of stakeholders, including some who are traditionally not included in building projects. An example would be the inclusion of community members in the charrette for the design of a corporate facility. Some recent projects have been including local building officials who will ultimately have to approve any innovative, out-of-the-box solutions that a high-performance building team may propose.

Due to its adversarial nature, the hard-bid delivery system is exceptionally difficult to employ for a green building project. The collaborative spirit needed for a successful high-performance green building project would be difficult to develop in this adversarial climate. The design-build delivery system has significant potential to deliver green buildings because, like negotiated work, it is designed to minimize adversarial relationships and simplify transactions among the parties. However, unlike conventional construction, the checks and balances provided by transparent interaction between the design team and the construction entity are virtually absent. And, as with other aspects of sustainable development, transparency is an important characteristic of green building projects. In spite of this potential problem, several successful green building projects have been executed using design-build, for example, the Orthopaedics and Sports Medicine Institute at the University of Florida in Gainesville, completed by the design-build team of URS and Turner Construction (see Figure 7.1).[3] IPD is a relatively new delivery system, but with its high emphasis on collaboration, it would seem to be an approach that will be highly compatible with green building delivery.

Figure 7.1 The project team for the Orthopaedics and Sports Medicine Institute at the University of Florida in Gainesville used the design-build delivery system to deliver this LEED-certified green building. (T. Wyman)

Executing the Green Building Project

Because the high-performance green building delivery system is distinctly different in many ways from conventional delivery systems, the project team needs to be aware of these differences and where they occur in the building design and construction process. After the programming and budgeting of the proposed building project has been accomplished by or on behalf of the owner, the execution of a high-performance green building project has the following phases:

1. Setting priorities for the green building project by the owner in collaboration with the project team.

2. Selecting the project team: the design team and the construction manager or the design-build firm.

3. Implementing an integrated design process (IDP) by orienting the project team to the concept of IDP and how it will be implemented during the design and construction processes. Note that the integrated design process, or IDP, is different from integrated project delivery, or IPD. The IDP is covered in more detail below.

4. Conducting a charrette to obtain input for the project from a wide variety of parties, including the project team, the owner and users, the community, and other stakeholders.

5. Executing the design process, consisting of schematic design, advanced schematic design, design development, construction documents, and documentation of green building measures for a project that is to be certified, all conducted using IDP. This involves full use of IDP in the development of the design, marked by extensive interdisciplinary interaction to maximize design synergies.

6. Constructing the building, to include implementing green building measures that address soil and erosion control, minimizing site disturbance,

protecting flora and fauna, minimizing and recycling construction waste, ensuring building health, and documenting the construction phase of green building measures.

7. Final commissioning and handover to the owner.

OWNER ISSUES IN HIGH-PERFORMANCE GREEN BUILDING PROJECTS

The decision to produce a high-performance green building brings with it a number of unique issues that have to be resolved by the owner prior to initiation of the design and construction of the building. Among the questions that must be answered are the following:

- Does the owner want the building to be a certified green building? Although the LEED approach is the predominant method in the United States for producing a green building, a green building based on a different philosophical and technical approach may be desirable. For example, the Green Globes building assessment protocol and the Living Building Challenge are alternative approaches that may be a good choice in some situations. In at least one state, Florida, there is a commercial green building assessment standard that can be used in lieu of the national standard.[4] The International Green Construction Code is being adopted by some jurisdictions, the result being that high-performance green construction may be required in those jurisdictions.[5]

- If the building is to be certified, what level of certification is desired (a platinum, gold, silver, or certified rating for LEED or one to four green globes in the case of Green Globes)? The building's owner may have a preconceived idea of the level of certification desired for the facility, in which case the task of the project team will be to design and build the facility to meet the owner's goals. Often the project team will have to address the cost/benefit issues involved in achieving different certification levels and provide life-cycle costing (LCC) analysis for each level to give the owner the data needed to make a decision.

- If the building need not be certified, what design criteria should be followed by the design team? The LEED and Green Globes assessment systems each provide a consistent framework that contains virtually all the criteria needed to produce a green building. If LEED or Green Globes is not to be the basis for creating the green building, the owner will have to provide the project team with a detailed description of the criteria the team members are to use in their work.

- What are the desired qualifications of the design team and construction manager with respect to the high-performance building? In the case of a design-build project, what background and training should the designers and construction professionals have? It is certainly advantageous for the owner to hire project team members who have green building experience. If certification is desired, significant documentation of numerous aspects of the project will be required. For example, if one of the credits being addressed is the recycled content of materials used in the project, the construction manager must be aware of the requirement to obtain information from most of the subcontractors about the quantity of recycled materials in the products they are using in the building, and then compile the data from all the subcontractors to determine the overall percentage of recycled content in the project.

- What level of capital investment, beyond that required for conventional construction, will the owner provide to make the facility a high-performance green building? And is the owner willing to consider trading off lower operational costs for higher front-end capital costs? Green buildings are specifically designed to have lower operational costs, which are often accompanied by higher front-end capital costs. An LCC analysis will provide a breakdown of costs versus savings on an annual basis and indicate where the break-even point, in years, for the investment occurs. It is up to the owner to decide whether the break-even point is satisfactory and, based on this information, whether the additional capital cost is warranted.

SETTING PRIORITIES AND MAKING OTHER KEY INITIAL DECISIONS

When the decision has been made to create a high-performance green facility, the owner must decide on the priorities for the building. For example, in water-short areas of the United States, water issues may be so important that the owner may decide to focus heavily on the building's hydrologic cycle (water conservation, water reuse, rainwater harvesting, graywater systems, and employment of reclaimed water) rather than, for example, to make an exceptional effort to reduce energy consumption. Another owner may opt for implementing an extensive and exceptional system of daylighting and lighting controls due to its energy-conserving possibilities and potential health benefits and, conversely, undertake minimal water conservation measures.

Another priority to be set and a decision to be made concern the financial investment the owner is willing to make in a high-performance building. Green buildings normally involve systems not commonly used in conventional buildings; for example, rainwater harvesting systems, with their associated piping, pumps, and cisterns, entail additional design effort. Many state governments are forced by law to operate within strict square footage cost guidelines. As a result, very simple, cost-effective measures must be considered. Other types of organizations may have revolving funds that can be used to invest in high-performance options that will pay back the fund over time. Harvard University, for example, has a $12 million revolving fund that can be used for investing in higher-capital projects that are repaid out of the savings. The federal government requires LCC to be employed to justify building investment decisions, a requirement that works in favor of high-performance building decisions. Private-sector owners have considerably more leeway, and their decisions can be based on LCC, as is the case for the federal government. Certified green buildings will have additional documentation requirements, requirements for commissioning, fees for registration and certification review, and other costs that must be allocated in the building budget.

SELECTING THE GREEN BUILDING TEAM

When an owner has decided to produce a high-performance green building, the next order of business is to select the design and construction teams. The actual selection process proceeds in the conventional fashion with the issuance of a request for proposal (RFP) or request for qualifications (RFQ) by the owner to announce the upcoming selection of the architect and construction manager. The RFP/RFQ should specify the additional qualifications required of the architect, interior designers, landscape architects, civil engineers, structural engineers, electrical engineers, and mechanical engineers supporting the design. One of the challenges in writing an RFP for a high-performance building is to ensure that the architects and construction managers understand the owner's green goals. To facilitate this effort, the Committee on the Environment (COTE) of the American Institute of Architects (AIA) has produced a guide to writing RFPs and RFQs for green buildings, "Writing the Green RFP."[6]

After reviewing the submissions by the architect and construction management firms or design-build firms that respond to the RFP/RFQ for the project, the owner typically creates a list of three to five firms in each category, and then organizes presentations by the short-listed design firms and construction management companies. The final selection is based on experience, qualifications, previous work, and demonstrated understanding of the owner's program and requirements, the building site, and the firm members' ability to work with other project team members. The architect and construction manager or design-build firm should be selected prior to the start of the design so that both will be on board during the entire project.

Clearly, it is important that the architect and engineers have a detailed understanding of the concept of green building and a commitment to investing creativity and energy to produce an exceptional building. At this point in the evolution of high-performance green buildings, even though the movement is relatively new, there are a large number of design professionals who have already engaged in the design of one or more green buildings. Detailed knowledge of the LEED building assessment standard or the Green Globes building assessment protocol is absolutely essential if the owner decides that the goal is green building certification. It is also important to note that there are some outstanding architects who have experience creating high-performance buildings that have not been submitted for LEED or Green Globes certification; thus, the owner must judge the ability of these firms to meet the owner's requirements.

If the building is to be certified, the construction manager should have strong familiarity with, or staff trained in, the requirements of the LEED or Green Globes assessment system. The certification process imposes enormous responsibility on the construction manager; lack of experience with the standards could compromise the certification of the project.

ROLE OF THE LEED ACCREDITED PROFESSIONAL OR GREEN GLOBES PROFESSIONAL IN THE PROCESS

The inclusion of building industry professionals trained in the use of green building certification systems helps facilitate decision making and the flow of information required to successfully navigate the relatively complex requirements of these systems. Both the US Green Building Council (USGBC) and the Green Building Initiative, the proponent of Green Globes, have training programs and designations for these individuals.

The USGBC offers training and testing which, if successfully passed, designates the individual as a LEED Accredited Professional (LEED AP). This designation provides the building owner with a high degree of assurance that the requirements of the USGBC certification programs will be understood and that the extensive documentation required for certification will be provided. The LEED AP examination is intended to test the individual's knowledge of green building principles, as well as familiarity with the LEED requirements. The following are the points covered on the LEED AP examination:[7]

- In-depth familiarity with the LEED building assessment system
- Understanding of LEED project registration/technical support/certification process
- Demonstrated knowledge of design and construction industry standards and process

- General understanding of the various designs referenced in LEED
- Understanding of green and sustainable design strategies and practices, and corresponding credits in the LEED rating system
- Familiarity with key green and sustainable design resources and tools

One of the other benefits of having a LEED AP on the project team is that one credit is awarded for the certification of the project. One of the drawbacks of the current system of awarding this credential is that it does not require in-depth knowledge of the building design and construction process, nor does it require professional experience. One of the challenges for the USGBC has been to create a rigorous accreditation process with requirements either for periodic recertification or for continuing education to maintain currency on green building issues and the LEED system.

In a similar fashion, the GBI offers training and testing designed to produce Green Globes Professionals (GGPs) who have detailed knowledge of Green Globes, the certification process, and documentation requirements. Like LEED, Green Globes provides credit for having a GGP as a member of the project team. Unlike the LEED AP, the GGP must be a qualified building industry professional prior to taking the accreditation examination to become a GGP. Green Globes also has a higher-level qualification, the Green Globes Assessor (GGA), who is the third-party advisor to the GBI regarding certification. The requirements to become a GGA far exceed those of becoming a GGP. The GGA provides the actual third-party certification for the project, assists the project team through the certification process, and makes the final judgment regarding achievement of certification and the level of certification. Chapter 6 covers the qualifications of GGAs and GGPs in more detail.

The Integrated Design Process

Although it is true that excellent teamwork is required for any building project, the level of interaction and communication needed to ensure the success of a green building project is significantly higher. Green buildings are a new concept to the industry, and it is generally necessary to orient all members of the project team to the goals and objectives of the project that are related to issues such as resource efficiency, sustainability, certification, and building health, to name a few. This orientation can serve three purposes. First, it can fulfill its primary purpose of informing the project team about all project requirements. Second, it can familiarize the project team with the owner's priorities for the high-performance green building aspects of the project. Third, it can provide an opportunity to accomplish team building in the form of group exercises for familiarizing the group with the building, the building program, and the building's green building issues.

Integrated building design or integrated design is the name given to the high levels of collaboration and teamwork that help differentiate a green building design from the design process found in a conventional project. According to the US Department of Energy, integrated design is

> [a] process in which multiple disciplines and seemingly unrelated aspects of design are integrated in a manner that permits synergistic benefits to be realized. The goal is to achieve high performance and multiple benefits at a lower cost than the total for all the components combined. This process often includes integrating green design strategies into conventional design criteria for building form, function, performance, and cost. A key to successful integrated building design is the participation of people from different specialties of design: general architecture, HVAC, lighting and

electrical, interior design, and landscape design. By working together at key points in the design process, these participants can often identify highly attractive solutions to design needs that would otherwise not be found. In an integrated design approach, the mechanical engineer will calculate energy use and cost very early in the design, informing designers of the energy-use implications of building orientation, configuration, fenestration, mechanical systems, and lighting options.[8]

IDP is characterized by early significant collaboration in the design process. In conventional design, the team begins their joint effort at the start of schematic design, whereas in a green building project employing integrated design, the collaboration starts at the very beginning of the project, and all team members have input on design decisions during the entire cycle of design (see Figure 7.2). The earlier integrated design is implemented, the greater the benefits (see Figure 7.3).

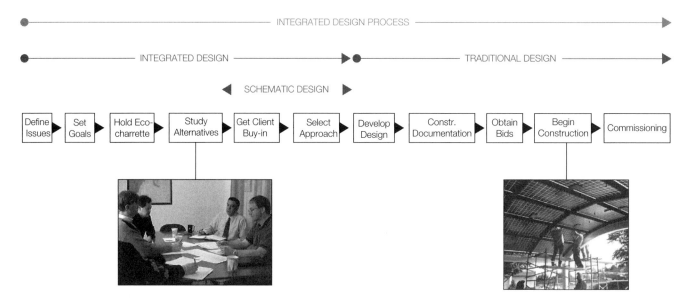

Figure 7.2   In green design, the integrated design starts much earlier in the project development process compared to conventional design, involving interaction with the owner to define issues and to set goals prior to schematic design and continuing through construction and commissioning. (Diagram courtesy of Interface Engineering, Inc.)

Figure 7.3   The earlier an integrated design is implemented, the greater the potential savings and the lower the cost of changes to the building design.

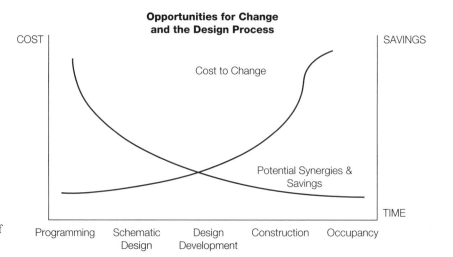

There are numerous potential areas for integrated design in any building project: the building envelope, the daylighting scheme, green roofs, minimization of light pollution, indoor environmental quality, and the building hydrologic cycle, to name but a few. The Green Globes building assessment protocol spells out the requirements for integrated design in its project management section, where a team can achieve 20 points for demonstrating that they have indeed implemented integrated design in the process. In addition to appointing a green design coordinator, the team must demonstrate how they interacted by documenting the results of their collaboration in the form of the minutes of goal-setting meetings and lists of items on which the team worked jointly for resolution.[9]

Another term that describes integrated design is *integrated design process* (IDP). Some of the foundational work on developing IDP occurred in Canada, and perhaps the most thorough definition was a result of a national workshop on IDP held in Toronto in 2001:[10]

> IDP is a method for realizing high performance buildings that contribute to sustainable communities. It is a collaborative process that focuses on the design, construction, operation and occupancy of a building over its complete life-cycle. The IDP is designed to allow the client and other stakeholders to develop and realize clearly defined and challenging functional, environmental and economic goals and objectives. The IDP requires a multi-disciplinary design team that includes or acquires the skills required to address all design issues flowing from the objectives. The IDP proceeds from whole building system strategies, working through increasing levels of specificity, to realize more optimally integrated solutions.

In addition to this extensive definition of IDP, the main elements of the IDP were identified as

- Interdisciplinary work between architects, engineers, costing specialists, operations people and other relevant actors right from the beginning of the design process.
- Discussion of the relative importance of various performance issues and the establishment of a consensus on this matter between client and designers.
- The addition of an energy specialist to test various design assumptions through the use of energy simulations throughout the process, to provide relatively objective information on a key aspect of performance.
- The addition of subject specialists (e.g., for daylighting, thermal storage, etc.) for short consultations with the design team.
- A clear articulation of performance targets and strategies, to be updated throughout the process by the design team.
- In some cases, a design facilitator may be added to the team, to raise performance issues throughout the process and to bring specialized knowledge to the table.

It was also noted that it may be useful to launch the IDP with a charrette, described in more detail in the following section.

Traditional design could be said to have three steps:

Step 1. The client and architect agree to a design concept that includes the general massing of the building, its orientation, its fenestration, and probably its general appearance and basic materials.

Step 2. The mechanical and electrical engineers are engaged to design systems based on the building design concept agreed to in step 1. The civil engineer and landscape architect develop a concept for landscaping, parking, paving, and infrastructure based on the building design concept and the owner's wishes.

Step 3. Each phase of design (schematic, design development, and construction documents) is carried out employing the same pattern, with minimal interaction between disciplines, little or no interdisciplinary collaboration, and great attention to the speed and efficiency of executing each discipline's design.

The result of traditional design is a linear, noncollaborative process with little attempt to set goals beyond meeting the owner's basic needs, and which meets the building code but is not optimized. Each discipline functions in isolation, with interdisciplinary communications kept to a minimum. As is the case with every other system, optimizing each subsystem of the project results in a suboptimal building. The most likely outcome is not only a suboptimal project but also a range of other potential problems caused by a lack of strong coordination among disciplines.

In contrast to traditional design, the point of IDP is to optimize the entire building project. The requirements for communication are intense, nonstop, and at all stages of the project, from design through construction, commissioning, handover to the owner, and postoccupancy analysis. Integrated design starts prior to the actual design process, with the project team articulating goals for the project and determining the opportunities for synergies in which design solutions have multiple benefits for the project. The following is a typical sequence of events that are indicative of integrated design:

- The project team establishes performance targets for a broad range of parameters, to include energy, water, wastewater, landscape performance, heat island issues, indoor environmental quality, and construction and demolition waste generation, to name a few. In conjunction with establishing these performance targets, the project team develops preliminary strategies to achieve the targets. IDP should bring engineering skills and perspectives to bear at the concept design stage, thereby helping the owner and architect to avoid becoming committed to a suboptimal design solution. It should also involve all members of the team bringing their skills to bear on designing the optimal building. Mechanical engineers are better placed in terms of their background in thermodynamics than the architect, and it makes sense to engage them in the design of the building envelope.

- The team should minimize heating and cooling loads and maximize daylighting potential through orientation; building configuration; an efficient building envelope; and careful consideration of the amount, type, and location of fenestration. A potentially wide variety of plug loads should be addressed due to the effects of large numbers of computers, printers, fax machines, sound systems, and other equipment on the performance of the building. Minimizing these loads and selecting equipment with the lowest possible energy consumption is needed so that the intent of the high-performance building is not compromised by neglecting to account for this consumption. The broad range of indoor environmental quality issues should be addressed, to include air quality, noise, lighting quality and daylighting, temperature and humidity, and odors. The team should also collaborate on site issues to maximize the use of natural systems; minimize hardscape; use trees to assist heating and cooling of the building; and integrate rainwater harvesting, graywater systems, and reclaimed water into the design of the building's hydrologic cycle.

- The team should maximize the use of solar and other renewable forms of energy, and use efficient heating, ventilation, and air conditioning (HVAC) systems, while maintaining performance targets for indoor air quality, thermal comfort, illumination levels and quality, and noise control.

- The result of the process should be several concept design alternatives, employing energy, daylighting, and other simulations to try out the alternatives, and then the selection of the most promising of these for further development.

The earlier IDP is instituted, the greater its effect on the design process. The maximum benefit occurs when the decision to employ IDP is made prior to the start of the design process and the project team has the opportunity to set goals for the project that guide the design process.

The result of IDP should be a full understanding of the potential design synergies and the connection of the project goals to the resulting building design. A truly collaborative process will use these project goals as the basis for wide-ranging, dynamic interaction among the project team members to capitalize on the potential for reducing resource consumption, limiting environmental impacts, and restoring the site to its maximum ecological potential. Figure 7.4 is a schematic that demonstrates how project goals can be used in conjunction with IDP to produce a wide range of benefits, both for the project and for the environment.

Another term related to integrated design is *whole-building design*, a concept advocated by the National Institute of Building Sciences (NIBS) and described as consisting of two components: an integrated design approach and an integrated team process.[11] Whole-building design has been adopted by a group of federal agencies as the core concept of high-performance green buildings, and the emphasis is on collaboration and life-cycle performance. The

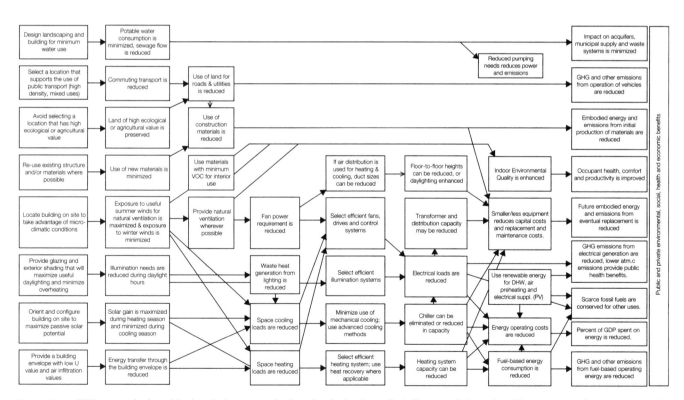

Figure 7.4 IDP can assist in achieving design synergies by stimulating interdisciplinary collaboration. The result can be green strategies such as those listed in the leftmost column of this example project being translated into benefits for the building owner and occupants as well as for the global environment. (Illustration courtesy of Nils Larsson, Natural Resources Canada, and the United Nations Environmental Program)

concept of collaboration is extended outside the project team to include all stakeholders in the building process. In the integrated team process, the design team and all affected stakeholders work together throughout the project phases to evaluate the design for cost, quality of life, future flexibility, and efficiency; overall environmental impact; productivity and creativity; and effect on the building's occupants. Whole-building design, as described by NIBS, draws from the knowledge pool of all the stakeholders across the life cycle of the project, from defining the need for a building, through planning, design, construction, building occupancy, and operations. The process does not conclude at the end of construction and handover to the owner. During operation, the building should be evaluated to ensure that it has met its high-performance design objectives. Furthermore, the building should be recommissioned periodically to maintain its high-performance character throughout its life cycle.

Role of the Charrette in the Design Process

Creating a green, sustainable building implies that the widest range of possible stakeholders will be engaged in the process because buildings ultimately affect a large variety of people and, in fact, affect future buildings. As the predominant artifacts of modern society, and due to their relative longevity, buildings are important cultural symbols; hence, they affect enormous numbers of people every day. Passersby are affected either positively or negatively by the appearance of a building based on its design, materials, color, location, and function. The stakeholders in a building will vary widely, depending on its type and its ownership. For example, a public building such as a library or city administration building will affect not only the employees who will directly use the building but virtually all persons in the local jurisdiction, who, as taxpayers, have contributed to its realization. In the case of a corporate building, although its impact may not be as widespread, a savvy owner would nevertheless engage a wide range of users, customers, local government, and citizens to obtain the maximum input. The process of gathering this input is referred to as a *charrette*. A general overview of the charrette concept is provided here. The detailed integration of the charrette into the design process is covered in the next section.

The word *charrette* is derived from the French term meaning "little cart." This concept has its roots in French architectural education when proctors at the École des Beaux-Arts in 19th-century Paris collected student projects on wheeled carts, literally pulling the drawings from the students' hands at the end of their final frenzied efforts on a design project. Today, the term is used to refer to an effort to create a plan. The National Charrette Institute (NCI) states that there are four guiding principles for a charrette (see Table 7.1). Note that these principles are meant to apply to a community planning charrette, not specifically to the design of a single building. Consequently, they are presented here in a modified form from the actual NCI guiding principles.[12]

The NCI has also proposed a four-step charrette process that, although designed for a community planning charrette, is also applicable to a building project charrette. These steps are outlined in Table 7.2.[13]

At the conclusion of the charrette, it is the responsibility of the project team to transform the results into a report that can be used to guide the design of the project. A final review of the outcome of the brainstorming sessions should be conducted to ensure that the measures selected for implementation meet cost and other criteria that may be important. Communications may need to be established with entities or groups external to the charrette to ensure that they act to maximize the high-performance aspects of the project. For example,

TABLE 7.1

Four Guiding Principles for a Built Environment Charrette

1. *Involve everyone from the start.* The identification and solicitation of stakeholders to provide input to a project is of the utmost importance because the participants in the charrette process will feel a sense of ownership for the outcomes. The broader the range of input, the more likely the project is to be successful and accepted by the community. It is also important to note that people or organizations that may potentially play a role in blocking a project should be invited to participate.

2. *Work concurrently and cross-functionally.* All disciplines engaged in a project should work together at the same time during the charrette and with the other stakeholders to generate alternative designs under the guidance of a facilitator. The level of design detail that emerges from the charrette will be a function of the time available and the complexity of the project. In general, a building design charrette produces a wide range of potential solutions and approaches that not only cover green issues but also address the function of the building and its relationship to the community. For larger, more complex projects, the participants can divide into groups to tackle specific issues, then return to a caucus or plenary meeting for each group to share its progress with other groups and to make decisions on how to proceed.

3. *Work in short feedback loops.* For a building project, proposed solutions and measures are laid out in a brainstorming session during which the participants, guided by a facilitator, cover all aspects of the building, its infrastructure, and its relationship to the community. This approach produces far more alternatives and engages far more creativity than a conventional design process. This is an advantage in that many more ideas and options are presented. That said, the information must also be processed efficiently and rapidly to provide useful input to the actual design process. The result of the brainstorming sessions must be distilled to the essential outcomes, and duplications must be eliminated and priorities established. For example, it would certainly be advantageous if all buildings had photovoltaics, but few owners have the resources at present to incorporate them into their facilities. The feedback loops between initial brainstorming sessions and design decisions should be as rapid as possible so that more than one iteration is possible during the charrette.

4. *Work in detail.* The more detail in a charrette the better. Alternatives for building appearance, orientation, massing, and electrical and mechanical systems should be sketched out in as much detail as possible. The NCI recommends working on problems at different scales during the charrette. Larger-scale issues of drainage, paving, and relationships to other buildings and the street should be addressed, as should details such as entrance location, window selection, and roof type.

Rinker Hall, a LEED gold building at the University of Florida in Gainesville, is connected to a central plant that provides its heating and cooling. The project team decided that the LEED Energy and Atmosphere point for eliminating hydrochlorofluorocarbon (HCFC) use could be justified only by obtaining a commitment from the university to implement a program to replace its older, HCFC-based chillers with efficient hydrofluorocarbon (HFC) refrigerant chillers. Another point, for maintaining open space, was acquired by obtaining a letter from the university administration stating that specific property contiguous to Rinker Hall would be maintained as open space for the life of the building. In the private sector, cooperation of municipal officials may be necessary to obtain points for proximity of mass transit.

The final version of the charrette report becomes one of the guiding documents for the launch of the schematic design phase of the project, and ultimately serves to help steer the project through design development, construction documents, and the actual construction process.

TABLE 7.2

Four Steps for a Built Environment Charrette

1. *Start-up.* In the context of a building project, the start-up for a charrette is very simple. It involves determining who the stakeholders are, engaging the stakeholders in the process, establishing the goals for the charrette, determining the time and place for holding the charrette, and notifying the participants of the details.

2. *Research, education, and concepts.* Prior to the charrette, the building owner, the charrette facilitator, and design team members should discuss the information needs for the charrette. The owner's directions, the building program, site details, utility information, and other pertinent data should be gathered and readied for the charrette. Information on specific technologies may be useful. For example, if a fuel cell is a strong-candidate technology for the project, technical information about the device, issues of connecting the fuel cell to the grid, and information about fuel and emissions should be gathered for use during the process. In some cases, the process of gathering information for the charrette may highlight the need to engage other organizations in the process. In the example of a fuel cell, the local utility company could provide valuable input as to how best to incorporate the fuel cell into the project. The location of the charrette should be selected to best facilitate its conduct. Generally, it is best to hold the charrette at the owner's location if adequate space and facilities are available. A large room with blackboards or whiteboards, space for large-paper tripods, and a projector and projector screen should be available.

3. *The charrette.* Generally, the charrette should be conducted by a facilitator familiar with the green building process. A typical building charrette might occur over several days and continue in phases until complete. The first step should be an effort to educate all the participants on the owner's requirements and the concept of high-performance green building. The second step would be to review the building program, previously generated architectural schemes, building siting, proposed construction budget, and construction schedule. The third step would be to lay out the goals of the project with respect to its green high-performance aspects. The owner may desire a specific level of certification, for example, a LEED gold certification, that will affect many of the decisions made during the charrette. When these steps have been completed and the project team and stakeholders understand the context of the project, the actual charrette begins. The facilitator conducts a guided brainstorming session that draws out input from the group about every aspect of the project, with a special emphasis on the sustainability of the building. During the conduct of the charrette, the team should keep a running scorecard on how the decisions made during the process are affecting the building assessment score. The economics of each decision also need to be taken into account, and the construction manager should ensure that enough data are available to provide a conceptual cost estimate for review by the owner.

4. *Review, revise, and finalize.* After the charrette is complete, the design team reviews the results with the owner, makes any appropriate adjustments and changes, and then produces a report of the charrette to guide the balance of the design process.

Green Building Documentation Requirements

To certify a green building using one of the major building assessment systems requires that a great deal of attention be paid to gathering information throughout the course of the design and construction of the project that ultimately will be reviewed as part of the certification process. The two main building assessment systems in the United States, LEED and Green Globes, have different approaches to how the documentation is ultimately reviewed, and the project team should carefully review the requirements for each approach.

LEED DOCUMENTATION

In the case of a project for which the owner is seeking USGBC certification, careful documentation of the efforts to achieve credits is needed. As noted earlier, the documentation requirements for the first versions of the LEED building assessment standard were relatively complex and difficult. For LEED-NC and other LEED products, the documentation requirements, while far simpler, are by no means easy to meet. The advent of LEED-Online, a sophisticated Web portal for posting of documentation and exchange of information, has made the entire process paperless. The documentation may be submitted in two batches: the design phase and the construction phase. The design phase submission is for those credits that are essentially complete during design and do not require any documentation during the construction phase. For instance, LEED Materials and Resources (MR) Prerequisite 1 requires that a space be set aside for the storage and collection of recyclables in the building. The required documentation is a drawing that shows this area and the location of the containers required for recycling. This prerequisite is completed during design and can be submitted with other design phase credits for review by the USGBC via LEED-Online. Most credits are documented at least in part by means of the LEED Online Templates. A LEED Online Template is to be filled out for each credit the project team is claiming for the building. For example, to demonstrate that the LEED-NC Prerequisite 1 for Construction Activity Pollution Prevention in the Sustainable Sites (SS) category has been adequately addressed, the civil engineer or other responsible party must fill out the LEED Online Template designated for this purpose, stating that the project followed US Environmental Protection Agency (EPA) Document EPA 832/R-92-005 (September 1992), "Storm Water Management for Construction Activities," or local erosion and sedimentation control standards and codes, whichever is more stringent. A brief list of the measures actually implemented must also be provided, along with a description of how they meet or exceed the local or EPA standards. The effort to document that this prerequisite has been met should be factored into the overall design and construction process to ensure that all the documentation has been prepared by the completion of construction.

Another example of required documentation is LEED Materials and Resources (MR) Credit 4 for Recycled Content. This credit is achieved if the project team can demonstrate that 10 percent of the value of the nonmechanical and nonelectrical materials in the building have a combination of postconsumer and preconsumer recycled content. Only one-half of the preconsumer content can be included in the calculation. For this MR credit, the architect, owner, or other responsible party must state that these requirements have been met and include details about products, product value, postconsumer and preconsumer recycled content, and the resulting overall recycled content for the project.

The project team also must decide at the start of the project how information will flow among the various parties and who will actually compile and produce the information for the appropriate LEED Online Template. For the MR credit, the calculations provided with the Online Template must clearly demonstrate that a requirement has been met by indicating the product or material, its value, and its postconsumer and preconsumer recycled content. The final computation should demonstrate that at least 10 percent of the total value of the materials, excluding mechanical and electrical systems, is recycled content, counting postconsumer content at its full percentage and preconsumer at half of its percentage in each product. This requirement can be challenging for products such as glass and aluminum storefronts, where part of the aluminum components may have recycled content but the glass will not. Additionally, because the product is likely to be assembled by a local glass subcontractor, that firm must research this information for its product. The contractor then compiles the information on the recycled content for all products to produce a final picture of the total recycled content of the project. Finally, either the contractor or the architect submits these data along with the MR Template for the project at LEED-Online.

It should also be noted that the USGBC audits submissions, meaning that much more extensive backup information may be required to verify the assertions made in the Templates. Therefore, it is good practice to ensure that full documentation is maintained throughout the design and construction processes and that all assumptions are clearly stated in the backup materials.

The LEED building assessment system is covered in detail in Chapter 5.

GREEN GLOBES DOCUMENTATION

Green Globes relies on an online questionnaire that the project team should utilize to guide the green aspects of the design and construction process. A careful review of the questionnaire should alert the team that, for example, as an indicator of Integrated Design, meetings should be held and documented to demonstrate that Integrated Design was indeed being fostered (see Section A1.4 of the Green Globes rating system in Table 6.1 of Chapter 6). Another indicator of integrated design is the appointment of a Green Globes Professional (GGP) who must be assigned duties such as

- Outlining the overall green design framework for the project
- Communicating the client's/user's intentions to the project team
- Developing measurable green design performance requirements
- Assisting in evaluating responses against the green design objectives

A careful review of the questionnaire will provide the project team with valuable information about the required documentation and what the Green Globes Assessor, who audits the project documentation in an on-site visit at the conclusion of construction, will be reviewing to determine if the documentation is adequate.

Case Study: Theaterhaus, Stuttgart, Germany

Theaterhaus represents some of the most advanced building engineering in Germany, a country where innovative design is the norm. It is a center for culture and arts in the Feuerbach area of Stuttgart and a meeting place for artists and literati to converse and collaborate about the state of music, literature, theater, and a wide range of other fine arts. In addition to being a theater complex and concert hall, the facility serves as a place for teaching music, as a gathering place for youth, and as a sports hall (see Figures 7.5–7.7). Theaterhaus is perhaps the largest naturally ventilated theater complex in the world and was designed with the dual goals of meeting the needs of the arts community while also producing an exemplar of ecologically responsive building and construction. It is a restoration of an industrial building known as the Rheinstahl-Werk, or Rhine Steelworks, which was built in the eastern section of Feuerbach in 1923. In spite of the industrial nature of the building, it was quickly recognized as an exceptional work of architecture and art, and in 1986, it was declared an official cultural monument of Stuttgart. In the process of

Figure 7.5 The entry to Theaterhaus in Stuttgart, Germany, is marked by an overhead shipping container with its name and logo.

Figure 7.6 The beautiful brickwork of the 1920s era Rhein Stahlwerk was preserved in its restoration and conversion into a cultural center.

Figure 7.7 The interior of Theaterhaus just inside the entrance, showing the staircase leading to the four performance halls. Extensive effort was made to preserve as much of the industrial character of the building as possible.

Figure 7.8 The key element of the Theaterhaus natural ventilation system is a 93-foot (30-meter) chimney, which induces airflow from outside the building, through the building interior space, to be exhausted by the stack, or chimney, effect. The chimney was a feature that was added to the building and which adds to the industrial appeal and appearance of the former steelworks building.

restoration and conversion to a cultural complex, the beautiful brick walls of the former Rheinstahl-Werk were meticulously preserved and have become a focal point upon approach. The renovation of the Rheinstahl-Werk into the Theaterhaus resulted in a total floor area of 122,000 ft^2 (12,200 m^2), with 82,000 ft^2 for the four performance spaces. About 10,000 ft^2 (1000 m^2) is provided for an organization called *Musik der Jahrhundert* (Music of the Century) and 30,000 ft^2 (2000 m^2) for an administrative area. In the theater zone, Hall 1 has a floor area of 7500 ft^2 (750 m^2) and seating for 1050 spectators. Hall 2 seats 450; Hall 3 seats 350; and Hall 4, the smallest performance space, seats 150. There are also several rehearsal spaces located throughout the theater zone in the building. The budget for the project, which was completed in 2003, was 17 million euros.

In addition to the remarkable restoration effort with respect to the building façade and interior structure, significant effort was invested in the design of a hyperefficient heating, cooling, and ventilation system that has resulted in remarkable energy performance. The energy required for moving air through the building was reduced 90 percent through the use of a natural ventilation system connected between the four theater spaces and the outside. A large exhaust chimney was appended to the top of the structure, and its shape and size induces airflow through the building by taking advantage of the buoyancy of warming air, the so-called *chimney effect.* Large intake louvers on the exterior of the building are connected to the chimney via a pathway that includes an earth-coupled canal, which cools the outside air in summer and warms it in winter. Air rising in the exhaust chimney induces this airflow, which, after passing through the earth canal, flows into each of the four theaters to meet their heating and cooling requirements (see Figures 7.8–7.13). In summer, no additional cooling is provided to temper the air flowing through the building, and the air cooled by ground contact is the sole medium provided for cooling. In winter, additional heating is provided as needed to boost the air warmed by the earth canal to suitable temperatures for conveyance into the building spaces. Additionally, in winter, a heat exchanger moves energy from the large exhaust chimney airstream to the intake air to warm it from an outside air temperature of 22 to 46°F (−5 to 8°C). If needed, an air heater boosts the temperature to about 68°F (20°C) before it is conducted into the theater spaces. If higher airflows are needed, fans can be used to move additional air

Figure 7.9 The grilles for the outside air intake for Theaterhaus are located on the side of the building. Air is induced to flow through the building by the rising warm air in the exhaust air chimney located on the top of the building.

Figure 7.10 After entering the building through the outside air intake grilles, air flows through the earth canal, which cools the air via ground contact in summer and heats it in winter.

Figure 7.11 Grilles located under the seats in the performance halls are the locations where air flows into the spaces for heating, ventilating, and cooling. The main mode of operation is natural ventilation, with airflow being induced by the buoyancy or rising warm air in the exhaust air chimney.

through the facility. Similarly, in summer, the outside air, which may be as warm as 90°F (32°C), is cooled to 77°F (25°C) before it is conducted into the spaces. Transsolar Energietechnik Engineering GmbH in Stuttgart, the designers of the natural ventilation–based HVAC system, predicts that, in addition to the ventilation energy savings of 90 percent, heating demand is reduced by about 20 percent and cooling demand by 100 percent.

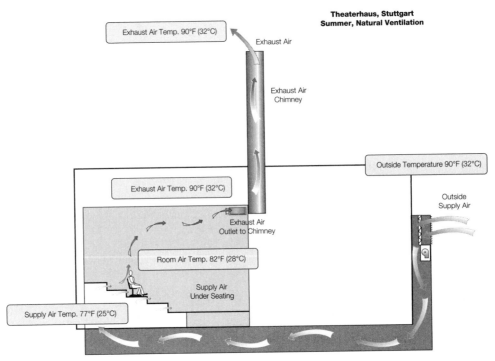

Figure 7.12 The summer natural ventilation scheme for Theaterhaus brings warm to hot air from outside and cools it in an underground tunnel by ground contact. The air then moves into the theater spaces and is exhausted to the large 20-meter chimney located on the roof of the building. When the natural ventilation system is active, the entire airflow is induced by the warm air rising in the chimney. (Illustration courtesy of Transsolar Energietechnik GmbH)

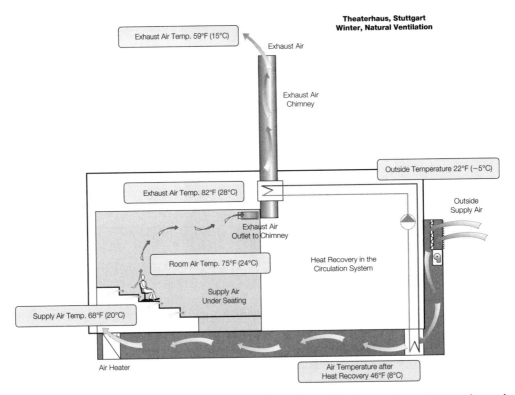

Figure 7.13 In the winter natural ventilation mode, air is conducted from the outside through the underground tunnel, which, together with a heat recovery system, warms the outside air. Additional heating is provided, if needed, to raise the supply air temperature into the theater to about 68°F (20°C). The heat recovery system is used to move energy from the air leaving the theater spaces to the outside air supply stream. (Illustration courtesy of Transsolar Energietechnik GmbH)

Summary and Conclusions

The process of green building and its delivery system are unique in that they provide not only improved buildings to owners but also an improved process. In a short time, this movement has developed several key elements that will undoubtedly find their way into mainstream construction, among them better teamwork among project team members, the use of the charrette to maximize input and creativity at the start of the design process, and the extensive use of building commissioning as a tool for ensuring that owners receive precisely the buildings they anticipated. In effect, this delivery system is based on the conventional construction management-at-risk delivery system, with significant improvements in the areas of collaboration and communication among the project team members. The design-build delivery system can also be modified to a green building delivery system by selecting a team with familiarity with green building and an orientation to environmentally friendly design and construction practices. IPD is a much newer construction delivery system with many attributes that make it highly compatible with the green building delivery process. The end result in either case should be a vastly superior end product, not only in its environmental attributes but also in the quality of design and construction, due to the improved working atmosphere fostered by the green building concept.

Notes

1. The definition of design-build is from the website of the Design-Build Institute of America, www.dbia.org.
2. A white paper on IPD (2011) written by Charles Thomsen of the Construction Management Association of America can be found at charlesthomsen.com/essays/ Managing Integrated Project Delivery.pdf.
3. As of January 2012, there were 55 green building projects at the University of Florida registered or certified by the USGBC LEED building rating system. More on these projects can be found at www.facilities.ufl.edu/leed/index.php.
4. The Florida Green Building Coalition Green Commercial Building Designation Standard can be found at www.floridagreenbuilding.org/standard.
5. On July 5, 2011, the city council of Scottsdale, Arizona, adopted the International Green Construction Code as the core component of its voluntary Commercial Green Building Program. This significant step makes it easier for developers of commercial and multifamily housing to be green certified. The new code provides flexibility to adapt to Scottsdale's geographic conditions and environmental quality of life while promoting uniformity and consistency from city to city. By integrating the voluntary code into the city's plan review and inspection process, green certification is streamlined and a Green Certificate of Occupancy is issued following the final building inspection. A report on this development can be found at www.scottsdaleaz.gov/ greenbuilding.
6. "Writing the Green RFP" can be found at the AIA COTE website, www.aia.org/ practicing/groups/kc/AIAS074658?dvid=&recspec=AIAS074658. The guide also provides examples of green RFPs/RFQs and highlights the experience of some people who have had a role in writing this type of document. It also contains "Sustainable Design Basics" and "Frequently Asked Questions (FAQs)" sections.
7. Current information about the LEED Accredited Professional Exam and the latest requirements can be found at the Green Building Certification Institute (GBCI) website, www.gbci.org/main-nav/professional-credentials/credentials.aspx#.
8. As found in the Building Toolbox section of the US Department of Energy's Building Technology Program at www1.eere.energy.gov/buildings/codes.html.
9. The potential Project Management points of Green Globes can be found at www .thegbi.org/assets/pdfs/Green-Globes-NC-Criteria-and-Point-Allocation.pdf.

10. The national workshop on IDP was held in Toronto, Canada, in October 2001. An excellent document describing the Canadian perspective on IDP is "Integrated Design Process Guide," written by Alex Zimmerman in 2006, available at www .cmhc-schl.gc.ca/en/inpr/bude/himu/coedar/upload/article_design_guide_en_aug23.pdf.

11. The concept of whole-building design and an online reference, "The Whole Building Design Guide," can be found at www.wbdg.org.

12. Adapted from the "Four Guiding Principles," proposed by the National Charrette Institute, available at www.charretteinstitute.org.

13. Adapted from the four-step charrette process proposed by the National Charrette Institute, available at www.charretteinstitute.org.

Chapter 8

The Sustainable Site and Landscape

Land use and landscape design are closely coupled—and offer perhaps the greatest opportunity for innovation in the application of resources needed to create the built environment. Buildings, while altering the local ecosystem, can become a contributory part of the ecosystem and function synergistically with nature. Carefully designed and executed work by architects, landscape architects, civil engineers, and construction managers is required to produce a building that optimizes the use of the site; that is highly integrated with the local ecosystem; that carefully considers the site's geology, topography, solar insolation, hydrology, and wind patterns; that minimizes impacts during construction and operation; and that employs landscaping as a powerful adjunct to its technical systems. Other members of the project team must also have a voice in the decisions made about land. The location of the facility on the site, the type and color of exterior finishes, and the materials used in parking and paving all affect the thermal load on the building and hence the design of the heating and cooling systems by the mechanical engineer. Minimizing the impact of light pollution requires the electrical engineer to carefully design exterior lighting systems to eliminate unnecessary illumination of the building's surroundings. Providing access to mass transportation, encouraging bicycling and alternative-fuel vehicles, or accommodating alternative-fuel vehicles ensures that the greater context beyond the building is not neglected. Collaboration among all these players marks high-performance green building as a distinct delivery system and is essential to make optimal use of the site and landscape.

Site and landscape also provide the opportunity to move beyond mere greening to the potential restoration of the land as an integral part of the building project. Until the advent of the green building movement, scant attention was paid to the impacts of construction on the environment, particularly on the land. Buildings alter the ecology, biodiversity, fecundity, and hydrology of the site, leaving it in a degraded state. Contemporary green building approaches call for the reuse of land, its cleanup in the case of contaminated land, and increasing density to minimize the need for greenfield development.

In the context of green buildings, the role of the landscape architect should perhaps be redefined from that of simply providing exterior amenities for the project to serving as the integrator of ecology and nature within the built environment. Because they are probably the best-equipped members of the project team to deal with natural systems, landscape architects should also provide expertise to the rest of the project team on the relationship between buildings and natural systems.

Historically, there has not necessarily been a recognizable connection between landscape architecture and the environment. As noted by Robert France in a 2003 critique of landscape architecture, "[T]he desire of planners to make their personal mark on the landscape, and of ecologists to understand the workings of nature, can be at odds with a desire to preserve, protect, and restore environmental integrity."[1] It might even be useful at this point in the evolution of the green building delivery system for members of this profession to review the term *landscape architect* and consider a more appropriate one, perhaps *ecological architect*. At present, there is no professional on the conventional project team with the knowledge of buildings, ecology, and the flow of matter and energy across the human-nature interface. New, emerging topics for landscape design include stormwater uptake, wastewater treatment, food production, contaminant remediation, and assisting in heating and cooling buildings. New approaches that include a robust role for natural systems in buildings are at the cutting edge of high-performance building and point to areas where their design must eventually evolve.

The appropriate use of land is a major issue in green building, if for no other reason than a building designed and constructed to the most exacting green building standards will be badly compromised if the users or occupants must drive long distances to reach it. Other land issues include building on environmentally sensitive property, in flood-prone areas, or on greenfields, or agricultural land, instead of on land already affected by human activities. Putting formerly contaminated land, or brownfields, back into productive use in a building project has the dual advantage of improving the local environment and recycling land, as opposed to employing greenfields. Contemporary green building approaches also require far more care in the use of the building site. In a green building project, the construction footprint is typically minimized, and the construction manager plans the construction process to minimize the destruction of plants and animal habitat from soil compaction. Erosion and sedimentation control are emphasized, and detailed planning of systems to minimize soil flows during construction is part of the green building delivery system. The potential for so-called heat islands, caused by the use of energy-storing materials in the building and on the site, is addressed. Likewise, the issue of light pollution from buildings is addressed in the design of a high-performance green building.

Land and Landscape Approaches for Green Buildings

Buildings require several categories of resources for their creation and operation: materials, energy, water, and land. Land, obviously, is an essential and valuable resource, so its appropriate use is a prime consideration in the development of a high-performance building. There are several general approaches to land use that fit in with the concept of high-performance green buildings:

- Building on land that has been previously utilized instead of on land that is valuable from an ecological point of view
- Protecting and preserving wetlands and other features that are key elements of existing ecosystems
- Using native and adapted, drought-tolerant plants, trees, and turf for landscaping
- Developing brownfields, properties that are contaminated or perceived to be contaminated
- Developing grayfields, areas that were once building sites in urban areas

- Reusing existing buildings instead of constructing new ones
- Protecting key natural features and integrating them into the building project for both amenity and function
- Minimizing the impacts of construction on the site by minimizing the building footprint and carefully planning construction operations
- Minimizing earth moving and compaction of soil during construction
- Fully using the sun, prevailing winds, and foliage on the site in the passive solar design scheme
- Maintaining as much as possible the natural hydroperiod of the site
- Minimizing the impervious areas on the site through appropriate location of the building, parking, and other paving
- Using alternative stormwater management technologies, such as green roofs, pervious pavement, bioretention, rainwater gardens, and others, which assist on-site or regional groundwater and aquifer recharge
- Minimizing heat island effects on the site by using light-colored paving and roofing, shading, and green roofs
- Eliminating light pollution through careful design of exterior lighting systems
- Using natural wetlands to the maximum extent possible in the storm-water management scheme and minimizing the use of dry-type retention ponds
- Using alternative stormwater management technologies such as pervious concrete and asphalt for paved surfaces

These approaches cover a wide range of possibilities. Their general purpose is to integrate nature and buildings, reuse sites that have already been impacted by human activities, and minimize disturbances caused by the building project.

Land Use Issues

The selection of a building site is generally the purview of the building owner, but often it may be affected by input from members of the project team. Rinker Hall at the University of Florida in Gainesville, a green building that achieved a gold certification under the US Green Building Council (USGBC) Leadership in Energy and Environmental Design (LEED) rating system, was originally slated for construction in an open space on the campus that had previously provided environmental amenity and recreation. However, following interaction between the project user group and the university administrators, the building was relocated to a plot of used land—in this case, a parking lot. The general population of the university benefited by this move in that it did not lose the environmental amenity of the open space, and, as it turned out, Rinker Hall's new location was a far more prominent site than its original location. One of the most important green measures in siting a new building is to locate it where the need for automobiles is minimized while conserving open space and amenities. Consequently, urban locations reasonably close to mass transit are highly desirable. In some cases, additional discussion with local government and the local transit authority may be required to articulate the need for bus service to what would otherwise be a good location for the facility.

In this section, several issues related to land use and siting are covered: the loss of prime farmland; building in 100-year-flood zones; using land that is

Figure 8.1 In the United States, farmland is being lost at the rate of 2 acres (0.8 hectare) per minute, with the most fertile, productive land being lost most rapidly. Farms abutting urban areas, as shown here, are especially threatened by land development and urban sprawl.

habitat for endangered species; and reusing brownfields, grayfields, and blackfields. These topics are addressed in the USGBC LEED and the Green Globes building assessment standards.

LOSS OF PRIME FARMLAND

In addition to addressing concerns over the loss of ecosystems, the green building effort considers the loss of agricultural land that, although impacted by human activities, is an important renewable resource (see Figure 8.1). Of the various categories of agricultural land, prime farmland is especially important to preserve. The US Department of Agriculture (USDA) defines prime farmland as follows:[2]

> Prime farmland is land on which crops can be produced for the least cost and with the least damage to the resource base. Prime farmland has an adequate and dependable supply of moisture from precipitation or irrigation and favorable temperature and growing season. The soils have acceptable acidity or alkalinity, acceptable salt and sodium content, and a few rocks. They are not excessively eroded. They are flooded less often than once in two years during the growing season and are not saturated with water for a long period. The water table is maintained at a sufficient depth during the growing season to allow cultivated crops common to the area to be grown. The slope ranges mainly from 0 percent to 5 percent. To be classified as prime, land must meet these criteria and must be available for use in agriculture. Land committed to nonagricultural uses is not classified as prime farmland.

In its publication "Farming on the Edge," published in 1997, the American Farmland Trust made the following observations about the impacts of development on the nation's farmland.[3]

- Every single minute of every day, America loses 2 acres of farmland. From 1992 to 1997, more than 6 million acres of agricultural land were developed, an area the size of Maryland.
- Farm and ranch land were lost at a rate 51 percent faster in the 1990s than in the 1980s. The rate of loss for 1992 to 1997, 1.2 million acres per year, was 51 percent higher than from 1982 to 1992.
- The best land, the most fertile and productive, is being lost the fastest. The rate of conversion of prime land was 30 percent faster, proportionally,

than the rate for nonprime rural land from 1992 to 1997. This results in *marginal* land, which requires more resources, like water, being put into production.

- Food is increasingly in the path of development: 86 percent of US fruits and vegetables, and 63 percent of our dairy products, are produced in urban-influenced areas.

- Wasteful land use is the problem, not growth itself. From 1982 to 1997, the US population grew by 17 percent, while urbanized land grew by 47 percent. Over the past 20 years, the acreage per person for new housing almost doubled; and since 1994, 10-plus acre housing lots have accounted for 55 percent of the land developed.

- Every state is losing some of its best farmland. Texas leads the nation in high-quality acres lost, followed by Ohio, Georgia, North Carolina, and Illinois. And for each of the top 20 states, the problem is getting worse.

Redirecting development away from prime farmland is addressed in the USGBC LEED-NC and Green Globes building assessment standards and by the Sustainable Sites Initiative™, indicating that preserving these valuable resources is high on the priority list for green building projects.

GREENFIELDS, BROWNFIELDS, GRAYFIELDS, AND BLACKFIELDS

Greenfields are properties that have experienced little or no impact from human development activities. Greenfields can also be defined to include agricultural land that has had no activity other than farming. Like recycling in general, recycling of land is an important objective in creating high-performance green buildings. *Land recycling* refers to reusing land impacted by human activities instead of using greenfields. There are at least three identifiable categories of potentially recyclable land: brownfields, grayfields, and blackfields.

The US Environmental Protection Agency (EPA) defines *brownfields* as abandoned, idled, or underused industrial and commercial facilities where expansion or redevelopment is complicated by real or perceived environmental contamination.[4] The official definition of a brownfield site, according to Public Law 107-118 (H.R. 2869), the Small Business Liability Relief and Brownfields Revitalization Act, signed into law January 11, 2002, is as follows: "With certain legal exclusions and additions, the term 'brownfield site' means real property, the expansion, redevelopment, or reuse of which may be complicated by the presence or potential presence of a hazardous substance, pollutant, or contaminant." The key word in the first definition is *perceived*; the key phrase in the second is *potential presence.* Former industrial properties are often thought to be contaminated because of the activities that occurred on the site—for example, metal plating or leather tanning. In fact, not infrequently these properties are fairly clean, requiring minimal cleanup. In many US cities, brownfields are now valuable real estate because of their proximity to extensive infrastructure and a potential workforce. Industries formerly fleeing to greenfields outside urban areas, thereby causing impoverishment of minority communities due to job loss, are returning to former industrial sites because the economics dictate the return to the city. A prime example of the potential success of a well-developed brownfields strategy is the Chicago Brownfields Initiative, which since 1993 has been assisting in the cleanup and transfer of 12 major former industrial sites in the city. An interesting aspect of the Chicago strategy has been to emphasize the return of these zones to industrial use, thus bringing jobs back into the city.[5]

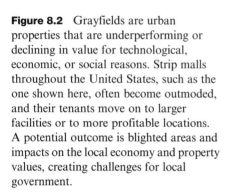

Figure 8.2 Grayfields are urban properties that are underperforming or declining in value for technological, economic, or social reasons. Strip malls throughout the United States, such as the one shown here, often become outmoded, and their tenants move on to larger facilities or to more profitable locations. A potential outcome is blighted areas and impacts on the local economy and property values, creating challenges for local government.

The LEED and Green Globes building assessment systems and the Sustainable Sites Initiative™ provide credit for the use of a former brownfield as a building site. According to a USGBC credit ruling on brownfields, the project team can consider a site not officially designated a brownfield by the EPA if the team can convince the EPA that the site fulfills the requirements of a brownfield and the EPA agrees in writing.

Grayfields, another form of urban property, can be defined as blighted or obsolete buildings sitting on land that is not necessarily contaminated (see Figure 8.2). The term *grayfield* is actually an expanded definition of *brownfield*. The state of Michigan, for example, embeds the term *grayfield* in the concept of *core community*, areas that are economically blighted and need investment to restore them to economic health. A grayfield could be a former machine shop that has become obsolete perhaps because it lacks a fire suppression system, had a septic system and old fuel tanks, or contains asbestos. Boarded-up housing can be an indication of a grayfield. The Congress for the New Urbanism (CNU) points out that former or declining malls can be classified as grayfields because they occupy impacted land that can be returned to productive use.[6] Declining malls are caused by a number of factors: population shifts, increasing numbers of big-box stores, changing demographics, and a failure of developers to reinvest in upgrades and modernization of older malls. Changes in the retail environment are also affecting the big-box stores as they continue to increase the size of their facilities. In June 2004, Wal-Mart Corporation listed 394 properties for sale, ranging in size from 2700 to 162,000 square feet (251 to 15,050 square meters).[7] Larger abandoned big-box stores are now referred to as *ghost boxes*. Some of the strategies communities are using to deal with these types of properties are as follows:[8]

- *Adaptive reuse.* Turning ghost boxes into office space, entertainment space, or space for light manufacturing.
- *Demalling.* Reversing storefronts to face the street; converting the property to give it a "Main Street" look; and making connections to nearby housing, using pedestrian-friendly planning.

- *Razing and reuse.* Older malls are being demolished to make room for new retail developments.

- *Passing community ordinances to prevent future grayfields and ghost boxes.* Some communities are setting a maximum size for big-box stores or requiring that an escrow account covering future demolition costs be established for the construction of a big-box store.

Both grayfields and brownfields are becoming valuable properties because of the presence of good infrastructure in urban areas; a trend toward urban living prompted by a perceived higher quality of life; and incentives offered by local and state governments in the form of tax rebates, tax credits, tax increment district financing, and other innovative strategies. In addition to access to infrastructure and a willing workforce for business, cities ultimately receive far greater tax revenues, creating a true win-win scenario. Though grayfields are not explicitly addressed in either LEED-NC or Green Globes, credits and points are awarded for building in a dense urban environment.

Yet another category of blighted land is *blackfields.* These properties are abandoned coal mines and are found in former coal-mining areas such as eastern Pennsylvania, where abandoned strip mines and subsurface mines comprise an area three times the size of Philadelphia and which will require an estimated $16 billion to clean up. Surface waters in these zones have a very low pH and are contaminated with iron, aluminum, manganese, and sulfates. The term *blackfields* also can be considered as an expanded definition of *brownfields.* There is a potential for obtaining LEED-NC points for using one of these properties for a building project.[9]

BUILDING IN 100-YEAR-FLOOD ZONES

Clearly, buildings should not be constructed in flood-prone areas due to the high potential for disasters that result not only in human suffering but also in enormous environmental and resource impacts caused by the cycle of destruction and rebuilding. This is such a vital matter that the Federal Emergency Management Agency (FEMA) has become deeply involved in issues of flood mapping and insurance. Specifically, in support of the National Flood Insurance Program (NFIP), FEMA has undertaken a massive program of flood hazard identification and mapping to produce Flood Hazard Boundary Maps, Flood Insurance Rate Maps, and Flood Boundary and Floodway Maps. Several areas of flood hazards are commonly identified on these maps. One of these areas is the *special flood hazard area* (SFHA), which is defined as an area of land that would be inundated by a flood having a 1 percent chance of occurring in any given year (previously referred to as the *base flood* or *100-year flood*). The 1 percent annual chance standard was decided after considering various alternatives. The standard constitutes a reasonable compromise between the need for building restrictions to minimize potential loss of life and property and the economic benefits to be derived from floodplain development. Development may take place within the SFHA, provided that it complies with local floodplain management ordinances, which must meet the minimum federal requirements. Flood insurance is required for insurable structures within the SFHA to protect federally funded or federally backed investments and assistance used for acquisition and/or construction purposes within communities participating in the NFIP.[10]

Before continuing with this discussion, it is important to point out that the term *100-year flood* is misleading. It is not the flood that will occur once every 100 years; rather, it is the flood *elevation* that has a 1 percent chance of being equaled or exceeded each year. Thus, the 100-year flood could occur more than once in a relatively short period of time. The 100-year flood, which is the standard used by most federal and state agencies, is also used by the NFIP as its standard for floodplain management and to determine the need for flood insurance. A structure located within an SFHA shown on an NFIP map has a 26 percent chance of suffering flood damage during the term of a 30-year mortgage.

To earn points when attempting to certify a green building using either LEED or Green Globes, the elevation of the building site must be at least 5 feet (1.52 meters) above the 100-year floodplain.

THREATENED AND ENDANGERED SPECIES

Passed in 1973 and reauthorized in 1988, the Endangered Species Act (ESA) regulates a wide range of activities affecting plants and animals designated as endangered or threatened. By definition, an *endangered species* is an animal or plant listed by regulation as being in danger of extinction. A *threatened species* is any animal or plant that is likely to become endangered within the foreseeable future. A species must be listed in the *Federal Register* as endangered or threatened for the provisions of the act to apply.

The ESA prohibits the following activities involving endangered species:

- Importing into or exporting from the United States
- Taking (which includes harassing, harming, pursuing, hunting, shooting, wounding, trapping, killing, capturing, or collecting) within the United States and its territorial seas
- Taking on the high seas
- Possessing, selling, delivering, carrying, transporting, or shipping any such species unlawfully taken within the United States or on the high seas
- Delivering, receiving, carrying, transporting, or shipping in interstate or foreign commerce in the course of a commercial activity
- Selling or offering for sale in interstate or foreign commerce

The ESA also provides for:

- Protection of critical habitat (habitat required for the survival and recovery of the species)
- Creation of a recovery plan for each listed species

The US Fish and Wildlife Service reported the following statistics for endangered and threatened species in the United States, as of July 2012:

- 600 US species of animals are listed.
- 794 US species of plants are listed.
- 30 US species of animals are currently proposed for listing.
- 5 US species of plants are currently proposed for listing.

As is the case with construction within a 100-year-flood zone, LEED and Green Globes do not provide credit if the project site is on land identified as habitat for species that are on state or federal lists of threatened or endangered species.

TABLE 8.1

Principles and Best Practices for Sedimentation and Erosion Control

Design the project to fit the site's context: its topography, soils, drainage patterns, and natural vegetation.

Minimize the area of construction disturbance and limit the removal of vegetative cover.

Remove viable topsoil for temporary stockpiling and reuse when the landscape is installed.

Reduce the duration of bare-area exposure by scheduling construction such that bare areas of the site are exposed only during the dry season or for as short a time as possible.

Decrease the amount of bare area exposed at any one time.

Shield soil from the impact of rain or runoff by using temporary vegetation, mulch, or groundcover on exposed areas.

Divert run-on and runoff water away from exposed areas.

Prevent off-site runoff from entering the site.

Inspect and maintain the erosion and sediment control practices that have been put in place.

Use vegetative buffer strips, mulching and temporary seeding, surface roughening, erosion control blankets, permanent vegetation, and gravel-surfaced construction areas for erosion control.

Use silt fences, fiber wattles, and logs; check dams in swales, sediment traps and basins, detention/retention ponds, and silt filters/inlet traps for sedimentation control.

Where high winds are likely to transport soil, use sand or wind fences as a barrier to soil movement.

When restoring or replacing soil, use native soil from nearby so that type, composition, microbes, and hydrologic characteristics are compatible with the region and are suitable for plants that will be used, especially native or adapted plants of the region.

SOIL EROSION AND SEDIMENT CONTROL

Sediment is eroded soil that is suspended, transported, and/or deposited by moving water or wind. *Erosion* is the process of displacing and transporting soil particles by the action of gravity. Some general principles and best management practices that should be used in sediment and erosion control are indicated in Table 8.1.

For high-performance green buildings, care must be taken to ensure that soil loss is minimized. The construction manager and subcontractors must pay attention to soil loss in the form of airborne dust and stormwater runoff. Additionally, the contemporary green building delivery system requires that measures be put in place to prevent the sedimentation of both stormwater systems and receiving water bodies. An erosion and sedimentation control plan is a prerequisite for certification under LEED-NC, meaning that this plan is required for the building to be considered for even the lowest level of certification. Green Globes awards a range of points for erosion and sedimentation control measures in Part B of this building assessment standard. Similarly the Sustainable Sites Initative[TM] awards credit for minimizing soil disturbance during construction.

Sustainable Landscapes

The advent of high-performance green buildings is causing noteworthy changes to the traditional notion of the constructed landscape. Landscape design has typically been an afterthought in the conventional building delivery system, and in many cases, it is given very low priority. As funding for a project becomes tighter near the end of construction, it will likely be the budget for the constructed landscape that will be reduced to the bare minimum. The outcome of such conventional thinking is that landscape design is given short shrift, treated *apart* from the building rather than *integral* to it. Today, the role of landscape design in high-performance building is in a state of transition; some projects treat it conventionally, while others realize that the role of the site is critical to the performance of the buildings, both individually and collectively. Among these new roles are to assist building heating and cooling, help control stormwater and eliminate stormwater infrastructure, treat waste, provide food, and contribute to biodiversity.

The concept of *sustainable landscape* predates the contemporary high-performance green building movement. The term emerged in the vocabulary of landscape architecture in 1988, when the Council of Educators in Landscape Architecture defined it as landscapes that contribute ". . . to human well-being and at the same time are in harmony with the natural environment. They do not deplete or damage other ecosystems. While human activity will have altered native patterns, a sustainable landscape will work with native conditions in its structure and functions. Valuable resources—water, nutrients, soil, etc.—and energy will be conserved, diversity of species will be maintained and increased."[11] The movement to reconsider the role of landscape architecture was initiated by John Tillman Lyle with the publication of his 1985 book, *Design for Human Ecosystems: Landscape, Land Use, and Natural Resources.* It was almost a decade, however, before more was heard on the subject of sustainable landscapes. In 1994, two volumes appeared, coincidentally at the onset of the American green building movement: Robert Thayer's *Gray World, Green Heart: Technology, Nature and the Sustainable Landscape*, and another book by Lyle, *Regenerative Design for Sustainable Development.*

In *Design for Human Ecosystems*, Lyle considered how landscape, land use, and natural resources could be shaped to make the human ecosystem function in the sustainable ways of natural ecosystems. He suggested that designers must understand ecological order and how it operates at a wide variety of scales, from minute to global. The understanding of ecological order has to be linked with human values in order to develop solutions that are long-lasting, beneficial, and responsible.

In *Gray World, Green Heart*, Thayer notes that landscape is the place where ". . . the conflict between technology and nature is most easily sensed." A sustainable landscape, according to Thayer, would have the following properties:

- An alternative landscape where natural systems are dominant.
- A landscape where resources are regenerated and energy is conserved.
- A landscape that allows us to see, understand, and resolve the battle between the forces of technology and nature.
- A landscape where essential life functions are undertaken, revealed, and celebrated.
- A landscape where the incorporated technology is sustainable, the best of all possible choices, and can be considered part of nature.
- A landscape that counters the frontier ethic of discovery, exploitation, exhaustion, and abandonment with one where we plant ourselves firmly, nurture the land, and prevent ecological impoverishment.
- A landscape that responds to the loss of place with reliance on local resources, celebration of local cultures, and preservation of local ecosystems.
- A landscape that responds to the view that landscape is irrelevant by making the physical landscape pivotal to our existence.

Thayer admits that this vision is utopian but suggests that such a vision is needed to give us direction. He goes on to provide five characteristics of a sustainable landscape that are based on the function and organization of natural landscapes:

1. Sustainable landscapes use primarily renewable, horizontal energy[12] at rates that can be regenerated without ecological destabilization.

2. Sustainable landscapes maximize the recycling of resources, nutrients, and by-products, and produce minimum waste or conversion of materials to unusable locations or forms.

3. Sustainable landscapes maintain local structure and function and do not reduce the diversity or stability of the surrounding ecosystems.

4. Sustainable landscapes preserve and serve local human communities rather than change or destroy them.

5. Sustainable landscapes incorporate technologies that support these goals and treat technology as secondary and subservient, not primary and dominant.

As a cautionary note, Thayer also tells us that "Without sustainable values, landscapes designed to be sustainable will be misused, become unsustainable, and fail." Contemporary American culture does not have a sense of, nor does it value, place, and it is oriented toward consumption, profit, and waste. Creating a sustainable landscape in the face of these values is challenging but necessary to at least launch a countermovement that values nature and ecosystems and that helps increase human awareness of their role in daily life (see Figure 8.3).

In *Regenerative Design for Sustainable Development*, Lyle introduced designers of the built environment to the concept of regenerative landscape, reminding them, as John Dewey did in 1916, that ". . . the most notable distinction between living and inanimate things is that the former maintain themselves by renewal."[13] He maintains that the developed landscape, the one created and built by humans, should be able to survive within the bounds of local energy and materials flows and that, in order to be sustainable, it must be *regenerative*, which, in the case of landscape, means being capable of *organic self-renewal*. Landscapes must be created using regenerative design, that is, design that creates cyclical flows of matter and energy within the landscape. According to Lyle, a regenerative system is one that provides for continuous replacement, through its own functional processes, of the energy and materials used in its operation. A regenerative system has the following characteristics:

- Operational integration with natural processes and, by extension, with social processes

- Minimum use of fossil fuels and man-made chemicals, except for backup applications

- Minimum use of nonrenewable resources, except where future reuse or recycling is possible and likely

- Use of renewable resources within their capacities for renewal

- Composition and volume of wastes within the capacity of the environment to reassimilate them without damage

Lyle gained considerable experience with regenerative landscapes as a professor at the 1-acre Center for Regenerative Studies that he founded at California State Polytechnic University in Pomona, where faculty and students worked with regenerative landscapes and technology to try to solve the daily

Figure 8.3 (A) The landscape design for NASA's Space Life Sciences Laboratory at Kennedy Space Center in Florida is self-maintaining and was envisioned as a model of environmental site design, with over 60,000 square feet of native grasses and wildflowers. (B) The building orientation reduces heat load and minimizes encroachment into isolated wetlands. (Photos from Zamia Design, Inc.)

problems of providing shelter, food, energy, and water and dealing with waste. He and his students took what was then a compacted cow pasture within sight of a large landfill and created what a former center director, Joan Stafford, described as a landscape that ". . . now yields armfuls of scented, exuberant lavender, sage, [and] rosemary, growing from rejuvenated soils."

TABLE 8.2

Principles of Sustainable Landscape Construction

Principle 1: Keep sites healthy. Ensure that biologically productive sites with healthy ecosystems are not harmed by the building project. Special attention must be paid to utility installation and road construction, which can be especially destructive to natural systems.

Principle 2: Heal injured sites. Using grayfields, brownfields, or blackfields reduces pressures on biologically productive sites and can result in restoration of blighted properties to productive ecosystems.

Principle 3: Favor living, flexible materials. Slope erosion can be controlled with living structures rather than artificial physical structures. Greenwalls, artificial structures that provide a support system for living matter, may be needed in especially steep terrain. Living materials on roofs create eco-roofs that provide additional green area and assist heating and cooling.

Principle 4: Respect the waters of life. Water bodies, including wetlands, should be protected and even restored. Rainwater can be harvested from roofs, stored in cisterns, and used for nonpotable applications. Landscape irrigation should be minimized and landscape designed to be durable and drought-tolerant.

Principle 5: Pave less. Paving destroys natural systems and should be minimized. Stormwater should be quickly infiltrated through the use of porous concrete and asphalt paving and through the use of pavers. Heat islands should be minimized by appropriate landscaping.

Principle 6: Consider the origin and fate of materials. Minimize the impact of landscape materials by carefully analyzing their embodied energy and other effects. Emphasize reused and recycled materials and avoid toxic materials.

Principle 7: Know the costs of energy over time. Landscape construction requires considerable energy in the form of work by machinery, the embodied energy of materials. The total energy consumption for all purposes, including maintenance, should be minimized.

Principle 8: Celebrate light, respect darkness. Landscape lighting should be accomplished such that plants are unaffected by lighting schemes, and lighting should be energy-efficient. Lighting should not spill over to areas where it is not wanted. Low-voltage lighting, fiber-optic lighting, and solar lighting should be considered.

In Sustainable Landscape Construction: A Guide to Green Building Outdoors, J. William Thompson and Kim Sorvig provide a set of principles to guide landscape design and construction for green buildings. These principles are outlined in Table 8.2.[14] In general, they are fairly straightforward and parallel the logic of LEED, which addresses many of these issues.

Some of the innovations emerging in today's high-performance green buildings include the application of landscaping directly to buildings in the form of green roofs, or living or eco-roofs, and the use of vertical landscaping, especially for skyscrapers. These two emerging landscaping concepts are described in the following sections.

GREEN, OR LIVING, ROOFS

A *green*, or *living, roof* is nothing more than an updated version of the ancient sod roof used in Europe that is making a comeback in today's green building movement. An alternative term used by some practitioners is *eco-roof.* The city

Figure 8.4 A roof garden on the Chicago City Hall containing over 20,000 plants and more than 150 species. (*Source:* City of Chicago)

of Portland, Oregon, provides tax breaks to motivate the creation of eco-roofs.[15] It approved a regulation in January 2001 that allows developers to expand their building plans if those plans include an eco-roof; it also waives certain code requirements for buildings with green roofs. Although Portland is the only city in North America to offer such an incentive, in other cities, including Chicago, Illinois; Toronto, Ontario; and Seattle, Washington, gardens are grown on the roofs of city halls and courthouses (see Figure 8.4). In Dearborn, Michigan, the Ford Motor Company has a 10-acre living roof on its Rouge Center assembly plant; county buildings in Anne Arundel County, Maryland, also have grassy roofs.

But Portland goes much further than most jurisdictions owing to the financial commitment the city has made to eco-roofs, in the form of tax breaks, grants, building code waivers, and the variety of private and public buildings with rooftop gardens.[16] Portland provides incentives for living roofs because they have been found to reduce building energy costs by 10 percent and to decrease summer roof temperatures by 70°F (21°C); furthermore, these roofs can reduce storm runoff by 90 percent and delay the flow of stormwater for several hours, thereby reducing the probability of stormwater and sewer system overflow. In an area like Portland, which suffers chronic stormwater and sewage system overflows that affect the Willamette and Columbia Rivers, an extensive array of eco-roofs on buildings may help mitigate this problem. Living roofs can also filter pollution and heavy metals from rainwater and help protect the regional water supply.

An eco-roof can fulfill several distinct roles: serving as an aesthetic feature, helping the building blend into its environment, and supporting climatic stabilization. An eco-roof is particularly useful in wet, snowy areas but has more limited potential in dry climates. Green roofs must be built on a sufficiently strong frame with carefully applied waterproofing, because it is very difficult to locate leaks once the growing medium is in place. The living aspect of the roof is a compost-based system, usually composed of a base of straw that is left to decompose, within which native or introduced plants can then take root. As might be expected, a living roof requires ongoing care; another disadvantage is that it could be a fire hazard in hot, dry climates. In contrast, it is advantageous in that it protects the waterproofing from damage by ultraviolet radiation, and it precludes the need for tiles or other shingles.

According to the Living Systems Design Guild, green roofs are generally classified as either *extensive* or *intensive*:

> *Extensive eco-roof systems.* Extensive landscaped roofs are defined as low-maintenance, drought-tolerant, self-seeding vegetated roof covers that incorporate colorful sedums, grasses, mosses, and meadow flowers that require little or no irrigation, fertilization, or maintenance (see Figure 8.5). The types of plants suitable for extensive landscaping are those native mainly from locations with dry and semidry grassy conditions or with rocky surfaces, such as an alpine environment. Extensive systems can be placed on low-slope and pitched roofs with up to a 40 percent slope.
>
> *Intensive eco-roof systems.* If there is adequate load-bearing capacity, it is possible to create actual roof gardens on many buildings. This type of eco-roof system may include lawns, meadows, bushes, trees, ponds, and terraced surfaces. Intensive systems are far more complex and heavy than extensive eco-roof systems and hence require far more maintenance.

Eco-roof systems are made up of 6 to 10 individual components, as shown in Table 8.3.[17] The soil substrate differentiates the extensive from the intensive system. The extensive system has a soil substrate of 4 to 6 inches (10 to 15 centimeters) of formulated, lightweight growing medium, whereas an intensive system may have as much as 18 to 24 inches (46 to 61 centimeters) of a heavier soil mix.

Figure 8.5 Cross section of an extensive eco-roof system that provides structure and drainage. (Image courtesy of American Hydrotech, Inc.)

TABLE 8.3

Components of Eco-Roof Systems

Plants. Extensive eco-roof systems include shallow root systems; regenerative qualities; and resistance to direct radiation, drought, frost, and wind. A much larger variety of plant selections are available for intensive roofscapes due to their greater soil depths.

Soil mix. The planting mix is a specially formulated, lightweight, moisture-retaining mix that is enriched with organic material.

Filter fabric. The filter fabric prevents fine particles from being washed out of the substrate soil, ensuring efficiency of the drainage layer.

Water retention layer. This layer is sometime used, commonly in the form of a fabric mat, to provide mechanical protection and retain moisture and nutrients. Profiled drainage elements retain rainwater for dry periods in troughs or cups on the upper side of this layer.

Drainage layer. Eco-roofs must have a drainage layer to carry away excess water; on very shallow, extensive eco-roofs, the drainage layer may be combined with the filter layer.

Root barrier. The root barrier prevents roots from affecting the efficiency of the waterproofing membrane in case it is not root-resistant.

Waterproof membrane. An eco-roof system may consist of a liquid-applied membrane or a specially designed sheet membrane.

Insulation layer. An insulation layer is optional and prevents water stored in the eco-roof system from extracting heat in the winter or cool air in the summer.

Figure 8.6 A vertical landscape at Universal City Walk in Universal City, California, provides a changing evergreen façade, which extends to a height of 75 feet (23 meters). (greenscreen®)

As should be obvious from this discussion, an eco-roof is far more complex than a conventional roof and requires significantly more research and planning. Additionally, eco-roofs generally cost twice as much as conventional roofs, or $10 to $15 per square foot ($107 to $162 per square meter). However, the payback due to energy savings alone can be fairly rapid, and the benefits of reduced stormwater infrastructure and natural water cleaning make eco-roofs an attractive option.

VERTICAL LANDSCAPING

The French designer, Patrick Blanc, is generally considered to have developed the notion of a *vertical garden* or *green wall* that takes advantage of vertical surfaces to provide buildings with some degree of ecological capacity. Skyscrapers are not normally thought of as candidates for landscaping. However, Ken Yeang, a Malaysian architect, has been advocating what he calls *vertical landscaping* to at least in part render these very large structures green. He also advocates vertical landscaping for reducing energy consumption, stating that a 10 percent increase in vegetated area can produce annual cooling load savings of 8 percent. He describes vertical landscaping as "greening the skyscraper," which he says involves introducing plants and ecosystem components at a high level, in addition to the ground-level landscaping.[18]

The vertical landscape creates a microclimate at the façade on each floor, can be used as a windbreak, absorbs carbon dioxide and generates oxygen, and improves the well-being of the occupants by providing greenery throughout the building (see Figure 8.6). This strategy also helps counterbalance the enormous mass of concrete, glass, and steel with plants and soil. In addition to these benefits, a vertical landscape that is well integrated with the building can provide architectural visual relief from otherwise uninteresting, nondescript surfaces. In order for the vertical landscaping to make visual sense, Yeang suggests that a series of stepped and linked planter boxes be designed into the building. The use of trellises also allows for vertical growth and interaction of the landscape from ground level to the roof. But because wind speeds at roof level will often be twice their ground-level speed, plants at upper levels may need protection, which can be provided by side louvers that allow the landscape to be seen, yet deflect the wind from around the plants (see Figure 8.6).

Enhancing Ecosystems

A desired outcome of any building project would be a landscape and an ecosystem that are regenerated and improved as a consequence of the project. *Environmental Building News* (*EBN*) provides a checklist for owners and designers to use in helping restore the vitality of natural ecosystems (see Table 8.4).[19] Although directed primarily at enhancing the presence of wildlife on a site, it is very useful for general ecosystem restoration or to regenerate or reconnect system components.

Stormwater Management

Transforming the natural environment by development dramatically affects the quantity and flows of stormwater across the surface of the earth. Covering natural landscapes with buildings and infrastructure replaces largely pervious

TABLE 8.4

EBN **Checklist for Wildlife Habitat Enhancement of Developed Land**

1. **Research and Planning**

   Hire a qualified consultant specializing in natural landscaping and ecosystem restoration. Test soils for contaminants. Inventory existing ecosystems.

   Research ecosystems that may have been on the site prior to European settlement. Inventory current landscape management practices. Develop an ecosystem restoration plan.

2. **Ecosystem Restoration**

   Reduce turf area.

   Eliminate invasive plants.

   Establish native ecosystems.

   Ensure diversity in plantings.

   Provide wildlife corridors.

   Use bioengineering for erosion control.

3. **Enhancements for Wildlife**

   Select native plant species that attract wildlife.

   Encourage birds to "plant" seeds of species they like.

   Provide edible landscaping. Provide "edge" areas.

   Establish a bird feeding program, if desired.

   Provide bird nesting boxes and platforms.

   Provide bat houses.

   Provide water features.

   Avoid chemical usage in the landscape.

4. **Helping People Appreciate Natural Areas and Wildlife**

   Provide wildlife viewing areas.

   Provide easy and inviting access to the outdoors.

   Provide for easy management of bird feeders and nesting boxes.

   Provide clear signage in public spaces.

   Provide features that will get people outside.

surfaces with impervious materials, thereby increasing the volume and velocity of horizontal water flows. Moreover, ecosystems, most prominently wetlands, which have a function of absorbing pulses of stormwater and returning it in a controlled manner to bodies of water and aquifers, are subject to modification or destruction by these same construction activities. One of the functions of green building is to address the issue of stormwater management by protecting ecosystems and the pervious character of the landscape, as well as to carefully consider how to affect as little as possible the natural hydroperiod of the site. *EBN* provides a useful checklist for dealing with stormwater issues; it is presented in Table 8.5.[20]

Low-Impact Development

Low-impact development (LID) is a relatively new strategy that integrates ecological systems with landscape design to effectively manage stormwater runoff. LID techniques minimize runoff to prevent pollutants from adversely impacting water quality and can decrease the required size of traditional retention and detention basins, resulting in cost savings over conventional stormwater control mechanisms. LID can be applied to new development, redevelopment, or as

TABLE 8.5

***EBN* Checklist for Stormwater Management**

Reduce the Amount of Stormwater Created

1. Minimize the impact area in a development.
2. Minimize directly connected impervious areas.
3. Do not install gutters unless rainwater is collected for use.
4. Reduce paved areas through cluster development and narrower streets.
5. Install porous paving where appropriate.
6. Where possible, eliminate curbs along driveways and streets.
7. Plant trees, shrubs, and groundcovers to encourage infiltration.

Keep Pollutants Out of Stormwater

8. Design and lay out communities to reduce reliance on cars.
9. Provide greens where people can exercise pets.
10. Incorporate low-maintenance landscaping.
11. Design and lay out streets to facilitate easy cleaning.
12. Control high-pollution commercial and industrial sites.
13. Label storm drains to discourage dumping of hazardous wastes into them.

Managing Stormwater Runoff at Construction Sites

14. Work only with reputable excavation contractors.
15. Minimize the impact area during construction.
16. Avoid soil compaction.
17. Stabilize disturbed areas as soon as possible.
18. Minimize slope modifications.
19. Construct temporary erosion barriers.

Permanent On-Site Facilities for Stormwater Control and Treatment

20. Rooftop water catchment systems
21. Vegetated filter strips
22. Vegetated swales for stormwater conveyance
23. Check dams for vegetated swales
24. Infiltration basins
25. Infiltration trenches
26. Dry detention ponds with vegetation
27. Retention ponds with vegetation
28. Constructed wetlands
29. Filtration systems

---

retrofits of existing development. LID has been applied to a range of land uses, from high-density ultra-urban settings to low-density development. An alternative terminology being used by the EPA for LID is *green infrastructure.*

In general, LID and green infrastructure refer to systems and practices of land development that use natural processes to infiltrate, evapotranspirate (return water to the atmosphere either through evaporation or by plants), or reuse stormwater runoff on the site where it is generated. LID employs principles such as preserving and re-creating natural landscape features, minimizing imperviousness, creating functional and appealing site drainage, and treating stormwater as a resource rather than as a waste product. There are many practices that can be used to implement LID, including bioretention facilities, rain gardens, vegetated roofs, rain barrels, and permeable pavements. By implementing LID principles and practices, water can be managed in a way that reduces the impact of built areas and promotes the natural movement of water within the system or watershed. Applied on an urban scale, LID can maintain or restore the watershed's hydrologic and ecological functions.

Another concept of LID in the urban setting is *artful rainwater design*. This idea is based on the premises that stormwater management techniques designed in conjunction with natural systems can provide site amenities with a strong aesthetic quality and even compel human interaction.

The LID approach to stormwater management is an enormous change from conventional practices, which historically divert stormwater through engineered conduit systems to natural water bodies or costly treatment plants. In contrast, the LID approach carefully considers the rate, volume, frequency, duration, and quality of discharge so as to allow for groundwater and aquifer recharge and the overall health of ecological systems.

The Nature Conservancy describes six principles for successful LID. These are stated below and should be used to guide the design of a LID system:[21]

1. *Use existing and valuable features.* Identify and work with all cultural and natural features that will immediately add value to the development: hedgerows, mature trees, wildlife habitats, streams, rural/architectural character, and heritage features.

2. *Let natural resources work for the project.* Use natural drainage by mimicking the existing systems and patterns. Minimizing construction disturbance and changes within the watershed will benefit the environment and reduce costs.

3. *Increase the value of the site with open spaces.* Clustering homes or buildings in the development enables the provision of open spaces and scenic views. Connections to open spaces within a larger city network can provide natural amenity, resulting in higher property values.

4. *Reduce the size of the water management needs on the site.* By limiting impervious surfaces, the amount of stormwater infrastructure can be greatly reduced. Buildings with smaller footprints, green roofs, permeable pavement, and narrow roads make stormwater management manageable.

5. *Treat stormwater close to the source.* Instead of expensive underground infrastructure, catch basins, piping, and stormwater ponds, use low-cost, low-maintenance, low-tech, nonstructural rain gardens and bioswales to infiltrate runoff.

6. *Smart landscaping can save money.* There is no question that good landscaping increases property values and that smart landscaping can also save money. Money is wasted on techniques such as clear-cutting, grading, and costly stormwater ponds that only address one problem. With a multifunctional landscape, it is possible to manage runoff, improve water quality, reduce power bills, increase property value, and save money.

The Nature Conservancy also describes 10 implementation measures that can help manage runoff while at the same time providing a landscape with natural amenity. These techniques are not meaningful individually, but as part of a larger LID strategy, they can be highly effective:[22]

1. *Impervious surface reduction.* Some techniques for reducing impervious surfaces include reducing the number of parking spaces, sharing parking with adjacent uses when possible, creating center landscape islands and cul-de-sacs, and reducing setbacks from the street to shorten driveway lengths.

2. *Tree preservation.* Not only do trees increase property value, but they are also excellent landscape features for the uptake of stormwater. Trees act as mini-reservoirs that absorb and store large quantities of water. They are excellent for controlling runoff at the source, reducing soil erosion, decreasing temperatures, absorbing carbon dioxide, and providing habitat for wildlife. A 12-inch-caliper oak tree can intercept roughly 2000 gallons of rain per year while a 30-inch-caliper maple can intercept as much as 12,000 gallons a year.

3. *Reduce lawn area/increase planted areas.* Lawns require a lot of watering, mowing, aerating, and chemicals and are not effective at absorbing water. Native grasses, shrubs, trees, and wildflowers are excellent species for absorbing stormwater. As is the case with trees, increasing the size of planting areas can result in higher property values while at the same time enhancing biodiversity.

4. *Bioswales and vegetated swales.* Bioswales assist in capturing rainwater runoff and then filter the runoff through prepared soil medium with suitable plants. They are ideal for median strips and parking lots along streets. The preferred depth of a bioswale is about 6 inches and with an overall size of 25×50 feet.

5. *Permeable pavement.* A wide range of permeable pavements are available, including interlocking pavers, grass pavers, porous asphalt, and porous concrete. These materials allow water to permeate the surface to an underlying stone or sand bed.

6. *Buffers and filter strips.* Buffers and filter strips are barriers between surfaces such as roads and parking lots, and waterways and sensitive aquatic environments. These buffers contain trees, bushes, wild grasses, and other natural plant species to remove particulates and other pollutants from stormwater crossing between the paved surfaces and sensitive bodies of water. Buffers can be linked to create a network of green infrastructure and provide opportunities and benefits to wildlife corridors. These buffers are also referred to as *conservation easements* and, like many other components of LID, can contribute to the present value of a property.

7. *Rain gardens.* A rain garden is a shallow depression planted with suitable trees, shrubs, flowers, and other species to capture stormwater runoff from impervious areas. They can be used as a buffer to capture runoff from landscaped areas before it enters a lake, pond, or river.

8. *Soil quality management.* Active soils can create standing water if the surface is impenetrable. To prevent soils from being overly compacted, driving on wet soils beyond the parking area and over tree roots should be prevented.

9. *Green roofs.* Green roofs can absorb rainwater, provide insulation, create wildlife habitat, and reduce the heat island effect. This chapter discusses the use of green roofs in high-performance green building projects.

10. *Rain barrels and rainwater harvesting systems.* A rainwater collection system can help capture and store stormwater from the roof for future use, reducing stormwater flows and decreasing water costs.

LID strategies inevitably save money. Table 8.6 provides four examples of LID projects and the economic effects of taking this approach. Illustrations of several LID projects are shown in Figures 8.7–8.9.

TABLE 8.6

Residential and Commercial Examples of LID Savings

Location	Description	LID Cost Savings
Madera Residential Subdivision Gainesville, Florida	44-acre, 80-lot development, used natural drainage depressions instead of new stormwater ponds	$40,000, or $500 per lot
Gap Creek Residential Subdivision Sherwood, Alaska	130-acre, 72-lot development, reduced street width and preserved natural topography and drainage networks	$200,021, or $4,819 per lot
OMSI Parking Lot Commercial Development Portland, Oregon	6-acre parking lot, incorporated bioswales and reduced piping and catch basin infrastructure	$78,000, or $13,000 per acre
Tellabs Corporate Campus Commercial Development Naperville, Illinois	55-acre site developed into office space, minimized site grading, preserved natural topography, eliminated storm sewer piping, and added bioswales	$564,473, or $10,623 per acre

Figure 8.7 Designed in partnership with the Housing Authority of Seattle, Washington, this natural drainage system for the High Point neighborhood of West Seattle will treat about 10 percent of the watershed feeding Longfellow Creek—one of Seattle's priority watersheds. The natural drainage system mimics nature in many ways by using features such as swales to capture and naturally filter stormwater and open, landscaped ponds or small wetland ponds to hold an overflow of stormwater. (Photo by Stuart Patton Echols)

Figure 8.8 The stormwater bioretention system for the Stata Center, a Frank Gehry–designed building on the Massachusetts Institute of Technology campus in Cambridge, Massachusetts, is a multifunctional constructed wetland that detains runoff to reduce peak downstream flow. The plants and planting medium of the wetland clean the runoff and allow some groundwater infiltration. (Photo by Stuart Patton Echols)

Heat Island Mitigation

An issue that is not normally considered in site and landscape design but that is a matter for consideration in high-performance green buildings is the *urban heat island effect*. Temperatures in cities are substantially higher than those in surrounding rural areas, usually in the range of 2 to 10°F (1 to 6°C) hotter (see Figure 8.10). The result is that cooling requirements for buildings in urban areas will be higher than those in a rural setting. The additional energy required to support the higher cooling loads results in more air pollution, greater resource extraction impacts, and higher costs. Reducing or mitigating urban heat islands can counter these negative effects and result in a more pleasant urban lifestyle.

Heat islands are caused by the removal of vegetation and its replacement with asphalt and concrete roads, buildings, and other structures. The shading effect of trees and the evapotranspiration, or natural cooling effect, of vegetation are replaced by human-made structures that store and release solar energy.

In addition to their negative energy impacts, heat islands are problematic for the following reasons:[23]

- Heat islands contribute to global warming by increasing fossil fuel consumption by power plants.
- Heat islands increase ground-level ozone pollution by increasing the reaction rate between nitrogen oxides and volatile organic compounds (VOCs).
- Heat islands adversely affect human health, especially that of children and older people, by increasing temperatures and ground-level ozone levels.

Heat island effects can be reduced by several measures:

- Installing highly reflective (or high-albedo) and emissive roofs that reflect solar energy back into the atmosphere

Figure 8.9 The stormwater system design for Chambers, Washington, employs a long water trail that exhibits a variety of water treatments. These include a wetland and a lined bed with river stone and plants interspersed with pieces of driftwood to emphasize the water theme. (Photo by Stuart Patton Echols)

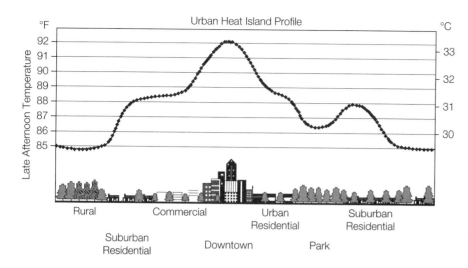

Figure 8.10 The removal of vegetation in urban areas and its replacement with buildings and infrastructure produces a heat island effect and results in urban temperatures that are 2 to 10°F (1 to 5°C) higher than those in nearby rural areas. (Illustration by Bilge Çelik)

- Planting shade trees near homes and buildings to reduce surface and ambient air temperatures
- Using light-colored construction materials where possible to reflect rather than absorb solar radiation

The EPA launched the Urban Heat Island Pilot Project in 1998 to quantify the potential benefits of reducing heat islands. For the city of Sacramento, California, a Lawrence Berkeley National Laboratory study showed the following:[24]

- Citywide energy bill reduction of $26.1 million per year, assuming high penetration of reduction measures
- Savings of 468 million watts (468 MW) of peak power and 92,000 tons (83,600 metric tons) of carbon annually
- An improvement in air quality caused by a decrease in ozone of 10 parts per billion
- Cooling-energy savings of 46 percent and peak power savings of 20 percent by increasing roof albedo, or reflectivity, on two school buildings

The LEED rating system provides points for mitigating heat islands in the Sustainable Sites category. For nonroof areas, LEED gives credit for creating shade or reducing heat islands for the site's impervious surfaces such as parking lots, walkways, and plazas. Credit is also given for providing a high-albedo (high-reflectivity) or vegetated roof. Similarly, Green Globes awards points on a sliding scale, depending on how much of the project hardscape and roof area include heat island mitigation measures.

Light Trespass and Pollution Reduction

Exterior lighting systems on buildings frequently emit light that, in addition to performing their primary role of illuminating the buildings and their walkways and parking areas, illuminate areas off-site. This condition is sometimes referred to as *light trespass*, defined as unwanted light from a neighboring property. This unwanted light poses a number of problems, ranging from being a nuisance to

Figure 8.11 The exterior lighting system for Rinker Hall, a LEED-NC gold certified building at the University of Florida in Gainesville, was designed to minimize light pollution. The result is a pleasant evening view of the building that enhances the experience of passersby. (Photograph courtesy of Gould Evans Associates and Timothy Hursley)

causing safety problems when it "blinds" pedestrians and automobile drivers. Nuisance light can also negatively affect wildlife, as well as human health, because it can interrupt normal daily light cycles that are needed for the average person's well-being. For example, chicken farmers have discovered that 24-hour lighting disturbs the growth of chicks. Bright lights can affect the migration patterns of birds and baby sea turtles.

Another negative lighting condition is *light pollution*, which prevents views of the night sky by the general population and astronomers. The solution to both light trespass and light pollution is proper lighting system design. The location, mounting height, and aim of exterior luminaires must all be taken into account to ensure that lighting energy is used efficiently and for its intended purposes. To prevent light pollution:

- Parking area and street lighting should be designed to minimize upward transmission of light.
- Exterior building and sign lighting should be reduced or turned off when not needed.
- Computer modeling of exterior lighting systems should be used to design exactly the level and quality of lighting needed to meet the project's requirements without straying off-site and causing undesirable conditions (see Figure 8.11).

Assessment of Sustainable Sites: The Sustainable Sites Initiative

Building assessment systems, such as LEED and Green Globes, focus on the building as the object of assessment. The building site and its location are typically evaluated as part of the building assessment, and the site ecology, stormwater, landscaping, and other factors are considered in this process. However, there are a wide range of projects that are not eligible for assessment by these well-known tools. For example, parking lots, athletic fields, plazas, streetscapes, and botanical gardens are just a few of the types of projects involving construction that do not necessarily involve a building. Also open space requirements for developments often result in requirements for easements, buffer zones, and transportation rights-of-way. The Sustainable Sites Initiative (SITES) was created to promote sustainable land development and management practices that can apply to sites with and without buildings, including the types of projects mentioned previously that are not routinely considered for environmental assessment along the lines of conventional building projects. SITES is a collaboration of the American Society of Landscape Architects (ASLA), the United States Botanic Garden, and the Lady Bird Johnson Wildflower Center. The USGBC is also now an active stakeholder in the development of SITES, with the notion that the performance guidance and benchmarks of SITES will be integrated with future versions of the LEED building assessment system.

SITES is developing tools for those who influence land development and management practices to assist them in addressing increasingly urgent global concerns such as climate change, loss of biodiversity, and resource depletion. These tools can be used by teams who design, construct, operate, and maintain landscapes, including planners, landscape architects, engineers, developers, builders, maintenance crews, horticulturists, governments, land stewards, and organizations offering building standards. The main objectives of SITES™ are as follows:[25]

- Elevate the value of landscapes by outlining the economic, environmental, and human well-being benefits of sustainable sites.
- Connect buildings and landscapes to contribute to environmental and community health.
- Provide performance benchmarks for site sustainability.
- Link research and practice associated with the most sustainable materials and techniques for site development construction and maintenance.
- Provide recognition for high performance in sustainable site design, development, and maintenance.
- Encourage innovation.

As part of a three-year-long stakeholder process, SITES engaged a wide variety of the country's leading sustainability experts, design professionals, and scientists, and gathered public input from hundreds of individuals and dozens of organizations. The latest version of this cumulative effort was issued in November 2009 in the form of an assessment tool, *The Case for Sustainable Landscapes* and *Guidelines and Performance Benchmarks 2009*. A number of pilot projects are testing the assessment tool for certification purposes, and, depending on the outcomes of this initial effort, it may be made available to qualified projects in the future.

The SITES guidelines and performance benchmarks offer four certification levels based on a four-star rating system, which works on a 250-point scale. In the pilot phase, a project achieving all 15 of the prerequisites and at least 100 credit points will become pilot certified with a rating of up to four stars as follows:

Certification Levels (250 Total Points)

One star (minimum points 40%)	100
Two stars (minimum points 50%)	125
Three stars (minimum points 60%)	150
Four stars (minimum points 80%)	200

SITES™ is an emerging assessment system that is especially important because it addresses construction projects that do not include buildings and for which there is no certification scheme. Additionally, it provides many areas of advanced thinking on site utilization, and its integration into the major building assessment systems in the United States will significantly improve the site and landscaping categories of these rating systems. An index of the SITES rating system, including its 15 prerequisites and 51 credits, can be found in Appendix B.

Summary and Conclusions

The most exciting and underutilized resources for creating high-performance green buildings are natural systems, and they should be employed as more than superficial components of the project. The ultimate green building will undoubtedly feature a much deeper integration of ecosystems with buildings, and exchanges of matter-energy between human systems and natural systems, in ways that are beneficial to both. The need to dramatically reduce building and infrastructure energy consumption will motivate designers to better understand the processing of waste by natural or constructed wetlands, which contribute to their sustainability and to that of the human systems with which they cooperate. Natural systems can shade and cool buildings, yet allow sunlight through for heating during appropriate seasons. They can also provide calories and nutrition and may be able to take up large quantities of stormwater, thus allowing the downsizing of conventional stormwater handling systems.

The high-level integration of ecosystems and the built environment is, at present, only a concept. But a future of high energy costs will inevitably force changes that decentralize many of the waste-processing functions that currently are performed at distant wastewater treatment plants to which building waste must be pumped, often through miles of piping, with motive energy provided by a series of lift stations. By integrating buildings with ecosystems, an alternative framework can be designed to ensure a future with a low energy profile. Though today's green building designers make only a minimal effort to use natural systems for anything other than amenities, in the future they will have a much more detailed knowledge of ecology and ecological systems, enabling them to successfully weave nature into the built environment.

Notes

1. From France (2003). The author provides an insightful analysis of how landscape architecture must change to participate in ecological design. He points out the possibility of landscape as "functional art," most prominently in the form of wetlands that, in addition to being pleasing to the human eye, provide numerous services, such as stormwater uptake and wastewater processing. He adds that the shift to multi-functional wetlands is a success story for sustainable landscape architecture.
2. As defined by the US Department of Agriculture and listed on the website of the American Farmland Trust, www.farmland.org.
3. Excerpted from "Farming on the Edge" (1997).
4. The US EPA brownfields website is www.epa.gov/brownfields.
5. The Chicago Brownfields Initiative is a partnership of private and public sector institutions that advocates and assists in the conversion of formerly contaminated industrial zones to productive use. The Chicago Department of the Environment hosts the website for this initiative at www.cityofchicago.org/city/en/depts/dgs/supp_info/chicago_brownfieldsinitiative.html.

6. The Congress of New Urbanism website is www.cnu.org.
7. Data on Wal-Mart stores is from the Wal-Mart Realty Company website, www.wal-martrealty.com.
8. Excerpted from an excellent article on the issue of grayfields, "Grayfields and Ghostboxes" (2003).
9. The Eastern Pennsylvania Coalition for Abandoned Mine Reclamation (EPCAMR) has an excellent website (www.orangewaternetwork.org) that describes the extent of the problem with blackfields or abandoned mine properties.
10. Detailed information about the NFIP, SFHA, and flood mapping can be found at the FEMA website, www.fema.gov/plan/prevent/fhm/index.shtm.
11. As defined in Thayer (1989). The Landscape Journal's website is www.wisc.edu/wisconsinpress/journals/journals/lj.html.
12. According to Robert Thayer, horizontal energy is low-intensity, widely dispersed, renewable energy in the form of sunlight, wind, water moving by tides or gravity, and energy fixed by plants.
   Horizontal energy is limited by its location and the rate of its natural generation, and landscape must exist within the limits of its availability.
13. From Dewey (1916).
14. Excerpted from Thompson and Sorvig (2000).
15. Portland's eco-roof program is described at the city's Bureau of Environmental Services website, www.portlandonline.com/bes/index.cfm?c=44422.
16. Excerpted from Flaccus (2002).
17. As described on the website of Living Systems Design Group, LLC, www.livingsystemsdesign.net/.
18. From Yeang (1996).
19. The "Checklist for Wildlife Habitat Enhancement of Developed Land" is excerpted from the February 2001 issue of *EBN*, pp. 8–12. The original checklist provides a detailed description of each of the points in the table. *EBN* is a publication of Building Green, Inc., www.buildinggreen.com.
20. The "Checklist for Stormwater Management Practices" is excerpted from the September/October 1994 issue of *EBN*, p. 1 and pp. 8–13. The original checklist provides a detailed description of each of the points in the table.
21. From Nature Conservancy (2010).
22. Ibid.
23. From the USEPA Heat Island Effect website, www.epa.gov/hiri/.
24. Lawrence Berkeley National Laboratory has a website devoted to heat island issues: http://eetd.lbl.gov/newsletter/nl08/eetd-nl08-5-meteorology.html.
25. From Sustainable Sites Initiative (2009).

References

American Farmland Trust. 1997. "Farming on the Edge: Sprawling Development Threatens America's Best Farmland." Available online from the American Farmland Trust website, www.farmland.org.
Campbell, Craig, and Michael Ogden. 1999. *Constructed Wetlands in the Sustainable Landscape.* Hoboken, NJ: John Wiley & Sons.
Dewey, John. 1916. *Democracy and Education.* New York: Free Press.
Flaccus, Gillian. 2002. "Portland at Forefront of Eco-Friendly Roof Trend." Available online at www.evesgarden.org/archives/2002/11/17/portland_at_forefront_of_ecofriendly_roof_trend.
France, Robert. 2003. "Grey World, Green Heart?" *Harvard Design Magazine*, no. 18, 30–36.
"Grayfields and Ghostboxes: Evolving Real Estate Challenges." 2003. *Let's Talk Business*, no. 81. Available online at www.uwex.edu/ces/cced/downtowns/ltb/lets/0503ltb.pdf.
Lyle, John T. 1985. *Design for Human Ecosystems: Landscape, Land Use, and Natural Resources.* Washington, DC: Island Press.

————. 1994. *Regenerative Design for Sustainable Development*. Hoboken, NJ: John Wiley & Sons.

Nature Conservancy. 2010. "Low Impact Development Principles and Techniques." Available at www.columbiatn.com/PDFs/LIDFinal.pdf.

Powell, Lisa M., E. Rohr, M. Canes, J. Cornet, E. Dzuray, and L. McDougle. 2005. *Low Impact Development Strategies and Tools for Local Governments: Building the Business Case*. Report LID50T1, LMI Government Consulting, September. Available at www.lowimpactdevelopment.org/lidphase2/pubs/LMI%20LID%20Report.pdf.

Sustainable Sites Initiative. 2009. "Guidelines and Performance Benchmarks 2009." Available at www.sustainablesites.org/report/Guidelines%20and%20Performance%20Benchmarks_2009.pdf.

Thayer, Robert. 1989. "The Experience of Sustainable Design." *Landscape Journal* 8: 101–110.

————. 1994. *Gray World, Green Heart: Technology, Nature, and the Sustainable Landscape*. Hoboken, NJ: John Wiley & Sons.

Thompson, Robert, and Kim Sorvig. 2000. *Sustainable Landscape Construction: A Guide to Green Building Outdoors*. Washington, DC: Island Press.

Yeang, Ken. 1996. *The Skyscraper Bioclimatically Considered*. London: Academy Editions.

Chapter 9

Energy and Carbon Footprint Reduction

Perhaps of all the challenges facing the development of high-performance green buildings, significantly reducing the energy and carbon footprints of the built environment is the most daunting. The environmental impacts of extracting and consuming nonrenewable energy resources such as fossil and nuclear fuels are profound. Pronounced land impacts from coal and uranium mining, acid rain, nitrous oxides, particulates, radiation, ash disposal problems, and long-term storage of nuclear waste are just some of the consequences of energy consumption by the built environment. Building energy consumption in the United States is at about the same scale as energy consumption by automobiles, with about 40 percent of primary energy being consumed by buildings and about the same quantity by transportation.[1] In fact, much automotive energy consumption is caused by the placement of buildings on the landscape.

The rollover point of oil production, the point at which the oil production rate reached its historical peak, likely occurred in 2006.[2] As discussed in Chapter 1, considerable additional energy and financial resources will be needed to extract the remaining oil resources. At the same time, economies around the world continue to grow, all of them dependent on abundant, cheap energy, none of them more so than the United States. H. T. Odum, the eminent ecologist who founded the branch of ecology known as *systems ecology*, forecasted that, at the rollover point, the energy required to extract the oil would be greater than its energy value.[3] Sounding another warning note, Odum and his colleagues calculated that some key technologies suggested as substitutes for a predominantly fossil fuel–powered energy system, among them photovoltaics and fuel cells, require more energy to produce than they themselves will ever generate. The point is that the technological optimists who believe that a technical solution will always be found to solve our energy, water, or materials problems have not found a cheap substitute for fossil fuel–derived energy. For the built environment, truly dramatic reductions in building energy consumption, accompanied by tremendous progress in passive design and the implementation of large-scale renewable energy systems, will be needed to meet a potentially costly energy future.

As we approach this day of reckoning, when energy costs are likely to rise dramatically as a result of fierce international demand and competition, we still have time to make some very important decisions with respect to how we live and the types of buildings we create. The green building movement and allied efforts to improve building energy performance are attempting to influence a major shift in the way buildings are designed. It is a fundamental transformation that must take place, one that does not just reduce energy consumption by a small percentage but that involves a total rethinking of building design. Advocates of just such a radical change believe that buildings should be *energy-neutral* or even net *exporters* of

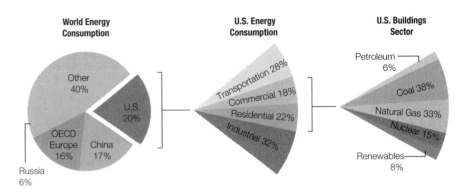

Figure 9.1 Energy consumption patterns worldwide, in the United States, and for US buildings. (*Source:* US Energy Information Administration)

energy. Advancing the use of solar energy, ground coupling, radiant cooling, and other radical approaches may indeed enable buildings to generate at least as much energy as they consume. In the interim, however, we must learn how to cut building energy use by a marked quantity, perhaps by as much as 90 percent—a daunting challenge, to be sure.

Building Energy Issues

US energy demand is staggering, with Americans consuming just over 100 quads of primary energy in 2011 with slightly more forecast for 2012. One quad alone is an enormous unit of energy, equaling 1 quadrillion, or 10^{15}, BTU, and American consumption represents 20 percent of the total primary energy consumed on the planet for a country with less than 5 percent of the world's population (see Figure 9.1). *Primary* energy is the best measure of energy consumption because it is the energy in fuel sources such as coal, oil, and natural gas before they are converted into electricity and other forms of *secondary* or *site* energy. Chinese consumption is growing rapidly, doubling in the six-year period from 2002 to 2008, while US consumption grew just 0.5 percent over the same time period. In the United States, buildings consume 40 percent of energy or about 8 percent of global primary energy (see Figure 9.2). Commercial buildings have the smallest fraction of energy consumption at 18 percent, but this number is growing the most rapidly. Although at first glance transportation and industrial energy appear to be unrelated to building energy, they are, in fact, coupled together. The relationships of buildings and the distances between them are a major contributor to transportation energy. Additionally, a considerable amount of industrial energy is invested in building products and infrastructure materials, and it is likely that the total energy, including the embodied energy of materials and the transportation energy attributable to buildings, is well over 60 percent of total primary energy. Energy consumed to support the built environment is dominated by coal at 33 percent of primary energy, but natural gas consumption is growing rapidly, and nuclear energy may possibly also be a growing fraction of electrical energy generation.

There is some good news with respect to energy, namely, energy use per capita and per unit of economic production is falling and will continue to decrease for the foreseeable future (see Figure 9.3). However, total energy consumption is still rising due to an increasing population and a growing economy, a problem at several levels. First, energy prices affect economic production, and higher demand drives prices higher, thereby putting a damper

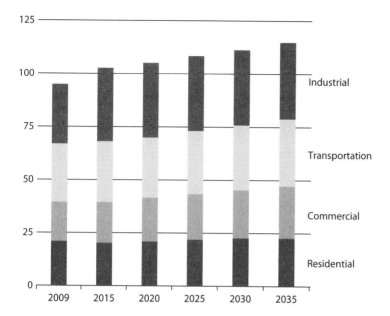

Figure 9.2 Primary energy use by end-use sector, 2009–2035, in quads. Energy use by buildings in the United States is growing, from about 40 percent in 2009 to a forecasted 48 percent in 2035. Commercial building energy consumption is growing at the fastest rate of the four sectors depicted in this diagram. (*Source:* US Energy Information Administration)

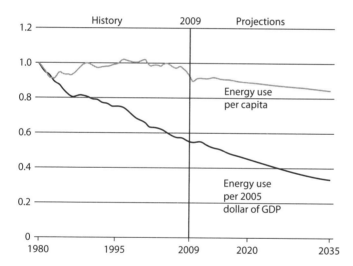

Figure 9.3 Index of energy use per capita and per dollar of GDP from 1980 to 2035 (the index for 1980 is set equal to 1). (*Source:* US Energy Information Administration)

on the economy. Second, the vast majority of energy consumption is via fossil fuel combustion, which has human health impacts. Finally, greater energy production generally means higher greenhouse gas production, contributing even more to climate change. To meet the challenges of the future, per capita and per gross domestic product (GDP) energy consumption must decline far more rapidly by developing more efficient manufacturing processes, reconfiguring cities to be more compact, and designing high-performance buildings and retrofitting existing buildings to consume far less energy. This last point is especially important because more than 99 percent of the building stock is composed of existing buildings. Shifting to renewable energy systems such as solar, wind, and biomass systems is also important because renewable energy is considered to be part of a clean energy production system without the negative health and climate change impacts.

Building energy consumption can be reduced through the use of better design backed up by more stringent national and state energy standards and codes. Truly low energy buildings are achievable by using passive

Efficiency gains for selected commercial equipment in three
cases, 2035 (percent change from 2009 installed stock efficiency)

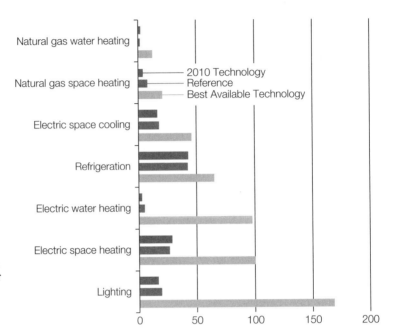

Figure 9.4 Building energy consumption
can be significantly lowered by employing
the best available technology for the major
energy-consuming systems in buildings.
(*Source:* US Energy Information
Administration)

energy strategies that take into account the orientation and mass of the
building to maximize daylighting and minimize heat gain except when
needed. Coupled with the best emerging technologies (see Figure 9.4),
significant reductions in building energy use can be realized. Energy consumption in US buildings is declining due to these very reasons, driven by a
combination of rising energy costs and more stringent standards (see Figure 9.5). A survey of buildings by the US Department of Energy (DOE),
known as the Commercial Buildings Energy Consumption Survey
(CBECS), was conducted in 2003 and found that building energy consumption averaged about 91,000 BTU/ft^2/yr (287 kWh/m^2/yr). Since this
survey, building consumption has been pushed lower by ever more stringent
standards such that the 2010 version of ASHRAE 90.1 sets a ceiling on
commercial building energy consumption of about 36,000 BTU/ft^2/yr (114
kWh/m^2/yr), a 60 percent reduction since 2003.

Recent programs in Germany indicate that buildings can be designed to use
far lower levels of energy than even the most ambitious US high-performance
buildings. As part of a 10-year demonstration program that ended in 2005, a
group of 23 office buildings throughout Germany were designed, built, and
monitored with a goal of using 32,000 BTU/ft^2 (100 kWh/m^2) of *primary energy*
annually. Primary energy is the source energy for the energy delivered to the
building, for example, the energy value of coal before it is combusted to produce
electricity. The efficiency of coal-fired electrical generating plants is such that
only one-third of the coal energy becomes electrical energy. Consequently, the
electrical energy used in the building, referred to as the *site energy*, is multiplied
by a factor of 3 to account for the primary energy. A code-compliant US office
building consumes on the order of 80,000 BTU/ft^2/yr (252 kWh/m^2/yr) of site
energy. For an all-electric building in the United States, this would equate to
240,000 BTU/ft^2/yr (756 kWh/m^2/yr). For a building that derives 80 percent of
its energy from electricity and the remainder from natural gas, the primary

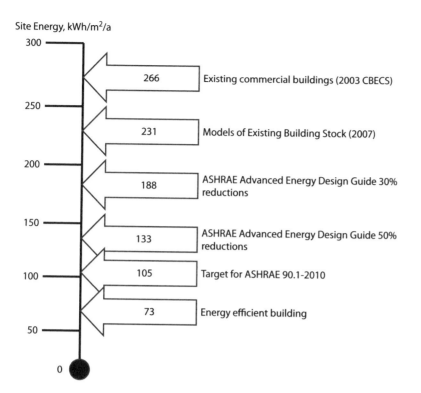

Site Energy, kWh/m²/a

- 266 — Existing commercial buildings (2003 CBECS)
- 231 — Models of Existing Building Stock (2007)
- 188 — ASHRAE Advanced Energy Design Guide 30% reductions
- 133 — ASHRAE Advanced Energy Design Guide 50% reductions
- 105 — Target for ASHRAE 90.1-2010
- 73 — Energy efficient building

Figure 9.5 Energy use in US buildings has been dropping rapidly since the DOE Commercial Buildings Energy Consumption Survey found an average of 287 kWh/m²/yr of operational energy in 2003. Since that time, standards such as ASHRAE 90.1-2010 are pushing energy far lower, to 114 kWh/m²/yr at present. (*Source:* US Department of Energy)

energy would be about 208,000 BTU/ft²/yr (656 kWh/m²/yr). Note that the best German buildings now have a primary energy target of 100 kWh/m²/yr. An energy-efficient, high-performance building in the United States would have to use one-fifth to one-seventh of the energy of a conventional US building to match today's best practices in Germany.[4] Even the best US practices, which cut energy consumption by 50 percent, result in the typical office building using at least twice the primary energy of a German building, pointing to a need for dramatic changes in the way buildings are designed in the United States.

A green building would ideally use very little energy, and renewable energy would be the source of most of the energy needed to heat, cool, and ventilate it. Today's green buildings include a wide range of innovations that are starting to change the energy profile of typical buildings. Many organizations are committed to investing in innovative strategies to help create buildings with Factor 10 performance, notably the federal government, which has been the leader in requiring life-cycle costing (LCC) analysis as the basis for decision making with respect to building procurement. Some state governments have followed suit, notably those of Pennsylvania, New York, and California; in contrast, others, such as Florida, have passed legislation requiring decisions based solely on the capital or first cost of a particular strategy. This latter, shortsighted approach will result in enormous expenditures of energy as we approach the rollover point.

Green building advocates often note that the strategies used to heat, cool, ventilate, and illuminate high-performance buildings allow a significant downsizing of the mechanical plant and a parallel reduction in the overall capital costs of the building. This is clearly the ideal outcome, wherein both capital and operating costs are lower than those of a comparable base-case building. However, there are very few of these buildings in typical US climactic zones for a variety of reasons, including building code constraints. LCC analysis of a building's performance is key for giving designers the creative freedom to optimize a given building's energy consumption.

High-Performance Building Energy Design Strategy

Over the past decade, a process for designing low-energy buildings has emerged that can produce 100 kWh primary energy buildings. The following are the steps in designing energy systems with low energy and low carbon footprints:

1. Use building simulation tools throughout the design process.
2. Optimize the passive solar design of the building.
3. Maximize the thermal performance of the building envelope.
4. Minimize internal building loads.
5. Maximize daylighting and integrate with a high-efficiency lighting system.
6. Design a hyperefficient heating, ventilation, and air conditioning (HVAC) system that minimizes energy use.
7. Select high-efficiency appliances and motors.
8. Maximize the use of renewable energy systems.
9. Harvest and use waste energy.
10. Incorporate innovative emerging strategies such as ground coupling and radiant cooling.

The design of an energy-efficient building is a complex undertaking, and these steps cannot just be performed in sequence; they are, in effect, part of an iterative process that starts with passive design. Trade-offs inevitably must be made, often because of the client's requirements and budget. Designed properly, a building with low energy and a low carbon footprint should provide greatly reduced operational costs for minimal or no increase in capital costs. In some cases, a well-executed passive design strategy can markedly reduce the costs of HVAC equipment due to the reduction in heating, cooling, ventilation, and lighting loads that can occur.

GOAL SETTING FOR HIGH-PERFORMANCE BUILDINGS

The design of the energy strategy for a high-performance building should involve an examination of energy targets for the building based on a combination of reviewing the performance of similar conventional buildings, an understanding of contemporary high-performance building best practices, and building energy simulations. The two major US building assessment systems, Leadership in Energy and Environmental Design (LEED) and Green Globes, take distinctly different approaches to energy goal setting. LEED relies on Illuminating Engineering Society of North America/American Society of Heating, Refrigerating and Air-Conditioning Engineers (IESNA/ASHRAE) 90.1-2007 (ASHRAE 90.1-2007) for direction on how to establish the baseline design for the building and compare it to the proposed design. The baseline design is generally thought of as a building designed to minimal building code requirements, with no special effort made to achieve energy efficiency. Green Globes relies on the US Environmental Protection Agency (EPA) Target Finder to determine the baseline design for comparison. Both of these approaches are described in more detail below.

ENERGY GOAL SETTING IN LEED

Building assessment systems, such as LEED, rely on ASHRAE 90.1-2007 to provide a standard set of instructions that dictate how the baseline design is to be defined and how the design of the high-performance building, referred to as the *proposed design*, is to be compared to the baseline design. The baseline design is simply a version of the building being designed but with minimal efforts to reduce its energy consumption below building code requirements. Appendix G of the ASHRAE 90.1-2007 standard describes the Performance Rating Method, which is a modification of the Energy Cost Budget Method. The baseline design is simulated for each of four orientations, with specified opaque assemblies, limits on vertical fenestration, and HVAC systems as defined in Appendix G. This approach uses energy cost as the basis for determining savings, with the cost of energy based on actual local utility rates or on state average prices published by the DOE's Energy Information Administration (EIA).[5]

ENERGY GOAL SETTING IN GREEN GLOBES

Green Globes uses a significantly different approach to setting targets for energy performance. Instead of developing an arbitrary baseline model based on ASHRAE 90.1-2007, Green Globes states that the building must surpass the 75 percent target as evaluated by the EPA Target Finder, meaning that the building has to be in the top 25 percent of buildings of that type, in the specific area, as contained in the Target Finder database. The 75 percent target also designates the building as meeting Energy Star standards. This target is the threshold performance for points in Green Globes, that is, it earns the minimum 10 points out of the maximum 100 points that can be awarded for minimizing energy consumption. The maximum number, 100 points, is achieved by buildings in the 96th percentile or higher. The advantage of this approach is that the target is based on actual buildings, and the designed building is compared to like structures in the immediate area. The drawback of Target Finder is that there is a limited range of building types listed in the database. Target Finder does have the capability of taking mixed-use buildings into account; for example, a building combining office and residential space can be analyzed to determine the appropriate target.[6]

BUILDING ENERGY SIMULATION AND DAYLIGHTING SIMULATION

Building energy simulation is an important tool in the design of a high-performance building. Contemporary building energy simulation tools allow the building to be modeled in great physical detail and to be operated on an hourly basis in a given configuration for an entire year. It is important to employ building energy simulation at a very early stage in the design process, when decisions about building shape, number of stories, and orientation are being made. Today's simulation tools allow the integration of active and passive building systems and can easily examine the interplay and trade-offs among heating and cooling systems, walls and roof choices, insulation, lighting, windows and doors, exterior and interior shading, and skylights. Perhaps the best-known whole-building energy simulation tool is DOE-2.2, which now has user-friendly interfaces and wizards to speed the energy simulation process.[7] Daylighting is a key component of an energy-efficient building, and performing simulations that optimize daylighting is important to understand the trade-offs among fenestration, envelope thermal resistance, and energy use for artificial lighting. Some building energy simulation tools, such as Energy-10, allow the integrated

TABLE 9.1

Comparison of Energy Models of LEED Buildings and Their Associated Base Cases Compared to Actual Energy Consumption

Building Type	Modeled Base Case	Modeled LEED Case	Savings of LEED Case Based on Modeling	Actual Energy Use*	Actual Energy Compared to Modeled[†]
Federal	131	117	21%	81	−30%
Nonfederal	105	61	42%	57	−7%

Note: For this study, nine federal and eight nonfederal, nonlaboratory buildings were compared.
*Thousands of BTU per square foot per year.
[†]A negative number indicates that the actual energy consumption was less than the modeled energy consumption.

evaluation of daylighting, passive solar heating, low-energy heating and cooling strategies, and envelope design.[8] Daylighting can also be evaluated with sophisticated software such as Radiance, developed by the Lawrence Berkeley National Laboratory. Radiance contains libraries of materials, glazings, electric lighting luminaires, and furniture to facilitate the daylighting analysis.[9] The simulation provides a quantitative check on the intuitive guesswork of the design team about the interrelationship of the building systems. Typical tools for whole-building energy simulation include eQUEST, DOE-2.2, and Energy-10.

To determine how well energy modeling represents the actual performance of buildings, Lawrence Berkeley National Laboratory conducted a study of 21 buildings certified under LEED 2.0 or LEED 2.1, with about half located in the Pacific Northwest and the others from areas throughout the United States. Part of the study separated federal from nonfederal buildings and considered only nonlaboratory buildings. A summary of this study is shown in Table 9.1.[10] It indicates a wide range of results when comparing modeling to actual performance. In some cases, the modeling is quite accurate, while in others it tends to be far off. The real value of modeling is to find the relative importance of changes to the building's envelope and energy systems; providing an accurate prediction of building energy performance is less important. The modeling of plug loads (computers, printers, fax machines, copiers, appliances) is notoriously inaccurate because the behavior of the building users is unpredictable. Actual plug loads are often substantially higher than those simulated in the energy model. Additionally, with the continual addition of new electrically powered devices in office buildings, plug loads tend to increase over time. The issues of plug loads and techniques for reducing them are addressed later in this chapter in the section "Plug Load Reduction."

VERIFYING BUILDING ENERGY PERFORMANCE

The International Performance Measurement and Verification Protocol (IPMVP) provides an overview of current best practices for verifying energy efficiency, water efficiency, and renewable energy performance for commercial and industrial facilities. It may also be used by facility operators to assess and improve facility performance. Energy conservation measures (ECMs) covered in the protocol include fuel-saving measures, water efficiency measures, load shifting, and energy reductions through installation or retrofit of equipment and/or modification of operating procedures. The IPMVP is maintained with the sponsorship of the DOE by a broad international coalition of facility owners/operators, financiers, energy services companies (ESCOs), and other stakeholders.

The IPMVP was first published in 1996 and contained methodologies that were compiled by a technical committee including hundreds of industry experts, initially from the United States, Canada, and Mexico. In 1996 and 1997, 20 national organizations from a dozen countries worked together to revise, extend, and publish a new version of the IPMVP in December 1997. The 2010 version, the latest edition, has been widely adopted internationally and has become the standard measurement and verification (M&V) document in countries ranging from Brazil to Romania. Volume 3 of the IPMVP applies to new construction, and its purpose is to provide a description of best practices for verifying the energy performance of new construction.[11]

The IPMVP requires the user to develop an M&V plan that includes defining the ECMs employed in the building, identifying the boundary conditions for measurement, establishing base year data, defining conditions to which all data will be adjusted for comparison, and meeting a range of other requirements that establish a standardized method for comparing information. The LEED-NC point that can be earned for M&V requires that the IPMVP be used for measuring both energy and water consumption data.

Passive Design Strategy

Due to the complexity of designing the energy systems for a high-performance green building, the starting point must be full consideration of *passive solar design*, or *passive design*. Passive design is the design of the building's heating, cooling, lighting, and ventilation systems, relying on sunlight, wind, vegetation, and other naturally occurring resources on the building site. Passive design includes the use of all possible measures to reduce energy consumption prior to the consideration of any external energy source other than the sun and wind. Thus, it defines the energy character of the building prior to the consideration of active or powered systems (chillers, boilers, air handlers, pumps, and other powered equipment). Randy Croxton, one of the pioneers of contemporary ecological design, describes a good passive design as one that allows a building to "default to nature." A building that has been well designed in a passive sense could be disconnected from its active energy sources and still be reasonably functional due to daylighting, adequate passive heating and cooling, and ventilation being provided by the chimney effect, cross-ventilation, operable windows, and the prevailing winds. A successful passive design scheme creates a truly climate-responsive, energy-conserving building with a wide range of benefits.

Passive design has two major aspects: (1) the use of the building's location and site to reduce the building's energy profile and (2) the design of the building itself—its orientation, aspect ratio, massing, fenestration, ventilation paths, and other measures. Passive design is complex, as it depends on many factors, including latitude, altitude, solar insolation,[12] heating and cooling degree days,[13] humidity patterns, annual wind strength and direction, the presence of trees and vegetation, and the presence of other buildings. An optimized passive design can greatly reduce the energy costs of heating, cooling, ventilation, and lighting.

Some of the factors that should be included in the development of a passive design strategy are as follows:

- *Local climate.* Sun angles and solar insolation, wind velocity and direction, air temperature, and humidity throughout the year
- *Site conditions.* Terrain, vegetation, soil conditions, water table, microclimate, relationship to other buildings

- *Building aspect ratio.* Ratio of the building's length to its width
- *Building orientation.* Long axis oriented east–west, room layout, glazing
- *Building massing.* Energy storage potential of materials, fenestration, color
- *Building use.* Occupancy schedule and use profile
- *Daylighting strategy.* Fenestration, daylighting devices (light shelves, skylights, internal and external louvers)
- *Building envelope.* Geometry, insulation, fenestration, doors, air leakage, ventilation, shading, thermal mass, color
- *Internal loads.* Lighting, equipment, appliances, people
- *Ventilation strategy.* Cross-ventilation potential, paths for routine ventilation, chimney effect potential

Like any concept, passive design can be improperly applied to building design. Its success is highly dependent on the wide range of factors just listed, and its application differs widely from New York to California, Colorado, or Florida. For example, using thermal mass as a passive design strategy, an excellent choice in the high desert altitudes found in New Mexico, with its abundant sunlight and wide daily temperature swings, would not be an appropriate choice in a hot, humid climate with generally narrow daily temperature differences, as would be found in Tampa, Florida. The optimum building orientation, the location and types of windows, the use of daylighting, and many other decisions must be based on a careful examination of the situation found in each locale.

SHAPE, ORIENTATION, AND MASSING

The classic passive design approach to orienting a building on its site is to locate the long side on a true east–west axis to minimize solar loads on the east and west surfaces, particularly during the summer. The *aspect ratio* is the ratio of a building's length to its width, which is an indicator of the general shape of a building. Passive design dictates that a building in the northern United States should have an aspect ratio close to 1.0; that is, it should be virtually square in shape. For buildings in the warmer southerly latitudes, the aspect ratio increases, with the building becoming longer and narrower. The reasoning behind this shift in aspect ratio is that a square building will have the minimum skin surface area compared to its volume. It is important in colder climates to minimize the surface area through which heat can be transmitted. Temperature differentials for heating are generally much greater than for cooling; thus, the total skin area of the building is more important in heating situations. The long, narrow building favored by passive design experts for warmer climates minimizes the relative exposure of east and west surfaces that experience the greatest sun load. Windows on east and west surfaces are typically minimized to eliminate as much as possible the potential high morning and afternoon solar loads. South-facing walls will experience a variable sun load during the day, and windows are easily protected from solar loads through the use of roof overhangs, shading devices, or recessed windows.

Thermal mass is an important aspect of passive design. In cases where passive solar heating is desired, the geometry of the building should be arranged to allow materials with high heat capacity and significant mass to store solar energy during the day. Materials such as brick, concrete masonry, concrete, and adobe, used for floors and walls, can absorb solar energy during the day and release it in the evening, when internal temperatures begin to drop. For passive solar cooling, buildings in climates such as that of Florida should have minimal mass for storing energy and should generally be lightweight and well insulated. Preventing solar energy transmission into the structure is the desired strategy for passive cooling. The ideal design, which would consider both passive heating and passive cooling, could provide heating in winter and promote cooling in summer. This requires careful consideration of orientation, fenestration, shading, and massing.

Because large commercial and institutional buildings are complex and are often restricted with respect to siting, trying out various passive design approaches using computer simulation is necessary to sort through the wide array of possibilities. The integration of landscaping with the building also has enormous potential for contributing to natural heating and cooling by shielding windows during the summer and allowing solar energy through in winter.

DAYLIGHTING

Using natural light or daylight for illumination is one of the hallmarks of a high-performance building. In addition to the benefits of supplying substantial light for free, natural lighting has been shown to provide great physical and psychological benefits to the building occupants. The first comprehensive scientific studies of the benefits of daylighting were conducted by the Pacific Gas and Electric Company in California in the late 1990s for two general types of buildings: retail stores and schools.[14] Daylighting in stores was shown to increase sales per square foot of retail space from 30 to 50 percent, while the learning rate of students was 20 to 26 percent higher in classrooms with daylighting compared to those with only artificial lighting.[15] Clearly, daylighting produces a win-win situation, marked by lower energy costs and better performance in classrooms. Most likely, the same is true in offices. Although not yet proven by scientific methods, it is thought that a 10 to 15 percent increase in office worker productivity can be expected as a consequence of daylighting. A 10 percent increase in employee productivity due to decreased illness and absenteeism or an improved sense of well-being translates into savings that far exceed the energy costs of a typical office building. If the connections between daylighting and human health could be proven with a high degree of certainty, this alone would cause an enormous transformation in the way buildings are designed and built. At present, productivity and health effects are not fully taken into account in the LCC analysis of high-performance buildings. However, if and when science catches up with speculation and the benefits are verified, daylighting will leap past its use as a green building strategy to near-universal incorporation. (Chapter 14 addresses LCC for green buildings in more detail.)

Developing an effective daylighting strategy can, however, be a complex undertaking due to the trade-offs that must occur between admitting light and cooling the building. The cost of windows, skylights, light shelves, and other features that function to transmit light, versus conventional construction where daylighting is not much of an issue, must also be factored in. Fortunately, experience with daylighting is growing at an exponential rate, along with the green building movement itself; consequently, the information from these efforts is becoming available to a wider audience of designers and owners. A list of key ideas for assessing daylight feasibility from Lawrence Berkeley National Laboratory is shown in Table 9.2.[16] An excellent checklist for daylighting from *Environmental Building News* (*EBN*) is shown in Table 9.3.[17]

TABLE 9.2

Key Ideas for Daylight Feasibility

Windows must see the light of day. A high-density urban site may make daylighting difficult if the windows will not see much sky.

Glazing must transmit light. A strong desire for very dark glazing generally diminishes the capacity to daylight in all but very sunny climates.

Install daylight-activated controls. To save energy, lights are dimmed or turned off with controls. Automated lighting controls in a daylit building can have other cost-saving applications (occupancy, scheduling, etc.) and benefits.

Design daylight for the task. If the occupants require very bright light, darkness, or a highly controllable lighting environment, tailor the design to meet their needs.

Assess daylight feasibility for each portion of the building. Spaces with similar orientation, sky views, ground reflectance, and design can be treated together. Within a single building, the feasibility and cost effectiveness of daylighting may vary greatly.

TABLE 9.3

Checklist for Daylighting

General Daylighting

1. Provide a daylighting scheme that will work under the range of sky conditions expected at that location.
2. Orient the building on an east–west axis.
3. Brighten interior surfaces.
4. Organize electric lighting to complement daylighting.
5. Provide daylight controls on electric lighting.
6. Commission the daylight controls.

Perimeter Wall Daylighting

1. Provide perimeter daylight zones.
2. Extend windows high on perimeter walls.
3. Provide light shelves on south-facing windows.
4. Minimize direct-beam sunlight penetration into work spaces.
5. Choose the right glazing.
6. Arrange interior spaces to optimize the use of daylighting.

Roof Daylighting

1. Provide roof apertures for daylighting.
2. Optimize skylight spacing.
3. Consider extending skylight performance with trackers.
4. Use reflective roofing on sawtooth clerestories.
5. Diffuse daylight entering the building through roof apertures.

Core Daylighting
Provide a central well or atrium for daylighting.

The energy and health benefits of daylighting are maximized in the design for Smith Middle School in Chapel Hill, North Carolina (see Figure 9.6). The strategy allows for multiple anidolic lighting systems to capture the south-facing sunlight and direct it deeply into the classrooms, gymnasium, media center, and main corridor, providing the most daylighting with the least amount of skylight glazing. The natural light is distributed through translucent,

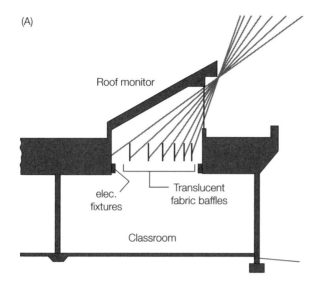

Figure 9.6   The daylighting strategy for Smith Middle School in Chapel Hill, North Carolina, employs (A) south-facing roof monitors with high-visible light transmission glazing and interior vertical baffles to provide for optimum controlled daylighting without glare throughout the entire classroom depth, (B) integrated lighting controls, and (C) exterior lighting shelves with high-visible light transmission glazing above to enhance daylighting and low-emissivity glass for view windows below. [(A) Image courtesy of Lighting Research Center/Rensselaer Polytechnic Institute; (B–C) Innovative Design]

ultraviolet-resistant cloth baffles to scatter direct rays and avoid glare. Recessed south-facing windows prevent glare by incorporating both anodized aluminum light shelves and low-emissivity double glazing. These light shelves reflect daylight onto the ceiling surface and deep into the room, while low-emissivity glazing reduces internal solar heat gain. The daylighting strategy is integrated with the lighting systems controls through occupancy sensors, passive infrared technology, and a manual light switch. To turn on internal lights, three conditions must be satisfied: the manual switch must be on, occupant motion must be detected, and the lighting level within the room must be below a predetermined set point. Once these conditions are met, the lights come on and a photosensor adjusts the lighting down to 10 percent in response to daylight levels. An energy simulation indicated that the incorporation of daylighting technologies makes energy efficiency achievable through smaller cooling system sizing, resulting in cost savings and an enhanced indoor experience. The emerging consensus is that learning environments enlivened by subtle and natural variation of light intensity, color, and direction throughout the day are healthier environments, leading to higher productivity.[18]

PASSIVE VENTILATION

Providing ventilation to building occupants is normally accomplished by using fans, dampers, and controls to move outside air into the building while at the same time removing an equal amount of interior air to the outside. In more advanced designs, an economizer cycle uses outside air for cooling, providing significant savings. Ventilation air using natural forces to move the air, rather than mechanical systems, can also be provided, greatly reducing the energy needed to move air. Passive ventilation can be accomplished by using a thermal chimney effect, whereby air normally rises due to heating, inducing airflow in a generally vertical direction; or a Venturi effect, whereby air movement is induced by the development of a low-pressure zone created by wind flow.

The Jubilee Campus of the University of Nottingham in the United Kingdom, designed by Sir Michael Hopkins and Partners and built in 1999, has one of the most advanced passive ventilation strategies among modern buildings. Wind catchers are used to position the air exhaust stacks for optimal ventilation. The wind catchers automatically turn in the direction of the wind, creating suction behind them and driving the ventilation system for the buildings. Cool, clean air is brought in at a high level and fanned down to the floor levels, where it starts to rise with the sunlight, body heat, and equipment. This intricate pattern of environmental cause and effect is echoed throughout the building's staircases and corridors. Thermal wheels are used in conjunction with the wind catchers to exchange energy between exiting exhaust air and incoming fresh air. The innovations in this design resulted in the Jubilee Campus winning the Royal Institute of British Architects (RIBA) sustainability award in 2001. Figures 9.7 and 9.8 show the passively ventilated building on the Jubilee Campus and depict the ventilation pattern through the building.

In a typical European passive ventilation design, the first determinant is the quantity of air required for ventilation. In England, the Chartered Institution of Building Services Engineers (CIBSE) publishes standards and guidelines requiring the following ventilation rates:

- Classrooms: 2 to 4 air changes per hour
- Offices: 4 to 6 air changes per hour
- Theaters: 6 to 10 air changes per hour
- Storage areas: 1 to 2 air changes per hour

Figure 9.7 Wind catcher (upper right) on the Jubilee Campus of the University of Nottingham in the United Kingdom. The wind catcher pivots in the wind, with the vane indicating the direction of airflow. Wind flowing past the vane induces the convection of air through the structure. (Photograph courtesy of Hopkins Architects and Ian Lawson)

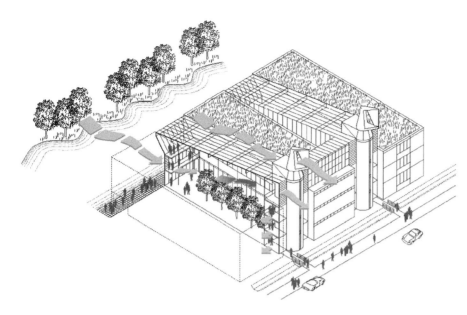

Figure 9.8   Schematic of the natural ventilation strategy for the Jubilee Campus of the University of Nottingham. Air flows from a low level at the rear of the building, moves gradually upward, and then exits through the pivoting wind catchers on the front of the building. A *Venturi effect* is induced in the wind catcher by the wind flowing past the vanes. (Illustration courtesy of Hopkins Architects)

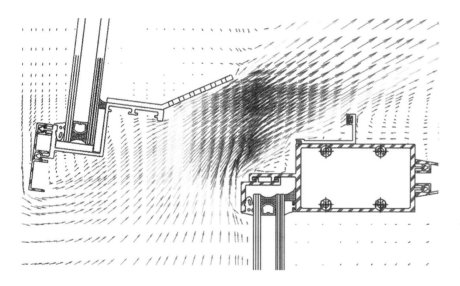

Figure 9.9   Design of passive ventilation systems requires the use of tools not traditionally used in building designs, such as the CFD modeling of wind and airflows around the Federal Building in San Francisco, California. The illustration shows a simulation of the design of an air deflector for the windows of the building that helps accelerate airflows, propelling them deep into the building's spaces. (Illustration courtesy of Natural Works)

The outside wind speed, which is generally in the range of 3 to 19 feet per second (1 to 6 meters/second) in England, is factored into the design, and the number of passive ventilation stacks required to move the calculated amount of ventilation air are designed into the structure. In the base of the stack, dampers connected to the building's energy management system, and possibly to carbon dioxide, humidity, and/or temperature sensors, control the rate of ventilation. Diffusers at ceiling level introduce the ventilation air into the occupied spaces. Solar tubes that bring in light, as well as air, are incorporated into some passive ventilation stacks.

In contrast to Europe, which has a wide range of examples of passive ventilation systems, the concept has not had much success in the United States. One of the best US examples is the Federal Building in San Francisco, California, for which a sample computational fluid dynamics (CFD) simulation is shown in Figure 9.9.

PASSIVE COOLING

Earlier in this chapter, it was noted that today's German office buildings achieve substantially better energy performance than their US high-performance counterparts. An obvious question is, How do the Germans achieve such exceptional energy performance in their buildings? The answer is that they are changing some of the basic assumptions of the past several decades about how buildings should operate. Rather than completely isolating the building occupants from outdoor conditions, designers now assume moderate interaction by means of natural ventilation, daylighting, and passive cooling. This concept, called *lean building*, results in smaller building service equipment for heating and cooling. In the German context, passive cooling is the interaction of all measures that reduce heat gains and render natural heat sinks—night air and the ground—accessible (see Figure 9.10).

Heat loads are transferred to the surrounding environment with some time delays, and heat storage in the building mass itself is substantial. The main design priority is to restrict the amplitude and dynamics of external heat gains. Limiting glazing while maintaining daylighting is the key to this strategy, and the ratio of glazing to façade area is less than 43 percent for the set of 23 German demonstration buildings that were mentioned previously. Almost all buildings use externally adjustable sun-shading devices, and total solar energy transmittance is kept below 15 percent. Cooling is accomplished by using night ventilation in which the building mass is cooled using earth-to-air heat exchangers, which are simply underground metal ducts through which the air is brought into the building, or by slab cooling in which groundwater is pumped through cavities in the slab. The coefficient of performance (COP) for mechanical and hybrid night ventilation ranges between 4.5 and 14, far higher than that of conventional cooling.[19] The earth-to-air heat exchangers have extremely high COPs, ranging from 20 to 280. Note that today's best-performing chillers, the heart of many air-conditioning systems, have a maximum COP of about 8. Figure 9.11 illustrates this strategy graphically.

Eliminating conventional cooling systems gives the project the resources to accomplish the technical analysis to design a lean building appropriate to the bioregion, one that transfers daytime internal energy to the structure and minimizes the intrusion of external heat energy into the building. Even if the outdoor conditions vary, the result is that indoor conditions remain within a well-defined comfort zone, meeting the needs of the occupants.[20]

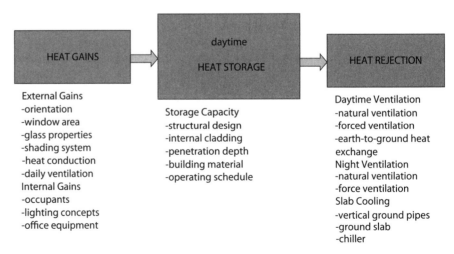

Figure 9.10   Passive cooling strategies use heat gain avoidance to minimize external thermal loads, minimize internal gains from occupants and electrical equipment, and use the building structure for storing residual heat gains, which are then removed by a combination of natural and forced ventilation with ground coupling.

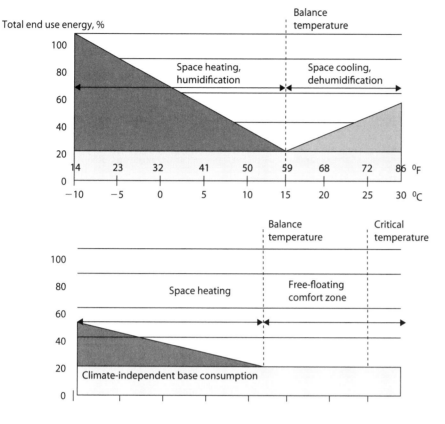

Figure 9.11 Low-energy German buildings use passive ventilation and cooling to eliminate much or all of the need for conventional mechanical cooling. Studies indicate that this strategy results in interior temperatures that rarely exceed acceptable space conditions for offices. As a consequence, buildings that use 100 kWh/m^2 (31,700 BTU/ft^2) of primary energy annually are achievable, a fraction of the energy of today's US high-performance green buildings.

The result of using this approach is an enormous reduction in the cooling capacity typically needed for office buildings. In monitoring these buildings, the researchers found that the upper desirable temperature limit of 77°F (25°C) was exceeded less than 10 percent of the working hours. During the unusually warm summer of 2002 in Germany, the naturally ventilated buildings exceeded the temperature criterion only 5 percent of the time, the equivalent of 1 hour every 2.5 days, a remarkable outcome. The one drawback of relying on a passive cooling strategy is that the mechanical plant will be unable to cope with the extreme weather conditions that can occur on occasion.

Building Envelope

After passive design is considered to minimize the need for external energy inputs, energy transmission through the building skin should be minimized through a tight, thermally resistant envelope. The building envelope must control solar heat gain, conduction or direct heat transmission, and infiltration or leakage heat transmission. The three major building envelope issues that need to be addressed are thermal resistance of the walls, window selection, and roof strategy. These are covered in the following sections. (The environmental impacts of materials selection are covered in Chapter 11.)

WALL SYSTEMS

The thermal conductance, or *U-value*, of building walls is an important factor in building energy efficiency because walls are generally the dominant component

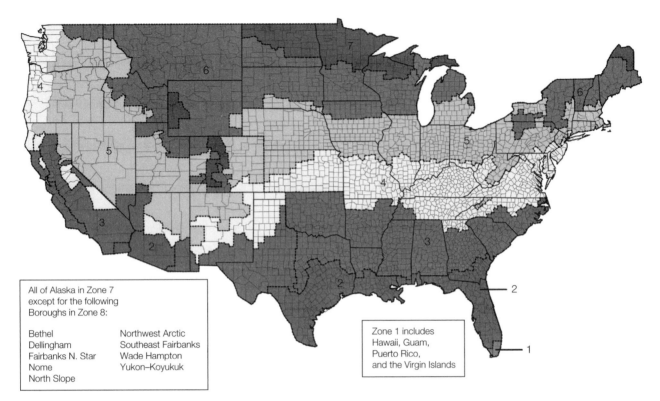

Figure 9.12 Climate zones by county in the United States based on the International Energy Conservation Code (IECC). (*Source: Pacific Northwest National Laboratory*)

of the envelope. U-values are measured in units of BTU/hr-ft^2-°F (W/m^2-°C). The lower the U-value of an assembly, the greater is its resistance to heat transfer. Maximum U-values are set by state building energy codes and by ASHRAE 90.1-2010. The maximum U-value is a function of the number of heating degree days (HDDs) and cooling degree days (CDDs) for the various climate zones in the United States (see Figure 9.12 and Table 9.4). Both HDD and CDD are measures of how much heating or cooling will probably be required in a given climatic zone. In general, wall thermal resistance becomes more important the farther north the building is located in the United States. Two other considerations in selecting wall systems are the thermal mass of the exterior surface that receives direct sunlight during the day and the placement of insulation with respect to the building façade. Placing insulation closer to the exterior nearest the outdoor conditions and having the thermal mass closest to the interior provides ideal conditions for using the mass beneficially and for minimizing the thermal loads transmitted into the building's interior that must be removed by air conditioning. In southern climates, it is generally important to design energy-shading façades that will reflect energy or that are ventilated to carry away energy that is absorbed on the building's skin.

WINDOW SELECTION

Windows play a variety of roles in the building envelope. They allow light into the room spaces, permit the occupants to admit air into the space in the case of operable windows, and provide a thermally resistant layer to energy movement. Windows must be installed so as to balance the amount of light admitted into the structure with the control of solar heat gain and conduction of energy through the window assembly. Window performance is a combination of

TABLE 9.4

US Climate Zones Defined by HDDs and CDDs

Zone Number	Thermal Criteria	
	IP Units	**SI Units**
1	9000 < CDD50°F	5000 < CDD10°C
2	6300 CDD50°F ≤ 9000	3500 < CDD10°C ≤ 5000
3A and 3B	4500 CDD50°F ≤ 6300 and HDD65°F ≤ 5400	2500 < CDD10°C ≤ 3500 and HDD18°C ≤ 3000
4A and 4B	CDD50°F ≤ 4500 and HDD65°F ≤ 5400	CDD10°C ≤ 2500 and HDD18°C ≤ 3000
3C	HDD65°F ≤ 3600	HDD18°C ≤ 2000
4C	3600 < HDD65°F ≤ 5400	2000 < HDD18°C ≤ 3000
5	5400 < HDD65°F ≤ 7200	3000 < HDD18°C ≤ 4000
6	7200 < HDD65°F ≤ 9000	4000 < HDD18°C ≤ 5000
7	9000 < HDD65°F ≤ 12600	5000 < HDD18°C ≤ 7000
8	12600 < HDD65°F	7000 < HDD18°C

Note: HDDs and CDDs are defined as the differences of daily average temperature from a base temperature, summed over the entire year. CDD50°F means the baseline for calculating cooling degree days (CDD) is 50°F. When the mean daily temperature is above 50°F, CDDs are calculated. A day with a mean temperature of 75°F would have (75°F − 50°F) = 25 CDDs. These are totaled for the full year in a given climate zone to establish the annual CDDs for that zone. Heating degree days (HDD) are based on a baseline of 65°F.

several factors: the *solar heat gain coefficient* (SHGC), the *visible transmittance* (VT) of the glass, the *thermal conductance* (U-value), and the infiltration or leakiness character of the window assembly.

Solar heat gain is largely a function of where windows are placed in the building and the types of glass used. SHGC and VT are used to express the radiation performance of windows in the building envelope. SHGC, with a value between 0 and 1, is the fraction of solar heat that enters the window and becomes heat; it includes both directly transmitted and absorbed solar radiation. The lower the SHGC, the less solar heat the window transmits through the glazing from the exterior to the interior and the greater its shading capability. In general, south-facing windows in buildings designed for passive solar heating should have windows with a high SHGC to allow beneficial solar heat gain in the winter. East- or west-facing windows encounter high levels of solar energy in the morning and afternoon and should generally have lower SHGC assemblies.

The VT, ranging in value from 0 to 1, refers to the percentage of the visible spectrum (380 to 720 nanometers) that is transmitted through the glazing. When daylight in a space is desirable, high-VT glazing would be the logical choice. However, lower-VT glazing may be more applicable for office buildings or where reduced interior glare is desirable. A typical clear, single-pane window has a VT of 0.90, meaning that it admits 90 percent of the visible light.

The ratio of SHGC to VT, known as the *light-to-solar-gain (LSG) ratio*, provides a gauge of the relative efficiency of different glass types in transmitting daylight while blocking heat gains. The higher the LSG, the brighter the room is without adding excessive amounts of heat. Table 9.5 shows average values of SHGC, VT, and LSG for typical windows.[21] Figure 9.13 is a diagram of the characteristics of a contemporary high-performance window that is optimized for heating climates. Windows that are filled with argon or krypton gas have more thermal resistance than those filled with air. Argon is inert, relatively abundant, and less costly than krypton, which provides higher thermal resistance but at a higher cost. Windows with a low SHGC were used in the relatively high temperature climate of northern Florida, as shown in Figure 9.14.

TABLE 9.5

Typical Values of SHGC, VT, and LSG for Total Window (Center of Glass) for Different Types of Windows

Window Type	Glazing	SHGC	VT	LSG
Single-glazed	Clear	0.79 (0.86)	0.69 (0.90)	0.97 (1.04)
Double-glazed	Clear	0.58 (0.86)	0.57 (0.81)	0.98 (1.07)
Double-glazed	Bronze	0.48 (0.62)	0.43 (0.61)	0.89 (0.98)
Double-glazed	Spectrally selective	0.31 (0.41)	0.51 (0.72)	1.65 (1.75)
Triple-glazed	Low-E	0.37 (0.49)	0.48 (0.68)	1.29 (1.39)

Figure 9.14 Low-E glazing on the Orthopaedics and Sports Medicine Institute at the University of Florida in Gainesville. The advent of high-technology glazing allows the design of buildings that admit visible light for daylighting but reflect infrared radiation. DOE-2.1 and daylighting simulations confirmed that daylighting and low-E glass produced greater savings than focusing solely on the thermal resistance of the building envelope. (T. Wyman)

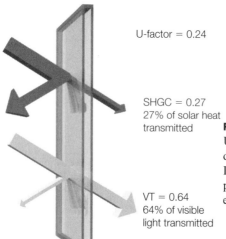

U-factor = 0.24

SHGC = 0.27
27% of solar heat transmitted

VT = 0.64
64% of visible light transmitted

Figure 9.13 Characteristics of a typical double-glazed window with a low SHGC, low-E glass, filled with argon gas. These windows are often referred to as *spectrally selective low-E glass* due to their ability to reduce solar heat gain while retaining high visible transmittance. Such coatings reduce heat loss and transmit less solar heat gain, making them suitable for climates with both heating and cooling concerns. (Illustration courtesy of Efficient Windows Collaborative)

Low-emissivity (low-E) and reflective coatings are applied to glazing to control the light passing through the glass and usually consist of a layer of metal a few molecules thick. The thickness and reflectivity of the metal layer (low-E coating) and the location of the glass to which it is attached directly affect the amount of solar heat gain in the room. Coating technology is advancing rapidly, and there are now low-E2 and low-E3 windows, with two and three silver coatings, respectively, that greatly improve glass performance. The low-E3 windows have a remarkably low SHGC of 0.30 or lower.

Any low-E coating is roughly equivalent to adding an additional pane of glass to a window. Low-E coatings reduce long-wave radiation heat transfer by 5 to 10 times. The lower the emissivity value (a measure of the amount of heat transmission through the glazing), the better the material reduces the heat transfer from the inside to the outside. Most low-E coatings also slightly reduce the amount of visible light transmitted through the glazing relative to clear glass. Representative emissivity values for different types of glass are as follows:

- Clear glass, uncoated: 0.84
- Glass with single hard-coat low-E: 0.15
- Glass with single soft-coat low-E2: 0.10

Increasing the window area to maximize daylighting has the effect of replacing a highly thermally resistant wall with far less thermally resistant glass, creating an opportunity for the infrared or heating component of light to enter the envelope and also creating the potential for infiltration around the window frame. In trading off daylighting to optimize the thermal envelope, controlling solar heat gain through windows is critically important. Prior to the development of today's window glazing and film technologies, 75 to 85 percent of infrared energy could pass through typical single- or double-paned glass.

A standardized national system for rating windows is important to enable performance comparisons of these important components of the building envelope. The National Fenestration Rating Council (NFRC) operates a uniform national rating system for measuring the energy performance of fenestration products, including windows, doors, skylights, and similar products.[22] The key to the rating system is a procedure for determining the thermal transmittance (U-factor) of a product. The U-factor rating procedure is supplemented by procedures for rating products for solar heat gain coefficient, visible transmittance, air leakage, and annual energy performance. Together, these rating procedures, as set forth in documents published by the NFRC, comprise the NFRC rating system. This system is expected to be supplemented by additional procedures for rating energy performance characteristics, including long-term energy performance and condensation resistance. The NFRC rating system employs computer simulation and physical testing by NFRC accredited laboratories to establish energy performance ratings for fenestration products and product lines (see Figure 9.15). The system is reinforced by a certification program under which a window and door manufacturer may label and certify its products to indicate those energy performance ratings.

Environmental Building News (*EBN*) suggests the approach to window selection depicted in Table 9.6.[23] *EBN* also provides the following additional recommendations:

- Use modeling software such as RESFEN and WINDOW to optimize the building fenestration system.
- As a minimum, select double-glazed, low-E, argon-filled windows for most US climate zones.
- For colder climates, select higher-performance windows with triple glazing, low-E2 coatings, and gas fill. In Germany, triple-glazed windows are now mandatory.
- Tune the windows to their orientation and climate. For east and west orientations when heat gain is not desirable, select low-SHGC windows. If passive solar heating is desired, high-SHGC windows on the south side may be desirable. For north-side windows in most climate zones, maximum thermal resistance is best, and SHGC is not necessarily important.

ROOF SELECTION: THERMAL RESISTANCE AND COLOR

The roof of a high-performance building is especially important because it is a major area for heat transmission due to its generally large area and exposure to

Figure 9.15 The NFRC adopted a new energy performance label in 2005. It lists the manufacturer, describes the product, provides a source for additional information, and includes ratings for one or more energy performance characteristics. (*Source:* National Fenestration Rating Council)

TABLE 9.6

Window Selection Approach Based on Climate and Solar Heat Gain

		Whole-Window U-factor	Whole-Window SHGC	Whole-Window SHGC for South Orientations When Solar Gain Is Desired
Hot Climate (Double or Triple Glazing)		0.16–0.30	0.25–0.37	0.36–0.63
		Lower is better	Lower is better	Very dependent on location
	Double Glazing	0.27–0.39	0.42–0.55	0.42–0.63
Cold Climate	Triple Glazing	0.17–0.26	0.33–0.49	0.42–0.63

Source: Environmental Building News.

the sun. According to Cool Communities, a nonprofit organization based in Rome, Georgia, the roofs of structures such as shopping malls, warehouses, and office buildings can reach 150°F (83°C) in the summer, enough to affect whole neighborhoods.[24] Using surfaces with high albedo (a measure of the reflectivity of solar radiation) for roofing can reduce the ambient air temperature so that the entire area is cooler. Light-colored roofs have high albedo, or high reflectivity, which helps reduce the thermal load on the building as well as the surrounding neighborhood. Both the Lawrence Berkeley National Laboratory and the Florida Solar Energy Center estimate that buildings with light-colored, reflective roofs use 40 percent less energy than similar buildings with dark roofs.

The Solar Reflectance Index (SRI), which measures how hot materials are in the sun, is used to easily describe the amount of solar energy reflected by roofing materials. A building with light-colored shingles and an SRI of 54 would

TABLE 9.7

Reflectance of Roof Materials and Air Temperatures above Roof

Material	Solar Reflectance	Temperature of Roof over Air Temperature ($°F/°C$)
Bright white coating (ceramic, elastomeric) on smooth surface	80%	15°/8°
White membrane	70%–80%	15°–25°/8°–14°
White metal	60%–70%	25°–36°/14°–20°
Bright white coating (ceramic, elastomeric) on rough surface	60%	36°/20°
Bright aluminum coating	55%	51°/28°
Premium white shingle	35%	60°/33°
Generic white shingle	25%	70°/39°
Light brown/gray shingle	20%	75°/42°
Dark red tile	18%–33%	62°–77°/34°–43°
Dark shingle	8%–19%	76°–87°/42°–48°
Black shingle or materials	5%	90°/50°

reflect 54 percent of incident solar energy, and would be very cool relative to a building with conventional dark shingles. Manufacturers have recently developed clean, "self-washing" white shingles with an even higher SRI, up to 62 percent. This is useful because the labor costs of maintaining the white color and high reflectivity of a conventional roof may exceed the worth of the energy being saved; consequently, a self-washing roof system will significantly reduce maintenance costs and improve energy performance. The reflectance of commonly used roofing materials is shown in Table 9.7.[25] As can be seen, dark-colored roofs have a tendency to absorb solar radiation, and they can be as much as 90°F (50°C) hotter than the air just above the roof. Because heat transmission is a function of temperature difference, a dark-colored, hot roof will have proportionately more heat conduction than a light-colored, relatively cool roof. Figure 9.16 shows the highly reflective, high-SRI roof used on a building in Hollywood, California.

Internal Load Reduction

Excellence in passive design and in the design of a high-performance building envelope needs to be combined with a significant effort to address the internal heat loads of the building. This is achieved in part by a good daylighting strategy, which has the dual benefit of reducing energy consumption for lighting and removing the lighting power saved from the total building cooling load. People constitute a major fraction of the building's internal heat load, and we can generally assume that reducing the number of people in a building is not a viable strategy. Reducing loads due to computers, peripherals, copiers, and other miscellaneous equipment is a promising strategy because it has been found that these loads constitute a substantial fraction of a building's energy consumption. Increasing wiring sizes beyond those required by code has the benefit of reducing energy losses in the wiring system and proportionately reducing the impact of these heat losses on the building's cooling system.

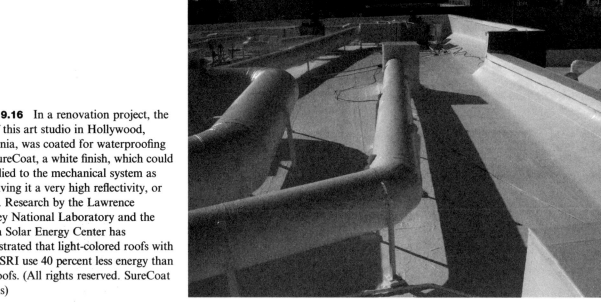

Figure 9.16 In a renovation project, the roof of this art studio in Hollywood, California, was coated for waterproofing with SureCoat, a white finish, which could be applied to the mechanical system as well, giving it a very high reflectivity, or albedo. Research by the Lawrence Berkeley National Laboratory and the Florida Solar Energy Center has demonstrated that light-colored roofs with a high SRI use 40 percent less energy than dark roofs. (All rights reserved. SureCoat Systems)

PLUG LOAD REDUCTION

Designers of high-performance buildings typically do not closely examine one of the major internal building loads, *plug loads*, a name for devices plugged into electrical outlets around the building that not only consume substantial energy but also increase cooling loads due to their heat emissions. The DOE estimates that office equipment loads make up about 18 percent of a US commercial building's electrical load, exceeded only by HVAC and lighting loads. In a study of plug loads for its newly renovated 6700-square-foot (622-square-meter) office building, IDeAs Z^2, a consulting engineering firm located in San Jose, California, estimated that plug loads would consume in excess of 40,000 kilowatt-hours per year, or almost 7 kilowatt-hours per square foot (75 kilowatt-hours per square meter) each year. In many office buildings, the largest plug loads are due to desktop computers, typically averaging about 160 watts per unit.[26] One alternative that should be considered for reducing desktop computer plug loads is to replace them with laptops designed for energy efficiency to maximize battery life. Laptops generally consume 40 watts, or about 25 percent of the power of desktops. The economics of replacing desktops with laptops will depend largely on the processing power required. In the case of IDeAs Z^2, which requires high-speed processors and large amounts of software to run its computational and graphics software, the cost difference between a desktop and a laptop with equivalent processing speed and storage was about $1835. An analysis indicated that the additional cost for high-end laptops would be greater than the cost of photovoltaics to offset the additional load; that is, it would be cheaper to buy photovoltaics than to replace desktops with laptops. In cases where exceptional computing capability is not needed, the cost differential and payback will be more favorable. Purchasing liquid crystal display (LCD) screens instead of cathode ray tube (CRT) screens results in a 50 percent energy savings, and LCD screens take up less space.

MISCELLANEOUS PLUG LOADS

Electrical loads for printers, scanners, copiers, and fax machines also contribute to higher building energy consumption. Energy Star–rated equipment costs 20 to 100 percent more than existing equipment, with energy savings of less than 10 percent. A typical high-efficiency, Energy Star–rated refrigerator saves 53 percent over a standard refrigerator but costs about $4.83 more per kilowatt-hour saved annually. There appears to be a very high premium for the highest-efficiency Energy Star–rated equipment. IDeAs Z^2 concluded that immediate replacement of existing devices was not cost-effective. Instead, its strategy was to replace old equipment with Energy Star–rated devices as much as possible and to evaluate the situation on an individual basis. It also researched energy-efficient dishwashers and concluded that, unlike refrigerators, which run 365 days a year, dishwashers run infrequently; therefore, the energy savings resulting from purchasing a high-efficiency dishwasher would be minimal. A final item that was evaluated was the existing coffeemaker, which stays on "warm" all day (and occasionally all night), holding half a pot of coffee. The firm will purchase a single-cup coffeemaker for daily staff use, which will heat one cup at a time and has no warming element. The old coffeemaker will only be used in conjunction with a thermos for large meetings.

PLUG LOAD CONTROL

Several types of control strategies are employed to reduce plug loads. Some equipment needs to be left on 24 hours a day, 7 days a week. This includes fax machines, main servers, and security systems. In these cases, high energy efficiency will be a key criterion when selecting equipment. Control of "phantom loads" in office equipment is another key strategy for conserving power. For equipment that has infrequent duty cycles, such as microwave ovens, the energy consumed during long hours of standby can be more than the energy consumed while in use. A second group of items, including printers, plotters, and copiers, need to be turned on only during working hours. These items typically have a long start-up time and would be inconvenient to turn on prior to each use, so they cannot be turned off between uses. However, there is no reason for these items to continue to run when the office is unoccupied, regardless of whether they are active or in sleep mode. IDeAs Z^2 found that the worst case was a laser plotter that consumed 1440 watts when plotting, 30 watts in the sleep mode, and 25 watts when manually switched to standby. Oddly enough, the plotter has no true off switch. To ensure that this equipment is not left on, the security system automatically turns off the electrical circuits to it when armed and turns on the circuits the next day when disarmed, reducing phantom loads. Occupancy sensor-controlled surge protectors are used at each workstation to turn off the power to task lights, computer monitors, speakers, and other nonessential peripherals when a user leaves his or her desk. Desktops are routinely left on all day but are set to go into sleep mode when not in use. Sleep mode saves energy and allows for fast restart times compared to the hibernate mode. However, if power is lost, data will also be lost. Hibernate mode saves data to the hard drive, so if power is lost, data will not be lost. However, computers in hibernate take much longer to restart when they come back to active mode. IDeAs Z^2 is currently working with EPA-sponsored researchers and experimenting with personal computer settings and individual occupancy sensors to determine how best to minimize energy consumption without significantly reducing productivity or creating inconvenience. There is a debate within IDeAs Z^2 as to whether it is wise to automatically shut off power to personal computers when the building is unoccupied. The argument against it is that risking the loss of

TABLE 9.8

Example of the Energy and Cost Benefits of Sizing Wiring in Electrical Circuits to Be Larger Than Code Minimum Requirements

	#8 AWG	#6 AWG
Conduit size	¾ in	1 in
Estimated loss (100% load, 75°C conductor temperature)	423 W	272 W
Wire cost*	$700	$800
Conduit cost*	$182	$259
Incremental cost		$177
Energy savings		604 kWh/yr
Dollar savings at $0.15/kWh, payback period		$90.60/yr, 2.0 years
Dollar Savings at $0.11/kWh, payback period		$66.45/yr, 2.7 years

*Wire and conduit costs in the above examples are based on those found at a large Nevada retailer in April 2009.

Figure 9.17 The IDeAs Z^2 office building in San Jose, California, is an NZE facility that was achievable through well-conceived approaches to reducing plug loads and upsizing electrical wiring. (Photograph courtesy of Integrated Design Associates, Inc.)

work left unsaved on the computer is not worth the attempt to save a few watts. However, most software has a built-in autosave feature, reducing the potential to lose a significant amount of work. IDeAs Z^2 is currently experimenting to determine if using the security system to turn off computer circuits is worth saving additional phantom losses.

UPSIZED ELECTRICAL WIRING

All circuits lose small amounts of energy through resistance as power flows through the wiring. Wire sizes recommended by code are based on keeping the heat generated from wiring losses below temperatures that would damage wire insulation. If wires are upsized, resistance in the wires is lower, and losses are reduced. IDeAs Z^2 estimated losses for sample circuits and compared the value of saved electricity with the additional cost to increase wire sizes one size above code recommendations. The paybacks were as low as four years for circuits that were highly loaded. All branch circuits carrying large, continuous loads were upsized to reduce wiring losses. In addition to reducing the electrical energy consumption of a circuit, this reduced the cooling loads associated with those losses. Table 9.8 is an example of the energy benefits found on a lighting circuit for a large retailer in Nevada. By upsizing one wire size—from #8 American Wire Gauge (AWG) to #6 AWG—rapid payback of just two to three years was achievable.[27] IDeAs Z^2 used upsized wiring in the design of its own LEED platinum, net zero energy (NZE) office building (see Figure 9.17).

Active Mechanical Systems

After the passive solar design of the building is optimized, the internal thermal loads in the building should be minimized. The thermal load of some buildings will be people-dominated; that is, the bulk of the load will be due to the number of people in the facility, so little can be done to reduce the load. A classroom building at a university is a good example of this situation. In other buildings, the load may be dominated by equipment, lighting, and other powered devices. In this situation, energy-efficient appliances, lighting, computers, and other energy-efficient systems can contribute to a significant reduction in cooling load.

Office buildings may have equipment-dominated loads if they have relatively large quantities of powered devices such as computers and copy machines and a moderate to low population.

A wide variety of HVAC systems can be used to meet the needs of a facility's occupants. The type of system selected is a function of the size of the building, the climatic conditions, and the load profile of the building. A typical building HVAC system will have an air side that delivers conditioned air into the spaces and a fluid side that creates chilled and hot water for use in the HVAC system, so equipment with the highest possible efficiency should be selected for all roles. The following sections contain information about selecting some of the major types of equipment in an HVAC system: chillers, air distribution system components, and energy recovery systems.

CHILLERS

According to Lawrence Berkeley National Laboratory's Environmental Energy Technologies Division, chillers are the single largest energy users in commercial buildings, consuming 23 percent of total building energy. Chillers also have the unfortunate characteristic of increasing their power consumption during the day, contributing to peak demand and forcing utilities to build new power plants to meet high daytime power demand. Consequently, chillers are responsible for large portions of peak power charges for commercial customers. In addition to these problems, most chiller plants tend to be oversized during the design process. Chillers operate at peak efficiency when they operate at peak load. However, chillers tend to operate at part load during much of the day. Even those that are correctly sized operate most of the time at low, part-load efficiencies.

Four types of chillers are commonly available today (note that 1 ton equals 12,000 BTU/hour of cooling capacity, or 3.4 kW):

- Centrifugal, primarily large tonnage above 300 tons (1000 kW)
- Screw [50 to 400 tons (170 to 1360 kW)]
- Scroll [up to 50 tons (170 kW)]
- Reciprocating [up to 150 tons (510 kW)]

Manufacturers of chillers have been working to produce high-efficiency chillers that meet the needs of high-performance green buildings. For example, the Trane CVHE/F EarthWise centrifugal chiller was awarded the EPA's Climate Protection Award. Rated at 0.45 kWh/ton of cooling, the EarthWise centrifugal chiller has the highest efficiency in this major category of HVAC equipment.

Chiller plant efficiency can be improved by more than 50 percent while improving reliability by combining new technologies such as direct digital control (DDC) and variable-frequency drives with improved design, commissioning, and operation. California tends to lead the nation in developing energy performance standards, and the latest version of California Title 24 Energy Efficiency Standards has substantially increased requirements for chiller efficiency.[28] Several different chiller technologies can be considered for a building. In general, water-cooled rotary screw or scroll chillers have the highest COP of all types of chillers. COP is the ratio of cooling power delivered by a chiller to the input power. A COP of 3.0, for example, indicates that the chiller provides 3 kWh of cooling for 1 kWh of input energy (see Figure 9.18). Note that a high-capacity screw or scroll water-cooled chiller has a COP of over 6, a very high level of performance and more than double the COP of 2.50 for an electrically

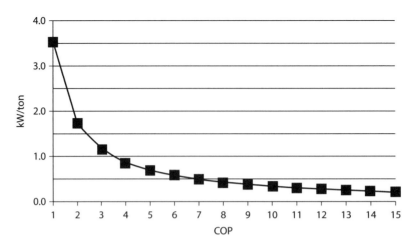

Figure 9.18 COP and kilowatt/ton are terms used to describe the performance of air-conditioning and cooling systems. COP is similar to the concept of efficiency, except it can be greater than 1.0 due to the nature of the refrigeration cycle. Specifically, COP is the ratio of cooling energy to input energy and has no units or dimensions. Kilowatt/ton is the inverse of COP; that is, it is the power required to produce 1 ton (12,000 BTU/hr) of cooling. A higher COP or a lower kilowatt/ton indicates a higher-efficiency system. A COP of 6 or higher represents a relatively high efficiency piece of air-conditioning equipment or a system. This corresponds to a kilowatt/ton value of 0.59 or lower. The highest-performing systems and equipment today have COPs of about 9, corresponding to about 0.40 kW/ton.

TABLE 9.9

Characteristics of a High-Efficiency Chiller Plant

Efficient design concept. Selecting an appropriate design concept that is responsive to the anticipated operating conditions is essential to achieving efficiency. Examples include using a variable-flow pumping system for large campus applications and selecting the quantity, type, and configuration of chillers based on the expected load profile.

Efficient components. Chillers, pumps, fans, and motors should all be selected for stand-alone as well as systemic efficiency. Examples include premium-efficiency motors, pumps that have high efficiency under the anticipated operating conditions, chillers that are efficient with both full and partial loads, and induced-draft cooling towers.

Proper installation, commissioning, and operation. A chiller plant that meets the first two criteria can still waste a lot of energy—and provide poor comfort to building occupants—if it is not installed or operated properly. For this reason, following a formal commissioning process that functionally tests the plant under all modes of operation can provide some assurance that the potential efficiency of the system will be realized.

---

operated, air-cooled chiller. One significant disadvantage of a water-cooled chiller is the need to provide a cooling tower to reject the energy absorbed from the building.

Absorption chillers tend to have a comparatively low COP, normally less than 1.0, which appears to indicate a very low level of performance. However, absorption chillers can use heat energy that would normally be wasted to provide cooling. A steam-driven screw chiller could reject its waste energy to an absorption chiller to provide additional cooling, thus increasing the COP of the overall system. Absorption chillers can also use relatively low temperature heat to produce chilled water and thus work well with solar thermal energy and waste heat from some varieties of fuel cells such as the phosphoric acid fuel cell (PAFC).

Table 9.9 describes the characteristics of a high-performance chiller plant,[29] and Table 9.10 provides several design strategies for achieving a relatively low-cost, high-efficiency chiller plant.[30]

TABLE 9.10

Design Strategies for a High-Efficiency Chiller Plant

1. Improve Chiller Plant Load Efficiency
Three methods for improving chiller plant load efficiency are: specify a chiller that can operate with reduced condenser water temperatures, specify a variable-speed drive (VSD) for the compressor motor, and select the number and size of chillers based on anticipated operating conditions.

2. Design Efficient Pumping Systems
Energy use in pumping systems may be reduced by sizing pumps based on the actual pressure drop through each component in the system, as well as the actual peak water flow requirements, accurately itemizing the pressure losses through the system and then applying a realistic safety factor to the total.

3. Properly Select the Cooling Tower
Proper sizing and control of cooling towers is essential to efficient chiller operation. Cooling towers are often insufficiently sized for the task. An efficient cooling tower should be specified based on using realistic wet-bulb sizing criteria; an induced-draft tower, if space permits; intelligent controls; and sequences of operation that minimize overall energy use.

4. Integrate Chiller Controls with Building Energy Management System
Although modern chillers are computer-controlled and have considerable intelligence to assist their operations, they should be integrated with the building's energy management system (EMS) to provide the capability to optimally operate the entire building energy plant. To accomplish this integration, the designers should specify an "open" communications protocol, use a hardware gateway, measure the power of ancillary equipment, and analyze the resultant data.

5. Commission the System
Commissioning a chiller system—that is, functionally testing it under all anticipated operating modes to ensure that it performs as intended—can improve efficiency and reliability and ensure that the owners are getting the level of efficiency they paid for.

AIR DISTRIBUTION SYSTEMS

Another major consumer of energy in modern buildings is the air distribution system, composed of air handlers, electric motors, ductwork, air diffusers, registers and grilles, energy and humidity exchangers, control boxes, and a control system. The air distribution system should be designed using much the same approach as for the chiller plant to deliver the precise capability needed and to do so efficiently across a wide range of operating conditions. According to *Greening Federal Facilities*[31], design options for improving air distribution efficiency include (1) variable air volume (VAV) systems, (2) VAV diffusers, (3) low-pressure ductwork design, (4) low–face velocity air handlers, (5) proper fan sizing with variable-frequency drive (VFD) motors, and (6) positive-displacement ventilation systems. VFD motors permit the speed of the motors to be matched to the exact amount of air required, which can produce enormous savings when the system is operating at less than peak load.

ENERGY RECOVERY SYSTEMS

Fresh air requirements for buildings mean that substantial quantities of fresh air are being brought into the facility while approximately the same amount of inside air is being exhausted to the outside. ASHRAE 62.1-2010 governs the quantity of fresh air that is required for building operation. The energy costs for this exchange of air can be considerable. For example, on a 90°F (32°C) summer day in New York City, with 80 to 90 percent relative humidity, the hot, humid outside air is being brought into buildings for ventilation purposes. At the same time, inside air at 72°F (22°C), with 50 percent relative humidity, is being exhausted to the outside. Clearly, it would be useful to have devices that could cool down the outside airstream in summer with the air being exhausted and, conversely, heat outside air being brought into the building during the winter by using the energy in the relatively warm air being exhausted. Another approach is to simply use outside air directly for conditioning the building when outside air conditions are just right for that purpose. Two technologies, economizers and energy recovery ventilators (ERVs), have been developed to use outside air for conditioning and to exchange energy between fresh intake air and exhaust air streams. Both approaches are described in the following sections.

Economizers

One rather obvious way to save energy in a typical building is to use outside air to cool the building when weather conditions are appropriate. The concept is quite simple: determine when the outside air temperature and humidity are in the same range as conditioned air delivered to the space would be and then duct the outside air to replace the conditioned airstream. The ductwork and dampers in the system are also designed so that all the return air can be exhausted from the building. Chillers and chilled-water pumps can be turned off, thus saving significant energy, as much as 20 to 30 percent of the energy that would ordinarily be invested in cooling (see Figure 9.18).

Unfortunately, economizers have a rather high rate of operational failure. Dampers become corroded and stick in place, temperature sensors fail, actuators fail, and linkages malfunction. Estimates of the failure rate of economizers vary widely, but the consensus of experts, according to Energy Design Resources, is that only 25 percent may be functioning properly within a few years. Malfunctioning economizers can actually cause significant energy waste. For example, a system mistakenly being operated in economizer mode in the middle of summer in a hot climate such as that of Florida or inland California can increase the cooling load by over 80 percent due to the large quantities of outside air that must be cooled. In spite of this, economizers have huge potential if properly installed, commissioned, and maintained.[32]

Energy Recovery Ventilators

Properly integrated desiccant dehumidification systems have become cost-effective additions to many innovative high-performance building designs. An Energy Recovery Ventilator (ERV) is an energy and humidity exchanger that employs desiccant technology for its functioning. ERV devices are placed between fresh air and exhaust airstreams, moving energy and humidity between the two streams to save significant quantities of energy. Additionally, indoor air quality can be improved by higher ventilation rates, and desiccant systems can help to increase fresh air makeup rates economically. In low load conditions, outdoor air used for ventilation and recirculated air from the building must be dehumidified more than cooled (see Figure 9.19).

Desiccants are materials that attract and hold moisture, and desiccant air-conditioning systems provide a method of drying air before it enters a conditioned space. With the high levels of fresh air now required for building ventilation, removing moisture has become increasingly important. Desiccant dehumidification systems are growing in popularity because of their ability to remove moisture from outdoor ventilation air while allowing conventional air-conditioning systems to deal primarily with control temperature (sensible cooling loads).

The Air-Conditioning and Refrigeration Institute (ARI) has developed a standard for ERVs, ARI 1060-2001, Rating Air-to-Air Heat Exchangers for Energy Recovery Ventilation Equipment, and ERV manufacturers should provide performance data in accordance with this standard. A typical ERV is shown in Figure 9.19. The device consists of a metal wheel coated with desiccant that rotates between the intake fresh air and exhaust airstreams. In summer, it

Figure 9.19 The ERV manufactured by Greenheck, Inc., houses a desiccant wheel that rotates between fresh and exhaust air streams, exchanging energy and humidity and providing enormous energy conservation benefits. (Photograph courtesy of Greenheck, Inc.)

dries and cools the hot, humid intake air with the cool, dry exhaust air from the building, saving significant quantities of energy, especially because the removal of moisture is accomplished via the desiccant, a very energy efficient strategy.

VENTILATION AIR AND CARBON DIOXIDE SENSORS

A healthy indoor environment is an important goal of green buildings. Creating a healthy interior requires that fresh outside air be brought into the building to dilute the buildup of potentially toxic components of indoor air. These toxic components include carbon dioxide from respiration, carbon monoxide from incomplete combustion of fuel, volatile organic compounds (VOCs) from building materials, and potentially others. The quantity of outside air required by ASHRAE 62.1-2010 for ventilation air is significant, and it must be either heated or cooled to allow it to remix with the supply airstream.

Contemporary US buildings have two basic methods for providing fresh or ventilation air for their occupants. First, the system can be designed to provide a constant quantity of fresh air based on a conservative evaluation of the number of occupants and the building's operating conditions. This approach has the advantage of being fairly simple, but the problem is that in a building with a variable population, substantial quantities of energy are wasted to condition the fresh air. A better approach would be to determine how many people are in the building and introduce the appropriate quantity of ventilation air based on the number of occupants. The concentration of carbon dioxide provides an indicator of how many people are in the building. Carbon dioxide is used as a surrogate ventilation index for diagnosing ventilation inefficiency or distribution problems. As the number of people in the space or the level of activity increases, so will the carbon dioxide concentration. Increased concentration of carbon dioxide in a space is also linked to discomfort and an increased perception of odors. Sensors are now available to detect the concentration of carbon dioxide in building spaces, and the data can be used as a surrogate for indoor air quality. The precise quantity of ventilation air needed to dilute the carbon dioxide to an appropriate level can be admitted to the space based on the measured carbon dioxide concentration. Buildings with populations that vary greatly can benefit from the use of this sensor technology because they can admit the exact amount of ventilation air needed, not the large quantities that would otherwise be required without this detection system.

Water-Heating Systems

In some types of buildings, water heating can consume large amounts of energy. In facilities with kitchens, cafeterias, health club facilities, or residences, there will be heavy demand for hot water. Solar water heating and tankless water heaters are technologies that can be used to reduce the hot water demand; these are described in the following sections.

SOLAR WATER-HEATING SYSTEMS

An estimated 1 million residential and 200,000 commercial solar water-heating systems have been installed in the United States. Although there are many different types of solar water-heating systems, the basic technology is very simple. Sunlight strikes and heats an absorber surface within a solar collector or an actual storage tank. Either a heat transfer fluid or the actual potable water to be used flows through tubes attached to the absorber and picks up the heat from it. Systems with a separate heat transfer fluid loop must utilize a heat exchanger to heat the potable water. The heated water is stored in a separate preheat tank or a conventional water heater tank until needed. If additional heat is needed, it is provided by electricity or fossil fuel energy by the conventional water-heating system. By reducing the amount of heat that must be provided by conventional water heating, solar water-heating systems directly substitute renewable energy for conventional energy, reducing the use of electricity or fossil fuels by as much as 80 percent.

Today's solar water-heating systems are well proven and reliable when correctly matched to climate and load. The current market consists of a relatively small number of manufacturers and installers that provide reliable equipment and quality system design. A quality assurance and performance rating program for solar water-heating systems, instituted by a voluntary association of the solar industry and various consumer groups, makes it easier to select reliable equipment with confidence.

Solar water-heating systems are most likely to be cost-effective for facilities with water-heating systems that are expensive to operate or with operations such as laundries or kitchens that require large quantities of hot water. A need for hot water that is relatively constant throughout the week and throughout the year, or that is higher in the summer, is also helpful for solar water-heating economics. Conversely, hard water is a negative factor, particularly for certain types of solar water-heating systems, because it can increase maintenance costs and cause those systems to wear out prematurely.

Although solar water-heating systems all use the same basic method for capturing and transferring solar energy, they use a wide variety of technologies. Systems can be either active or passive, direct or indirect, pressurized or nonpressurized. As a rough guide, the solar system should have 10 square feet (1 square meter) of collector area for every 14 gallons (50 liters) of daily hot water usage, and the storage tank should have 1.4 gallons per square foot (50 liters per square meter) of collector area. This corresponds to 40 square feet (4 square meters) of collector for every apartment suite in multiunit residential buildings and 10 square feet (1 square meter) of collector for every five office workers in an office building.

TANKLESS (INSTANTANEOUS) WATER-HEATING SYSTEMS

Tankless, or instantaneous, water heaters eliminate the need for hot water storage by supplying energy at the point of demand to heat water as it is being used. Clearly, this takes high energy input, either electric or gas, at the point of use, but energy losses from storage tanks are eliminated. Unlike storage water heaters, tankless water-heating systems can theoretically provide an endless supply of hot water. The actual maximum hot water flow is limited by the size of the heating element or thermal input of the gas heater.

Demand water heaters, common in Japan and Europe, began appearing in the United States about 30 years ago. Unlike conventional tank water heaters, tankless water heaters heat water only as it is used, or on demand. A tankless unit has a heating device that is activated by the flow of water when a hot water valve is opened. Once activated, the heater delivers a constant supply of hot water. The output of the heater limits the rate of flow of the heated water.

Gas tankless hot water units typically heat more gallons per minute than electric units, but in either case, the rate of flow is limited. Electric tankless heaters should use less energy than electric storage systems. But gas-fired tankless heaters are only available with standing pilot lights, which lower their efficiency. In fact, the pilot light can waste as much energy as is saved by eliminating the storage tank.

Tankless heaters have either modulating or fixed output control. The modulating type delivers water at a constant temperature, regardless of flow rate. The fixed type adds the same amount of heat, regardless of flow rate and inlet temperature.

Electrical Power Systems

In addition to the building's air-conditioning and heating systems, the lighting system and electric motors are major consumers of electrical energy. Major advances have been made in lighting fixture and lighting control technologies that can dramatically reduce energy consumption. Because electric motors in buildings drive fans, pumps, and other devices, using the most energy-efficient motor can result in substantial energy savings. The following sections describe advances in lighting and motor technology that can produce substantial energy savings in buildings.

LIGHTING SYSTEMS

Lighting is a voracious consumer of electrical energy, consuming on the order of 30 percent of total building electrical energy in the United States; thus, a primary goal of all designs should be to reduce dependence on artificial light and to maximize the use of daylighting. These efforts should become an integrated strategy, that is, combining natural lighting and powered lighting to provide high-quality, low-energy illumination for the building's spaces.

When specifying lighting, several technical terms are used for selecting the most energy efficient and effective system for the application: efficacy, Color Rendering Index (CRI), and color temperature. These three terms are defined below.

> *Efficacy* is used as the measure of lighting efficiency, and it is measured in lumens per watt (lm/W) or light output per energy input. Clearly, higher efficacy means a more energy efficient lighting system. Fluorescent lamps have efficacies that range from 80 to 93 lm/W, while the emerging LED technology has a maximum efficacy of 130 lm/W.

Color Rendering Index (CRI) describes how a light source affects the appearance of a standardized set of colored patches under standard conditions. A lamp with a CRI of 100 will not distort the appearance of the patches in comparison to a reference lamp, while a lamp with a CRI of 50 will significantly distort colors. The minimum acceptable CRI for most indoor applications is 70; levels above 80 are recommended.

Color temperature influences the appearance of luminaires and the general "feel" in the space and is expressed in kelvins (K). Low color temperature (e.g., 2700 K) provides a warm feel similar to that of light from incandescent lamps; 3500 K provides a balanced color; and 4100 K emits "cooler," bluish light. Standardizing the color temperature of all lamps in a room or facility is recommended.

Fluorescent Lighting

Fluorescent lighting is the best source for most building lighting applications because it is very efficient and can be switched and controlled easily. Modern linear fluorescent lamps have good color rendering and are available in many styles. Lamps are classified by length, form (straight or U-bend), tube diameter (e.g., T-8 or T-5), wattage, pin configuration, electrical type (rapid or instant-start), CRI, and color temperature. When specifying a lighting system, it is important that the lamp and ballast be electrically matched and the lamp and fixture optically matched.

Fluorescent lamp diameters are measured in ⅛-inch (0.3-centimeter) increments—for example, T-12s are 12/8 inch (3.2 centimeters) or 11/2 inch (3.8 centimeters) in diameter, and T-8s are 1 inch (2.5 centimeters) in diameter. Typical linear fluorescent lamps are compared in Table 9.11; note that efficacy (lumens per watt) is higher with smaller-diameter lamps.[33]

T-5 lamps are designed to replace T-8 fluorescent lamps. The T-5 lamp operates exclusively with electronic ballasts and offers continuous dimming. It has an efficacy of about 93 lm/W, compared to the 89 lm/W achievable with T-8 lamps. Most manufacturers use internal protective shield technology to minimize light depreciation to a predicted 5 percent over the life of the lamp. This technology has also made it possible to reduce the mercury content of lamps to about 3 mg, compared to the previous 15 mg.

Color rendering of fluorescent lamps is very important. Modern, efficient fluorescent lamps use rare-earth phosphors to provide good color rendition. T-8 and T-5 lamps are available only with high-quality phosphors that provide CRIs greater than 80. Electronic ballasts with linear fluorescent lighting should be specified. These are significantly more energy efficient than magnetic ballasts and eliminate the hum and flicker associated with older fluorescent lighting. Dimming electronic ballasts are also widely available.

TABLE 9.11

Fluorescent Light Fixture Characteristics

Lamp Type	T-12	T-12 ES	T-8	T-5*
Watts	40	34	32	54
Initial lumens	3200	2850	2850	5000
Efficacy (lumens/watt)	80	84	89	93
Lumen depreciation†	10%	10%	5%	5%

*High-output T-5 in metric length.
†Change from initial lumens to design lumens.

Luminaires should be selected based on the tasks being performed. Reflectorized and white industrial fixtures are very efficient and good for production and assembly areas but are usually inappropriate for office applications. Lensed fluorescent fixtures (prismatic lens style) typically result in too much reflected glare off computer screens to be a good choice for offices. In areas with extensive computer use, the common practice is to install parabolic luminaires, which minimize high-angle light that can cause reflected glare off computer screens; however, these may result in unpleasant illumination in the presence of dark ceilings and walls. Instead, for tall ceilings, over 9 feet (2.7 meters) in height, direct/indirect pendant luminaries should be used. For lower ceilings, 8 feet, 6 inches (2.6 meters) in height, parabolic luminaires with semispecular louvers should be considered.

Luminaires should not be selected solely on the basis of efficiency. A very high efficiency luminaire can have inferior photometric performance. The most effective luminaires are usually not the most efficient, but they deliver light where it is most needed and minimize glare. The Luminaire Efficiency Rating (LER) used by some fluorescent fixture manufacturers makes it easier to compare products. Since the LER includes the effect of the lamp and ballast type, as well as the optical properties of the fixture, it is a better indicator of the overall energy efficiency than simple fixture efficiency. An LER of 60 is good for a modern electronically ballasted T-8 fluorescent fixture; 75 is very good and is close to state of the art.

Fiber-Optic Lighting

Fiber-optic lighting utilizes light-transmitting cable fed from a light source in a remote location. A fiber-optic lighting system consists of an illuminator (light source), fiber-optic tubing, and possibly fixtures for end-emitting uses. When light strikes the interface between the core and the cladding of the cable, total internal reflection occurs and light bounces or reflects down the fiber within the core. Two types of fiber are used: small-diameter strands bundled together or a solid core (the latter being more limited in application). The lighting source is generally a halogen or metal halide lamp. Fiber-optic lighting is generally energy-efficient and provides illumination over a given area. The only electrical connection needed for the system is at the illuminator. No wiring or electrical connection is required along any part, either at the fiber-optic cable or at the actual point source fixture.

Fiber-optic lighting systems provide many benefits and eliminate many problems encountered with conventional lighting systems. Infrared and ultraviolet wavelengths produced by a given light source are undesirable by-products, and fiber-optic systems can filter these out, eliminating the damaging effect of ultraviolet and infrared radiation. Fiber-optic lighting requires no voltage at the fixture, is completely safe, emits no heat, and is virtually maintenance-free. This lighting technology is especially useful for retail settings, supermarkets, and museums because it emits no heat or ultraviolet radiation (see Figure 9.20).

Light-Emitting Diodes

Light-emitting diodes (LEDs) for lighting systems are evolving very rapidly, and white-light LEDs are now being produced that can be used in many building applications. LEDs are based on semiconductors that emit light when current is passed through them, converting electricity to light with virtually no heat generation. Until the early 1990s, red, yellow, and green LEDs were being produced. In the early 1990s, blue LEDs and then white LEDs were developed.

Figure 9.20 Fiberstars' EFO lighting, shown in Trammell Crow's office in Houston, Texas, is low-energy, lightweight, and ultrasafe because it does not conduct electricity. The manufacturer claims that a single EFO lamp uses 68 watts and replaces about 400 watts of halogen lamps. (Photograph courtesy of Fiberstars, Inc.)

LEDs have an efficacy of 130 lm/W, compared to about 52 lm/W for incandescent bulbs. At present, linear fluorescent lights produce 80 to 93 lm/W and compact fluorescent lightbulbs about 65 lm/W. The latest laboratory versions are producing 150 lm/W. LEDs are also very tough and durable and able to absorb large shocks without malfunctioning. The color temperature appearance of LEDs has improved, with warm-white (2700 to 3000 K) and neutral-white (3500 to 4000 K) versions now available. The CRI cited by leading LED manufacturers is now at least 80, which is the minimum recommended for indoor applications. With a projected lifetime of 50,000 hours, LEDs last 20 times longer than incandescent lightbulbs and 2 to 3 times longer than fluorescent lights. However, LEDs become dimmer over time, and the actual useful life, which is defined as when the LED is emitting just 70 percent of its initial light output, is about 30,000 hours. This dimming phenomenon is known as *LED depreciation* and is caused by the heat generated at the internal junction in the LED. Costs of LED lights declined by half from 2009 to 2010, from $36 to $18 per thousand lumens (kilolumens, or klm), and prices are expected to be about $2/klm by 2015. Some forecasts are that, by 2020, LED technology will be in 70 to 80 percent of all building lighting applications (see Figures 9.21 and 9.22).

LIGHTING CONTROLS

Ideally, lighting controls should comprise an integrated system that performs two basic functions:

1. Detects occupancy and turns lights on or off in response to the presence or absence of occupants.
2. Throttles lights up and down or turns lights on and off to compensate for levels of natural light provided by the daylighting system.

Figure 9.21   An intelligent LED lighting system by Color Kinetics and 4 Wall Entertainment illuminates the exterior of the Hard Rock Hotel and Casino in Las Vegas, Nevada, cutting annual energy costs by 90 percent. (Photograph courtesy of 4Wall Entertainment)

Figure 9.22   LED lighting is now available as architectural grid lay-in lighting that measures 2 feet by 2 feet. (Photograph courtesy of Lunera Lighting, Inc.)

Research has shown that daylight-linked electrical lighting systems—such as automatic on/off and continuous dimming systems—have the potential to reduce the electrical energy consumption in office buildings by as much as 50 percent.

There are two basic types of daylighting control systems: dimming and switching. Dimming controls vary the light output over a wide range to provide the desired light level. Switching controls turn individual lamps off or on as required. In a conventional two-lamp fixture, there are three settings: both lamps off, one lamp on, both lamps on. The same strategy can be used with three- and four-lamp fixtures. Dimming systems, which require electronic dimmable ballasts, are more expensive than switching systems; however, they achieve greater savings and do not have the abrupt changes in light level characteristic of switching systems. Dimming systems are best suited to offices, schools, and those areas where deskwork is being performed. Switching systems can be used in areas with high natural light levels (e.g., atria and entranceways) and where noncritical visual tasks are being performed (e.g., cafeterias and hallways).

Of course, neither system is appropriate in nondaylit areas. The lighting control zones and number of sensors need to be carefully designed. At least one sensor is required for each building orientation. The lighting control zone should only be as deep into the building as is effectively daylit—about 16 feet (5 meters) from windows in conventional office plans. Light shelves can extend the daylit zone deeper into the building's interior.

In addition to energy savings, electric light dimming systems offer two other advantages over conventional lighting systems. First, conventional lighting systems are typically designed to initially overilluminate rooms, to account for the 30 percent drop in lighting output over time. Electric light dimming systems automatically compensate for this reduced output to give a constant light level over time. Second, daylighting controls can be adjusted to give the desired light level for any space. Thus, when floor plans are changed, it is easy to adjust the light levels to meet the lighting needs of each area, provided that the system is properly zoned and has adequate lighting capacity.

The cost of switching controls is quite modest, and these systems should be considered in all applications where changes in light level can be tolerated. Dimming lighting controls are approximately twice the price of switching controls and require electronic dimmable ballasts.

ELECTRIC MOTORS

Electric motors are important components of modern buildings, as they drive fans, pumps, elevators, and a host of other devices. Over half of all electrical energy in the United States is consumed by electric motors. Motors typically consume 4 to 10 times their purchase cost in energy each year, so energy-efficient models often make economic sense. For example, a typical 20-horsepower, continuously running motor uses almost $8000 worth of electricity annually at 6 cents per kilowatt-hour, about nine times its initial purchase price. Improving the efficiency of electric motors and the equipment they drive can save energy and reduce operating costs.

The construction materials and the mechanical and electrical design of a motor dictate its final efficiency. Energy-efficient motors utilize high-quality materials and employ optimized design to achieve higher efficiencies. Large-diameter copper wire in the stator and more aluminum in the rotor reduce resistance losses of the energy-efficient motor. An improved rotor configuration and an optimized rotor-to-stator air gap reduce stray load losses. An optimized cooling fan design provides ample motor cooling with a minimum of windage loss. Thinner and higher-quality steel laminations in the rotor and stator core allow the energy-efficient motor to operate with substantially lower magnetization losses. High-quality bearings result in reduced friction losses.

Innovative Energy Optimization Strategies

At least partially because of the green building revolution, a wide variety of innovations in building systems are emerging. Four of the more innovative approaches are described here: radiant cooling, ground coupling, renewable energy systems, and fuel cells. Each of these is a cutting-edge strategy that can have a marked effect on energy consumption if used properly in a building.

RADIANT COOLING

In the United States, cooling is generally delivered to conditioned spaces using air that is pressurized by fans and delivered via ductwork to the various spaces. Air has a very low heat capacity, and the result is that rather large quantities of air must be delivered to a space to provide the needed cooling effect. Additionally, air, a compressible medium, is relatively energy-intensive to move, compared to water, which is incompressible, has very high heat capacity, and can be moved comparatively cheaply via pumping. That is why, in Europe, radiant cooling is frequently used for cooling spaces. These systems use water, which has 3000 times the energy transport capacity of air, as the medium for delivering cooling to the space. In Germany, radiant cooling systems have become the new standard.

Radiant cooling systems circulate cool water through tubes in ceiling, wall, or floor elements or panels. The water temperature does not differ noticeably from the room temperature, so care must be exercised to ensure that the temperature of the circulated water does not reach the dew point of the air in the space. Otherwise, condensation will occur, resulting in moisture problems. The cost of a radiant cooling system is approximately the same as that of a VAV system, but the life-cycle savings are 25 percent higher compared to those of a VAV system. Moreover, the energy required for circulating water is only about 5 percent of the energy needed to circulate a comparable capacity of air.

There are three main types of radiant cooling systems (see Figure 9.23):

1. *Concrete core.* Plastic tubes are buried in concrete floor and ceiling slabs.
2. *Metal panels.* Metal tubes are connected to aluminum panels.
3. *Cooling grids.* Plastic tubes are embedded in plaster or gypsum.

The metal panel system is the most commonly used radiant cooling system and, due to its metal construction, has a relatively fast response time to changing conditions. Cooling grids are generally the choice for retrofit projects because the grid of plastic cooling tubes is readily placed in plaster or gypsum in existing walls. As a guide to system sizing, the total heat transfer rate (combined radiation and convection) is about 11 $W/m^2/°C$ (0.7 $W/ft^2/°F$) temperature difference for cooled ceilings.

Design guidelines for radiant cooling systems are as follows:

1. The building should be well sealed.
2. In humid areas, the intake fresh air should be dehumidified prior to its entry into conditioned spaces.
3. Radiant cooling requires a large surface area due to the relatively small temperature difference between the cooling surface and the room air.
4. The set points for cooling and heating must be carefully considered to deliver maximum conditioning without causing moisture problems. For instance, for a typical system in Germany, during the cooling season, the

Figure 9.23 Radiant cooling panels provide a low-energy solution for cooling, requiring only a fraction of the energy of a conventional system based on air handling and ductwork. (A) A dropped-panel installation showing a radiant cooling panel installation and ease of maintenance. (B) Installing a cooling mat: grids of plastic tubing carrying chilled water are placed under the ceiling drywall. (Photographs courtesy of Juan Rudek, Karo Systems)

room temperature set point is about 80°F (27°C), with cold or chilled water entering the radiant cooling panels at 61°F (16°C) and leaving at 66°F (19°C). For heating, the room set point is 68°F (20°C), with heated water delivered at 95°F (35°C) and leaving the radiant panels at 88°F (31°C).

5. Humidity sensors should be used to detect when the temperature of the supply water is approaching the dew point to activate valves that will prevent condensation from occurring.

GROUND COUPLING

One innovative method for reducing energy consumption in a building is ground coupling, in which the thermal characteristics of the earth and groundwater in the vicinity of the building are used for cooling and heating purposes. There are two major methods for applying ground coupling for building conditioning: direct and indirect. In the direct approach, groundwater is employed in radiant cooling systems, and fresh air is cooled through ground contact. The indirect approach employs heat pumps in conjunction with the ground or groundwater to move heating and cooling energy between the building and the earth. It is feasible, for example, to use groundwater in the 60°F (16°C) range in a radiant cooling system for a building and virtually eliminate the need for a chilled-water plant. The following sections describe these two approaches.

Ground Source Heat Pumps

Ground source heat pump (GSHP) systems use the ground as a heat source in the heating mode and as a heat sink in the cooling mode. The ground is an attractive heat source or sink compared to outdoor air because of its relatively stable temperature. In many locations, the soil temperature does not vary significantly over the annual cycle below a depth of about 6.5 feet (2 meters). For example, in Louisiana, outdoor air temperatures may range from wintertime lows of 32°F (0°C) or lower to summertime highs of about 95°F (33°C), while the soil temperature at depths greater than 6.5 feet (2 meters) never falls below about 64°F (18°C) or rises to approximately 77°F (25°C), averaging around 68°F (20°C). A number of different methods have evolved for thermally connecting, or coupling, the heat pump systems with the ground, but the two major methods are vertical systems and horizontal systems. These systems depend on how the piping that makes ground contact is laid out.

The *horizontal* ground-coupling system uses plastic piping placed in horizontal trenches to exchange heat with the ground. Piping may be placed in the trenches either singly or in multiple-pipe arrangements. The primary advantage of horizontal systems is lower cost. This is a result of fewer requirements for special skills and equipment, combined with less uncertainty about subsurface site conditions. The disadvantages of the horizontal ground-coupling system are its high land area requirements, its limited potential for heat exchange with groundwater, and the wider temperature swings of the soil at typical burial depths.

Vertical ground coupling is the most common system used in commercial-scale systems. Vertical U-tube plastic piping is placed in boreholes and manifolded in shallow trenches at the surface. Vertical ground coupling has several advantages: low land area requirements, stable deep-soil temperatures with greater potential for heat exchange with groundwater, and adaptability to most sites. Among the disadvantages of vertical ground coupling are potentially higher costs, problems in some geological formations, and the need for an experienced driller/installer. The regulatory requirements for vertical boreholes used for ground-coupling heat exchangers vary widely by state. One note of caution to the designer is that some regulations, installation manuals, and/or local practices call for partial or full grouting of the borehole. The thermal conductivity of materials normally used for grouting is very low compared to the thermal conductivity of most native soil formations. Thus, grouting will tend to act as insulation and hinder heat transfer to the ground.

In addition to ground-coupled heat pumps, systems that use both surface water and groundwater have been successful. In fact, for commercial-scale applications, if groundwater is available in sufficient quantities, it should be considered as the first alternative, as it will often turn out to be the least costly.

Direct Ground Coupling for Fresh Air and Chilled Water

It is also possible to heat and cool fresh air being introduced into a building by bringing it in underground through large-diameter, 1- to 2-meter (3- to 7-foot)

Figure 9.24 Ground-coupled system showing an air intake tube, 1.8 meters (6 feet) in diameter, under a Mercedes showroom in Stuttgart, Germany. The galvanized steel tube, which is 100 meters (325 feet) long, heats cold air via ground contact in winter and cools hot air in summer. The tube is located in a zone under the building where the temperature is a constant 60°F (16°C).

galvanized steel tubes, known as *earth-to-air heat exchangers*. Additionally, groundwater can sometimes be used as the source of chilled water, reducing or eliminating the need for mechanical chilled-water systems. Both of these practices are becoming common in Germany, where buildings are now routinely conditioned using a comprehensive ground-coupling scheme. For example, a 50,000-square-meter (538,000-square-foot) Mercedes showroom in Stuttgart has all of its fresh air brought in through a corrugated steel tube with a diameter of 1.8 meters (6 feet), the top of which is located 2 meters (6.6 feet) underground at a velocity of about 155 meters (500 feet) per minute (see Figure 9.24). The ground temperature at this depth is a stable 60°F (16°C). In winter, cold outside air is warmed up to approximately the ground temperature prior to introduction into the building. In summer, hot outside air is significantly cooled prior to its introduction into the facility.

Groundwater, where permitted by local jurisdictions, can be used directly in a radiant cooling system, where the temperature is adjusted by mixing valves or by employing a relatively small heat pumping system to move energy to and from the groundwater stream. The groundwater is pumped into the radiant cooling system and discharged back to the ground with only a few degrees of temperature change.

It is also feasible to design and install a ground-coupling system that both conditions the fresh air being brought into the building and uses groundwater for a radiant cooling system. A well-designed ground-coupled HVAC system can provide significant savings by greatly reducing the requirements for equipment, ductwork, and air handlers.

Renewable Energy Systems

Renewable energy can be generated on-site by three different techniques: photovoltaics, wind energy, and biomass. Each of these has advantages and disadvantages and varying levels of complexity. A brief summary of each is provided in Table 9.12.

TABLE 9.12

Advantages and Disadvantages of Renewable Energy Systems

Renewable Energy Type	Advantages	Disadvantages
Photovoltaics (PVs)	New technologies allow integration into building façade	Remains relatively expensive
	Price of PV modules is dropping as demand increases	Potential metering problems with local utility
Wind	Lowest kilowatt-hour cost of any renewable energy source	Generally large, unsightly generators
		Significant annual wind speed needed
Biomass	Can use local vegetation for fuel	Systems for buildings are not readily available
	Potentially low-cost energy source	

PHOTOVOLTAICS AND BUILDING-INTEGRATED PHOTOVOLTAICS

Photovoltaic (PV) cells are semiconductor devices that convert sunlight into electricity. They have no moving parts. Energy storage, if needed, is provided by batteries. PV modules are successfully providing electricity at hundreds of thousands of installations throughout the world. Especially exciting are building-integrated photovoltaic (BIPV) technologies that integrate PV cells directly into building materials, such as semitransparent insulated glass windows, skylights, spandrel panels, flexible shingles, and raised-seam metal roofing. PV elements can be fabricated in different forms. They can be used on or be integrated into roofs and façades as part of the outer building cladding, or they can be used as part of a window, skylight, or shading device. PV laminates provide long-lasting weather protection. Their expected life span is in excess of 30 years. Warranties are currently available for a 20-year period.

PV systems are modular in nature; hence, they can be adapted to changing situations. They can usually be added, removed, and reused in other applications. Typical modules consist of glass laminates, plastic Tedlar bounding material, and silicon cells with trace amounts of boron and phosphorus. Their disposal or recycling after the end of their life span should not create any environmental problems.

A variety of attractive BIPV products are available that allow building surfaces, such as the roof, walls, skylights, and sunshades, to double as solar collectors. Integrating these products into the building envelope creates a large solar collection area, enabling solar power to displace more of the electricity used in the building. The cost of the PV system is offset by the fact that the BIPV products displace standard building envelope components. PVs can be integrated into the roofing system through PV roof shingles, roof tiles, and metal roof products, all of which can replace the standard roof. Alternatively, framed PV modules can be incorporated into the roofing system. BIPV glazing systems are available that allow sloped and overhead glazing to capture solar energy. These glazing systems are insulated and can be specified to provide the desired level of light transmission for daylighting, typically as needed per kilowatt of capacity.

Curtainwall offers considerable potential for BIPVs. A wide variety of PV products can be used in place of architectural spandrel glass and vision glass. Sunshades and skylights, common BIPV applications, have become popular in Europe. BIPV systems are available for sunshades and skylights that are

Figure 9.25 BIPVs in the Solaire building in New York City's Battery Park. The BIPVs are the specked surfaces between the windows. (Photograph courtesy of the Albanese Development Corporation)

visually transparent or provide partial shading. It is an easy upgrade to substitute preengineered BIPV sunshades for conventional sunshades. The look and color of these building-integrated PV products vary with the application and the type of solar collector technology. The most efficient solar collectors are deep blue to black in color, although BIPV products are also available in dark gray and medium blue; some manufacturers also may produce custom-colored BIPV products for large orders.

Depending on the type of collection medium, BIPV can generate approximately 5 to 10 watts of power per square foot (50 to 100 watts per square meter) of collector area in full sunlight. That means that a collector area of 100 to 200 square feet (10 to 20 square meters) typically is needed per kilowatt of capacity. The annual power output varies with the latitude and climate, as well as with the orientation of the building surface that comprises the PV material. The annual energy output ranges from 1400 to 2000 kWh per kilowatt of installed system capacity. Figure 9.25 shows the BIPV system installed on the façade of Solaire, a 27-story luxury residential tower in New York City's Battery Park.

WIND ENERGY

Wind energy is the fastest-growing form of energy production, with an estimated year-on-year growth of 25 percent. According to the DOE's National Renewable Energy Laboratory (NREL), the cost of wind energy has declined from $0.40/kWh in the 1980s to less than $0.05/kWh today. In 2002, the United States doubled the 1700 megawatts (MW) of additional wind-generating capacity brought online in 2001. By the beginning of 2002, installed capacity in this country was over 4200 MW, reaching over 10,000 MW in 2006. As of the end of 2011, the United States had 46,916 MW of cumulative installed wind capacity, 20 percent of the world's installed wind power. In the past few years, wind energy constituted 30 percent of all new installed capacity, and there are over 8300 MW of wind energy projects under construction. The American Wind Energy Association (AWEA) has estimated that, with the support of the government and utilities, wind energy could provide at least 6 percent of the nation's electricity supply by 2020. The AWEA estimates that 20 percent of US electricity demand could ultimately be met by wind energy. Texas has the highest wind energy capacity in the United States, about 10,400 MW in late 2011, adding new capacity at the rate of about 400 MW per year. The reasons for this high installation rate is the Texas Renewable Portfolio Standard, which mandates installation of wind energy by utilities, and federal tax credits, which provide a $0.015/kWh write-off for the first 10 years of a project's operation.

Small wind turbines (those with less than 100 kW output) suitable for building-scale applications are available, and there are innovative programs that can make their incorporation into a building project financially feasible (see Figure 9.26).

BIOMASS ENERGY

The term *biomass* refers to any plant-derived organic matter available on a renewable basis, including dedicated energy crops and trees, agricultural food and feed crops, agricultural crop wastes and residues, wood wastes and residues, aquatic plants, animal wastes, municipal wastes, and other waste materials. Handling technologies, collection logistics, and infrastructure are important aspects of the biomass resource supply chain.

According to the American Bioenergy Association, the United States has the available land and agricultural infrastructure to produce adequate biomass in a sustainable way to replace half of the country's gasoline usage or all of its nuclear power without a major impact on food prices.[34] Shifting part of the $50 billion now spent for oil imports and other petroleum products to rural areas would have a profoundly positive effect on the economy in terms of jobs created (for production, harvesting, and use) and industrial growth (facilities for conversion into fuels and power). David Morris of the Institute for Local Self-Reliance refers to this as moving partway from a "hydrocarbon economy to a carbohydrate economy."[35]

Fuel Cells

Fuel cells are devices that generate electricity in a process that can be described as the reverse of electrolysis. In electrolysis, electricity is input to electrodes to decompose water into hydrogen and oxygen. In a fuel cell, hydrogen and oxygen molecules are brought back together to create water and generate electricity. The principle behind fuel cells was discovered in 1839, but it took almost 130 years before the technology began to emerge, first in the US space program and more recently in a host of new technologies and applications. Fuel cells provided power for onboard electronics for the Gemini and Apollo spacecraft and electricity and water for the space shuttles.

Fuel cells generally consist of a fuel electrode (anode) and an oxidant electrode (cathode) separated by an ion-conducting membrane. Fuel cells must take in hydrogen as a fuel, but any hydrogen-rich fuel can be processed to extract its hydrogen for fuel cell use. A device called a *reformer* is used to process nonhydrogen fuels to extract the hydrogen. This device reformulates non-hydrogen fuels such as gasoline, methane, diesel fuel, and ethanol to turn them into hydrogen. Due to their complexity, reformers are still very expensive. Some of the higher-temperature fuel cells can directly process some nonhydrogen fuels—methane, gasoline, and ethanol—without using the reformer.

There are several different types of fuel cells, including phosphoric acid, alkaline, molten carbonate, solid oxide, and proton exchange membrane (PEM) fuel cells. PEM fuel cells are of great interest because they operate at relatively low temperatures (below 200°F/93°C), have high power density, and can vary their output quickly to meet shifts in power demand. Current-generation fuel cells last anywhere from one to six years before they wear out or need an overhaul. Fuel cells are currently expensive to manufacture and depend on ongoing technological innovations to ensure their eventual economic viability. And, as noted, unless hydrogen is available as a fuel cell, a reformer must be utilized to process hydrogen-rich fuels to extract the hydrogen gas, an expensive additional component that adds considerable complexity to the fuel cell system.

For buildings and utilities, fuel cell power plants are beginning to make economic sense. The potential for home and commercial building power systems to use fuel cells, particularly in the United States in an era of utility deregulation, is quite high. An additional positive feature is that heat produced by some types of fuel cells can be used for thermal cogeneration in building power systems (see Figure 9.27).

Fuel cells specifically designed for building use are beginning to emerge. Plug Power is developing the GenSys fuel cell, which will produce electricity by using the hydrogen contained in natural gas or liquid petroleum gas (LPG).[36] For most building applications, this system has three major components:

- A reformer that extracts hydrogen from the natural gas or LPG
- A fuel cell that changes the hydrogen to electricity
- A power conditioner that converts the fuel cell's electricity to the type and quality of power required for use in the building

Figure 9.26 Three wind turbines, each 29 meters in diameter, provide 10 to 15 percent of the power needed to operate the Bahrain World Trade Center, which opened in 2007 in Manama, Bahrain. (Courtesy of Atkins)

Figure 9.27 An array of five PC25 PAFCs manufactured by UTC Power, Inc., powering a postal facility in Anchorage, Alaska. The PC25 provides about 200 kW of power, using natural gas as the fuel source. Waste heat generated by the PC25 can be used for heating applications or to create cooling using and absorption cycle chiller. (Photograph courtesy of UTC Power, Inc.)

TABLE 9.13

Building Systems Typically Found in a Smart Building

Fiber-optics capability

Built-in wiring for Internet access

Wiring for high-speed networks

Local area network (LAN) and wide area network (WAN) capability

Satellite accessibility

Integrated digital services network (IDSN)

A redundant power source

Conduits for power/data/voice

High-tech, energy-efficient HVAC system

Automatic on/off sensor in the lighting system

Smart elevators that group passengers by floor designation

Automatic sensors installed in faucets/toilets

Computerized/interactive building directory

Smart Buildings and Energy Management Systems

In its simplest form, a building's energy management system (EMS) is a computer with software that controls energy-consuming equipment to ensure that the building operates efficiently and effectively.[37] Many EMS are also integrated with fire protection and security systems. A newer innovation, *smart buildings*, uses the concept of information exchange to provide a work environment that is productive and flexible. In each building zone, a building automation system (BAS) and high-bandwidth cabling connect all building telecommunications; heating, ventilation, air-conditioning, and refrigeration (HVAC&R) components; fire, life, and safety (FL&S) systems; lighting; emergency or redundant power; and security systems. The smart building concept is important for consideration in green buildings because of the enormous demand for flexible layout and responsiveness, both afforded by smart buildings. A survey of building owners conducted by the Building Owners and Managers Association (BOMA) found that there were 13 systems desired by tenants of smart buildings (see Table 9.13).[38] In addition to the items on this list, also in demand today is the capability for wireless technologies to enable telecommunications and Internet connections. EMS can produce substantial energy savings, on the order of 10 percent of building energy consumption.

Modern smart buildings also use digital controls, referred to as *direct digital controls* (DDCs), to control the growing variety of devices and control systems in the building's HVAC&R systems. In addition to controlling systems based on temperature and humidity, DDC permits the integration of information about air quality and carbon dioxide levels. Digital systems can process and store information and manage complex interrelationships between components and systems. Control of lighting systems can be accomplished with DDC systems that allow occupant control of lighting, a prime feature of smart buildings.

Ozone-Depleting Chemicals in HVAC&R Systems

Of the many building systems, mechanical systems used for generating cooling and for fire protection employ the largest quantity of ozone-depleting chemicals. Removing these chemicals from building inventories and replacing systems with new products that use non-ozone-depleting chemicals are priorities. This section describes the replacement of refrigerants in air-conditioning systems with newer technologies that do not impact the ozone layer. HVAC&R systems or equipment constitute the majority of mechanical engineering systems.

Before 1986, the chemicals known as chlorofluorocarbons (CFCs) were commonly used as refrigerants in chillers, mechanical devices that are used to generate cooling; CFC-11 and CFC-12 were the two most common. Then, in 1986, their release into the atmosphere was found to be a major cause of destruction of the ozone layer, and international treaties soon called for their phaseout. The impact of CFCs on the ozone layer is indicated in terms of a quantity called the *ozone depletion potential* (ODP). The ODP is defined as 1.0 for CFC-11, meaning that a substance with an ODP of 10.0 depletes ozone at 10 times the rate of CFC-11. Other typical CFCs have a value of 1.0. For hydrochlorofluorocarbons (HCFCs), the ODP ranges from about 0.02 to 0.11, or about 10 to 50 times less impact than that caused by CFCs.

Several families of chemicals have been used to replace CFCs, among them HCFCs and hydrofluorocarbons (HFCs). Although HCFCs are a great improvement over CFCs, they still have relatively high ODPs. HFCs, on the other hand, have a zero ODP and, as a result, have no impact on the ozone layer. HFC-containing equipment is available from all major manufacturers. HFC-134a has become the dominant refrigerant, replacing HCFC-123 and HCFC-22 in most chillers designed for building use. HCFC-22 is currently used in a large proportion of positive-displacement compressor-based chillers and in some larger-tonnage centrifugal chillers. These uses predate the Montreal Protocol but will be phased out as part of the overall HCFC phaseout. In the United States, HCFC-22 was phased out for use in new equipment as of January 1, 2010.

According to the Carrier Corporation, HFC-134a has proven to be an optimal refrigerant in chiller applications because it has no chlorine molecules and does not contribute to ozone depletion. HFC-134a is a highly efficient thermodynamic refrigerant in application. Current centrifugal chillers using HFC-134a are 21 percent more efficient than chillers sold just six years ago and 35 percent more efficient than the chillers installed during the 1970s and 1980s. Because HFC-134a is a positive-pressure refrigerant, pressure vessels using it must conform to the American Society of Mechanical Engineers (ASME) pressure code, and every step in their construction must be inspected by third-party insurance companies. As a result of the stringent testing and applied technology, chiller leak rates can be lowered to less than 0.1 percent annually. Existing chillers have a leak rate of 2 to 15 percent. HFC-134a also has a smaller molecular mass than the past CFCs and HCFCs. This is an important feature, as it results in an overall product size that is 35 to 40 percent smaller, a size reduction that helps offset the cost of construction and facilitates the use of smaller interconnecting pipes. This advantage has led to the addition of isolation valves to the chiller piping connection so that the HFC-134a can be stored in the chiller during service. This feature gives the end user the option of never having to remove the refrigerant from the vessels once charged, a real "no emissions" feature. An additional advantage of HFC-134a chillers is their smaller size, requiring much less plant space than the CFC-11 chillers they replaced.

Reducing the Carbon Footprint of the Built Environment

Significantly reducing built environment energy consumption is a very important goal of sustainable construction. More recently, the issue of how to reduce the climate change impacts of the built environment has become equally important. Climate change is being caused by human activities that are increasing the concentrations of heat-trapping, carbon-containing gases, particularly CO_2, in the atmosphere. The quantity of these gases being released is a function of both the quantity of energy being consumed and the source of the energy. The term *carbon footprint* is commonly used to describe the quantity of CO_2 and other carbon gases being released by an activity, for example, electrical energy generation or the manufacture of drywall. The carbon footprint of the built environment has four major components: (1) the output of carbon gases due to building operation (operational energy); (2) the carbon invested in the materials and products of construction (embodied energy); (3) the carbon emissions from transportation energy; and (4) the output of carbon gases associated with processing and moving water, wastewater, and stormwater. As noted earlier in this chapter, the total energy associated with the built environment is probably on the order of 65 percent of total US energy consumption, or about 100 quads. Although there are some differences in energy sources for building, transportation, and industry, the carbon footprint of the built environment is likely about this percentage of the total human carbon footprint. Of the 100 quads of energy being consumed annually in the United States, 40 quads are consumed by building operations, and another 25 quads are consumed by transportation energy, the embodied energy of the materials and products of construction, and water pumping and processing. Consuming 100 quads of energy annually produces 6600 million metric tons of carbon dioxide equivalent (CO_2e), with the built environment contributing about 4300 million metric tons of CO_2e. The equivalent notation is used because each greenhouse gas has different impacts. For example, each mass of methane has 21 times the impact of the same mass of CO_2. Thus, each gram of methane has 21 g CO_2e/kWh of impact (see Table 9.14). This is sometimes referred to as the *greenhouse multiplier* for a given gas. To reverse course with respect to climate change requires that the built environment be the main focus of activities addressing this, the most serious issue of the 21st century.

There are several strategies that can be used to reduce the built environment carbon footprint, among them:

1. Dramatically reducing energy consumption
2. Shifting to renewable energy sources
3. Emphasizing compact forms of development
4. Shifting to mass transportation
5. Designing buildings for durability and adaptability
6. Restoring natural systems
7. Designing low-energy built environment hydrologic systems
8. Designing buildings for deconstruction and material reuse
9. Selecting materials for their recycling properties
10. Including the carbon footprint of buildings in building assessment systems

TABLE 9.14

Greenhouse Multiplier for Various Atmospheric Gases

Atmospheric Gas	Greenhouse Multiplier
CO_2 (carbon dioxide)	1
CH_4 (methane)	21
NO_2 (nitrous oxide)	310
CFC-11 (CCl_3F)	1320
CFC-12 (CF_2Cl_2)	6650
HCFC-22 ($CHClF_2$)	1350
Surface ozone	100

Note: The multiplier indicates how many grams of CO_2 equivalent impact each gram of gas causes. Although some gases have large multipliers, the vast mass of CO_2 being emitted dwarfs the mass of other gases, causing over 99 percent of the climate change impact.

Reducing atmospheric carbon will require a concerted effort on the part of all stakeholders to the built environment to shift building design onto a course that focuses on long-term strategies that rebalance the emissions of carbon into the atmosphere by greatly reducing the carbon associated with building construction and operation and with the distribution of buildings in communities. This latter point addresses the problem of how buildings and their location drive energy consumption and carbon emissions and transportation systems. Additionally, enormous efforts must be made to restore the quantity of biomass on the planet to help in the reabsorption of carbon. Although many technical fixes have been proposed to reduce and absorb carbon in the atmosphere, thus far no technical fix has been proven to work at large scale to remove the enormous quantities of carbon that would be required to stabilize the atmosphere.

OPERATIONAL ENERGY

Operational energy is the energy required to power the built environment. All industrial systems and the electrical power systems that support them have carbon footprints. Carbon dioxide and other greenhouse gases are emitted over the life cycle of the power plant, and their climate change impact is expressed as grams of CO_2 equivalent per kilowatt-hour of generation. For electrical energy generation, there are both direct and indirect emissions of carbon dioxide and other greenhouse gases. Direct emissions are those arising from the operation of the power plant, and indirect emissions arise from other phases of the life cycle such as fuel extraction, transportation of fuel, processing of the fuel, construction of the power plant, maintenance, and power plant decommissioning. A life-cycle assessment (LCA) of a power plant to determine its carbon footprint is carried out in exactly the same way as an LCA for products of any kind, and it is based on the International Organization for Standardization (ISO) 14000 series of standards. For any type of electrical generation plant, whether it be a fossil fuel plant, hydropower installation, nuclear power plant, or solar photovoltaic array, the exact same analysis is used to determine the carbon footprint (see Figure 9.28).[39]

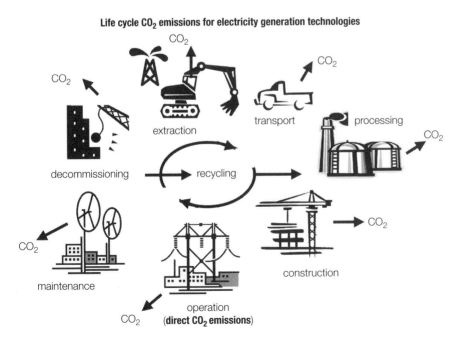

Life cycle CO_2 emissions for electricity generation technologies

CO_2

CO_2

CO_2

CO_2

extraction

transport

processing

CO_2

decommissioning

recycling

CO_2

CO_2

maintenance

construction

CO_2

operation
(direct CO_2 emissions)

Figure 9.28   The carbon footprint of an electrical power generating station includes the emissions from all phases of the life cycle in extracting and transporting the fuel required for the plant, the materials and products from which the plant is constructed, plus all emissions associated with the extraction and transportation of these resources. The carbon footprint also includes the maintenance and decommissioning of the power plant at the end of its useful life. (*Source:* Parliamentary Office of Science and Technology)

Fossil fuel–burning power plants have by far the largest carbon footprint of all forms of power generation. This is due to the long-chain, carbon-containing molecules from which fossil fuels are derived. A typical coal-fired plant, for instance, will have emissions on the order of 1000 g CO_2e/kWh. Oil-fired power plants also have a significant contribution of carbon for each kilowatt-hour of electricity generated, on the order of 600 g CO_2e/kWh. Gas-fired power generation, which is becoming an increasing fraction of power generation, contributes on the order of 400 g CO_2e/kWh. Renewable forms of energy have significantly lower carbon footprints, as shown in Figures 9.29 and 9.30.

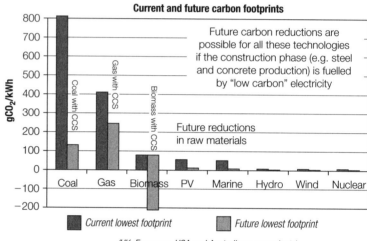

Figure 9.29 The carbon footprint of various electrical power generation technologies. Carbon capture and storage (CCS) technologies that remove carbon dioxide from combustion gases are being developed but have not been proven and tested at large scale. Coal-fired power plants not only dominate the electrical power generation industry but also have by far the highest production of carbon dioxide at about 1000 g CO_2e/kWh. Renewable energy systems range from about 50 g CO_2e/kWh for photovoltaics to about 5 g CO_2e/kWh for hydropower and wind power. Nuclear power plants also have a relatively small carbon footprint at about 5 g CO_2e/kWh. (*Source:* Parliamentary Office of Science and Technology)

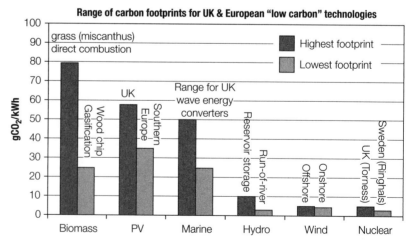

Figure 9.30 Contributions of renewable and nuclear power systems to climate change. Biomass, at about 80 g CO_2e/kWh, has just 1/10 the climate change contribution of coal-fired power plants. At the lower end of the spectrum, wind energy and nuclear have just 5 g CO_2e/kWh. (*Source:* Parliamentary Office of Science and Technology)

Biomass has a carbon footprint ranging from 25 to about 80 g CO_2e/kWh, while hydropower has a fairly low carbon footprint of 10 g CO_2e/kWh or lower. Nuclear power has the lowest carbon footprint of all forms of energy generation at 5 g CO_2e/kWh and lower. Photovoltaic power production has a lower but still surprisingly high carbon footprint compared to wind energy and hydropower, ranging from about 30 to 60 g CO_2e/kWh.[40]

EMBODIED CARBON OF MATERIALS AND PRODUCTS

The amount of energy and carbon associated with the materials and products that comprise the built environment is quite large and can be equivalent to from 5 to 20 years of operational energy, depending on the type of building and its energy profile. Oddly enough, for high-performance green buildings with low energy profiles, the years of operational energy that are equivalent to the embodied energy of the building may be much longer due to the lower annual energy consumption. In Germany, the Deutsche Gesellschaft für Nachhaltiges Bauen (DGNB) and Bewertungssystem Nachhaltiges Bauen für Bundesgebäude (BNB) building assessment systems do have provisions for rating the building based on its total embodied carbon footprint per square meter. DGNB/BNB has an LCA tool that requires that the total mass of all building materials be input into the tool to determine the embodied carbon per square meter as well as other impacts. The team designing the building can then try various trade-offs, for example, more insulation and shading devices to reduce the size of the mechanical plant and reduce the annual operational energy requirements to determine if the total carbon footprint can be reduced. The database tool contains historic data on German buildings, and any new designs can be compared to this database to determine how the proposed design rates with respect to its carbon footprint. Although this level of detailed information is not yet available in the United States, both LEED and Green Globes have provisions for comparing the impacts of alternative building assemblies such as wall sections. In any event, the life-cycle emissions of carbon are being determined for a wide range of materials, including carpet (see Figure 9.31). These data are

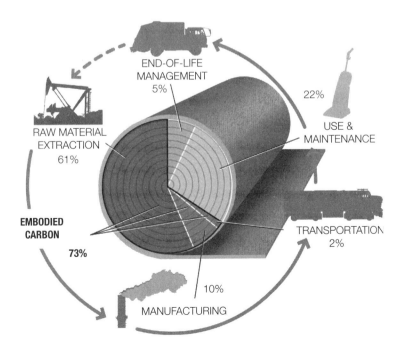

Figure 9.31 The embodied carbon of carpet is determined by examining the entire life cycle of the materials from extraction through removal. (Peter Harris/BuildingGreen, Inc.)

TABLE 9.15

Embodied Energy and Embodied Carbon of Common Materials Used in Construction as Indicated in the Inventory of Carbon and Energy Developed by the University of Bath

Material	Embodied Energy (MJ/kg)	Embodied Carbon (kg CO_2e/kg)
Aluminum—virgin	218	12.79
Aluminum—recycled	29	1.81
Asphalt 6% binder	3.93	0.076
Brick	3.00	0.24
Concrete 0% fly ash	0.55	0.076
Concrete 15% fly ash	0.52	0.069
Concrete 30% fly ash	0.47	0.061
Copper tube—virgin	57.0	3.81
Copper tube—recycled	16.5	0.84
Glass	15	0.91
Paint	70	2.91
Plastics (general)	80.5	3.3
Steel—virgin	35.40	2.89
Steel—recycled	9.40	0.47
Timber	10	0.72

being compiled into detailed databases such as the Inventory of Carbon and Energy (ICE) developed by researchers at the University of Bath in the United Kingdom. The information displayed in Table 9.15 was extracted from ICE.[41] Note that both the embodied energy and the embodied carbon of recycled metals are about one-third to one-seventh the levels for virgin metals. This is true for all materials, whether they be concrete, plastics, glass, paper, or wood products.

The embodied carbon footprint of buildings can be greatly reduced by creating facilities that are durable, low maintenance, and adaptable. Doubling the lifetime of a building from 50 to 100 years in effect cuts the embodied carbon footprint in half, a major effect. Clearly, good planning with a long time horizon will ensure that frequent redesign of urban areas that requires removal of large numbers of buildings is unnecessary.

TRANSPORTATION CARBON FOOTPRINT

A recent National Academy of Sciences report calls for a shift to compact development to reduce carbon emissions.[42] The report states that if 75 percent of new development were built at twice the current density norms, vehicle-miles traveled would drop 25 percent and greenhouse gas emissions by 8 percent by 2050. A report by the Brookings Institution backed up these findings with an analysis of the carbon footprint of metropolitan America compared to the nation as a whole. The study reported the following conclusions:[43]

> **Large metropolitan areas offer greater energy and carbon efficiency than nonmetropolitan areas.** Although they house 67 percent of the nation's population and have 75% of its economic activity, the country's 100 largest metropolitan areas emitted just

56% of US carbon emissions from highway transportation and residential buildings in 2005.

Carbon emissions increased more slowly in Metropolitan America than the rest of the country between 2000 and 2005. The average per capita carbon footprint of the 100 largest metro areas in the US grew by only 1.1% during this five-year time frame. The US footprint as a whole grew twice as rapidly, by 2.2%, during the same timeframe.

Per capita carbon emissions very substantial by metropolitan area. For example in 2005, per capita carbon emissions were highest in Lexington, Kentucky and lowest in Honolulu. Lexington emitted 3.46 metric tons per capita compared with 1.36 metric tons in Honolulu. This is at least in part due to the metropolitan area's economic output, or gross metropolitan product (GMP), an indicator of carbon intensity. For example, Youngstown, Ohio, a heavy industrial area, had a carbon footprint of 97.6 million metric tons of carbon per dollar GMP while San Jose, California, more of a high tech area, had just 22.5 million metric tons per dollar GMP.

Development patterns and rail transit play important roles in determining carbon emissions. Many of the older, denser cities in the northeast, midwest, and California (for example, Boston, New York, Chicago, and San Francisco) are low emitters. These cities also have some of the highest annual rail ridership in the nation, ranging from 296 to 757 miles per capita and carbon footprints ranging from 1.5 to 2.0 tons of carbon per capita, much lower than the average of 2.2 tons for the 100 metropolitan areas.

Other factors, such as weather, the fuels used to generate electricity, and electricity prices are also important. Areas in the northeast, which have a much greater reliance on carbon intensive home heating fuels such as fuel oil, have higher carbon footprints. Similarly, warm areas in the south often have large residential footprints because of their heavy reliance on carbon intensive air-conditioning. The fuel mix used to generate electricity matters for the size of residential carbon footprints. For example, the Washington, DC Metro area's residential electricity footprint was 10 times larger than Seattle's in 2005. Seattle draws its energy primarily from essentially carbon-free hydropower while Washington, DC relies largely on coal-fired energy power plants.

In general, the carbon footprint of transportation varies greatly with development density and the availability of modes of transportation other than the automobile. Rail travel is particularly important due to its low carbon footprint for both commuter and long-distance travel (see Table 9.16).

TABLE 9.16

Carbon Emissions for Various Modes of Transportation

Mode of Travel	CO_2 Generation
Vehicle	8.91 kg (19.6 lb) per gallon of gasoline 0.44 kg (0.88 lb) per passenger-mile*
Air travel	0.40 to 0.60 kg (0.88 to 1.32 lb) per passenger-mile
Rail travel (commuter and subway)	0.16 kg (0.35 lb) per passenger-mile
Rail travel (long distance)	0.19 kg (0.42 lb) per passenger-mile
Bus travel (inner city)	0.30 kg (0.66 lb) per passenger-mile
Bus travel (long distance)	0.18 kg (0.18 lb) per passenger-mile

*Assumes an automobile performance of 20 miles/gallon.

Case Study: River Campus Building One, Oregon Health and Science University, Portland

Medical facilities pose significant challenges and provide enormous opportunities for high-performance building project teams. While they are generally far more complex than other building types and have special requirements for controlling the movement of viruses and other pathogens that can result in the transmission of disease, they also have the potential for contributing to health and well-being by virtue of their design. In expanding its main campus in Portland, the Oregon Health and Science University required a new 400,000-square-foot, 16-story medical office and wellness building, which was named River Campus Building One. In addition to a two-story wellness center, the building houses several different types of university operations, including biomedical research, clinical space, an outpatient surgery, and educational space. The medical offices are built atop a three-level, below-grade parking structure (see Figure 9.32).

The project team included the Portland office of Interface Engineering, Inc., a multidisciplinary engineering firm that provided HVAC, plumbing, electrical, power and backup power distribution, lighting, security, energy, telecommunications, data, and fire alarm systems design, as well as all tenant improvements and basic commissioning. Interface Engineering's project team was instrumental in River Campus Building One's receiving a LEED platinum certification from the US Green Building Council (USGBC). How this project team helped achieve this rating through a holistic approach to the design of these technical systems is an excellent case study on how to create a low-energy building.

THE PROJECT BUDGET

The total project was initially budgeted at $145.4 million, with $30 million allocated for the building's mechanical, electrical, and plumbing (MEP) systems. Interface's MEP design approach resulted in savings of nearly $4 million of the initial $30 million

Figure 9.32 The Oregon Health and Science University's River Campus Building One, shown in the late stages of construction, is a LEED-NC 2.1 platinum certified building. (Courtesy of Interface Engineering)

budget. What was truly remarkable in this project is that energy consumption was reduced about 60 percent compared to the baseline model, while at the same time the capital cost of the MEP systems was reduced 10 percent. Conventional wisdom is that high-performance MEP systems will cost more than their code-compliant alternatives. For River Campus Building One, Interface's engineering team did indeed "tunnel through the cost barrier," as suggested was possible by Paul Hawken, Amory Lovins, and L. Hunter Lovins in their 1999 book *Natural Capitalism.*

THE STRATEGY: INTEGRATED DESIGN

Achieving a win-win combination of lower energy consumption and lower capital costs for energy systems is a difficult but clearly achievable strategy. Interface was able to reach this holy grail of sustainable design using an integrated design approach. As articulated for the River Campus Building One project, integrated design is different from conventional design in two key respects: (1) goal setting starts early in the sustainable design process, during the programming and conceptual design phases, and (2) the entire design team is involved in the process much earlier than is usually the case so that engineers can provide inputs to architectural decisions that affect energy and water consumption, as well as indoor air quality. For River Campus Building One, this meant that several disciplines were able to collaborate early in the design regarding the green roof, PV, and rainwater harvesting system. This early collaboration started with an eco-charrette in which participants and stakeholders from diverse backgrounds helped craft ambitious goals for the project. One of the goals that emerged was a 60 percent reduction in energy consumption relative to that of a comparable building (see Table 9.17).

Making key decisions early in the design process allowed the design team to focus on collaboration to ensure their implementation. The abundant rainfall in Portland and the facility's large roof area meant that rainwater could be used for nonpotable water uses, including cooling tower makeup water. Moderate temperatures allowed the use of outside air to flush and precondition the building at night. Due to Oregon's generous tax credits for renewable and alternative energy systems, the team also opted for PV panels on the south side of the building and a microturbine system in the central utility plant. Integrated design also allowed the design team to eliminate solutions that were not feasible early on—for example, roof-mounted, vertical-axis wind turbines.

TABLE 9.17

The Interface Team's "Back of the Envelope" Goals for River Campus Building One

Load	Oregon Energy Code kBTU/SF/yr*	Percent	Target Savings, kBTU/SF/yr*
Heating	35	27	22
Cooling	10	7.7	5
Fans	6	4.6	2
Hot water	30	23	28
Lighting	30	23	15
Equipment	15	11.5	5
Exterior lighting	4	3	1
Totals	**130**	**100%**	**78**

*Thousands of BTU per square foot per year.

THE DETAILS—HOW INTERFACE APPROACHED THE MECHANICAL SYSTEMS DESIGN

The Interface team had two core principles guiding the mechanical design: (1) optimum health and (2) reduced energy use. To achieve this, the engineers followed the basic sustainable engineering dictum laid out by Amory Lovins: optimize the system, not the subsystems. Doing otherwise, that is, optimizing the subsystems without considering the system as a whole, will inevitably produce suboptimal results. Applied to buildings, the system includes all components of the building that affect energy consumption: the mass and orientation of the building, its envelope (thermal resistance, fenestration, roof, infiltration, shading), its plug loads (computers, printers, copiers, and other plug-in devices), its air delivery systems, the lighting system (lights and lighting controls), the cooling and heating plant, fans, motors, pumps, piping and duct sizing and layout. In many cases, it calls for challenging conventional wisdom. For example, mechanical engineers use tables that assume an embedded level of friction loss for fluids such as air and water circulated in pipes or ductwork. Lowering the acceptable friction loss may result in the use of larger-diameter pipes of larger cross-sectional ducts or the selection of smoother pipes with less friction per unit length.

Early on, the team examined the building's energy profile and worked with the architects to optimize the building's envelope. The team used the BetterBricks Integrated Design Lab in Portland to study year-round shading, including the shading effects of adjacent buildings. As a result, River Campus Building One was designed so that windows were shaded in the summer, allowing sunlight to warm the interior during the winter. Sunshades and building PV panels were used to assist in the shading above the fourth floor.

Plug loads from computers, printers, and other devices were examined to ensure that the selection of these components contributed to the 60 percent energy reduction goal. Similarly, all fans, water heaters, pumps, and motors were selected to support the energy-conserving goals of the team.

Computational fluid dynamics (CFD) models were used extensively to explore approaches to natural ventilation and the building's air distribution approach. A whole-building CFD model allowed the team to determine wind pressures on each face and optimize a natural ventilation strategy. CFD models were also used in making the decision to select a positive-displacement ventilation system for patient examination rooms (see Figure 9.33). Similarly, the supply air temperature for the examination rooms was selected based on CFD modeling, allowing the temperature to be raised from a typical 55 to 60°F (13 to 16°C). In addition to lower energy costs, this permits the more extensive use of the typically temperate outside air in the Pacific Northwest to cool the building (see Figure 9.34).

In short, the Interface team, by collaborating with the project architects, was able to design a downsized mechanical system. As part of this process, the engineers also bought into the notion of right-sizing the mechanical system rather than oversizing the system to accommodate hypothetical unknowns. The team accomplished this by (1) eliminating excessive safety factors, (2) calculating heating and cooling demands using basic physics rather than simply applying conventional HVAC rules of thumb, (3) assuming nothing and proving everything, (4) building in expansion capabilities rather than trying to accomplish everything at the beginning, and (5) challenging restrictive codes that add cost without benefit by making successful appeals.

Right-sizing is just one of eight design points articulated by Andy Frichtl, PE, the lead engineer for River Campus Building One. The other seven design points he advocates are (1) transfer savings in HVAC systems to other important aspects of

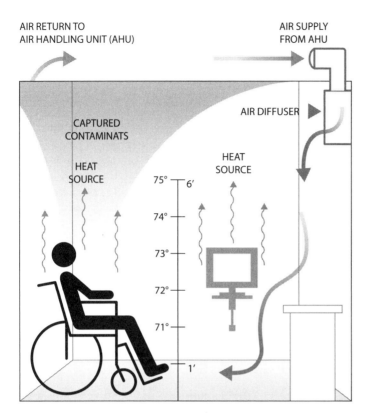

Figure 9.33   CFD modeling of patient examination rooms indicates the waterfall effect of cool incoming air falling down the walls, pooling on the floor, and then rising as it is heated by people, computers, and lights. (Courtesy of Interface Engineering)

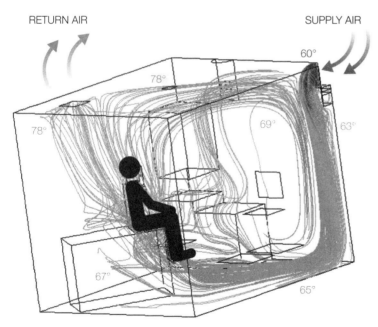

Figure 9.34   Modeling of patient examination room temperatures aided the Interface team in deciding to raise the supply temperature to 63°F (17°C), creating a more comfortable examination room with less air movement. (Courtesy of Interface Engineering)

Figure 9.35 Chilled-beam systems are aluminum-finned copper assemblies through which water circulates, providing radiative cooling or heating and inducing airflow by convective effects. (Courtesy of Interface Engineering)

the project; (2) use free resources such as the sun, wind, ground temperature, and groundwater to reduce building energy consumption; (3) reduce the demand for heating and cooling by superior envelope design, reduction in plug loads, and providing high-efficiency appliances and other devices; (4) shift loads from peak to off-peak periods by using energy storage strategies; (5) challenge standard practice by emphasizing comfort and health, which may also involve challenging building codes; (6) utilize radiant space conditioning, which uses radiative rather than convective heat transfer, with significantly lower energy consumption; and (7) relax comfort standards by allowing temperature and humidity set points to float within a specified comfort zone.

The result of applying these strategies was a variety of energy-efficient design measures to achieve the high-performance goals of the project:

- Radiant cooling of the atrium and lobby ground floor using reclaimed rainwater and groundwater in the concrete slab
- Radiant cooling with an overhead chilled beam (see Figure 9.35)
- High-efficiency boilers and chillers
- Double-fan VAV air handlers and VFDs on most pumps and motors
- Demand-controlled ventilation (DCV) using carbon dioxide sensors and occupancy sensors to prevent overventilating and overlighting unoccupied spaces
- Heat recovery systems, including laboratory and general exhaust
- Displacement ventilation for core exam and office areas
- Load shifting using a system of hot and cold water storage to reduce peak demand
- Energy-efficient lighting fixtures and controls, incorporating daylighting where feasible
- Night-flush precooling with outside air
- Economizers for free cooling using outside air when outside temperatures permit

Chilled beams represent a potential breakthrough strategy for conditioning buildings. The HVAC systems employing this technology can be one-third the size of systems using forced air as the heat transfer medium. While relatively new to the United States, radiant cooling systems are fairly standard practice in Germany. They can function passively using only radiant effects for cooling or, with the assistance of a fan passing air through the beam, provide convective cooling. The compact size of the chilled beams allows reduced floor-to-floor heights because larger ductwork is eliminated and the space required for mechanical rooms and shafts can be reduced. Although the beams cost $100 to $250 a lineal foot ($328 to $820 per meter), the net result is reduced HVAC system costs and lower costs for architectural and structural elements.

INTEGRATING LIGHTING AND DAYLIGHTING SYSTEMS

A properly designed lighting system for a high-performance building should integrate daylighting, lighting fixtures, and lighting controls to provide a low-energy lighting solution. For River Campus Building One, the Interface team's goal was to reduce the typical lighting system's 23 percent share of the total energy use by 50 percent. They managed to achieve a 45 percent reduction in the actual building, a savings of 16 percent in total energy use. In the exam rooms, the standard two 1- to 4-foot (0.3- to 1.2-meter) lensed fluorescent luminaires were replaced by a single lensed skydome, 48 inches (122 centimeters) in diameter, that mimics natural light. Combined wall switch/occupancy sensors turn on only half of the exam room lights, permitting the other half to switch on automatically when needed. Reduced lighting levels were specified for lobbies and other pass-through spaces. When there is adequate natural light, hallway daylight sensors switch off normal and emergency lighting. Outdoor lighting was significantly reduced using cutoff fixtures that also eliminate unnecessary light pollution. In the high-bay athletic club, lighting levels automatically switch down as more daylight becomes available. Occupancy sensors in stairwells switch lighting on and off to follow an occupant up or down, allowing the lighting to stay on for the minimum time needed for passage. Perimeter offices also have occupancy sensors and daylighting sensors.

Figure 9.36 The PV panels used in River Campus Building One were assembled at Benson Industries, a major supplier of curtainwalls and exterior cladding systems for larger buildings. (Courtesy of Interface Engineering)

INNOVATIVE SOLAR ENERGY APPLICATIONS: BIPV AND SOLAR AIR HEATER

The project team for River Campus Building One specified sunshades in the design of the south façade and used the sunshade surface for PV panels (see Figure 9.36). In addition to using renewable energy for the building, PVs are subsidized by generous federal and many state incentives such as tax credits, accelerated depreciation, and, in the case of Oregon, bonuses from the Oregon Energy Trust. These BIPV panels have a peak of 60 kW and produce about 66,000 kWh annually.

On the 15th and 16th floors of the building, the façade serves as a giant solar heater, 190 feet (58 meters) long by 32 feet (9.8 meters) high. Sheets of low-iron glass are located 4 feet (1.2 meters) from the building skin. The air between the skin and glass is warmed by solar energy and then moved by air-handling units across a heat exchanger for use in preheating water for use in bathroom sinks and exam rooms. The integrated design approach used by the project team allowed the fusion of architecture and engineering to create this innovative water heating system. This system has the added benefit of serving as a Trombe wall, warming clinic and lab spaces in winter and reducing the amount of total heating energy. It requires almost no maintenance and has no replacement costs over time.

Acknowledgment

The River Campus Building One Case Study is used with the permission of Interface Engineering, Inc. It is also available from Interface Engineering in a comprehensive booklet, *Engineering a Sustainable World*, published in October 2005.

THOUGHT PIECE: ADVANCING THE STATE OF THE ART IN BUILDING ENERGY MODELING

One of the key elements in developing very low energy and carbon buildings is being able to accurately and dynamically simulate the energy performance of a facility and to be able to use this model to optimize its design. The emergence of building information modeling (BIM) as the best current construction documents tool for building design and construction is bringing with it a new era with new potential for creating plug-in energy models that use what amounts to the design drawings and data to create the model. In this thought piece, Ravi Srinivasan, an international expert on both BIM and energy modeling, discusses this new direction and the exciting outcomes that can be expected as a result of the fusion of these two ideas.

Building Energy Analysis: The Present and Future

Ravi Srinivasan, College of Design Construction and Planning, University of Florida, Gainesville

Building energy analysis (BEA) is widely realized for both new and existing buildings nowadays. Several BEA tools are available to develop building energy conservation measures (ECMs) for greater energy efficiency. BEA requires extensive data gathering of model inputs. It is essential to bypass arbitrary and/or incorrect inputs when using BEA tools. Quality inputs to BEA tools are central to energy estimating. This may be achieved through integrating quality control mechanisms in BEA procedures. The possibility of erroneous inputs increases when modeling large buildings as it involves tedious, oftentimes iterative and repetitive data inputs. Among other model inputs, plug load density values and building occupancy schedules are important. Plug load density relates to equipment energy use per unit area. Plug load densities can be calculated by using equipment nominal power data and diversity (or utilization) factors. Benchmarking plug load densities is not as easy as it may seem. The reason is that not all equipment peaks at the same time as some may be in idle mode. Only a few building energy standards, guidelines, and technical reports discuss such densities. As more *simulationists* play a decision-making role for the design team, they tend to lean on building energy standards and guidelines for plug load densities. However, the recommended values of standards and guidelines vary, posing a challenge for early design decision making. Such discrepancy may lead to unrealistic determination of energy use. Benchmarking of plug load densities will pave the way for instituting targets for trimming plug load densities in new and retrofit building projects. Recently, plug load densities for K−12 schools were benchmarked under two new categories—*classrooms with computers* and *classrooms without computers* (Srinivasan et al. 2011b). Eighteen K−12 schools, including nine elementary, two middle, and seven high schools, were assessed for actual plug load densities. Additionally, for the same case study buildings, four existing approaches—National Renewable Energy Laboratory (NREL), Commercial Energy Services Network (COMNET), ASHRAE 90.1-1989, and California Title 24—were evaluated for plug load densities. Results show under- and overestimation of plug load densities over actual densities calculated.

Similarly, the importance of building operating schedules cannot be understated. Any changes in operating schedules will significantly change the results. Among other building types, convention centers are complex to model with BEA tools owing to both their mix of spaces and their occupancy patterns. For one such BEA, the building operating schedules were developed based on the convention center's event calendar (Srinivasan, Lakshmanan, and Srivastav 2011c). The model adapted adjusted ASHRAE hourly operating schedules for event, nonevent, and move in, move out (MIMO) days, and used the event calendar and actual occupancy data. This drilldown approach of replicating the event calendar proved effective in model calibration. Calibration revealed that the energy model had a monthly variance of less than 8 percent for electricity. The calibrated model was then used to evaluate an array of energy efficiency measures (EEMs).

Although several BEA tools are available on the market, a single tool with up-to-date algorithms representing new, state-of-the-art technologies for building systems and controls is not available. Currently, it is the modeler who selects the "right" tool that closely *attempts* to represent the building systems and controls. Workarounds are developed to represent unavailable systems and controls wherever applicable. These workarounds are also limited to the capability of the selected tool. Moreover, rapid prototyping of new building systems and controls using current BEA tools is cumbersome as the entire simulation code needs to be executed rather than just portions of it. Wetter's (2011) argument of component-based modeling using *Modelica* (Mattson and Elmqvist 1997), an open-source language, offers a solution to this inherent modeling problem. The concept behind this type of modeling approach is the use of equation-based object-oriented modeling that allows the design and analysis of building energy and controls systems.[44] The Buildings library contains dynamic and steady-state component models that are applicable for analyzing control algorithms to assess energy performance. Using this library, rapid prototyping and improved representation of advanced building energy and control systems can be achieved. The Building Controls Virtual Test Bed (BCVTB), developed by the Building Technologies Department at Lawrence Berkeley National Laboratory (LBNL), may be used for enhanced collaboration. This test bed enables data exchange between simulation programs such as EnergyPlus and Radiance, allows integration with physical sensors polling real-time data, and accesses the Modelica-based Buildings library. Using this platform, manufacturers and advanced *simulationists* can develop new building energy and controls systems. The BCVTB platform can also be used to update the simulation algorithms using simple state machines. A few experiments were conducted to utilize the power of the Buildings library, BCTVB, and BEA tools by LBNL. Notable is the implementation of model predictive control (MPC) of the University of California's Merced chilled-water plant to reduce peak demand reduction (Haves et al. 2010). With the use of physical sensors, MPC predicts optimal solutions in real time. Results show improvement in chiller performance over the baseline policy. This investigation also revealed the significance of rapid development of new control algorithms and their implementation in real-world scenarios to improve actual performance.

Yet, in today's building design-construction-operation realm, there is still an impasse in sharing project files. One may recall two notable developments this past decade—the Industry Foundation Classes (IFCs), developed by the International Alliance for Interoperability (IAI), to describe building and construction industry data, and the green building XML (gbXML), originally developed by Green Building Studio, to facilitate the transfer of building properties stored in building information models (BIMs) to engineering analysis tools. In spite of such developments, the transfer of data from BIM to BEA tools has not materialized in its entirety. In other words, gbXML data exported from BIM tools are not fully compatible for executing whole-building energy simulation as one would develop and conduct in a BEA tool directly. At present, gbXML exported from BIM software such as Revit Architecture 2012, Revit MEP 2012, and ArchiCAD 10 can be directly imported to BEA tools such as Ecotect Analysis and Trane Trace 700. However, the gbXML exported from BIM software is not robust enough to populate all necessary model inputs to run a BEA tool without additional involvement of the designer. Well, then, whatever happened to the goals of interoperability? It is more than a decade since IFC and gbXML have been in development, and yet we notice this partial disconnect—a crucial component for any green building integrated project design and delivery. This enormously affects the seamless work process from design to analysis, documentation, construction, and measurement and verification. What is fundamentally required is not only a seamless and effective project data transfer between project team members but also a unified approach toward sustainability that deals not only with building operative energy but also with information related to the overall building life-cycle, including emissions, embodied energy, carbon, renewable energy balance (Srinivasan et al. 2011a), and so forth. Rather than work in silos, such a unified approach will allow us to effectively simulate sustainability scientifically.

Summary and Conclusions

As might be expected, energy receives the most emphasis in both the LEED and Green Globes building assessment systems. Clearly, improving building performance through the application of passive solar design techniques that use the materials, fenestration, and orientation of the building to maximize the amount of free energy that can be used is the key. Passive solar design addresses heating, cooling, daylighting, and ventilation of the building to minimize the employment of active mechanical and electrical systems, especially those powered by nonrenewable energy systems. The other measures called for in the energy categories of building assessment systems help round out the concept of a building that is both energy-efficient and environmentally responsible. The elimination of atmospheric ozone-depleting chemicals is a very worthwhile objective of any building rating scheme, and reducing energy consumption helps to lower the incidence of a wide range of power plant emissions.

One innovation in building assessment is the incorporation of strict requirements for building commissioning, ensuring that the building not only functions as designed but is also built to the highest-quality standards. Both LEED and Green Globes also provide impetus for the development of renewable energy sources on a large scale by providing a possible credit for using energy from renewable energy power plants.

Notes

1. Primary energy refers to raw energy in the form of oil, coal, and natural gas that is input to a process. It does not refer to electricity leaving a generating plant, which accounts for only a fraction of the input, primary energy.
2. The oil rollover point is described in more detail in Chapter 1.
3. Systems ecology was developed into a full-fledged ecological theory by H. T. Odum during his five decades at the University of Florida. The current program in systems ecology in the Department of Environmental Engineering at the University of Florida is described at www.ees.ufl.edu/research/area.asp?AID=3.
4. From Löhnert et al. (2006).
5. The EIA website is www.eia.gov.
6. The EPA Target Finder website is www.energystar.gov/index.cfm?c=new_bldg_design.bus_target_finder.
7. The best website for information about DOE-2.2 and the eQUEST interface is www.doe2.com.
8. Energy-10 was developed by the National Renewable Energy Laboratory and is available from the Sustainable Buildings Industry Council under license to the Midwest Research Institute. Detailed information about Energy-10 is available at www.sbicouncil.org/store/e10.php.
9. Detailed information about the capabilities of Radiance can be found at radsite.lbl.gov/radiance.
10. Information adapted from Diamond et al. (2006).
11. The January 2006 version of the IPMVP Protocol, "Concepts and Practices for Determining Energy Savings in New Construction," Volume 3, Part 1, plus other IPMVP references are available from the Efficiency Valuation Organization website, www.evo-world.org.
12. *Insolation* is an acronym for incoming solar radiation.
13. An HDD or a CDD is a measure of the deviation of the site's temperature profile from the average temperature in a building. For heating, the average temperature is 65°F (18°C); for cooling, the average temperature used for calculations is 75°F (24°C). For example, a day with an average temperature of 60°F (16°C) would result in five Fahrenheit-based (two Celsius-based) HDDs [(65°F − 60°F) (18°C − 16°C) × 1 day]. The number of HDDs or CDDs is an indicator of how extreme the temperature profile of a site is and how much energy may be required to provide heating or cooling.
14. The study of the effects of skylights on retail sales is in the report "Skylighting and Retail Sales" (1999).
15. Data on student performance are from "Daylighting in Schools" (1999).
16. Excerpted from "Tips for Daylighting with Windows" (1997).
17. From "Daylighting: Energy and Productivity Benefits" (1999).
18. A description of the daylighting and other strategies employed to make Rinker Hall a high-performance building can be found at the American Institute of Architects

(AIA), Committee on the Environment (COTE) website, www.aiatopten.org/hpb/energy.cfm?ProjectID286.

19. COP is a measure of the performance of heat pumps and air-conditioning systems and is defined as the ratio of energy removed or added to the energy input to the system. Both energy removed and energy input must have the same units—for example, BTUs per hour or kilowatts. Unlike efficiency, which has a maximum value of 1, COP can be greater than 1 and indeed should be much greater than 1. For example, efficient screw chillers can have a COP of 7 or higher. Another related term is *Seasonal Energy Efficiency Ratio* (SEER), which describes the ratio of energy removed, in BTU to watts of input power, and is used to describe the performance of smaller residential-scale air-conditioning systems. An air-conditioning unit with a SEER 14 rating would have an equivalent COP of 4.

20. From Löhnert et al. (2006).

21. Excerpted from "Solar Heat Gain Control for Windows" (2006).

22. The National Fenestration Rating Council's website is www.nfrc.org.

23. The original table upon which this is based is from "Choosing Windows: Looking through the Options" (2010).

24. The Cool Communities network advocates for measures that prevent urban heat islands. Its website is www.coolcommunities.org.

25. From Florida Solar Energy Center (2000).

26. Excerpted from Kaneda et al. (2006).

27. From IAEI (2009).

28. Excerpted from "Chiller Plant Efficiency" (2000).

29. Ibid.

30. Ibid.

31. *Greening Federal Facilities* is a guide developed for use by federal government facility managers to use in greening their buildings during the course of routine operations and maintenance. It is downloadable from www.nrel.gov/docs/fy01osti/29267.pdf.

32. Excerpted from "Economizers" (2000).

33. From Philips Lighting; excerpted from *Greening Federal Facilities* (2001).

34. The American Bioenergy Association no longer exists and its activities have been taken on by the Environmental and Energy Study Institute whose website is www.eesi.org.

35. The Carbohydrate Economy Clearinghouse was sponsored by the Institute for Local Self-Reliance (ILSR) and covered the broad range of issues associated with shifting to biobased renewables.

36. Information about fuel cell applications can be found at www.fuelcells.org.

37. An excellent overview of building EMS is available from Energy Design Resources in the form of a design brief, "Energy Management Systems" (1998).

38. Excerpted from "What Office Tenants Want" (2000).

39. From POST (2006).

40. Ibid.

41. The Inventory of Carbon and Energy was developed by Geoff Hammond and Craig Jones at the University of Bath (UK) and is in the form of an Excel spreadsheet. Available at perigordvacance.typepad.com/files/inventoryofcarbonandenergy.pdf.

42. From NRC (2009).

43. From Brown, Southworth, and Sarzynski (2008).

44. A Buildings library following the Modelica Fluid library (Elmqvist, Tummescheit, and Otter 2003; Casella et al. 2006) is available for download at http://www.modelica.org/libraries/Buildings.

References

Brown, Marilyn A., Frank Southworth, and Andrea Sarzynski. 2008. "Shrinking the Carbon Footprint of Metropolitan America." Metropolitan Policy Program at the Brookings Institution. Available at www.brookings.edu/~/media/Files/rc/reports/2008/05_carbon_footprint_sarzynski/carbonfootprint_report.pdf.

"Building Simulation." 2000. Energy Design Brief, Energy Design Resources. Available at www.energydesignresources.com/resources/publications/design-briefs.aspx.

Casella, F., M. Otter, K. Proelss, C. Richter, and H. Tummescheit. 2006. "The Modelica Fluid and Media Library for Modeling of Incompressible and Compressible Thermo-Fluid Pipe Networks." Proceedings of the Fifth International Modelica Conference, Austria.

"Chiller Plant Efficiency." 2000. Energy Design Brief, Energy Design Resources. Available at www.energydesignresources.com/resources/publications/design-briefs.aspx.

"Choosing Windows: Looking through the Options." 2010. *Environmental Building News* 20 (2), 1–14.

"Concepts and Practices for Determining Energy Savings in New Construction." 2006. Volume 3, Part 3, EVO 30000-1: 2006. Prepared by the New Construction Sub-committee of the IPMVP. Available at the Energy Valuation Organization website, www.evo-world.org.

"Daylighting: Energy and Productivity Benefits." 1999. *Environmental Building News* 8 (9): 1, 10–14.

"Daylighting in Schools: An Investigation into the Relationship between Daylighting and Human Performance: A Condensed Report." 1999. Conducted by the Heschong Mahone Group for the Pacific Gas and Electric Company. Available at www.pge.com/includes/docs/pdfs/shared/edusafety/training/pec/daylight/Schools Condensed820.pdf.

Diamond, R., M. Opitz, T. Hicks, B. Vonneida, and S. Herrera. 2006. "Evaluating the Site Energy Performance of the First Generation of LEED-Certified Commercial Build-ings." *Proceedings of the 2006 Summer Study on Energy Efficiency in Buildings*, LBNL-59853. Washington, DC: American Council for an Energy Efficient Economy. Available at http://epb.lbl.gov/homepages/Rick_Diamond/LBNL59853-LEED.pdf.

"Economizers." 2000. Energy Design Brief, Energy Design Resources. Available at www.energydesignresources.com/resources/publications/design-briefs.aspx.

Elmqvist, H., H. Tummescheit, and M. Otter. 2003. "Object-Oriented Modeling of Thermo-Fluid Systems." Proceedings of the Third Modelica Conference, Sweden.

"Energy Management Systems." 1998. Energy Design Brief, Energy Design Resources. Available at www.energydesignresources.com/resources/publications/design-briefs.aspx.

Florida Solar Energy Center. 2000. "Laboratory Testing of the Reflectance Properties of Roofing Materials," by D. S. Parker et al. Available at www.fsec.ucf.edu/en/publications/html/FSEC-CR-670-00/.

Fruits, Eric. 2011. "Compact Development and Greenhouse Gas Emissions: A Review of Recent Research." *Center for Real Estate Quarterly* 5(1), 2–6. Available at http://pdx.edu/sites/www.pdx.edu.realestate/files/01%20Fruits%20Quarterly%202011-02.pdf.

Greening Federal Facilities, 2nd ed. 2001. DOE/GO-102001-1165. Washington, DC: Department of Energy. Available at www.nrel.gov/docs/fy01osti/29267.pdf.

Haves, P., B. Hencey, F. Borrelli, J. Elliot, Y. Ma, B. Coffey, S. Bengea, and M. Wetter. 2010. "Model Predictive Control of HVAC Systems: Implementation and Testing at the University of California, Merced Campus."

Hawken, P., A. Lovins, and L. H. Lovins. 1999. *Natural Capitalism.* New York: Little, Brown.

IAEI. 2009. "Energy Loss. Global Warming and Voltage Drop." *IAEI Magazine*, July–August. Available at www.iaei.org/magazine/2009/07/energy-loss-global-warming-and-voltage-drop.

Kaneda, David, Scott Shell, Peter Rumsey, and Mark Fisher. 2006. "IDeAs Z^2 Design Facility: A Case Study of a Net Zero Energy, Zero Carbon Emission Office Building." Proceedings of Rethinking Sustainable Construction 2006, September 18–22, Sarasota, FL.

Löhnert, Günther, Sebastian Herkel, Karsten Voss, and Andreas Wagner. 2006. "Energy Efficiency in Commercial Buildings: Experiences and Monitoring Results from the German Funding Program Energy Optimized Building, ENOB." Proceedings of Rethinking Sustainable Construction 2006, September 18–22, Sarasota, FL.

Mattson, S. E., and H. Elmqvist. 1997. "Modelica—An International Effort to Design the Next Generation Modeling Language." Proceedings of the Seventh IFAC Symposium on Computer Aided Control Systems Design, Belgium.

NRC. 2009. "Driving and the Built Environment: The Effects of Compact Development on Motorized Travel, Energy Use, and CO_2 Emissions." Transportation Research Board Special Study 298, National Research Council. Available at www.nap.edu/catalog.php?record_id=12747.

Parker, D. S., J. E. R. McIlvaine, S. F. Barkaszi, D. J. Beal, and M. T. Anello. 2000. "Laboratory Testing of the Reflectance Properties of Roofing Material." FSEC-CR-670-00, Florida Solar Energy Center, Cocoa, FL. Available at www.fsec.ucf.edu/en/publications/html/FSEC-CR-670-00/.

POST. 2006. "Carbon Footprint of Electricity Generation." Number 268, Parliamentary Office of Science and Technology. Available at www.parliament.uk/documents/post/postpn268.pdf.

"Radiant Cooling." 2000. Energy Design Brief, Energy Design Resources. Available at www.energydesignresources.com/resources/publications/design-briefs.aspx.

Reference Package for New Construction and Major Renovations (LEED-NC). 2005. Washington, DC: US Green Building Council.

Skanska. 2010. "Carbon Footprinting in Construction: Examples from Finland, Norway, Sweden, UK and US." Case Study 76, Skanska AB. Available at skanska-sustainability-case-studies.com/pdfs/76/76_Carbon_Footprinting_v001.pdf.

"Skylighting and Retail Sales: An Investigation into the Relationship between Daylighting and Human Performance: A Condensed Report." 1999. Conducted by the Heschong Mahone Group for the Pacific Gas and Electric Company. Available at www.pge.com/includes/docs/pdfs/shared/edusafety/training/pec/daylight/Retail Condensed820.pdf.

"Solar Heat Gain Control for Windows." 2002. *EREC Reference Briefs.* Washington, DC: Office of Energy Efficiency and Renewable Energy, Department of Energy.

Srinivasan, R. S., W. W. Braham, D. E. Campbell, and C. D. Curcija. 2011a. "Re(de)fining Net Zero Energy: Renewable Energy Balance in Environmental Building Design." *Building and Environment* 47: 300–315.

Srinivasan, R. S., J. Lakshmanan, E. Santosa, and D. Srivastav. 2011b. "Plug-Load Densities for Energy Analysis: K–12 Schools." *Energy and Buildings* 43 (11): 3289–3294.

Srinivasan, R. S., J. Lakshmanan, and D. Srivastav. 2011c. "Calibrated Simulation of an Existing Convention Center: The Role of Event Calendar and Energy Modeling Software." Proceedings of 2011 Building Simulation Conference, November 11–14, Sydney.

"Tips for Daylighting with Windows: The Integrated Approach." 1997. LBNL-39945, Lawrence Berkeley National Laboratory, Berkeley, CA.

Wetter, M. 2011. "A View on Future Building System Modeling and Simulation." In *Building Performance Simulation for Design and Operation*, edited by J. L. M. Hensen and R. Lamberts. London: Routledge.

"What Office Tenants Want: 1999 BOMA/ULI Office Tenant Survey Report." 2000. BOMA International Foundation.

Chapter 10

Built Environment Hydrologic Cycle

Of the various resources needed for the built environment, water is arguably the most critical. In his book *The Bioneers*, Kenny Ausubel notes that biologists occasionally refer to this resource as "Cleopatra's water" because, like all other materials on the planet, water stays in a closed loop. The water you sip from a drinking fountain may have once been used by the Egyptian queen in her bath. The human body is 97 percent water, and water is more crucial to survival than food. It serves as a buffer in human metabolism for the transfer of oxygen at small scale, as a damper on rapid changes in the planet's environment at large scale, and as a shock absorber in cellular function at microscopic scale. Water plays a role in most of the world's spiritual traditions and religions, from baptism in the Christian faiths to sweat lodges in Native American rituals to the cleanliness traditions of the Baha'i faith. Water is the source of life for both humans and other species, yet it also has the power to destroy. It is used as a metaphor for truth and as a symbol for redemption and the washing away of sin. Water serves as habitat for a substantial fraction of the earth's living organisms, and the remainder are totally dependent on it for their survival.

In spite of water's symbolic and practical values, water resources throughout the planet are badly stressed. In July 2010, the United Nations passed a resolution affirming the right of all people to safe and clean water and sanitation.[1] At present, nearly 2 billion people live in water-stressed areas of the world, and 3 billion have no running water within about 0.6 mile (1 kilometer) of their homes. Every eight seconds, a child dies of a waterborne disease, which would be preventable if their families had adequate financial resources. The world is running out of water, and the future will likely be grim for populations that cannot afford the technology and energy needed to produce clean water from seawater or polluted water. A recent McKinsey & Company report stated that, by 2030, global demand for water will exceed supply by more than 40 percent, a foreshadowing of the dire predicament that the human population of the planet will face in the near future.[2] The McKinsey report also forecasted that of the new demand between now and 2030, about 42 percent would be from just four countries: China, India, Brazil, and South Africa.

It is important to note the actual amount of water needed by a population because this defines the limits of supply and consumption for a region. For bare survival, the World Health Organization (WHO) suggests that 0.5 to 1 gallon (2 to 4.5 liters) of water is needed per person for drinking and another 1 gallon (4 liters) for cooking and food preparation. The US Agency for International Development (USAID) states that 26.4 gallons (100 liters) a day per person are required to maintain a reasonably good quality of life. In the United States, direct per capita daily water use is approximately four times higher, about 100 gallons (400 liters); and if agricultural and industrial water use is included, the amount per person per day is approximately 1800 gallons (7000 liters)—an enormous quantity of a limited and precious resource.

In addition to problems of water supply, public health and hygiene are important issues. Waterborne diseases, including diarrhea, typhoid, and cholera, are responsible for 80 percent of the illnesses and death in developing countries. Some 15 million children per year die from these diseases. Raw sewage and toxic materials, including industrial and chemical wastes, human waste, and agricultural waste, are dumped into water systems at the rate of 2 million tons per day. About 300,000 gallons (1.1 million liters) of raw sewage are dumped every minute into the Ganges River in India, which is also a primary source of water for many Indians. Wastewater treatment lags in most of the world: only 35 percent is treated in Asia and approximately 14 percent in Latin America.

Global Water Resource Depletion

Of all the earth's water, only 2.75 percent is freshwater, and of that, three-quarters, or about 2 percent, is sequestered, or locked up, in glaciers and permanent snow cover. Only a tiny fraction of planetary water, about 0.01 percent, is surface water found in rivers and lakes and thus readily accessible (see Table 10.1). The remainder is buried deep in the ground, and in some cases, once removed, it can be replenished only over hundreds of years. In much of the world, freshwater removed from both ground and surface sources is being used up far faster than it is being replenished. Western Asia has the most severe water supply problem in the world, with over 90 percent of its population experiencing severe water stress. In Spain, over half of its approximately 100 aquifers are overexploited. In the United States, the situation is better but not significantly and perhaps not for long. In Arizona alone, more than 520 million cubic yards (400 million cubic meters) of water are removed from aquifers each year, double the replenishment rate from rainwater.

Perhaps the best known case of water supply depletion is the Aral Sea, which in the 1960s began supplying water to Soviet collective farms for the production of cotton. Formerly, it was a source of large fish; by the early 1980s, they had been virtually eliminated. By the 1990s, the Aral Sea occupied half of its original area, and it had shrunk in volume by 75 percent. A once beautiful, large, rich, and deep lake with complex ecosystems had been largely destroyed in about 40 years due to human activities (see Figure 10.1).

TABLE 10.1

Inventory of Water on the Earth's Surface

Reservoir	Volume (cubic km × 1,000,000)	Percentage of Total
Oceans	1370	97.25
Ice caps and glaciers	29	2.05
Groundwater	9.5	0.68
Lakes	0.125	0.01
Soil moisture	0.065	0.005
Atmosphere	0.013	0.001
Streams and rivers	0.0017	0.0001
Biosphere	0.0006	0.00004

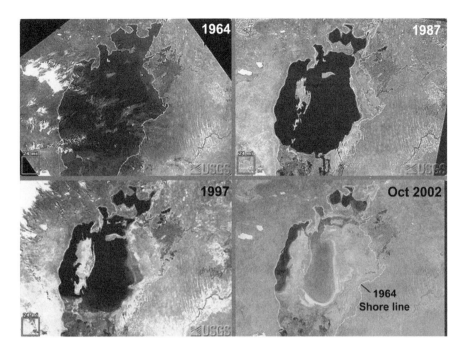

Figure 10.1 The Aral Sea has all but disappeared and its ecosystems totally destroyed in the 40-year period from the 1960s to the 1990s, a victim of withdrawals for growing cotton and industrialization. (*Source:* US Geological Survey)

Water Distribution and Shortages in the United States

In the United States, water crises are occurring almost everywhere. The Florida panhandle's ecologically significant Apalachicola, located at the southern end of a complex watershed comprising the Apalachicola, Flint, and Chattahoochee Rivers, is under threat due to issues far from Apalachicola Bay into which the system flows. At the far north of this watershed lies Atlanta, Georgia, a growing city of 5 million that draws most of its water from the Chattahoochee River and competes with the sparsely populated rural and fishing communities further south along the Alabama border and into Florida for the limited water in this system. A three-year drought ending in 2009 resulted in a three-state water war that pitted the urban interests of Atlanta against the rural needs of Georgia in a conflict that is being mirrored many times over in the United States alone (see Figure 10.2). In October 2007, Georgia governor Sonny Perdue declared a state of emergency for the northern third of the state of Georgia and asked President George W. Bush to declare it a major disaster area. At that time, Georgia officials warned that Lake Lanier, a 38,000-acre reservoir that supplies more than 3 million residents with water, was less than three months from depletion. Smaller reservoirs were dropping even lower. The competition for the limited water is refereed by the US Army Corps of Engineers, which releases more than a billion gallons of water from Lake Lanier every day. The water releases are based on two requirements that the Corps of Engineers is mandated to meet: the minimum flow needed for a coal-fired power plant in Florida and mandates to protect two mussel species in a Florida river. Consequently, the needs of Atlanta are pitted against the downstream needs of a largely rural region and the protection of natural species that support the livelihood of Gulf Coast fishermen. Governor Perdue asked a federal judge to significantly reduce the outflows from the lake and set aside more water for the residents of northern Georgia. Similar dramas have reoccurred several times, and the three-state water war among Florida, Georgia, and Alabama continues.

Figure 10.2 Lake Lanier, northeast of Atlanta, Georgia, supplies water to its burgeoning population of 5 million, competing with, among others, Gulf of Mexico oystermen, for critical and increasingly scarce water. The picture shows Lake Lanier during the October 2007 drought when water levels were 14.4 feet below normal. (Dick McMichael, dicksworld.wordpress.com)

Figure 10.3 The enormous growth of Las Vegas, Nevada, has contributed to significant aquifer depletion in less than 30 years. The satellite imagery of Las Vegas illustrates the spatial patterns and rates of change resulting from the city's urban sprawl. (*Source:* United Nations Environment Programme)

Water crises are also apparent in the moratoriums imposed on development and growth because of either a shortage of water supplies or insufficient wastewater treatment capacity. A growth moratorium in Las Vegas, Nevada, currently one of the fastest-growing municipalities in the United States, has been under active discussion several times since 2004. In the Diamond Valley, near Las Vegas, water levels dropped over 100 feet (30 meters) during the 1970s and 1980s and have never recovered (see Figures 10.3 and 10.4).[3] In January 2004, the town commissioners of Emmitsburg, Maryland, passed an ordinance that invokes a growth moratorium for lots not already approved for development until the maximum design capacity of the city's wastewater treatment plant, which is 800,000 gallons (3 million liters) per day, is not exceeded for 180 days.[4]

The sheer scale of water consumption is enormous but has flattened out over time, with over 410 billion gallons extracted each day in the United States for all uses in 2005 with the same level of consumption estimated for 2010.[5]

Figure 10.4   Most areas in Las Vegas, Nevada, require water irrigation for golf courses, country clubs, and other landscaping to further attract people to this region of the Mojave Desert. (Paul Francis, www.lasvegasrealestatehome.com)

Figure 10.5   The water demands of the United States are causing significant water level drops in various aquifers throughout the region. Southern Arizona is one of many areas that extract extensive quantities of water from the aquifer, causing land subsidence. These areas are vulnerable to runoff contaminating basin aquifers. (*Source:* Arizona Department of Water Resources)

The rate of water consumption is over 40 times that of gasoline, and some argue that one day in the not too distant future, water may be more expensive than gasoline. In fact, the equivalent price of bottled water in a convenience store is already at least $5.80 per gallon ($1.50 per liter). The bright side of this picture is that in 1980 US water use was even higher, 450 billion gallons a day, meaning that total and per capita water use dropped in spite of an additional 70 million people and a doubling of US gross domestic product. Thus, less water was used in an economy of $14 trillion than in an economy of $6 trillion. Although direct consumption by people in buildings is not a large fraction of total water use in the United States, water shortages in many areas of the country are having an impact on development and construction (see Figures 10.5 and 10.6).

Figure 10.6 Sinkholes are an example of land subsidence due to groundwater extraction. Since groundwater serves partly as a structural component to the rock, its depletion results in voids and eventual collapse, sometimes sudden and unpredictable, creating a substantial hazard to people and infrastructure. (Photograph courtesy of Ildar Sagdejev)

Agriculture is the cause of serious water supply problems because it is responsible for over 80 percent of water consumption, and 60 percent of irrigation water is wasted because of leaky canals, evaporation, and mismanagement. Similar problems occur in the cities of many developing countries, with about 40 percent of the water in large cities being lost to leaky systems.

Buildings account for about 12 percent of freshwater withdrawals. The built environment hydrologic cycle, characterized by the input of high-quality potable water and the release of used, contaminated water, is inefficient, wasteful, and illogical. In its more extended context, the built environment hydrologic cycle also includes the irrigation of landscaping and the handling of stormwater (see the discussion in Chapter 8 on stormwater, which is generally included with the general topic of the building site). As pointed out by Hawken, Lovins, and Lovins, the invention of the water closet by Thomas Crapper was perhaps the start of an unfortunate trend in decision making with respect to building water use.[6] In order to dispense with the human waste generated in buildings, water closets mix high-quality potable water with disease-ridden feces and relatively clean urine for the purpose of diluting this mix. Consequently, enormous quantities of water are wasted, and a potentially useful source of fertilizer is released into sanitary sewer systems to combine with industrial waste. The end result is a complex, chemically intense, energy-consuming, pollution-producing system of wastewater treatment plants. Major rethinking of the built environment hydrologic system is clearly needed to make better use of increasingly scarce and expensive potable water and to reduce the impact and cost of treating effluent from buildings.

In this chapter, we address how high-performance buildings can help contribute to reducing pressure on the increasingly scarce water resource and to improving the health of local ecosystems. We also discuss strategies for selecting water sources, employing recent technological improvements in plumbing fixtures, evaluating alternative wastewater strategies, implementing sustainable stormwater management, and optimizing landscape water consumption. Additionally, this chapter covers the subject of setting targets for water use and modeling building water consumption to assess progress in meeting these targets.

Hydrologic Cycle Terminology

Before discussing a high-performance building hydrologic strategy, it is important to define common terms used in the context of the built environment water cycle. The following are definitions of the more important concepts that should be understood in any discussion of the design of high-performance building water systems.

> **Hydrologic cycle.** The continuous cycling of water between planetary reservoirs such as the ground, water bodies, and the atmosphere. The hydrologic cycle is also referred to as the *water cycle.* Table 10.1 shows the distribution of water on the earth's surface and in the atmosphere. The residence time for water on the earth's surface varies from as little as 1 month for soil moisture to as much as 10,000 years for deep groundwater (see Table 10.2).[7]

> **Built environment hydrologic cycle.** The flow and storage of all types of water on sites altered from their natural state for the purpose of building and infrastructure. Water types include potable water, rainwater, stormwater, graywater, blackwater, and reclaimed water that are used, processed, stored, and moved by employing a variety of technologies that, in the case of high-performance buildings, are coupled with natural systems.

> **Potable water.** Water that is safe for human consumption, that is, has high quality and low risk of harm. Potable water is generally obtained from groundwater or surface water sources and then processed to increase its quality to drinking water standards.

> **Groundwater.** Water that is found underground in rock formations such as aquifers and in soils. Groundwater is extracted for human consumption using shallow wells or deep, artesian wells. Water that seeps into the ground to add to the supply of groundwater is referred to as *recharge water.*

> **Surface water.** Water that collects on the earth's surface in rivers, streams, lakes, and ponds, and which serves as a source of replenishment for groundwater.

> **Fossil water.** Deep groundwater that has a long residence time, sometimes on the order of thousands of years. In spite of the long period of existence of this water source in underground aquifers, fossil water is being rapidly depleted because it is not readily replenished and is essentially a non-renewable resource. In the United States, the US Department of Agriculture reported that in parts of three leading grain-producing states that draw water from the Ogallala Aquifer—Texas, Oklahoma, and Kansas—the underground water table has dropped by more than 30 meters (100 feet); as a result, wells have gone dry on thousands of farms in the southern Great Plains.

> **Stormwater.** Water that does not infiltrate into the ground and either runs off into bodies of water or enters the stormwater system. Includes water from the precipitation of rain and snow, water from melting snow, and water from overwatering.

> **Rainwater.** Water from liquid precipitation, excluding water from snow, hail, and sleet, that has not entered a stream, lake, or other body of water.

> **Rainwater harvesting.** The collection, storage, and use of rainwater. Most systems use the roof surface as the collection area and a large galvanized steel, fiberglass, polyethylene, or ferrocement tank as the storage cistern. When the water is to be used just for landscape irrigation, only sediment

TABLE 10.2

Typical Residence Times of Water Found in Various Reservoirs

Reservoir	Average Residence Time
Glaciers	20–100 years
Seasonal snow cover	2–6 months
Soil moisture	1–2 months
Groundwater: shallow	100–200 years
Groundwater: deep (fossil)	10,000 years
Lakes	50–100 years
Rivers	2–6 months

filtration is typically required. When water is being collected and stored for potable uses, additional measures are required to purify the water and ensure its safety. Rainwater harvesting offers several important environmental benefits, including reducing pressure on limited water supplies and reducing stormwater runoff and flooding. It can also be a better-quality source of water than conventional sources. After purification, rainwater is usually very safe and of high quality.

Reclaimed water. Water from a wastewater treatment plant that has been treated and can be used for nonpotable purposes such as landscape irrigation, cooling towers, industrial process uses, toilet flushing, and fire protection. In some areas of the United States, reclaimed water may be referred to as irrigation quality (IQ) water, but potential uses can extend well beyond irrigation.

Blackwater. Water containing human waste. Wastewater from kitchen sinks and dishwashers is sometimes considered blackwater because it contains oil, grease, and food scraps, which can burden the treatment and disposal processes.

Graywater. Water from showers, bathtubs, bathroom sinks, washing machines, and drinking fountains. Graywater may also include condensation water from refrigeration equipment and air conditioners, hot tub drainwater, pond and fountain drainwater, and cistern drainwater. Graywater contains a minimum amount of contamination and can be reused for certain landscape applications. Although the issue is still being debated by public health officials, no case of illness has ever been traced to graywater reuse. Both graywater and blackwater contain pathogens—humans should avoid contact with either—but blackwater is considered a much higher risk medium for the transmission of waterborne diseases. Although they are not blackwater, the following water sources should not be included in graywater that is to be used for irrigation: garden and greenhouse sinks, water softener backflush, floor drains, and swimming pool water. In buildings served exclusively by composting toilets, and thus producing no true blackwater, it may be useful to include kitchen wastewater in graywater by taking special precautions to eliminate organic matter.

Xeriscaping. A landscaping strategy that focuses on using drought-tolerant native and adapted species that require minimal to no water for their maintenance. The term is derived from the Greek word *xeri*, meaning "dry"; the strategy is also referred to as *enviroscaping*.

Living Machine. A trademark and brand name for a form of ecological wastewater treatment designed to mimic the cleaning functions of wetlands. The system is an intensive bioremediation system that can also produce beneficial by-products, such as reuse quality water, ornamental plants, and plant products usable for building materials, energy, or animal feed.

High-Performance Building Hydrologic Cycle Strategy

One of the key issues that the green building movement is attempting to include in the dialogue about the future direction of high-performance buildings is the interaction of the natural water cycle with the built environment. The built environment hydrologic cycle involves the handling and use of water both internal and external to buildings. Water is imported into the built environment for consumption and other uses and then exported as wastewater. Water used inside the building can be potable water from the municipal water system or from wells; rainwater from cisterns; or, when permitted, graywater recycled within the building. Outside the building, the built environment hydrologic cycle can be extraordinarily complex with the challenges of handling sometimes large volumes of rainwater and providing water for landscape irrigation. Rainwater falling on the building site can have several fates. On a greenfield or previously undeveloped site, most rainwater infiltrates into the ground and the remainder flows off into streams or other bodies of water. On a developed site, the situation can be reversed, with relatively little water infiltrating into the ground. Buildings and hardscape such as sidewalks and roads cover the ground, preventing infiltration of rainwater and inducing water flow across parking lots and roads. Rainwater must be either collected and conducted into municipal stormwater systems that receive and process this water or stored on-site in retention and detention ponds.

Designers of high-performance buildings have developed novel built environment hydrologic strategies that are having significant positive impacts on water consumption. The focus of these approaches is threefold: (1) to minimize the consumption of potable or drinking quality water from wells or the municipal wastewater system, (2) to minimize wastewater generation, and (3) to maximize rainwater infiltration into the ground.

These strategies, together with the emergence of several key technologies, are resulting in high-performance buildings with enormous reductions in their water consumption and wastewater generation profiles. These innovative strategies and technologies are described below.

THE BENEFITS OF WATER EFFICIENCY

Reducing building water consumption and rethinking the wastewater strategy employed for the built environment can dramatically extend the available supply of water, improve human health, and reduce threats to ecological systems. In addition to these benefits, the Rocky Mountain Institute (RMI) suggests that water efficiency can have these other tangible and calculable benefits:[8]

- *Energy savings.* More money can be saved by reducing the energy needed to move, process, and treat water than the actual value of the saved water.

- *Reduced wastewater production.* Reducing water consumption also reduces wastewater generation, lowering costs for building owners. Wastewater costs are significantly higher than the cost of potable water.

- *Lower facilities services investments.* Designing water-efficient buildings reduces the costs of water and wastewater infrastructure.

- *Industrial processes.* Innovations in water use in production systems can result in new processes and approaches.

- *Higher worker productivity.* Facilities that incorporate resource efficiency measures are known to have a more productive workforce.

- *Reduced financial risk.* Implementing water efficiency can be accomplished as needed, thus reducing costs and risks for large facilities.

- *Environmental benefits.* Lowering water consumption results in reduced impact on natural systems.

- *Public relations value.* Protecting the environment is looked upon favorably by the general public and clients.

The building hydrologic cycle and energy use are tightly coupled, with very little of the impact being apparent to the building owner. Complex and expensive systems extract potable water from surface water and groundwater sources, then pump it for treatment and distribution, requiring large quantities of energy that are generally subsidized by the low cost of water. Similarly, wastewater must be pumped through an extensive system of sanitary sewers and lift stations to central wastewater treatment plants, consuming relatively large amounts of energy. (The term *watergy* is sometimes used to describe the tightly intertwined relationship of water and energy.) The good news is that reducing water consumption reaps numerous positive benefits, not only by reducing flows through the system but also by lowering overall energy consumption and associated pollution from energy sources.

STEPS IN DEVELOPING A HIGH-PERFORMANCE HYDROLOGIC STRATEGY

The following logical steps can be used to develop a hydrologic strategy for high-performance buildings:

1. *Select the appropriate water sources for each consumption purpose.* Potable water must be used only for those applications that involve human consumption or ingestion. In addition to potable water, other water sources include rainwater, graywater, and reclaimed water. These alternative sources of water can be used for landscape irrigation, fire protection, cooling towers, chilled and hot water, toilet and urinal flushing, and other applications for which valuable potable water can be minimized. In each case, the availability of each alternative water source should be analyzed to determine which mix is optimum for the particular project and its forecasted water use profile.

2. *For each purpose, employ technologies that minimize water consumption.* This strategy can include a combination of low-flow fixtures (toilets, urinals, faucets, and showerheads), no-flow fixtures (composting toilets, waterless urinals), and controls (infrared sensors). For cooling towers, chemical-free electromagnetic technology can reduce scaling caused by biological contaminants and corrosion, both of which can reduce the system performance. For landscaping, highly efficient drip irrigation systems use far less water and deliver the water to the plant roots with more than 90 percent efficiency. Additionally, drought-tolerant native and adapted species can be employed in the landscape scheme, an approach that can often eliminate the need for an irrigation system.

3. *Evaluate the potential for a dual wastewater system.* Such a system separates lightly contaminated water from sinks, drinking fountains, showers, dishwashers, and washing machines from human waste —contaminated sources such as toilets and urinals. This dual piping system separates graywater from blackwater, thus providing the capability for water recycling within the building.

4. *Analyze the potential for innovative wastewater treatment strategies.* For example, constructed wetlands or Living Machines can be employed to process effluent. These approaches are rapidly evolving and beginning to appear in more high-performance building projects each year as the practice of using nature in symbiosis with the building process becomes more refined.

5. *Apply life-cycle costing* (LCC) to analyze the costs and benefits of adapting practices that reduce water flow through the building and its

landscape beyond the levels mandated by the Energy Policy Act of 1992 (EPAct 1992). A simple LCC that examines nothing more than the cost of potable water will generally provide long payback times, perhaps in the 10- to 20-year range. Including reductions in wastewater generation and the costs associated with its treatment will provide an accelerated payback. A more liberal interpretation of costs, such as the actual energy cost of moving water and wastewater, emissions associated with energy generation, worker productivity improvements, and general environmental benefits, would also shorten the payback time of the initial investment. Finally, it can be reasonably expected that the price of potable water in most regions will increase at a greater rate than the general inflation rate and perhaps dramatically faster. Including this in the LCC evaluation, along with other indirect cost factors, should bring the paybacks into the same range as those for good energy conservation measures, namely, seven years or less.

6. *Design landscaping to use minimal water* for its maintenance and upkeep and consider the restoration of ecological systems as an important part of the building design.

7. *Design parking, paving, roads, and landscaping to maximize the infiltration of stormwater.* Prior to buildings being present on the site, a natural hydrologic cycle functioned to move water between the atmosphere, ground, bodies of water, and ecological systems. Restoring the natural hydrologic cycle can benefit both natural systems as well as reduce the need for complex and expensive stormwater infrastructure.

8. *Incorporate green roofs* into buildings to store and naturally process stormwater and contribute to the regeneration of the ecology of the building location.

ESTABLISHING WATER CONSUMPTION TARGETS

Limits on water consumption in buildings are set by building codes, which are, in turn, based on legislation. Table 10.3 shows the progress in setting maximum water consumption levels for typical building plumbing fixtures. One of the landmark pieces of legislation concerning potable water consumption is the Energy Policy Act of 1992 (EPAct 1992). EPAct 1992 requires all plumbing fixtures used in the United States to meet ambitious targets for reducing water consumption; as a result, building codes now mandate these dramatically lower levels of water consumption. Additional requirements for water efficiency for prerinse spray valves used in commercial kitchens were set by the Energy Policy Act of 2005 (EPAct 2005). In 2007, California passed legislation that established even more stringent requirements for toilets and urinals, reducing toilet water consumption from 1.6 gallons (6 liters) per flush to 1.28 gallons (4.8 liters) per flush and urinal water consumption from 1.0 gallon (3.8 liters) per flush to 0.5 gallon (1.9 liters) per flush.

Beyond legislation and code requirements, the US Environmental Protection Agency (EPA) created the voluntary WaterSense label in 2006, which required that WaterSense-certified fixtures use at least 20 percent less water than the requirements of EPAct 1992. The WaterSense label is awarded based on third-party certification that the fixture meets EPA requirements (see Figure 10.7).

Setting goals for a building's water consumption that exceed code requirements is an important first step in designing a strategy that makes sense. If the Factor 10 concept described in Chapter 2 is applied to the issue of water consumption, potable water—and, by inference, wastewater—should be reduced by 90 percent for the purpose of producing a sustainable future. This

Figure 10.7 The EPA created the WaterSense label to stimulate the development of technologies that improve on the EPAct 1992 requirements by at least 20 percent. (Source: US Environmental Protection Agency)

TABLE 10.3

Water Efficiency Standards and Best Technology for Typical Building Plumbing Fixtures

Fixture Type	Units	EPAct 1992	EPAct 2005	WaterSense 2006	California 2007	Best Technology
Water closet, flushing	gpf	1.6		1.28	1.28	0.8[*]
	lpf	6.0		4.8	4.8	3.0
Urinal	gpf	1.0		0.8	0.5	0.0/0.13[†]
	lpf	3.8		3.0	1.9	0.0/0.47
Showerhead	gpm at 80 psi	2.5		2.0		0.57
	gpm at 60 psi	2.2		1.8		0.49
	lpm at 450 kPa	9.5		7.6		2.2
	lpm at 410 kPa	8.5		6.8		1.9
Faucet	gpm at 80 psi	2.5		2.0		0.5
	gpm at 60 psi	2.0		1.6		
	lpm at 450 kPa	9.5		7.6		1.9
	lpm at 410 kPa	7.6		6.1		
Replacement aerator	gpm	2.5				0.5
	lpm	9.8				1.9
Metering faucet	gpc	0.25				0.09
	lpc	0.98				0.34
Prerinse spray valves	gpm at 60 psi		1.6			
	lpm at 410 kPa		6.0			

Key:
gpf = gallons per flush, gpm = gallons per minute, gpc = gallons per cycle
lpf = liters per flush, lpm = liters per minute, lpc = liters per cycle
psi = pounds per square inch pressure, kPa = thousand pascals pressure
Notes:
[*] The best technology in this case is the water closet with the lowest water flush rate. Composting toilets use no water but generally have limited application.
[†] For urinals the best technology is the waterless urinal, which has no water use. Ultra-low-flow urinals use about one-eighth of a gallon per flush and can be selected in cases where a waterless urinal is not appropriate or desirable.

means that typical per capita household consumption of potable water in this country must be reduced from 100 gallons (380 liters) per day to about 10 gallons (40 liters) per day. To accomplish this remarkable reduction requires that water be reused and recycled at high rates. For example, per capita consumption of water is almost evenly divided between outdoor and indoor uses. If only recycled water were used outdoors for irrigating landscaping, per capita consumption of potable water would drop to 50 gallons (190 liters) per day. Indoors, almost half of the water consumed is for toilet and urinal flushing, and using only recycled water for this purpose would further reduce water consumption to 25 gallons (85 liters) per day. These relatively straightforward measures produce an immediate Factor 4 reduction. Additional measures that incorporate low-flow fixtures and electronic controls can nearly produce the desired Factor 10 reduction.

As an alternative to using Factor 4 or Factor 10 strategies to set targets for building water consumption, a more recent approach known as the *net zero built environment* is emerging. In addition to addressing energy, the net zero strategy addresses water consumption by setting limits to water usage based on annual precipitation and water recycled within the building.[9] This is referred to as *net zero water* and is required for certification under the Living Building Challenge. A number of US military installations such the Aberdeen Proving Grounds in Maryland and Fort Hood in Texas are participating in net zero water pilot programs. As an example, if a net zero water target were established as a criterion for a building in Gainesville, Florida, where the average annual rainfall is 36 inches, each square foot of roof would provide 3 cubic feet, or about 22.5 gallons, of water. For Rinker Hall, a Leadership in Energy and Environmental Design (LEED) gold certified building at the University of Florida in Gainesville, with three stories and a 15,000-square-foot roof, the water budget would be about 330,000 gallons per year. The following section on water modeling delves further into the issue of water modeling and budgeting.

WATER SUPPLY STRATEGY

The basic strategy for the water supply of a high-performance building is to reduce potable water consumption to the maximum extent possible. Thus, the first two steps in the high-performance building hydrologic cycle strategy just given also apply to the water supply strategy. The first step is to assess the potential for using nonpotable water sources to replace potable water in a wide range of applications. In this context, nonpotable water includes rainwater, graywater, and reclaimed water. When the feasibility of using each of these nonpotable sources has been assessed, the next step is to ensure that consumption of both potable and non-potable water is minimized. A wide range of high-efficiency fixtures are now available that provide flow rates well below the EPAct 1992 requirements. Waterless plumbing fixtures are becoming more widely available and price-competitive as manufacturers begin offering more alternatives. EPAct 1992 set relatively ambitious limits on water use for water fixtures. However, water use by high-performance green buildings normally exceeds the EPAct 1992 require-ments. For example, the LEED requires at least a reduction of 20 percent in potable water consumption over the EPAct 1992 requirements.

BUILDING PLUMBING FIXTURES AND CONTROLS

The following sections describe the main types of plumbing fixtures currently in use and their low-flow/high-efficiency alternatives.[10] Note that, in this context, *low flow* refers to fixtures that meet the EPAct 1992 requirements, and *high efficiency* refers to fixtures that meet the EPA requirements of using 20 percent less water than the EPAct 1992 requirements.

Toilets and Urinals

Toilets account for almost half of a typical building's water consumption. Americans flush about 4.8 billion gallons (18.2 billion liters) of water down toilets each day, according to the EPA. According to the Plumbing Foundation, replacing all existing toilets with models that use 1.6 gallons (6 liters) per flush would save almost 5500 gallons (25,000 liters) of water per person each year. A widespread toilet replacement program in New York City apartment buildings found an average 29 percent reduction in total water use for the buildings studied. The entire program, in which 1.3 million toilets were replaced, is estimated to be saving 60 to 80 million gallons (230 to 300 million liters) per day. However, there is a common perception that low-flow toilets do not perform adequately. The reason is that a number of early 1.6-gallon (6-liter) per flush gravity flush toilets that were adapted from the 3.5-gallon (16-liter) per flush model (rather than being engineered to operate effectively with the lower volume) performed very poorly, and some low-flow toilets may still suffer from this problem. But studies show that most 1.6-gallon (6-liter) per flush toilets work very well.

Several technologies of 1.6-gallon (6-liter) toilets are available:

- *Gravity tank toilets.* Use basically the same design as for older toilets, but with steeper sides to allow more rapid cleaning during the flush cycle.

Figure 10.8 Waterless urinals save about 40,000 gallons (151,400 liters) of water per year per fixture. (Courtesy of Sloan Valve Company)

- *Dual-flush toilets.* Have two handles for flushing, one for minimal needs such as urine, which uses 1.0 gallon (3.8 liters) per flush; the second for a maximum flow of 1.6 gallons (6 liters).

- *Flushometer toilets.* Capture pressure developed in the flush cycle to assist in the subsequent flush.

- *Vacuum-assisted toilets.* Use the reverse principle of a flushometer toilet by employing a vacuum, which is regenerated by flushing action, to pull the wastewater from the toilet.

For toilets, a high-efficiency toilet (HET) fixture would consume 20 percent less water than a toilet that uses 1.6 gallons (6 liters) per flush, that is, less than 1.28 gallons (4.8 liters) per flush. Where flush performance is a particular concern or where water conservation beyond that of a model that uses 1.28 gallons (4.8 liters) per flush is required, electromechanical flush toilets and dual-flush toilets should be considered. Electromechanical toilets use electrically powered mechanical devices such as pumps and compressors to assist the removal of wastewater from toilets and use less than 1.0 gallon (3.8 liters) of water per flush.

Even greater water conservation can be achieved in certain (limited) applications with composting toilets. Because of the size of composting tanks, lack of knowledge about performance, local regulatory restrictions, and higher first costs, composting toilets are rarely an option except in certain unique applications, such as national park facilities. Composting toilets are being used very successfully, for example, at Grand Canyon National Park.

For urinals, water conservation well beyond the standard 1.0 gallon (4.5 liters) per flush can be obtained using high-efficiency urinals (HEUs) or waterless urinals that use no water. HEUs use at least 20 percent less water than a code-compliant urinal and typically use about 0.5 gallon (1.9 liters) per flush, or 50 percent less than the federal requirements. Waterless urinals use a special trap with a lightweight biodegradable oil that allows urine and water to pass through but prevents odors from escaping into the restroom; there are no valves to fail, and clogging does not cause flooding. The water and wastewater savings that can be achieved are truly remarkable. For example, Falcon Waterfree Technologies cites an annual net savings of $12,600 for a 75-unit installation, or about $168 per installed urinal. The payback for this rate of savings is less than three years at today's water and wastewater prices, and will be far greater in the future as pressure mounts to optimize the use of increasingly scarce sources of potable water (see Figure 10.8).

Showers

EPAct 1992 requires that showerheads deliver a maximum of 2.5 gallons (9.5 liters) per minute at 80 pounds per square inch (psi). Prior to this legislation, showerheads used 3 to 7 gallons (11 to 27 liters) per minute at normal water pressure, about 80 psi (550 kPa). A 5-minute shower now uses about 12.5 gallons (47 liters) of water while an older showerhead typically consumed 15 to 35 gallons (60 to 130 liters). High-quality replacement showerheads that deliver 1.0 to 2.5 gallons (3.8 to 9.5 liters) per minute can save many gallons per shower when used to replace conventional showerheads. Products vary in price from $3 to $95, and many good models are available for $10 to $20. A variety of spray patterns are also available, ranging from misty to pulsing and massaging. These showerheads typically have narrower spray jets and a greater mix of air and water than conventional showerheads, enabling them to provide what feels like a full-volume shower while using far less water.

Flow regulators on the shower controls and temporary cutoff buttons or levers incorporated into the showerhead reduce or stop water flow when the individual is soaping or shampooing, further lowering water use. When the water flow is reactivated, it emerges at the same temperature, eliminating the need to remix the hot and cold water. Flow restrictors are washer-like disks that fit inside existing showerheads, and they are tempting retrofits. Flow restrictor disks were given away by many water conservation programs; however, they provide poor water pressure in most showerheads, leading to poor acceptance of water conservation in general. Permanent water savings are better provided through the installation of well-engineered showerheads.

Faucets

Faucets are generally found in bathrooms, kitchens, and workrooms. Bathroom faucets need no more than 1.5 gallons (5.7 liters) per minute, and residential kitchens rarely need more than 2.5 gallons (9.5 liters) per minute. Institutional bathroom faucets may include automated controls and premixed temperatures. Institutional kitchen faucets may include special features such as swivel heads and foot-activated on/off controls. Older faucets with flow rates of 3 to 5 gallons (11 to 19 liters) per minute wasted tremendous quantities of water. Federal guidelines mandated that all lavatory and kitchen faucets and replacement faucet tips (including aerators) consume no more than 2.5 gallons (9.5 liters) per minute at 80 psi (550 kPa).

Metered-valve faucets are restricted to a discharge rate of 0.25 gallon (0.95 liter) per cycle after this date. Metered-valve faucets usually have push buttons and deliver a preset amount of water and then shut off. For water management purposes, the preset amount of water can be reduced by adjusting the flow valve. The Americans with Disabilities Act (ADA) requires a 10-second minimum on-cycle time.

Variations in water pressure can occur in buildings, and pressure-compensating faucets can be used to automatically maintain 2.5 gallons (9.5 liters) per minute at varying water pressures. For kitchens, devices are available to maintain the water pressure at 2.2 to 2.5 gallons (8.3 to 9.5 liters) per minute. In washrooms, 0.5 to 1.25 gallons (1.9 to 4.7 liters) per minute will often prove adequate for personal washing purposes.

Foot controls for kitchen faucets provide both water savings and hands-free convenience. The hot water mix is set, and the foot valve turns the water on and off at the set temperature. Hot water recirculation systems reduce water wasted while users wait for water to warm up as it flows from the faucet. To prevent these water-saving systems from wasting large amounts of energy, hot water pipes should be well insulated.

Drinking Fountains

Drinking fountains can be metered or nonmetered. Due to the design of water supply systems, drinking fountains vary with respect to discharge rate. In order to meet EPA WaterSense requirements, metered drinking fountains are limited to 0.25 gallon per cycle and nonmetered to 0.7 gallon per cycle. Self-contained drinking fountains have an internal refrigeration system. Adjusting the exit water temperature to 70°F (21°C) versus the typical 65°F (18°C) will result in substantial energy savings. Insulating the piping, chiller, and storage tank will save energy. If appropriate, adding an automatic timer to shut off the unit during evenings and weekends will add to the savings. Remote chillers or central systems are used in some facilities to supply cold drinking water to multiple locations. Sensor faucets require either electrical wiring for the connection of alternating current (AC) power or regular replacement of battery power supplies.

ELECTRONIC CONTROLS FOR FIXTURES

Automated controls for faucets, toilets, and urinals can dramatically lower water consumption and potentially eliminate disease transmission via contact with bathroom surfaces and fixtures. These controls are rapidly gaining popularity in all types of commercial and institutional facilities, although the driver is generally hygiene rather than water or energy savings.

Electronic controls can be installed with new plumbing fixtures or retrofitted onto many types of existing fixtures. Although water savings depend greatly on the type of facility and the particular controls used, some facilities report 70 percent water savings. This type of on-demand system can also produce proportional savings in water heating (for faucets) and sewage treatment. Electronic controls for plumbing fixtures usually function by transmitting a continuous beam of infrared (IR) light. With faucet controls, when a user interrupts this IR beam, a solenoid is activated, turning on the water flow. Dual-beam IR sensors or multispectrum sensors are generally recommended because they perform better for a wider range of users. With toilets and urinals, the flush is actuated when the user moves away and the IR beam is no longer blocked. Some brands of no-hands faucets are equipped with timers to defeat attempts to alter their operation or to provide a maximum on cycle—usually 30 seconds. Depending on the faucet, a 10-second handwash typical of an electronic unit will consume as little as 1⅓ cups (0.3 liter) of water.

Electronic controls can also be used for other purposes in restrooms. Sensor-operated hand dryers are hygienic and save energy by automatically shutting off when the user steps away. Soap dispensers can be electronically controlled. Electronic door openers can be employed to further reduce contact with bathroom surfaces. Even showers are now sometimes being controlled with electronic sensors—for example, in prisons and military barracks. Electronic fixtures are particularly useful for handicapped installations and hospitals, greatly reducing the need to manipulate awkward fixture handles and removing the possibility of scalding caused by improper water control. No-touch faucets are available with (1) the sensor mounted in the wall behind the sink, (2) the sensor integrated into the faucet, or (3) the sensor mounted in an existing hot or cold water handle hole and the faucet body in the center hole. For new installations, the first or second option is usually best; for retrofit installations, the last option may be the only one feasible. At sports facilities where urinals experience heavy use, the entire restroom can be set up and treated as if it were a single fixture. Traffic can be detected and the urinals flushed periodically based on traffic rather than per person. This can significantly reduce water use. Computer controls can be used to coordinate water usage to divert water for fire protection when necessary. Thermostatic valves can be used with electronic faucets to deliver water at a preset temperature. Reducing hot water consumption saves a considerable amount of energy. A 24-volt transformer operating off a 120-volt AC power supply is typically used for electronic controls, at least with new installations. The transformer should be listed by Underwriters Laboratories (UL), and for security reasons, the transformer and the solenoid valve should be remotely located in a chase.

NONPOTABLE WATER SOURCES

Rainwater Harvesting

Rainwater has been considered a crucial source of water for survival for all of human existence. For building applications, rain was typically collected from the roofs of homes and other buildings and conducted into a storage tank or cistern. With the advent of centralized potable water systems, rainwater systems all but disappeared until the emergence of the modern high-performance green building movement. *The Texas Guide to Rainwater Harvesting* cites three factors that are propelling rainwater back into the picture as a viable water source:[11]

1. The escalating environmental and economic costs of providing water by centralized water systems or by well drilling

2. Health concerns regarding the source and treatment of polluted waters

3. The perception that there are cost efficiencies associated with reliance on rainwater

Rainwater systems are appropriate when one or more of the following factors is present:

■ Groundwater or aquifer water supplies are limited or fragile. Fragile aquifer systems are those that, when pumped, can threaten ecologically valuable surface waters and springs.

■ Groundwater supplies are polluted or significantly mineralized, requiring expensive treatment.

■ Stormwater runoff is a major concern.

A rainwater harvesting system generally has the following key components:

■ *Catchment area.* With most rainwater harvesting systems, the catchment area is the building's roof. The best roof surface for rainwater harvesting does not support biological growth (e.g., algae, mold, moss), is fairly smooth so that pollutants deposited on the roof are quickly removed by the roof wash system, and should have a minimal number of overhanging tree branches above it. Galvanized metal is the roofing material most commonly used for rainwater harvesting.

■ *Roof wash system.* This is a system for keeping dust and pollutants that have settled on the roof out of the cistern. It is necessary for systems used as a source of potable water but is also recommended for other systems, as it keeps potential contaminants out of the tank. A roof wash system is designed to purge the initial water flowing off a roof during rainfall.

■ *Prestorage filtration.* To keep large particulates, leaves, and other debris out of the cistern, a domed stainless steel screen should be secured over each inlet leading to the cistern. Leaf guards over gutters can be added in areas with significant windblown debris or overhanging trees.

■ *Rainwater conveyance.* This is the system of gutters, downspouts, and piping used to carry water from the roof to the cistern.

■ *Cistern.* This is usually the largest single investment required for a rainwater harvesting system. Typical materials used include galvanized steel, concrete, ferrocement, fiberglass, polyethylene, and durable wood (e.g., redwood or cypress). Costs and expected lifetimes vary considerably among these options. Tanks may be located in a basement, buried outdoors, or located aboveground outdoors. Light should be kept out to prevent algae growth. Cistern capacity should be sized to meet the expected demand. Particularly for systems designed as the sole water supply, sizing should be modeled on the basis of 30-year precipitation records, with sufficient storage to meet the demand during times of the year having little or no rainfall (see Figure 10.9).

Figure 10.9 The rainwater harvesting system for Rinker Hall at the University of Florida in Gainesville has a cast-in-place cistern (shown here under construction) located under the south stairwell of the building. The rainwater is used for flushing the building's toilets. (Photograph courtesy of Centex-Rooney, Inc.)

- *Water delivery.* A pump is generally required to deliver water from the cistern to its point of use, although gravity-fed systems are occasionally possible with appropriate placement of system components.

- *Water treatment system.* To protect plumbing and irrigation lines (especially with drip irrigation), water should be filtered through sediment cartridges to remove particulates, preferably down to 5 micrometers. For systems providing potable water, additional treatment is required to ensure a safe water supply. This can be provided with microfiltration, ultraviolet sterilization, reverse osmosis, or ozonation (or a combination of these methods). With some systems, higher levels of treatment are provided only at a single faucet where potable water is drawn.

Rainwater harvesting systems have immense potential for reducing potable water consumption by introducing a water source that is readily obtainable in many regions of the United States (see Figures 10.10 and 10.11). In spite of this advantage, there are no standard designs or approaches to designing a rainwater harvesting system; hence, currently, each system designed for a building is unique. Factors to include in the design include the roof material and slope, rainfall intensity, airborne pollutants (such as smoke, dust, and automobile exhaust), and debris generated from trees and other nearby vegetation. As a consequence, these systems can be prone to failure and unreliable, resulting in a potential erosion of interest in rainwater as a substitute for potable water. The creation of clear standards, designs, and standard components would go a long way toward resolving this problem and making the implementation of these systems standard practice.

Graywater Systems

Graywater is generally considered to comprise the nonhuman waste fraction of wastewater. Graywater collection involves separating graywater from blackwater, which, as defined previously, is the human waste−contaminated water from toilets and urinals. Graywater is generally used for landscape irrigation, but it can also be used to flush toilets and urinals.

Buildings with graywater systems must have a dual waste piping system, one for each type of water. Graywater waste lines should run to a central

Figure 10.10 Rainwater is harvested in a cistern system consisting of the following components: (1) nontoxic, noncorrosive roofing material; (2) nontoxic, noncorrosive gutters and downspouts; (3) first-flush diverter with a cleanout trap and bleed valve; (4) debris traps and sediment filter; (5) easily accessible but locked passageway; (6) engineered cistern that avoids direct-sunlight exposure to the collected rainwater; (6) automatic water refill with air gap supplied from the building; (7) cistern; (8) pump electrical supply; (9) pump start/stop relay; (10) backflow prevention valve; and (11) water distribution area. The submersible pump is located inside the cistern while the overflow system is located behind the cistern. (H_2Options, Inc.)

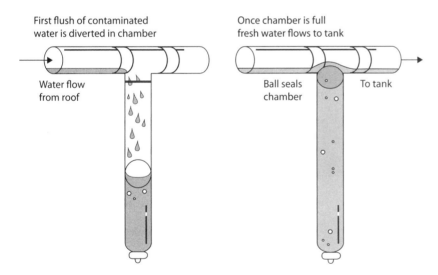

First flush of contaminated water is diverted in chamber

Water flow from roof

Once chamber is full fresh water flows to tank

Ball seals chamber   To tank

Figure 10.11 The simplest first-flush diverter is a standpipe that captures and diverts contaminants washed from the roof. Rainwater fills the standpipe, backs up, and then allows water to flow into the main collection piping after the contaminants have been flushed out. (*Source:* Texas Water Development Board)

location where a surge tank can collect and hold the water until it drains or is pumped into an irrigation system or for other appropriate end uses. An overflow for the graywater collection system should be provided that feeds directly into the sewer line. If excess graywater fills the system due to a mismatch between supply and outflow, or due to a filter or pumping malfunction, the overflow conducts the excess flow to the sewer system. A controllable valve should also be included so that graywater can be shunted into the sewer line when the area(s) being irrigated become too wet or other reasons preclude the use of graywater (see Figure 10.12).

Graywater should not be stored for extended periods of time before use. Decomposition of the organic material in the water by microorganisms will quickly use up available oxygen, and anaerobic bacteria will take over,

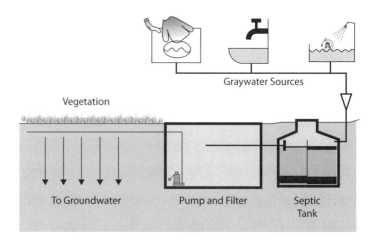

Figure 10.12 A graywater system collects water from showers, sinks, and washing machines into a septic system. The water is then filtered, pumped, and reclaimed for irrigation before seeping into the groundwater. (D. Stephany)

producing unpleasant odors. Some graywater systems are designed to dose irrigation pipes with a large, sudden flow of water instead of allowing the water to trickle out as soon as it enters the surge tank. For a dosing system, holding the water for some amount of time will be necessary, but this should be limited to no more than a few hours. If a filter is used in the graywater system, it should be one that is easy to clean or self-cleaning. Filter maintenance is a major problem with many graywater systems. For complete protection from pathogens, graywater should flow by gravity or be pumped to a belowground disposal field (subsurface irrigation). Perforated plastic pipe—with a minimum diameter of 3 inches (76 millimeters)—is called for in California's graywater regulations, although, with filtering, smaller-diameter drip irrigation tubing can also be used. The California standards require that untreated graywater be disposed of at least 9 inches (about 230 millimeters) below the surface of the ground. Some graywater systems discharge into planter beds—sometimes even beds located inside buildings. Some ready-made systems are available by mail order, but these should be modified for specific soil and climate conditions. As a general rule, graywater can be used for subsurface irrigation of lawns, flowers, trees, and shrubs, but it should not be used for vegetable gardens. Drip irrigation systems have not yet proven to be effective for graywater discharge because of clogging or high maintenance costs.

Reclaimed Water

Reclaimed water is wastewater that has been treated for reuse. The use of reclaimed water for nonpotable purposes can greatly reduce the demand on potable water sources. Municipal wastewater reuse now amounts to about 4.8 billion gallons (18 million cubic meters) per day (about 1 percent of all freshwater withdrawals). Industrial wastewater reuse is far greater—about 865 billion gallons (3.2 billion cubic meters) per day.

In areas of chronic water shortage, the design team should check with the local water utility and inquire whether it has a program to provide reclaimed water to the building's location. Reclaimed water programs are particularly popular in California, Florida, Arizona, Nevada, and Texas.

There are a host of potential applications for reclaimed water: landscaping; golf course or agricultural irrigation; decorative features, such as fountains; cooling tower makeup; boiler feed; once-through cooling; concrete mixing; snowmaking; and fire main water. Making use of reclaimed water is easiest if

Figure 10.13 (A) Reclaimed water is former wastewater that is cleaned and redistributed through a clearly coded system of bright purple pipes. Reclaimed water is used when the application does not require potable water. (B) Posting is mandated where the water comes in contact with the public so as to prevent human consumption. (*Source:* City of Clermont, Florida)

this is planned for at the outset of building a new facility, but major renovations or changes to a facility's plumbing system provide opportunities as well. For certain uses, such as landscape irrigation, required modifications to the plumbing system may be quite modest. It is important to note, however, that the use of reclaimed water may be restricted by state and local regulations. For locations such as universities or military bases that often have their own wastewater treatment plant (WWTP), there may be an opportunity to modify the plant to provide on-site reclaimed water (see Figure 10.13).

To consider using reclaimed water for a building, one or more of the following situations should be present: (1) high-cost water or a need to extend the drinking water supply, (2) local public policy encouraging or mandating water conservation, (3) availability of high-quality effluent from a WWTP, or (4) recognition by the building owner of environmental benefits of water reuse.

Technologies vary with end uses. A modern WWTP has three stages of treatment—primary, secondary, and tertiary—with each succeeding stage requiring more energy and chemicals than the previous stage. In general, tertiary or advanced secondary treatment is required, either of which usually includes a combination of coagulation, flocculation, sedimentation, and filtration. Virus inactivation is attained by granular carbon adsorption plus chlorination or by reverse osmosis, ozonation, or ultraviolet exposure. Dual water systems are beginning to appear in some parts of the country where the water supply is limited or where water shortages may constrain development. Buildings may have two water lines coming in, one for potable water and the other for reclaimed water. The former is for all potable uses, the latter for nonpotable uses. Piping and valves used in reclaimed water systems should be color-coded with purple tags or tape. This minimizes piping identification and cross-connection problems when installing systems. Liberal use of warning signs at all meters, valves, and fixtures is also recommended. (Note that potable water mains are usually color-coded blue, while sanitary sewers are green.) Reclaimed water should be maintained at 10 psi (70 kPa) lower pressure than potable water mains to prevent backflow and siphonage in the event of accidental cross-connection. Although it is feasible to use backflow prevention devices for safety, it is imperative never to directly connect reclaimed and potable water piping. One additional precaution is to run reclaimed water mains at least 12 inches (30 centimeters) lower (in elevation) than potable water mains and to separate them from potable or sewer mains by a minimum of 10 feet (3 meters) horizontally.

Although water prices vary greatly throughout the country, reclaimed water costs significantly less than potable water. For example, in Gainesville, Florida, the price of potable water is now $3.40 per 1000 gallons ($0.92 per cubic meter) versus $0.60 per 1000 gallons ($0.07 per cubic meter) for reclaimed water. Similar pricing differences occur wherever reclaimed water is available.

WASTEWATER STRATEGIES

Reducing potable water consumption is relatively straightforward compared to the effort needed to change wastewater treatment strategies. Contemporary WWTPs are large, centralized, energy- and chemical-intensive operations designed to ensure that public health is protected. However, future high energy costs and increasing public resistance to chemical use are motivating building owners to consider other options for treating wastewater. The fundamental approaches being used today rely on nature, either directly or indirectly, for these alternative approaches. In the direct approach, effluent from buildings is treated by surface or subsurface wetlands. In the indirect approach, nature is brought into the building and enclosed in tanks and vats through which wastewater is passed and cleaned up by plants, light, and bacteria. The following sections describe two natural system–based approaches to wastewater treatment, constructed wetlands and the Living Machine concept.

Constructed Wetlands

One of the ultimate goals of green building is the application of ecological design to the greatest extent possible, including a synergistic relationship among natural systems, buildings, and the humans occupying them. Using nature to perform tasks that would otherwise be accomplished by energy-intense mechanical and electrical systems has four distinct advantages:[12]

1. Nature is self-maintaining, self-regulating, and self-organizing.
2. Nature is powered by solar energy and chemical energy stored in organic materials.
3. Natural systems can degrade and absorb undesirable toxic and metal compounds, converting them into stable compounds.
4. Natural systems are easy to build and operate.

The use of wetlands to treat wastewater from buildings provides precisely this type of opportunity because these ecological systems can break down organic waste, minimizing the need for complex infrastructure and creating nutrients that benefit the species performing these services. Constructed wetlands can be characterized as passive systems for wastewater treatment. They mimic natural wetlands by using the same filtration processes to remove contaminants from wastewater (see Figure 10.14A). In addition to removing organic nutrients, constructed wetlands have the ability to remove inorganic substances; thus, they can be used to treat industrial wastewater, landfill leachate, agricultural wastewater, acid mine drainage, and airport runoff. Constructed wetlands also provide the added benefit of an environmental amenity and can blend into natural or rural landscapes. Moreover, in addition to treating wastewater, constructed wetlands can provide surge areas for stormwater and treat this often contaminated runoff.[13]

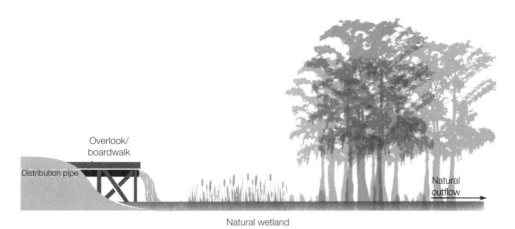

Natural wetland

Figure 10.14 (A) Wetlands, sometimes referred to as nature's kidneys, are natural habitats with distinct characteristics of soil percolation, vegetation, and wildlife habitat. They play an important role in the ecosystem while providing a filtering process whereby contaminants in stormwater runoff are degraded before entering the groundwater. Wastewater can be purposely directed to natural wetlands for the highest cost benefit in terms of both conventional economic and natural capital. (T. Wyman)

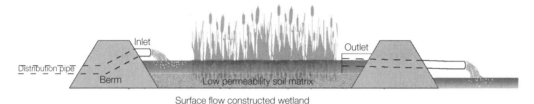

Surface flow constructed wetland

Figure 10.14 (B) Surface flow in constructed wetlands mimics natural wetlands because the water flows aboveground as sheet flow. Wetland plants are selected to provide attachment areas for microbes, which are essential for water quality improvement. The outlet receives water from the wetland cell and directs it either to downstream wetland cells or to a natural water system. (T. Wyman)

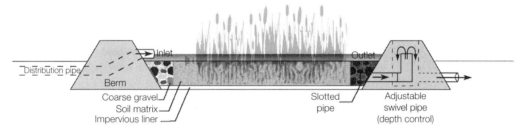

Subsurface flow constructed wetland

Figure 10.14 (C) Subsurface flow constructed wetlands closely resemble wastewater treatment plants and must initiate and maintain all surface flow through the bed media to the outlet where water is collected from the base of the media. (T. Wyman)

Wetlands remove contaminants from water by several mechanisms, including nutrient removal and recycling, sedimentation, biological oxygen demand, metals precipitation, pathogen removal, and toxic compound degradation.

A number of site-specific factors must be taken into account when considering the use of a constructed wetland for wastewater treatment: hydrology (groundwater, surface water, permeability of ground), native plant species, climate, seasonal temperature fluctuations, local soils, site topography, and available area. Constructed wetlands are built for either surface or subsurface flow. Surface flow systems (see Figure 10.14B) consist of shallow basins with wetland plants that are able to tolerate saturated soil and aerobic conditions. The wastewater entering the surface system slowly moves via sheet flow through the basin and is released as clean water. Subsurface systems (see Figure 10.14C), where the wastewater flows through a substrate such as gravel, have the advantages of higher rates of contaminant removal, compared to surface flow systems, and limited contact for humans and animals. They also work especially well in cold climates due to the earth's insulating properties. Cost is an important factor in deciding which approach is best for a particular situation. The good news about constructed wetlands is that both the capital and operating costs are far lower than for conventional wastewater treatment plants, with the added benefit of reduced direct and indirect environmental impacts associated with materials extraction, processing, and manufacturing.

Living Machines

In addition to using constructed wetlands to treat wastewater from the built environment, nature can be brought directly into a building in order to break down the materials in the wastewater system. Although there are several approaches, the best known is the Living Machine, created by John Todd, a pioneer in the development of natural wastewater processing systems. The Living Machine differs from a conventional WWTP in four basic respects:[14]

1. The vast majority of the Living Machine's working parts are live organisms, including hundreds of species of bacteria, plants, and vertebrates such as fish and reptiles.

2. The Living Machine has the ability to design its internal ecology in relation to the energy and nutrient streams to which it is exposed.

3. The Living Machine can repair itself when damaged by toxics or when shocked by interruption of energy or nutrient sources.

4. The Living Machine can self-replicate through reproduction of the organisms in the system.

The concept of the Living Machine can be applied not only to an alternative WWTP but also to a range of other systems that can generate fuel, grow food, restore degraded environments, and even heat and cool buildings. Several successful examples of the Living Machine have been integrated into buildings. An example of a Living Machine is the one located in the Lewis Center for Environmental Studies at Oberlin College in Oberlin, Ohio, which processes wastewater from the occupants of this 14,000-square-foot (1400-square-meter) building (see Figure 10.15).

Designing the High-Performance Building Hydrologic Cycle

Designing the water, wastewater, and stormwater systems for a high-performance building is a challenging task. In general, the first objective is to minimize potable water consumption. To determine how well a given strategy is working and to meet the requirements for most green building certification schemes, a baseline model of the building water and wastewater systems is created to allow comparisons. Both the types of plumbing fixtures and the alternative sources of water (rainwater, reclaimed water, and graywater) can be varied in the baseline model to determine how much potable water has been saved. Table 10.4 shows the flow rates for flush and flow fixtures. Flush fixtures, as the names implies, are plumbing fixtures that use a fixed quantity of water for their function, while flow fixtures use a quantity of water that depends on the length of time during

Figure 10.15 The Living Machine built into the Lewis Center for Environmental Studies at Oberlin College in Oberlin, Ohio, contains biological organisms that break down wastewater components into nutrients that are then fed into a constructed wetland inside the building. (Photograph courtesy of Oberlin College)

TABLE 10.4

Water Use by Various Types of Plumbing Fixtures

Flush Fixture Type	Water Use (gpf)
Conventional low-flow water closet	1.60
High-efficiency toilet (HET), single-flush gravity	1.28
HET, single-flush pressure assist	1.00
HET, dual-flush (full-flush)	1.28
HET, dual-flush (low-flush)	1.00
HET, foam flush	0.05
Waterless toilet	0.00
Composting toilet	0.00
Conventional low-flow urinal	1.00
High-efficiency urinal (HEU)	0.50
Waterless urinal	0.00

Flow Fixture Type	Water Use (gpm)
Conventional low-flow lavatory faucet	2.20
High-efficiency lavatory faucet	1.80
Conventional low-flow kitchen sink faucet	2.20
High-efficiency kitchen sink faucet	1.80
Conventional low-flow showerhead	2.50
High-efficiency showerhead	Max 2.00
Low-flow janitor sink faucet	2.50
Low-flow handwash fountain	0.50
Conventional low-flow self-closing faucet	0.25 gallons/cycle
High-efficiency self-closing faucet	Max 0.20 gallons/cycle

TABLE 10.5

Uses per Day for Plumbing Fixtures by Gender and Type of Building Occupants

Fixture Type Gender, Duration, Application	FTE	Student Visitor	Retail Customer	Resident
Uses per Day				
Water closet				
Female	3.0	0.5	2.0	5.0
Male	1.0	0.1	0.1	5.0
Urinal	2.0	0.4	0.1	n/a
Lavatory faucet	3.0	0.5	0.2	5.0
Commercial at 15 sec, 12 sec with autocontrol; residential at 60 sec				
Shower	0.1	0.0	0.0	1.0
Commercial at 300 sec; residential at 480 sec				
Kitchen sink	1.0	0.0	0.0	4.0
Commercial at 15 sec; residential at 60 sec				

use of the fixture. The water use and flow rates shown in this table are the starting point for determining the water use of a building.

Establishing the population and occupant type is the next step in identifying a baseline. Along with the population, the female-to-male ratio must be known in order to quantify water consumption through fixture types. For projects that have either unknown ratios or ratios that are relatively the same, it is best to model with an even gender distribution. Occupant type classifies the people who use the facility's plumbing system. The largest occupant type difference is between a full-time equivalent (FTE) and a transient or someone temporarily visiting the facility. An FTE refers to a person who occupies the building in the equivalent of an eight-hour day. Table 10.5 shows the typical daily use patterns of plumbing fixtures based on gender and type of occupant.

BASELINE WATER MODEL EXAMPLE

To best understand how to generate a baseline water model, we start with a few pieces of information. As an example, we will use an academic building designed to have a total of 50 full-time male and 30 full-time female occupants. The building is assumed to have 300 transient male visitors and 200 transient female visitors per day. The baseline water model assumes that the fixtures used in the building meet the EPAct 1992 requirements for maximum flow rates for plumbing fixtures. Fixture performances are selected from Table 10.4, and the number of uses per person per day can be found in Table 10.5. Each fixture type must be modeled in order to identify a total water use (in gallons) per day. This value is determined by identifying the product of multiplying the appropriate occupant type by the number of daily uses per person and by the total water consumed per fixture use. This calculation can be found in Table 10.6. The rightmost column indicates the estimated fixture water use per day, which is then summed for both flush and flow fixtures.

To accurately model the annual water consumption from the facility, the number of workdays must be multiplied by the daily total water use; in this case, 260 days was determined. In this particular example, the total annual potable

TABLE 10.6

Example of a Baseline Water Model

Occupant Type	Flush Fixture	Daily Uses	Potable Water (gpf)	No. of Occupants	Water Use (gal)
FTE	Conventional low-flow water closet (male)	1.0	1.6	50	80
FTE	Conventional low-flow water closet (female)	3.0	1.6	30	144
FTE	Conventional low-flow urinal	2.0	1.0	50	100
Transient	Conventional low-flow water closet (male)	0.1	1.6	300	48
Transient	Conventional low-flow water closet (female)	0.5	1.6	200	160
Transient	Conventional low-flow urinal	0.4	1.0	300	120
	Total Flush Fixture Potable Water Use (gal)				*652*

Occupant Type	Flow Fixture	Daily Uses	Potable Water (gpm)	Duration (sec)	No. of Occupants	Water Use (gal)
FTE	Conventional low-flow lavatory faucet	3.0	2.2	15	80	132
FTE	Conventional low-flow kitchen sink faucet	1.0	2.2	15	80	44
FTE	Conventional low-flow showerhead	0.1	2.5	300	80	100
Transient	Conventional low-flow lavatory faucet	0.5	2.2	15	500	138
Transient	Conventional low-flow kitchen sink faucet	0.0	2.2	15	500	0
Transient	Conventional low-flow showerhead	0.0	2.5	300	500	0
	Total flow fixture potable water use (gal)					*414*
	Total daily potable water use (gal)					**1,066**
	Annual workdays					**260**
	Total annual potable water use (gal)					**277,030**
	Total annual wastewater generation (gal)					**277,030**

water use is predicted to be 277,030 gallons per year. Note that the estimated water quantity for a conventional water system can be used to estimate the wastewater quantity for the building.

USE OF LOW-FLOW FIXTURE STRATEGY

The most straightforward strategy for reducing potable water consumption in buildings is to incorporate plumbing fixtures that use significantly less water than code-compliant fixtures. For example, while a code-compliant urinal has a maximum water use of 1 gallon per flush, a HET is required to have a maximum potable water use of 0.5 gallon per flush; furthermore, a waterless urinal uses no water at all. Table 10.7 shows the same baseline water consumption calculations with the modification of installing HETs and HEUs instead of their conventional counterparts. The results of this modification indicate a water use reduction of almost half. It is now possible to determine the feasibility of such a retrofit by associating costs savings in both water consumption and wastewater treatment.

USE OF ALTERNATIVE WATER SOURCES STRATEGY

Further significant savings in potable water consumption can be achieved by substituting other suitable water sources for potable water. Table 10.8 shows the impact of incorporating rainwater catchment and graywater systems. In this case, we are assuming that the size of the rainwater harvesting system is

TABLE 10.7

Water Model for a Low-Flow Fixture Scenario

Occupant Type	Flush Fixture	Daily Uses	Potable Water (gpf)	No. of Occupants	Water Use (gal)
FTE	HET, single-flush gravity (male)	1.0	1.28	50	64
FTE	HET, single-flush gravity (female)	3.0	1.28	30	115
FTE	Waterless urinal	2.0	0.0	50	0
Transient	HET, single-flush gravity (male)	0.1	1.28	300	38
Transient	HET, single-flush gravity (female)	0.5	1.28	200	128
Transient	Waterless urinal	0.4	0.0	300	0
	Total flush fixture potable water use (gal)				*346*

Occupant Type	Flow Fixture	Daily Uses	Water Use (gpm)	Duration (sec)	No. of Occupants	Water Use (gal)
FTE	High-efficiency lavatory faucet	3.0	1.8	15	50	68
FTE	High-efficiency kitchen sink faucet	1.0	1.8	15	30	14
FTE	High-efficiency showerhead	0.1	1.8	300	50	45
Transient	High-efficiency lavatory faucet	0.5	1.8	15	300	68
Transient	High-efficiency kitchen sink faucet	0.0	1.8	15	200	0
Transient	High-efficiency showerhead	0.0	1.8	300	300	0
	Total flow fixture potable water use (gal)					*194*
	Total daily potable water use (gal)					**539**
	Annual workdays					**260**
	Total annual potable water use (gal)					**140,166**
	Total annual wastewater generation (gal)					**140,166**
	Potable water savings compared to the Baseline Model					**49.4%**

sufficient enough to supply graywater to be used for flushing toilets and urinals. By comparing the potable water consumption in this scenario to the baseline model, a water use reduction of 82 percent can be achieved.

Water Budget Rules of Thumb (Heuristics)

Based on the three models shown in the previous sections, it is now possible to develop some rules of thumb, sometimes called *heuristics*, to set targets for potable water consumption. For the low-flow fixture strategy, high-efficiency fixtures are aggressively used, and the result is about a 50 percent, or Factor 2, reduction in potable water consumption compared to code requirements. For a combination of low-flow fixtures and alternative water strategies, an 80 percent reduction in potable water consumption was achieved. Consequently, it is possible to develop water reduction strategies that are in excess of Factor 4 for an aggressive strategy that includes alternative water sources and low-flow fixtures and at least Factor 2 for a less aggressive strategy that uses a simple low-flow fixture strategy.

As mentioned earlier in this chapter, the concept of net zero water is receiving serious consideration, and it actually provides a sensible approach based on one of the core ideas of sustainability; that is, resource use should be constrained to what nature provides. If we assume the building in the three

TABLE 10.8

Water Model for a Combination of Alternative Water and Low-Flow Fixture Strategy

Occupant Type	Flush Fixture	Daily Uses	Potable Water (gpf)	No. of Occupants	Water Use (gal)
FTE	HET, single-flush gravity (male)	1.0	0.0	50	0
FTE	HET, single-flush gravity (female)	3.0	0.0	30	0
FTE	Waterless urinal	2.0	0.0	50	0
Transient	HET, single-flush gravity (male)	0.1	0.0	300	0
Transient	HET, single-flush gravity (female)	0.5	0.0	200	0
Transient	Waterless urinal	0.4	0.0	300	0
	Total flush fixture potable water use (gal)				*0*

Occupant Type	Flow Fixture	Daily Uses	Potable Water (gpm)	Duration (sec)	No. of Occupants	Water Use (gal)
FTE	High-efficiency lavatory faucet	3.0	1.8	15	50	68
FTE	High-efficiency kitchen sink faucet	1.0	1.8	15	30	14
FTE	High-efficiency showerhead	0.1	1.8	300	50	45
Transient	High-efficiency lavatory faucet	0.5	1.8	15	300	68
Transient	High-efficiency kitchen sink faucet	0.0	1.8	15	200	0
Transient	High-efficiency showerhead	0.0	1.8	300	300	0
	Total flow fixture potable water use (gal)					*194*
	Total daily potable water use (gal)					**194**
	Annual workdays					**260**
	Total annual potable water use (gal)					**50,310**
	Total annual wastewater generation (gal)					**50,310**
	Potable water savings compared to the Baseline Model					**81.8%**

water models in the preceding sections were located in a climate zone with 24 inches of annual rainfall and that the facility had a roof area of 15,000 square feet, then 30,000 cubic feet, or 224,000 gallons, of water would be available for all uses. The baseline model shows that 277,030 gallons are required, and the result is that a net zero water strategy would require about a 20 percent reduction in water consumption, which matches up to using high-efficiency WaterSense fixtures throughout the facility.

Sustainable Stormwater Management

Stormwater management has long been a challenging issue for built environment development. Replacing plants and trees that naturally uptake large quantities of water with buildings and covering porous soils with impermeable surfaces result in large quantities of water flowing horizontally across parking and paving and picking up particles and chemicals along the way. The result has been an enormous headache for municipalities that then have to build large stormwater management facilities at high cost to taxpayers and with additional costs to the environment. Water supplies are threatened by polluted stormwater, and the health of ecosystems into which the stormwater is discharged is often compromised.

One of the results of adopting sustainable construction approaches has been the emergence of innovative, effective schemes that attempt to maintain the

Figure 10.16 (A) Disconnecting downspouts from storm sewer systems prevents roof runoff from overloading these systems by dispersing it to vegetated areas. (*Source:* City of Gresham, Oregon)

Figure 10.16 (B) Rain barrels collect roof runoff and store it for later nonpotable use. (Maxine Thomas, Florida Master Gardener/University of Florida–IFAS Extension Realtor®)

Figure 10.16 (C) Cisterns are similar to rain barrels except they are more permanent and constructed with more durable material. They can be installed above or beneath the ground, with sizes ranging from 100 to 10,000 gallons. (*Source:* City of Portland, Oregon)

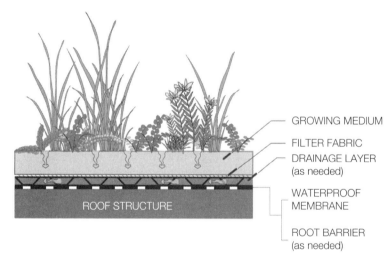

Figure 10.16 (D) Eco-roofs are an extensive green roof system. These roofs are typically constructed with layers of waterproof membrane, drainage material, and a lightweight soil and planted with shallow-root plant material. This application is appropriate for conventional roofs that are flat or low-sloped. (*Source:* City of Portland, Oregon)

natural hydrology of the area. Sometimes referred to as *sustainable stormwater management*, this strategy, according to the Portland Bureau of Environmental Services, " . . . mimics nature by integrating stormwater into building and site development to reduce the damaging effects of urbanization on rivers and streams. Disconnecting the flow from storm sewers and directing runoff to natural systems like landscaped planters, swales and rain gardens or implementing an ecoroof reduces and filters stormwater runoff."[15]

Sustainable stormwater management recognizes that there is a relationship between the natural and built environments and treats them as integrated

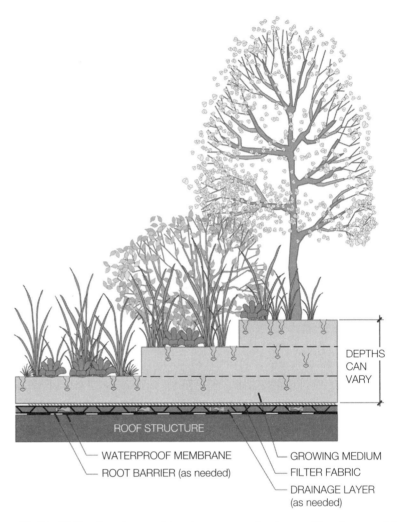

DEPTHS
CAN
VARY

ROOF STRUCTURE

WATERPROOF MEMBRANE

ROOT BARRIER (as needed)

GROWING MEDIUM

FILTER FABRIC

DRAINAGE LAYER
(as needed)

Figure 10.16   (E) Roof gardens are intensive green roof systems, with a deeper soil layer that allows for deeper-rooted and thus larger plant material than extensive systems. Some green roofs will have access points and walkways for occupants to enjoy. (*Source:* City of Portland, Oregon)

components of the watershed. Instead of the traditional approach of using piping and extensive and expensive collection systems, it focuses on on-site collection and conveyance of stormwater from roofs, parking lots, streets, and other services, to promote the infiltration of water into the ground. Vegetated natural systems slow and filter the water and enhance the intersection and evaporation of rainfall through their leaves and roots. Vegetation also reduces stormwater runoff and removes pollutants in the process. Studies have shown that this approach can reduce stormwater runoff volume by as much as 65 percent. It can also remove 80 percent of suspended solids and heavy metals, and as much as 70 percent of nutrients such as phosphate and nitrogen.

Sustainable stormwater management integrates natural components such as landscape swales and infiltration basins with structural devices such as cisterns, planters, pervious pavers, and pervious concrete and asphalt surfaces. Figure 10.16A–N illustrates some of the components that may be part of a sustainable stormwater management system.

Pervious asphalt and pervious concrete are particularly interesting materials because they allow stormwater to rapidly infiltrate through the hard

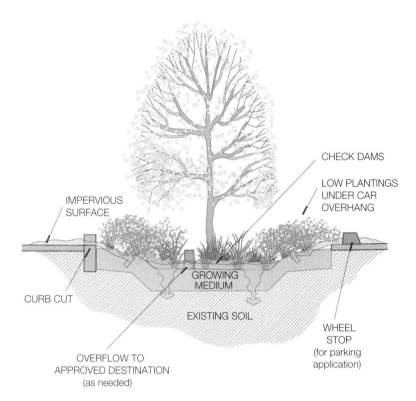

Figure 10.16 (F) Vegetated swales, or bioswales, are gently sloping depressions planted with dense vegetation or grass to divert and treat stormwater runoff. The plant material slows and filters the water as it seeps into the ground. (*Source:* City of Portland, Oregon)

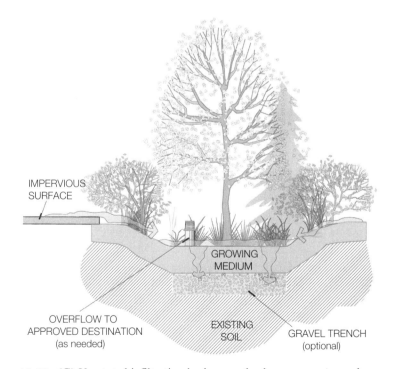

Figure 10.16 (G) Vegetated infiltration basins are also known as *rain gardens* or *detention ponds.* The basin is either excavated or created with berms, then landscaped to temporarily store runoff until it infiltrates into the ground. These designs temporarily detain water during a large storm and usually incorporate an overflow system for safety purposes. (*Source:* City of Portland, Oregon)

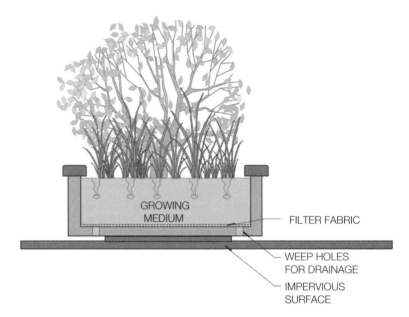

Figure 10.16 (H) Contained planters are filled with soil and plants that absorb fallen rainwater. Excess water infiltrates to the bottom of the planter and drains through weep holes. (*Source:* City of Portland, Oregon)

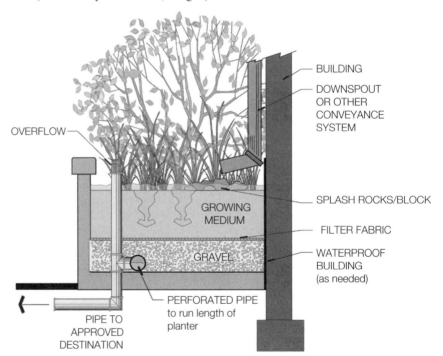

Figure 10.16 (I) Flow-through planters are used in areas where the water cannot infiltrate into the ground. They are sealed and filled with gravel, soil, and vegetation to absorb and filter the rainwater. Excess water escapes through a perforated pipe located at the bottom of the planter or to an overflow system. (*Source:* City of Portland, Oregon)

exterior surface and then into the ground. Pervious asphalt consists of coarse stone aggregate and asphalt binder, with very little fine aggregate. Water percolates through the voids caused by the absence of fine aggregates. A thick layer of gravel underneath allows water to drain quickly through the surface. Pervious asphalt is similar to conventional asphalt, although with a rougher service,

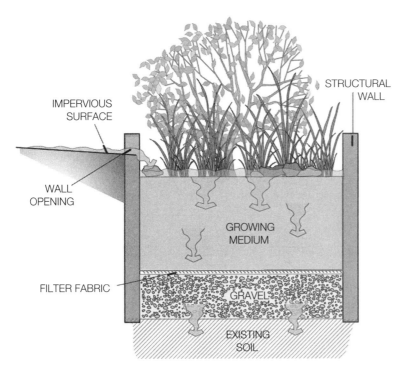

Figure 10.16 (J) Infiltration planters have open bottoms to allow stormwater to collect in the topsoil and slowly infiltrate into the ground. Materials and sizes range depending on the application. (*Source:* City of Portland, Oregon)

Figure 10.16 (K) Pervious pavers or unit pavers replace impervious surfaces and allow stormwater to soak into the ground. They typically are made out of precast concrete, brick, stone, or cobbles and form interlocking patterns with the gaps filled with either sand or gravel. (Photograph courtesy of Holly Piza)

Figure 10.16 (L) Pervious pavement is made from concrete or asphalt, with coarse aggregates that create air voids, allowing water to pass through the system. (*Source:* City of Fairway, Kansas)

Figure 10.16 (M) Turf block, also known as *grass grid* or *open-cell unit pavers*, has gaps filled with soil and grasses allowing water to pass through. This option only accepts precipitate and not stormwater runoff; thus, suited for low traffic and infrequent car parking. Applications include patios, walkways, emergency access roads, street shoulders, and residential driveways. (Photograph courtesy of Western Interlock, Inc.)

which accounts for its name, "popcorn mix." Pervious concrete consists of specially formulated mixtures of portland cement, coarse aggregate, and water. Owing to the absence of fine aggregate, it has enough void space to allow the rapid percolation of water. Due to the lack of fine aggregates. the pervious concrete has a rough surface and resembles exposed aggregate concrete.

Applying stormwater management strategies (see Figure 10.17) from a tool chest of available technologies provides the best solution to minimize stress on water treatment plants and maximize groundwater and aquifer recharge.[16]

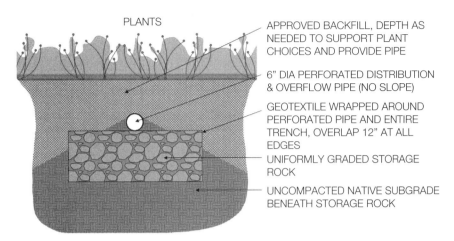

Figure 10.16   (N) Soakage trenches, or infiltration trenches, are shallow trenches lined with perforated pipe. The pipe collects rainwater from roofs or other impervious surfaces and disperses it underground to the backfill material in which it can infiltrate to the ground. (D. Stephany)

Figure 10.17   Several components are integrated into this sustainable stormwater management system in Portland, Oregon. All constraints and benefits are considered when selecting which components to use for any given site. (Source: City of Portland, Oregon)

Landscaping Water Efficiency

Approximately 30 percent of residential water use, or about 32 gallons (121 liters) per person per day, is used for exterior uses; and the bulk of this, as much as 29 gallons (110 liters) per person per day, is used for maintaining landscaping, with wide variations depending on the climatic region. Most of the water used for this purpose is wasted due to overwatering. Water-intensive turfgrass creates the major demand for irrigation. In the United States, more than 16,000 golf courses consume 2.7 billion gallons (10.2 billion liters) of water per day.[17]

Several forms of sustainable landscaping are emerging after several decades of evolution. The best known is *xeriscaping*, which emphasizes the use of drought-tolerant native and adapted species of plants and turfgrass. (Note that the terms *enviroscaping* and *water-wise landscaping* are sometimes used interchangeably with *xeriscaping*.) Seven principles can be used to ensure a well-designed, water-efficient landscape:

1. Proper planning and design
2. Soil analysis
3. Appropriate plant selection
4. Practical turfgrass areas
5. Efficient irrigation
6. Use of mulches
7. Appropriate maintenance

Perhaps an even more sustainable form of landscaping than xeriscaping is *natural* or *native landscaping*. Using restorative landscaping principles, natural landscaping supports the use of indigenous plants that, once established, virtually eliminate the need for watering. Even turfgrass, the most ubiquitous consumer of water, can be replaced with indigenous species because there are thousands of native species in the United States. The restoration of native landscapes has other benefits as well. Animal species that live in native landscapes are reestablished, natural landscapes filter stormwater effectively, and the natural beauty of the landscape is restored. In 1981, Darrel Morrison, a professor at the University of Georgia and a member of the American Society of Landscape Architects (ASLA), defined three characteristics necessary for natural landscape design:[18]

1. Regional identity (sense of place)
2. Intricacy and detail (biodiversity)
3. Elements of change

Opposition to natural landscaping was initially strong because many people, after having grown accustomed to manicured turfgrass lawns, had difficulty accepting landscaping that appeared wild and unconventional. In fact, numerous people were prosecuted for attempting to implement natural landscaping; they were accused of violating weed laws. Fortunately, natural landscaping is now far more widely accepted, and the beauty and aesthetics of this approach are winning over most skeptics. Natural landscaping can include butterfly gardens, native trees and shrubs that attract birds, small ponds, native groundcovers in lieu of turfgrass, and gardens composed of native plants. Native plants have several environmental advantages that fit in with the concept of a high-performance green building: they survive

Figure 10.18   The Environmental Nature Center in Newport Beach, California, is landscaped with a water-saving design and a diverse selection of native species. Selection of native plant material in the landscape requires little or no irrigation and helps to maintain not only the ecosystem but also the indigenous character of the site. (LPA, Inc./ Costea Photography, Inc.).

without fertilizers or synthetic pesticides and rarely need watering, they provide food and habitat for wildlife, and they contribute to biodiversity (see Figure 10.18).

Case Study: LOTT Clean Water Alliance, Olympia, Washington

The new Regional Services Center for Olympia's wastewater treatment facility fosters active engagement of the public in the wastewater treatment process. This multiuse facility contains water quality laboratories and offices, as well as an educational and technology center (see Figure 10.19A–D). One of the goals of the facility is to create a strong community outreach program emphasizing water conservation while providing the highest-quality reclaimed water to four counties with a population totaling approximately 85,000. Visitors to the facility are quickly surrounded by water being processed as they approach the building—a design that promotes community education at many levels, from the reclaimed water in the front plaza to the hands-on children's museum. Once inside, the technology center continues to communicate the importance of water and the process by which the facility meets the demands of the region.

The success of this project can be attributed to the early collaborative effort among the owner, design team, construction manager, facilities and management staff, and other stakeholders. The project goals were clearly identified and communicated to everyone involved with the project. Some of the high-performance features of this building include polished concrete floors, a former brownfield site, the use of reclaimed timber, daylighting of offices, and louver shading to reduce solar gain.

Figure 10.19 (A) The new Regional Services Center for Olympia's LOTT Clean Water Alliance stands as an icon in the neighborhood and welcomes the public for hands-on water treatment education. A reclaimed water pond edges two sides of the building, enticing visitors to become engaged in the process. (© Nic Lehoux)

Figure 10.19 (B) Visitors enter the building by way of a bridge, which puts them in close contact with the center's reclaimed water pond. (© Nic Lehoux)

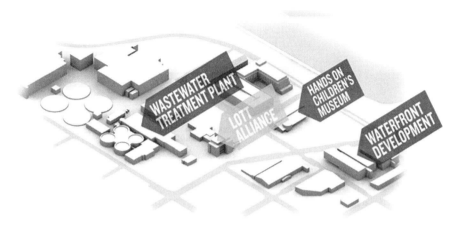

Figure 10.19   (C) The administrative offices of the LOTT Clean Water Alliance are situated in the center of the region's wastewater treatment facility. The public is welcomed and frequents the offices, technology center, and adjacent children's museum to gain a hands-on understanding of water issues. (Susan Kelly)

Figure 10.19   (D) Daylighting of the offices and the use of reclaimed timber are just two examples of the many sustainable features of the building's design. (© Nic Lehoux)

Summary and Conclusions

Much of the attention to high-performance green building design has focused on superior energy performance because there are demonstrable, easy-to-document savings that can be used to justify investments in energy conservation. But for the building hydrologic system, the savings for water conservation and innovative handling of wastewater are not so easy to document because, in the United States, water has been a heavily subsidized resource, as has been the treatment of wastewater effluent and stormwater. However, it is water, not energy, that can be the limiting resource for development as demonstrated by several growth moratoriums that have been imposed to limit or stop development and construction activity until the shortage of water or lack of an adequate wastewater treatment system can be resolved.

Water is, in fact, such a critical issue that project teams in many areas of the United States should consider making extraordinary efforts to reduce potable water consumption to exceptionally low levels. Extensive recent experience has shown that the use of rainwater harvesting, reclaimed water, graywater systems, and new waterless fixture technologies are eliminating the need for water use in urinals. HETs are also available, requiring only about one-half of the water needed by toilets meeting current plumbing codes. One area where progress still needs to be made is landscape irrigation where about 50 percent of the total potable water for the built environment is consumed.

If the construction industry does not make significant reductions in the consumptive water profile of the built environment, growth moratoriums, often instituted because of water or wastewater limitations, will reduce the volume of its business. It is now apparent that finding more appropriate ways of using potable water and treating wastewater will result in a win-win situation for both the public and the construction industry.

Notes

1. The resolution affirming the human right to water was passed by the UN General Assembly on July 28, 2010; the text of the resolution can be found at www.un.org/News/Press/docs/2010/ga10967.doc.htm.
2. The title of the McKinsey report is *Charting Our Water Future: Economic Frameworks to Inform Decision-Making* (2009). It was produced on behalf of the 2030 Water Resources Group, an alliance of concerned bodies, including the World Bank Group, and private interests such as the Coca-Cola Company, SAB Miller, and Standard Chartered Bank.
3. From the Letters to the Editor section of the online version (www.elytimes.com) of the *Ely Times*, April 7, 2004.
4. From the January 31, 2004, online edition of Emmitsburg.net (http://emmitsburg.net), a nonprofit Internet source for information about the Emmitsburg area.
5. The US Geological Survey publishes a detailed report on US water consumption at five-year intervals for a time frame five years earlier. The title of the 2010 report is *Estimated Water Use in the United States in 2005* and can be found at pubs.usgs.gov/circ/1344/.
6. Hawken, Lovins, and Lovins (1999), chapter 11, describe the faulty logic of contemporary building water and wastewater systems and suggest remedies that can ensure the sustainability of the world's potable water supply.
7. Tables 10.1 and 10.2 are from the *Fundamentals eBook* at www.physicalgeography.net/fundamentals/contents.html.
8. RMI is a nonprofit organization that provides a wide range of consulting services on energy, water, development, and green building issues. It provides many valuable resources at www.rmi.org.
9. The online publication *Toward Net Zero Water: Best Management Practices for Decentralized Sourcing and Treatment*, by Sisolak and Spataro (2011), provides current information on net zero water approaches in the United States.
10. The description of alternative water systems is from sections 6.1–6.6 of *Greening Federal Facilities* (2001).
11. *The Texas Guide to Rainwater Harvesting* (2005) provides an excellent overview of rainwater harvesting and addresses health issues, materials, and safety concerns in a succinct, informative manner.
12. In the early 1970s, the EPA began investigating alternatives to centralized, technically complex WWTPs; part of this effort was the creation of an alternative technology program to encourage the development of systems that employ ecological systems to break down their own waste. This program and the advantages noted here are discussed by Campbell and Ogden (1999).

13. An excellent summary of the state of the art of constructed wetlands technology can be found in Lorion (2001).
14. Todd (1999) describes the concept of the Living Machine in great detail in chapter 8 of *Reshaping the Built Environment*.
15. The definition of sustainable stormwater management is from "A Sustainable Guide to Stormwater Management" and can be found on the Portland Bureau of Environmental Services website at www.portlandonline.com/bes/index.cfm?c=34598. Portland, Oregon, has many award-winning sustainable stormwater projects and its website contains a wide range of resources for supporting sustainable stormwater management.
16. Adapted from the *Stormwater Solutions Handbook* developed by the Bureau of Environmental Services of the city of Portland, Oregon (2011).
17. A detailed explanation of landscape water conservation practices can be found in Vickers (2001), chapter 3. Information in this section is from this source, which also contains extensive information about water conservation in general, and technical and policy information on the subject of reducing potable water use.
18. An excellent source of information about natural landscaping is the nonprofit organization Wild Ones: Native Plants, Natural Landscapes. Information and free downloads are available at its website, www.for-wild.org.

References

Ausubel, Kenny. 1997. *The Bioneers: A Declaration of Interdependence.* White River Junction, VT: Chelsea Green Publishing.

Campbell, Craig S., and Michael H. Ogden. 1999. *Constructed Wetlands in the Sustainable Landscape.* Hoboken, NJ: John Wiley & Sons.

"Graywater Guide." 1995. Sacramento, CA: California Department of Water Resources, Publications Office; www.dwr.water.ca.gov.

Greening Federal Facilities: An Energy, Environmental and Economic Resource Guide for Federal Facility Managers and Designers, 2nd ed. 2001. DOE/GO-102001-1165. Washington, DC: Federal Energy Management Program, Office of Energy Efficiency and Renewable Energy, US Department of Energy. Available at www.nrel.gov/docs/fy01osti/29267.pdf /.

Hawken, Paul, Amory Lovins, and L. Hunter Lovins. 1999. *Natural Capitalism: Creating the Next Industrial Revolution.* Boston: Little, Brown.

Kibert, Charles J., ed. 1999. *Reshaping the Built Environment: Ecology, Ethics, and Economics.* Washington, DC: Island Press.

Lorion, Renee. 2001. "Constructed Wetlands: Passive Systems for Wastewater Treatment." Technology Status Report for the US Environmental Protection Agency's Technology Innovation Office. Available at www.epa.gov/swertio1/download/remed/constructed_wetlands.pdf.

Ludwig, Art. 1999. *Builder's Greywater Guide.* Santa Barbara, CA: Oasis Design; www.oasisdesign.net/.

McKinsey & Company. 2009. *Charting Our Water Future: Economic Frameworks to Inform Decision-Making.* Available at www.mckinsey.com/App_Media/Reports/Water/Charting_Our_Water_Future_Exec%20Summary_001.pdf.

Sisolak, Joel, and Kate Spataro. 2011. *Toward Net Zero Water: Best Management Practices for Decentralized Sourcing and Treatment.* Cascadia Green Building Council. Available at ilbi.org/education/Resources-Documents/Reports-Docs/WaterDocs/toward-net-zero-water-report.

Stormwater Solutions Handbook. 2011. City of Portland, Oregon. Available at www.portlandonline.com/bes/index.cfm?c=43110.

Texas Guide to Rainwater Harvesting, 3rd ed. 2005. Austin, TX: Texas Water Development Board, in collaboration with the Center for Maximum Potential Building Systems. Available at www.twdb.state.tx.us/publications/reports/rainwaterharvestingmanual_3rdedition.pdf or from the Texas Water Development Board.

Todd, John. 1999. In *Reshaping the Built Environment: Ecology, Ethics, and Economics*, edited by Charles J. Kibert. Washington, DC: Island Press.

Vickers, Amy. 2001. *Handbook of Water Use and Conservation.* Amherst, MA: WaterPlow Press. Book details available at www.waterplowpress.com.

Chapter 11
Closing Materials Loops

The selection of materials and products for high-performance green building projects has historically been a major challenge for project teams. The characteristics that make materials and products acceptable for application in high-performance buildings include high recycled content, reused building materials, locally and regionally available materials, certified wood products, and wood products made from rapidly renewable resources. However, coming to a common understanding of how to prioritize and combine these attributes into a decision system for product selection has been lacking. The good news is that significant progress is now being made in crafting a widely accepted approach for determining the environmental efficacy of materials and products used in construction. The advent of *environmental product declarations* (EPDs) and *environmental building declarations* (EBDs) promises to ease past problems of determining the impacts of both products and whole buildings based on a commonly accepted approach. In short, an EPD is the equivalent of a nutrition label for products and materials and is issued by independent third-party organizations that ensure uniformity and transparency in the process. An EBD can be considered to be the sum total of EPDs for all the products and materials in a building and represents its total impact. Some building rating systems, notably the Deutsche Gesellschaft für Nachhaltiges Bauen (DGNB) certification, include whole-building impact assessment as part of the scoring system for the certification. This process has historically been very difficult to develop because it is dependent on information supplied by independent third-party organizations and because it is very data intensive. Additionally, for what they perceive as competitive reasons, manufacturers have historically been resistant to participating in a transparent scheme of product declarations. However, the old days of manufacturers being unwilling to provide information to third-party certification organizations is drawing to a close as a competitive atmosphere created by early adopters is forcing others to share their information as part of the overall green building process. Noteworthy among the early adopters is InterfaceFLOR Corporation, a manufacturer of carpet tiles, which pledged to obtain third-party-validated EPDs on all InterfaceFLOR products globally by 2012. Ultimately, at a point in time when EPDs are available for all products that comprise a building, including complex items such as air handlers and lighting systems, then EBDs, or whole-building declarations, will be possible. At present, EPDs simply allow a comparison between products being used for the same purpose, for example, steel versus concrete structural systems. Whole-building declarations, or EBDs, will allow trade-offs between systems in order to minimize total impact. For example, the impacts of significant additional insulation and triple-pane, gas-filled windows can be compared to the effects of reducing the size and mass of heating, ventilation, and air-conditioning (HVAC) components, thus supporting a holistic approach to decision making about products and materials.

In addition to the advent of EPDs and EBDs, another significant trend is the emergence of independent organizations that produce standards, conduct testing, and provide important input to building assessment systems. These third-party organizations include Green Seal, GreenGuard, and the Forest Stewardship Council. They provide the equivalent of environmental labels, or ecolabels, for products that meet the requirements of green building rating systems. Meeting their standards is often mentioned as a requirement for gaining points toward building certification for the US building assessment systems. A second tier of organizations that are industry-related, or "second party," are contributing to the atmosphere of openness and to easing the selection process for a wide variety of materials used in construction. These organizations include the Carpet and Rug Institute (CRI) and the Sustainable Forestry Initiative (SFI). Although not without their critics due to their industry ties, many of these groups provide a useful service while at the same time setting high standards.

Another issue that is highly important when discussing the closing of materials loops is the fate of building material at the end of service life for a facility. Building assessment systems address the recycling and reuse of materials from building demolition (see Figure 11.1) and the reduction of waste in the construction process.

This chapter addresses the issues of green building materials and products, the criteria for defining environmentally friendly products, the application of life-cycle assessment (LCA) in decision making for materials selection, and the subject of EPDs and their role in high-performance green building design and construction. The role of third-party standards organizations in high-performance building assessment and certification is addressed. Finally, this chapter provides information about specific materials and product groups where new technologies and approaches are beginning to take hold in support of the green building movement.

Figure 11.1 Partial demolition of the Levin College of Law library at the University of Florida in Gainesville for a building expansion project. Truly green buildings of the future should be designed for deconstruction to maximize the reuse and recovery of building components and materials. (Photograph courtesy of M. R. Moretti)

The Challenge of Materials and Product Selection

The selection of building materials and products for a high-performance green building project has historically been the most difficult and challenging task facing the project team. In the third edition of *Green Building Materials: A Guide to Product Selection and Specification*, one of the first books about the subject of green building materials, Ross Spiegel and Dru Meadows defined green building materials as "those that use the Earth's resources in an environmentally responsible way."[1] At present, however, there is no clear consensus about the criteria for materials and products that would characterize them as *environmentally preferable, environmentally responsible*, or *green*. As a matter of fact, alternative terminologies are rapidly infiltrating the language of high-performance building green materials and products. For example, the label *environmentally preferable products* (EPPs) is commonly used and can be found in US government specifications for building materials and products. As a result, the question of what is or is not environmentally preferable is still being settled and is still open to controversy. For example, some organizations promote green products based on a narrow range of attributes they specify as being important for this purpose. The Forest Stewardship Council (FSC), represented in the United States by the SmartWood Program and Scientific Certification Systems, defines green products as wood products derived from a sustainably managed forest. The Greenguard Environmental Institute instead relies on levels of chemical emissions that affect indoor environmental quality (IEQ) to describe what constitutes a green product.[2] As noted above, the advent of EPDs is changing this situation somewhat by providing third-party-verified information on the impacts of products using a transparent process.

Clearly, it would be advantageous for green products to carry a certification, or ecolabel, to designate them as being preferable on the basis of consensus standards that address each type of building product. EPDs, while providing detailed information about products, are not certifications attesting to the environmental friendliness or "greenness" of a product. Ecolabels, in contrast, designate the superior performers in a given class of products. In Europe, several ecolabels cover at least some building materials. The Blue Angel ecolabel in Germany, the Nordic Swan ecolabel of the Nordic countries, and the European Union ecolabel all have programs for labeling some types of building products. For example, the Blue Angel Standard RAL-UZ-38 addresses the requirements for certification of wood panels.[3] Unfortunately, the range of products covered by these labeling programs is very limited; consequently, they provide minimal assistance in identifying those that might be considered green. Thus, the project team must rely on their own best judgment in deciding which materials fit the criteria for environmental friendliness.

On the positive side, several tools are available to assist this process, the most familiar being LCA. LCA provides information about the resources, emissions, and other impacts resulting from the life cycle of materials use, from extraction through disposal, and incorporates a high degree of rigor and science in the evaluation process. LCAs are also important because they are the tool used in crafting EPDs, which will likely become the commonly accepted approach for comparing products in the decision-making process. Two readily available LCA programs, Athena[4] and Building for Environmental and Economic Sustainability (BEES),[5] apply to North American projects and can provide the project team with a decision system for materials selection that is based on science. These are covered later in this chapter.

ISSUES IN SELECTING GREEN BUILDING MATERIALS AND PRODUCTS

As discussed above, determining how building materials and products will affect the environment is the central unresolved problem of the green building movement. Even evaluating the relative worth of using recycled versus virgin materials—which should be a relatively simple matter—can result in controversy. One school of thought, here referred to as the *ecological school*, maintains that keeping materials in productive use, as in an ecological system, is of primary importance, and that the energy and other resources needed to feed the recycling system are of secondary importance. Nature, after all, does not use energy *efficiently*, but it does employ it *effectively*; that is, it matches the energy needed to the available energy sources. Another school of thought, here referred to as the *LCA school*, suggests that if the energy and the emissions due to energy production are higher for recycling than for the use of virgin materials, then virgin materials should be used. The LCA school also generally contends that too much attention is given to solid waste and that greater emphasis should be put on climate change.[6]

Nothing, in fact, is obvious when it comes to using renewable resources in construction. Consider wood from old-growth forests. Although these forests are certainly a renewable resource, extracting resources from them is generally frowned upon by environmental groups, and the green building movement is in favor of protecting the biodiversity of these beautiful and increasingly rare natural assets. Rather, it is generally agreed, wood should come from plantation forests and, even better, from rapidly renewable species. The US Green Building Council (USGBC) Leadership in Energy and Environmental Design (LEED) standard defines a class of materials known as *rapidly renewable resources*, which are species with a growth and harvest cycle of 10 years or less. However, in spite of this strategy to shift extraction from old-growth forests to other resources, plantation forestry, which produces rapidly renewable resources, must be called into question, because it can require large quantities of water, fertilizer, pesticides, and herbicides to support the rapid growth cycle and protect the company's financial investment, not to mention that monoculture forestry runs counter to the notion of biodiversity. The definition of *rapidly renewable* as 10 years or less is itself arbitrary, and any number of other definitions are equally applicable.

Besides determining which materials are environmentally preferable or green, one must decide which products or materials will have low environmental impact. Many building products are selected to help reduce the overall environmental impact of the building, not for their own low environmental impacts. Using an energy recovery ventilator (ERV), for example, a relatively complex device containing desiccants, insulation, wiring, an electric motor, controls, and other materials, contributes to an exceptionally low energy profile for the building, but it cannot be considered inherently green because its constituent materials cannot be readily recycled. Today, one of the greatest challenges in designing a high-performance green building is selecting materials and products that lower the overall impact of the building, including the impact on its site. As time progresses, a hoped-for outcome is the development of more products that both have a low environmental impact and are inherently green—that is, can be disassembled into their recyclable constituent materials.

Distinguishing between Green Building Products and Green Building Materials

The terms used to refer to the materials and products used in high-performance building can be contradictory and confusing. *Green building products* generally refer to building components that have any of a wide range of attributes that make them preferable to the alternatives. For example, low-emissivity (low-E) glass is a spectrally selective type of glass that allows visible light to pass through but rejects a substantial part of the heat-producing infrared portion of the light spectrum. As a product, it is preferable to ordinary float glass in windows because of its energy performance. *Green building materials* refer to basic materials that may be the components of products or used in a stand-alone manner in a building. Green building materials have low environmental impacts compared to the alternatives. As noted earlier, an example of a classic green building material is wood products certified by the FSC as having been grown using sustainable forestry practices. Wood is a renewable resource, the forest is managed to produce wood at a replenishable rate, and the biodiversity of the local ecosystems is protected. In short, wood meets all the criteria for a green building material as a raw input to the production process. However, the processing of the sustainably harvested wood may produce significant waste, requires large quantities of energy and water, and may contribute to the degradation of the environment. Consequently, although the raw material may be ideal from an environmental point of view, the entire life cycle must be considered to fully assess the environmental performance of a product.

The point is, depending on how they are defined, green building products may not even be made of green building materials. For example, the glass in the low-E window may be difficult or impossible to recycle because of the films utilized to provide spectral selectivity, which are glued to the glass. In contrast, ordinary float glass can be readily recycled; therefore, with respect to materials, it may be considered greener than the low-E product. This example illustrates the complexity of the product and materials selection process for high-performance buildings.

GREEN BUILDING MATERIALS

The basic materials of construction and construction products have changed over time from relatively simple, locally available, natural, minimally processed resources to a combination of synthetic and largely engineered products, especially for commercial and institutional buildings. Vernacular architecture—design rooted in the building's location—evolved to take advantage of local resources such as wood, rock, and a few low-technology products made of metals and glass. Today's buildings are made from a far wider variety of materials, including polymers, composite materials, and metal alloys. A side effect of these evolving building practices and materials technology is that neither buildings nor the products that comprise them can be readily disassembled and recycled. There is some controversy over the relative merits of materials from natural resources versus those of synthetic materials made from a wide variety of materials, some of which do not even exist in nature. Most ecologists would, in fact, agree that there is nothing fundamentally wrong with synthetic materials. For example, it could be argued that recyclable plastics can be more environmentally friendly than cotton, whose cultivation requires large quantities of energy, water, pesticides, herbicides, and fertilizer. Nonetheless, debate continues in the contemporary green building movement about the efficacy of synthetic materials versus materials derived from nature.

GREEN BUILDING PRODUCTS

A basic philosophical approach to selecting materials for building design is sorely lacking in today's green building movement. Consequently, there are many different schools of thought, many approaches, and abundant controversy. It is not obvious, for example, that building products made from postcommercial, postindustrial, or postagricultural waste are, in fact, green. Many of the current green building products contain recycled content from these various sources.

To shed light on this topic, this section describes three philosophies or points of view about what constitutes a green building product: the Natural Step, the Cardinal Rules for a Closed-Loop Building Materials Strategy, and a pragmatic approach suggested by *Environmental Building News.*

The Natural Step and Construction Materials

One philosophical approach to designing the built environment is to use the well-known Natural Step, a tool developed to assess sustainability, as guidance for materials, product, and building design. The Natural Step, which is based on four scientifically based "system conditions," was developed in the 1980s by Dr. Karl-Henrik Robèrt, a Swedish oncologist. These conditions are as follows:[7]

1. In order for a society to be sustainable, nature's functions and diversity are not systematically subjected to increasing concentrations of substances extracted from the Earth's crust. In a sustainable society, human activities such as the burning of fossil fuels and the mining of metals and minerals will not occur at a rate that causes them to increase systematically in the ecosphere. There are thresholds beyond which living organisms and ecosystems are adversely affected by increases in substances from the Earth's crust. Problems may include an increase in greenhouse gases leading to global climate change, contamination of surface water and groundwater, and metal toxicity, which can cause functional disturbances in animals. In practical terms, the first condition requires society to implement comprehensive metal and mineral recycling programs and decrease economic dependence on fossil fuels.

2. In order for a society to be sustainable, nature's functions and diversity are not systematically subjected to increasing concentrations of substances produced by society. In a sustainable society, humans will avoid generating systematic increases in persistent substances such as DDT (dichlorodiphenyltrichloroethane), PCBs (polychlorinated biphenyls), and freon. Synthetic organic compounds such as DDT and PCBs can remain in the environment for many years, bioaccumulating in the tissue of organisms, causing profound deleterious effects on predators in the upper levels of the food chain. Freon and other ozone-depleting compounds may increase the risk of cancer due to added ultraviolet radiation in the troposphere. Society needs to find ways to reduce economic dependence on persistent human-made substances.

3. In order for a society to be sustainable, nature's functions and diversity are not systematically impoverished by overharvesting or other forms of ecosystem manipulation. In a sustainable society, humans will avoid taking more from the biosphere than can be replenished by natural systems. In addition, they will avoid systematically encroaching upon nature by destroying the habitat of other species. Biodiversity, which includes the great variety of animals and plants found in nature, provides the foundation for ecosystem services that are necessary to sustain

life on this planet. Society's health and prosperity depend on the enduring capacity of nature to renew itself and rebuild waste into resources.

4. In a sustainable society, resources are used fairly and efficiently in order to meet basic human needs globally. Meeting this System Condition is a way to avoid violating the first three System Conditions for sustainability. Considering the human enterprise as a whole, we need to be efficient with regard to resource use and waste generation in order to be sustainable. If 1 billion people lack adequate nutrition while another 1 billion have more than they need, there is a lack of fairness with regard to meeting basic human needs. Achieving greater fairness is essential for social stability and the cooperation needed for making large-scale changes within the framework laid out by the first three System Conditions. To achieve this fourth System Condition, humanity must strive to improve technical and organizational efficiency around the world and to use fewer resources, especially in affluent areas. System Condition 4 implies an improved means of addressing human population growth. If the total resource throughput of the global human population continues to increase, it will be increasingly difficult to meet basic human needs, as human-driven processes intended to fulfill human needs and wants are systematically degrading the collective capacity of the Earth's ecosystems to meet these demands.

Applying the system conditions to new building construction, with a particular focus on building materials, produces a matrix, as shown in Table 11.1.[8] The matrix indicates the relationship between the system conditions and the various major types of materials used or generated in construction: durables, consumables, and solid waste. It also shows which system conditions are violated when contemporary practices are used.

TABLE 11.1

Violation of Natural Step Conditions in the Application of Construction Materials

Item	Violation Examples	1	2	3	4
Item	Use of less abundant mined metals and minerals (copper, chromium, titanium)		X		X
	Use of heavy metals (mercury, lead, cadmium)				
Durables	Use of persistent synthetic materials [polyvinyl chloride (PVC), hydrochlorofluorocarbon (HCFC), formaldehyde]	X	X	X	X
	Use of wood from rainforests and old-growth timber that is harvested unsustainably				
	Use of petroleum-based products (solvents, oils, plastic film)				
Consumables	Excessive packaging and other disposables		X	X	X
Solid Waste	Landfill disposal of construction and demolition waste, including toxic components such as lead and asbestos	X	X	X	X

In practical terms, applying the Natural Step to the employment of building materials would result in the following materials practices:[9]

1. All materials are nonpersistent and nontoxic and procured either from reused, recycled, renewable, or abundant (in nature) sources.
   a. *Reused* means reused or remanufactured in the same form, such as remilled lumber, in a sustainable way.
   b. *Recycled* means that the product is 100 percent recycled and can be recycled again in a closed loop in a sustainable way.
   c. *Renewable* means able to regenerate in the same form at a rate greater than the rate of consumption.
   d. *Abundant* means that human flows are small compared to natural flows—for example, aluminum, silica, and iron.
   e. In addition, the extraction of renewable or abundant materials has been accomplished in a sustainable way, efficiently using renewable energy and protecting the productivity of nature and the diversity of species.

2. Design and use of materials in the building will meet the following criteria in order of priority:
   a. Material selection and design favor deconstruction, reuse, and durability appropriate to the service life of the structure.
   b. Solid waste is eliminated by being as efficient as possible; or,
   c. Where waste does occur, reuses are found for it on-site; or,
   d. For what is left, reuses are found off-site.
   e. Any solid waste that cannot be reused is recycled or composted.

On a systemwide—in this case, planetary—scale, the Natural Step contends that, unless we are willing to severely compromise human health, we ultimately need to eliminate the extraction of ores and fossil fuels mined and extracted to produce energy and materials. Additionally, the Natural Step calls for the ultimate elimination of synthetic materials whose concentration in the biosphere is compromising not only human health but also the very health of the biosphere in which we reside. The Natural Step also cautions against the degradation of the biosphere by human activities because it is the very source of the resources needed to sustain life. And, finally, it addresses the social aspects of sustainability by noting that human needs in all parts of the world must be met. In sum, the message of the Natural Step is to reduce resource extraction, increase reuse and recycling, and minimize emissions that affect both ecosystems and human systems.

Cardinal Rules for a Closed-Loop Building Materials Strategy

A truly green building product should ideally be composed of several different materials that are also green. As pointed out earlier in this chapter, currently there are many green building products that are not themselves inherently green: for example, low-E windows, T-8 lighting fixtures, and ERVs. Although there are many arguments about what constitutes a green building product, perhaps the primary question relates to the ultimate fate of the product and its constituent materials. Presuming that ecology is the ideal model for human systems, and that in nature there is said to be no waste, it follows that the building materials cycle should be closed and as waste-free as the laws of thermodynamics permit. A *closed-loop* building product and materials strategy must address several levels of materials use in its implementation: the building, the building products, and the materials used in the building products and in construction. Ideally, the building materials system should follow the Cardinal Rules for a Closed-Loop Building Materials Strategy listed in Table 11.2.

The cardinal rules state that the complete dismantling of the building and all of its components is required so that materials input at the time of the building's

TABLE 11.2

Cardinal Rules for a Closed-Loop Building Materials Strategy

1. Buildings must be deconstructable.
2. Products must be disassemblable.
3. Materials must be recyclable.
4. Products/materials must be harmless in production and in use.
5. Materials dissipated from recycling must be harmless.

construction can be recovered and returned to productive use at the end of the building's useful life. These rules also establish the ideal conditions for materials and products used in building. It is, however, important to point out that very few materials and products today can adhere to these five rules, meaning that the behavior of materials is far from its ideal state. As it stands, devising a system of materials, products, and buildings to support closed-loop behavior is in the distant future. Nonetheless, this thought process can be used as a touchstone for making decisions about the development of new products, materials, and technologies that support the high-performance green building movement.

Pragmatic View of Green Building Materials

In order to take a pragmatic view of green building materials, it is useful to examine contemporary efforts to wrestle more directly with these issues based on our current understanding, capabilities, and technologies. As noted several times in previous chapters, *Environmental Building News* (*EBN*) is an excellent source of well-reasoned approaches to most matters concerning high-performance buildings, and the subject of building materials and products is no exception. According to *EBN*, green building products can be broken down into five major categories:[10]

1. Products made from environmentally attractive materials
   a. Salvaged products
   b. Products with postconsumer recycled content
   c. Products with postindustrial recycled content
   d. Certified wood products
   e. Rapidly renewable products
   f. Products made from agricultural waste material
   g. Minimally processed products

2. Products that are green because of what is not there
   a. Products that reduce material use
   b. Alternatives to ozone-depleting substances
   c. Alternatives to products made from PVC and polycarbonate
   d. Alternatives to conventional preservative-treated wood
   e. Alternatives to other components considered hazardous

3. Products that reduce environmental impacts during construction, renovation, or demolition
   a. Products that reduce the impacts of new construction
   b. Products that reduce the impacts of renovation
   c. Products that reduce the impacts of demolition

4. Products that reduce the environmental impacts of building operation
   a. Building products that reduce heating and cooling loads
   b. Equipment that conserves energy
   c. Renewable energy and fuel cell equipment
   d. Fixtures and equipment that conserve water
   e. Products with exceptional durability or low maintenance requirements
   f. Products that prevent pollution or reduce waste
   g. Products that reduce or eliminate pesticide treatments

5. Products that contribute to a safe, healthy indoor environment
   a. Products that do not release significant pollutants into the building
   b. Products that block the introduction, development, or spread of indoor contaminants
   c. Products that remove indoor pollutants
   d. Products that warn occupants of health hazards in the building
   e. Products that improve light quality

This pragmatic view of building materials and products is a useful starting point because it deals with the contemporary supply chain and with today's technologies and practices. The question, then, is, How do we evolve closer to the ideal of green building materials and products espoused by the Natural Step and the Cardinal Rules for a Closed-Loop Building Materials Strategy?

PRIORITIES FOR SELECTING BUILDING MATERIALS AND PRODUCTS

There are three priorities in selecting building materials for a project:

1. As with energy and water resources, the primary emphasis should be on reducing the quantity of materials needed for construction.
2. The second priority is to reuse materials and products from existing buildings; this is a relatively new strategy called *deconstruction*. Deconstruction is the whole or partial dismantling of existing buildings for the purpose of recovering components for reuse.
3. The third priority is to use products and materials that contain recycled content and that are themselves recyclable or to use products and materials made from renewable resources.

TECHNICAL AND ORGANIC RECYCLING ROUTES

There are two general routes for recycling: technical and organic. The *technical recycling route* is associated with synthetic materials, that is, materials that do not exist in pure form in nature or are invented by humans. These include metals, plastics, concrete, and nonwood composites, to name a few. As noted earlier, only metals and plastics are fully recyclable; hence, they can potentially retain their engineering properties through numerous cycles of reprocessing. Materials in the technical or synthetic category require major investments of energy, materials, and chemicals for their recycling. Materials recyclable through the *organic recycling route* are described in the previous section under renewable resources. Composting is the best-known organic recycling route. This route is designed to allow nature to recycle building materials and turn them back into nutrients for ecosystems. Although feasible in theory, organic recycling has not been attempted on a large scale in the United States. For the organic route to work, it would have to incorporate products from a wide range of applications, including agricultural waste and landscape clearing debris, as well as organic waste from construction.

GENERAL MATERIALS STRATEGY

Assuming that a building is, in fact, needed for a given function, minimizing the environmental impacts of building materials suggests the following strategy, in general order of priority:

1. *Reuse existing structures.* By modifying an existing building and reusing as much of its structure and systems as possible, one can minimize the use of new materials, with their accompanying impacts of resource extraction; transportation; and processing energy, waste, and other effects. Clearly, trade-offs must be made when considering a building for

reuse. For example, a building that, historically, has been inefficient and would need significant changes to its envelope and mechanical/electrical systems might incur significant waste, as well as require enormous quantities of new materials, in order for the original structure to be retrofit for its new use.

2. *Reduce materials use.* Using the minimal amount of materials required for a building project also lowers the environmental impact of introducing products manufactured from virgin resources. In a typical building, however, the opportunities for *dematerialization* are few, and center on the possible elimination of systems that are not absolutely necessary. Rejecting floor finishes in favor of finished concrete is an example of reducing materials use, but probably at the cost of aesthetic appeal. Materials waste caused by handling and conventional construction processes also contributes to unnecessary materials use. In general, dematerializing a building is difficult because of building code provisions, the desires of the users, and, sometimes, the need for new systems that are becoming standard in high-performance green buildings. An example of a relatively new system frequently used in green buildings is a rainwater harvesting system that requires cisterns, piping, pumps, power, and controls, which are not present in a conventional building. Fortunately, building performance can often be enhanced by the introduction of more systems and materials that may offset the impacts caused by increasing the mass of materials in the building project. The building materials cycle also can be enhanced by modifying existing building designs so that they incorporate design for deconstruction (DfD) as a component of the overall building design strategy and by using materials that will have future value for recycling. DfD is addressed in more detail later in this chapter.

3. *Use materials created from renewable resources.* Materials created from renewable resources offer the opportunity to close materials loops via an organic recycling process. The organic route involves recycling by biodegradation, that is, by composting or aerobic/anaerobic digestion, either by nature itself or by processes that mimic the decomposing action of nature. This approach applies to all products made of wood or other organic materials such as jute, hemp, sisal, wool, cotton, and paper. Recycling of renewable resources or organic products via the organic recycling route can be accomplished with low to zero energy, additional materials, and chemicals. Note, however, that some materials are composites of organic and technical materials and hence would fall into the technical class for purposes of recycling. Other emerging materials such as polylactic acid (PLA) polymers are hybrid synthetics. PLA is a polymer made from the lactic acid that results from cornstarch fermentation; it is used in plastics that are competitive with and often superior to hydrocarbon-based polymers and is completely renewable. PLA can be engineered to be biodegradable in controlled compost situations, so although it is a synthetic material, it can be recycled through the organic route.

4. *Reuse building components.* Reusing intact building components from deconstructed buildings reduces the environmental impacts of building materials because these components require minimal resources for reprocessing. Progress in the techniques for deconstructing existing buildings, instead of demolishing them, means that used building components are

becoming more widely available; likewise, businesses that specialize in the sale of components salvaged from deconstructed buildings are becoming more commonplace. One problem that remains to be solved, however, is how to recertify most used building products. That said, good progress has been made in developing visual regrading standards for some types of dimensional lumber—for example, western cedar and southern yellow pine.

5. *Use recyclable and recycled-content materials.* To close the materials loop in construction, all materials must have the capacity for recycling. Currently, this remains a very ambitious objective simply because few building materials are recyclable and many others can be recycled only into a lower-value application. For example, recycled concrete aggregate can be used as a subbase material but not—at least not readily, in the United States—as an aggregate in new concrete. Metals and plastics are perhaps the only materials that are fully recyclable without loss of their basic strength and durability properties. A wide range of recycled-content materials is available for the green building market. These generally contain either postindustrial or postconsumer waste. *Postindustrial waste* refers to materials recycled within the manufacturing plant. For example, during the extrusion of plastic lumber made from high-density polyethylene (HDPE), sprools of the HDPE, which peel off during the process, can be recycled back into the plastic being input to the process. *Postconsumer waste* refers to materials that are recycled from home or business use into new products. Plastic lumber made entirely of HDPE from recycled milk bottles would be considered to have 100 percent postconsumer content. Postconsumer waste recycling is far more difficult than postindustrial waste recycling. This fact is reflected in the LEED-NC building assessment system, which weights postconsumer content as double postindustrial content for the purpose of awarding points.

6. *Use locally produced materials.* Examining the resources and emissions associated with transporting materials between the various sites of extraction, materials production, product manufacture, and installation is one of the steps in an LCA evaluation. There is no doubt that minimizing transportation distances by using locally produced materials and locally manufactured products can greatly reduce the overall environmental impacts of materials. Defining what is meant by *local* can, however, be a challenge. The LEED-NC building assessment standard sets 500 miles (806 kilometers) as the radius within which a product is considered local for the purpose of obtaining points. Another difficulty with assigning weight to locally produced products is that improved technologies may be passed over. A classic example is the introduction of Japanese automobiles to the US marketplace in the late 1970s, when the quality and workmanship of these cars compelled a rapid shift away from American products. The subsequent bailout by the US government of the Chrysler Corporation in 1979–1980, coupled with higher energy prices, forced US companies to rethink their products. Ultimately, fundamental changes took place in the design and production of American cars. Today, American cars are almost on a par with Japanese automobiles and exceed many European cars in terms of quality. In short, products not considered local may be far superior, result in lower life-cycle environmental impacts, and encourage improvements in local products (see Figure 11.2).

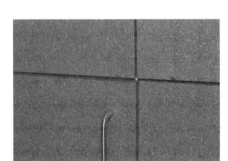

Figure 11.2 Typical of new materials emerging to serve the green building market is compressed wheatboard, made from wheat straw, which can be used in millwork or for cabinetry, as shown here in a laboratory at Rinker Hall at the University of Florida at Gainesville. (T. Wyman)

The materials selection process may be summarized as follows: rely on the three Rs—reduce, reuse, and recycle (with the meaning of *recycle* being extended to address products and materials with recycled content or from renewable resources).

LCA of Building Materials and Products

As stated previously, the most important tool currently being used to determine the impacts of building materials is LCA. LCA can be defined as a methodology for assessing the environmental performance of a service, process, or product, including a building, over its entire life cycle.[11] LCA comprises several steps, which are defined in the International Organization for Standardization (ISO) 14000 series of standards that address environmental management systems.[12] These steps include inventory analysis, impact assessment, and interpretation of the impacts.

Put simply, LCA is a methodology for assessing the environmental performance of a product over its full life cycle, often referred to as *cradle-to-grave or cradle-to-cradle* analysis. Environmental performance is generally measured in terms of a wide range of potential effects, for example:

- Fossil fuel depletion
- Other nonrenewable resource use
- Water use
- Global warming potential
- Stratospheric ozone depletion
- Ground-level ozone (smog) creation
- Nutrification/eutrophication of water bodies
- Acidification and acid deposition (dry and wet)
- Toxic releases to air, water, and land

Comparing these effects for a building takes careful analysis. For example, the total energy for a building's life cycle is composed of the embodied energy invested in the extraction, manufacture, transport, and installation of its products and materials, plus the operational energy needed to run the building over its lifetime. For the average building, the operating energy is far greater than the embodied energy, perhaps 5 to 10 times higher. Consequently, the operational stage has far more energy impacts than those up through the construction stage. For other effects, however, the impacts of the stages up through construction can be far greater. Toxic releases during resource extraction and the manufacturing process can be far greater than those occurring during building operation. The net result is that the designer using these tools must keep in mind the entire life cycle of the building, not just the stages leading to construction.

ATHENA ENVIRONMENTAL IMPACT ESTIMATOR

The Athena Environmental Impact Estimator (EIE) is an LCA tool that focuses on the assessment of whole buildings or building assemblies such as walls, roofs, or floors. It was created and is maintained by the nonprofit Athena Institute and is intended to assist project team members make decisions about product selection early in the design stage. The EIE has a regional character, meaning that the user can select the project site from among 12 different North American locations. It accounts for materials maintenance and replacement over an assumed building life and distinguishes between owner-occupied and rental facilities, if relevant. If an energy simulation for a design has been completed, it can be entered into the EIE to take account of operating energy impacts and the impacts of generating that energy. The EIE has a database of generic products covering 90 structural and envelope materials. It can simulate more than 1000 different assembly combinations and can model the structure and envelope systems for over 95 percent of the building stock in North America. The output of the EIE provides cradle-to-grave and region-specific results of a design in terms of detailed flows from and to nature. It also provides summary measures for embodied energy use, global warming potential, solid waste emissions, pollutants to air, pollutants to water, and natural resources use. Graphs and summary tables show energy use by type or form of energy, and emissions by assembly group and life-cycle stage. A comparison dialogue feature allows the side-by-side comparisons of up to five alternative designs. Similar projects with different floor areas can be compared on a unit floor area basis.[13]

A typical array of information produced by EIE, version 3.0, is shown in Table 11.3. The information in this table is not meaningful unless it is compared to alternative strategies. For example, the building depicted is an 18-story office building with five levels of underground parking; it has a concrete structure and an exterior curtainwall. An alternative would be a steel structure with masonry walls. The purpose of making these comparisons is to determine the building systems that have the lowest life-cycle impact—within the construction budget. An LCA program such as the EIE has a very complex array of outputs, as shown in Figure 11.3.

TABLE 11.3

Example of an LCA Output

Building Components	Embodied Energy (GJ)	Solid Waste (metric tons)	Air Pollution* (index)	Water Pollution* (index)	GWP† (equivalent CO_2 metric tons)	Weighted Resource Use (metric tons)
Structure	52,432	3,273	859.0	147.0	13,701	34,098
Cladding	17,187	281	649.8	24.7	5,727	2,195
Roofing	3,435	145	64.8	5.8	701	1,408
Total	73,054	3,554	1,573.6	177.5	20,129	37,701
Per square meter	2.36	0.11	0.05	0.006	0.65	1.21

*The air and water pollution indices are based on the critical volume measure (method).
†GWP is global warming potential. Energy and emission estimates do not include operating energy.

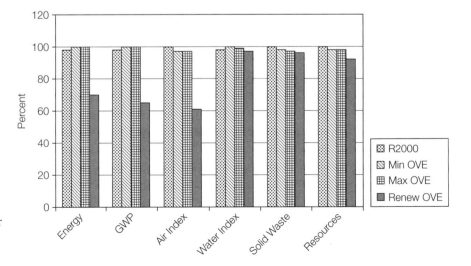

Figure 11.3 Sample output screen from the Athena EIE, version 3.0, program showing energy use, various impacts, and other resource use for a comparison of four products. (Courtesy of the Athena Sustainable Materials Institute)

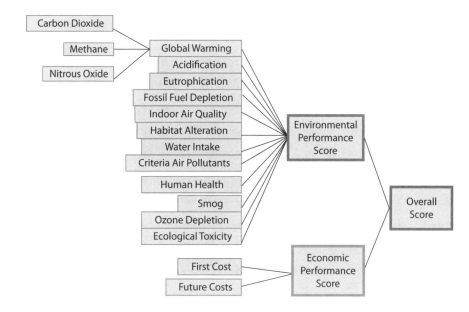

Figure 11.4 The BEES model combines environmental and economic performance into a single score for use in comparing product selection options. (*Source:* National Institute of Standards and Technology)

BUILDING FOR ENVIRONMENTAL AND ECONOMIC SUSTAINABILITY (BEES)

As noted earlier, BEES is the other prominent North American tool for LCA of building materials and products; it is specific to the United States. It was developed by the National Institute of Standards and Technology (NIST) with support from the US Environmental Protection Agency (EPA) Environmentally Friendly Purchasing Program. BEES allows side-by-side comparison of building products for the purpose of selecting cost-effective, environmentally preferable products, and includes both LCA and life-cycle costing (LCC) data (see Figure 11.4). The result is that the user obtains both environmental performance and economic comparisons.

In addition to the typical measures of performance, BEES provides data about air pollutants, indoor air quality, ecological toxicity, and human health for each material or product. BEES can compare building elements to determine where the greatest impacts are occurring and which building elements need the most improvement. The user assigns weights to categories, then combines the environmental and economic performance into a single performance score. For example, the user first decides how to weigh environmental versus economic performance, say, 50–50 or 40–60. The user then selects from among four different weighting schemes for the environmental performance measures. The latest version of BEES is available online and has a database of over 200 building products, including 80 brand-name products.[14] As an example, for floor coverings, there are 18 brand-name products and 17 distinct generic products. A sample output screen for a BEES LCA analysis is shown in Figure 11.5.

Environmental Product Declarations

An environmental product declaration (EPD) presents quantified environmental data for products or systems based on information from an LCA that was conducted using a standard approach defined by the International Organization for Standardization (ISO). Specifically, the LCA approach is defined

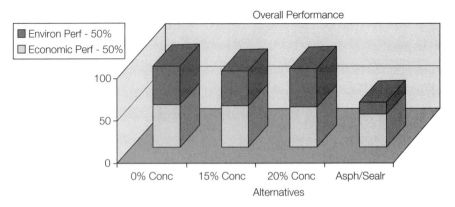

Figure 11.5 Sample output from BEES 3.0 showing comparative environmental performance for concrete with various levels of fly ash and for sealed asphalt. BEES is a free LCA program available from the National Institute of Standards and Technology (NIST). (*Source:* National Institute of Standards and Technology)

Note: Lower values are better

Category	0% Conc	15% Conc	20% Conc	Asph/Sealr
Economic Perf - 50%	50	49	48	39
Environ Perf - 50%	45	41	44	14
Sum	95	90	92	53

by ISO 14040, Environmental Management—Life Cycle Assessment—Principles and Framework. An EPD is based on the output of an LCA, and its format is governed by ISO 14025, Environmental Labels and Declarations—Type III Environmental Declarations—Principles and Procedures. The LCA includes information about the environmental impacts associated with a product or service, such as raw material acquisition; energy use and efficiency; content of materials and chemical substances; emissions to air, soil, and water; and waste generation. It also includes product and company information. An EPD is a voluntarily developed set of data that provides third-party, quality-assured, and comparable information regarding the environmental performance of products based on an LCA. It is a statement of product ingredients and environmental impacts that occur during the life cycle of a product, from resource extraction to disposal. An EPD is similar to a nutrition label on a box of cereal, but instead of nutrients and calories, it indicates raw material consumption; energy use; air, soil, and water emissions; water use and waste generation; and other impacts. An EPD is not a certification, a green claim, or a promise; it simply shows product information in a consistent way, certified to a public standard and verified by a credible third party.

Although a potential breakthrough in the arena of selecting high-performance building materials and products, EPDs are not especially helpful in isolation. A significant number of EPDs must be available in each product class, for example, ceiling tiles, to allow comparison among like products. And, as noted earlier, the ultimate goal is whole-building comparisons that combine EPDs into an EBD to allow trade-offs between systems. InterfaceFLOR is now the North American leader in issuing EPDs and, as noted earlier, has made a commitment to issuing EPDs for all of its products by the end of 2012. Extracts from its EPD for GlasBac nylon carpet tiles are shown in Figures 11.6–11.8.[15]

Materials and Product Certification Systems

One of the means of selecting green building materials and products for high-performance buildings is to rely on certification programs that are well recognized, especially by building assessment systems such as LEED. For example,

Layer	Component	Material	Availability	Mass %	Origin
Wear Layer	Face Cloth/Yarn	Nylon 6 Post Industrial & Post Consumer Recycled	Recycled material, abundant	17%	IT
Carrier	Tufting Primary	Polyester	Fossil resource, limited	3%	US
Backing	Latex	Ethylene vinyl acetate	Fossil resource, limited	5%	US
	Filler	CaCO3	Mineral resource, non-renewable, abundant	15%	US
Stabilization	Fiberglass	Silica	Mineral resource, non renewable, abundant	1%	US
Structural Backing	GlasBac® Backing	Polyvinyl chloride copolymer	Ethylene – Fossil resource, limited and Salt – Mineral resource, non-renewable, abundant	10%	US
		di-isononyl phthalate	Fossil resource, limited	10%	US
		Calcium alumina glass spheres, post industrial	Recycled material, abundant	39%	US

Figure 11.6 Extract from the EPD for InterfaceFLOR's GlasBac type 6 nylon carpet tiles showing materials selection. (InterfaceFLOR Commercial, Inc.)

Yarn Weight	Unit	Total Life Cycle	Production			Installation	Use*	End of Life
Low (441 grams/square meter)	MJ	132.04	120.67			2.68	6.56	2.13
			Primary Material 101.65	Secondary Material 6.58	Internal Processing 12.44			
Medium (712 grams/square meter)	MJ	144.64	132.95			2.85	6.56	2.28
			Primary Material 109.39	Secondary Material 11.12	Internal Processing 12.44			
High (949 grams/square meter)	MJ	155.68	143.72			2.99	6.56	2.41
			Primary Material 128.23	Secondary Material 3.05	Internal Processing 12.44			

Figure 11.7 Extract from the EPD for InterfaceFLOR's GlasBac type 6 nylon carpet tiles showing primary energy use. (InterfaceFLOR Commercial, Inc.)

wood products certified by the Forest Stewardship Council (FSC), low-emission paints certified as meeting Green Seal standards, and low-emission carpets certified by the Carpet and Rug Institute (CRI) are typical of programs that both provide assurance that the products comply with their standards and are referenced in building assessment systems. Like ecolabels, product certification by programs established by reputable organizations greatly simplifies the search for environmentally friendly products. Figure 11.9 is a list of the major materials certification programs generally applicable in the United States. It is useful to distinguish between certification organizations and the standards they develop.[16] For example, Green Seal is a certification organization while Green Seal Standard 13 for Paints and Coatings is a standard produced by Green Seal that specifies maximum volatile organic compound (VOC) content for specific classes of paints and coatings. Green Seal is a third-party certifier; that is, it is an independent entity. The CRI, because it was set up by the carpet industry,

	Yarn Weight			Units
	441	712	949	grams/square meter
	13	21	28	ounces/square yard
PCR Impact Category		**Impact**		**Units**
US TRACI				
TRACI, Acidification Air	1.7	1.9	2.1	mol H+ Equiv.
TRACI, Eutrophication Water & Air	0.003	0.003	0.003	kg N-Equiv.
TRACI, Global Warming Air	9.28	10.28	11.34	kg CO2-Equiv.
TRACI, Ozone Depletion Air	1.2×10^{-6}	1.3×10^{-6}	1.4×10^{-6}	kg CFC 11-Equiv.
TRACI, Smog Air	1.6×10^{-5}	1.8×10^{-5}	1.9×10^{-5}	kg NOx-Equiv.
CML				
CML, Abiotic Depletion (ADP elements)	1.1×10^{-5}	1.1×10^{-5}	1.1×10^{-5}	kg Sb-Equiv.
CML, Acidification Potential (AP)	0.034	0.040	0.044	kg SO2-Equiv.
CML, Eutrophication Potential (EP)	0.007	0.008	0.008	kg Phosphate-Equiv.
CML, Global Warming Potential (GWP 100 Years)	9.46	10.57	11.54	kg CO2-Equiv.
CML, Ozone Layer Depletion Potential (ODP, steady state)	1.2×10^{-6}	1.3×10^{-6}	1.3×10^{-6}	kg R11-Equiv.
CML, Photochem, Ozone Creation Potential (POCP)	0.005	0.006	0.006	kg Ethene-Equiv.

Figure 11.8   Extract from the EPD for InterfaceFLOR's GlasBac type 6 nylon carpet tiles showing environmental impacts generated by the product LCA. (InterfaceFLOR Commercial, Inc.)

is considered a second-party certifier. Although not totally independent, they are considered to be fair and reliable with respect to their testing and certification. First-party certification is provided directly by the manufacturer, and the data provided have not been verified by an outside organization. An example is a material safety data sheet (MSDS) provided by the manufacturer for the purposes of complying with Occupational Safety and Health Administration (OSHA) safety requirements on the construction site.

Key and Emerging Construction Materials and Products

For many conventional building materials, admirable progress is being made in rethinking their extraction and application in construction. Perhaps the most notable and successful effort has been the inclusion of sustainable forestry as the key criterion for wood products used in construction. LEED-NC, for example, provides a point if a minimum of 50 percent of the building's wood-based materials and products are certified in accordance with the FSC's Principles and Criteria. Green Globes provides points for wood products certified by the FSC, the Sustainable Forestry Initiative, under the Canadian Standards Association Standard for Sustainable Forest Management (CAN/CSA Z809), and the American Tree Farm System. For metal products, the emphasis is on their recycled content, and organizations such as the Steel Recycling Institute ensure that the benefits of their member companies' products are well known.

New technologies are being developed to improve performance or to provide new capabilities of building products. But in part because there is no commonly accepted vision of what constitutes a green building material or product, there are many different approaches to product development. Therefore, one of the most effective ways to track progress is to examine the products emerging to serve the

Program	Managing Organization	Product Range	Levels	Used in LEED*	Type of Standard or Certification	Comments
			Sustainable Forestry			
Forest Stewardship Council	Forest Stewardship Council (FSC)	Forest products	Variety of labels for pure and percentage content	Y	Third-party certification to regionally specific standards	Only forestry program in LEED; roots in the environmental movement. More prescriptive than SFI.
Sustainable Forestry Initiative	Sustainable Forestry Initiative (SFI)	Forest products	Variety of labels for pure and percentage content	N	Third-party certification	Thorough system but less prescriptive than FSC. Historically close to the forest industry.
American Tree Farm System	American Forest Foundation	Forest products	Single standard for small US landowners	N	Third-party certification	Very nonprescriptive standard. Does not label products itself, but the SFI label applies.
CSA Sustainable Forest Management System	Canadian Standards Association (CSA)	Forest products	Single standard applied to specific forest areas	N	Third-party certification	Industry- and government-backed standard used in Canada.
			Emissions Certifications			
California Section 01350	California Department of Health Services	Wide range of interiors products	n/a	Y	Specification guidance on which other certifications are based	Designed to reduce pollutant concentrations in classrooms and offices.
Greenguard	Greenguard Environmental Institute	Wide range of interiors products	Greenguard Indoor Air Quality, Greenguard for Children & Schools	Y	Third-party certification	Uses ASTM test methods.
FloorScore	Scientific Certification Systems, Resilient Floor Coverings Institute	Nontextile flooring	FloorScore	Y	Third-party certification	Based on California Section 01350 Specification. Equivalent to Indoor Advantage Gold.
Indoor Advantage	Scientific Certification Systems (SCS)	Wide range of interiors products	Indoor Advantage, Indoor Advantage Gold	Y	Third-part certification based on a variety of standards	Indoor Advantage meets LEED requirements; Indoor Advantage Gold also meets stricter California Section 01350 limits.
Green Label	Carpet & Rug Institute (CRI)	Carpet, pad, adhesive	Green Label, GreenLabel Plus	Y	Second-party certification	Green Label Plus meets California Section 01350 limits.

(Continued)

Program	Managing Organization	Product Range	Levels	Used in LEED*	Type of Standard or Certification	Comments
Energy						
Energy Star	US EPA and US Department of Energy	Range of products	n/a	Y	Government label based on manufacture data	Popular program with wide impact. Moderate standards capture wide market share.
Multiattribute Standards and Certifications						
Sustainable Choice	Scientific Certification Systems (SCS)	Carpet; others expected	Silver, Gold, Platinum	ID	Third-party certification, based on both consensus and proprietary standard	Similar to SCS's EPP standard but with social considerations. Respected as a leader in the field.
Cradle to Cradle (C2C)	McDonough Braungart Design Chemistry (MBDC)	Wide range of products	Biological, Technical Nutrients; Silver, Gold, Platinum	ID	Second-party certification, based on a proprietary standard	Developed by respected industry leaders, but key pieces are not transparent.
SMaRT Consensus Sustainable Product Standards	Institute for Market Transformation to Sustainability (MTS)	Wide range of products	Sustainable, Silver, Gold, Platinum	ID	Third-party certification	Works with outside auditors to verify performance.
NSF-140 Sustainable Carpet Assessment	NSF International	Carpet	Bronze, Silver, Gold, Platinum	ID	Standard, requiring third-party certification	California Gold Sustainable Carpet Standard has merged with NSF-140 Platinum.
Sustainable Furniture Standard	Business and Institutional Furniture Manufacturer's Association (BIFMA)	Furniture	Silver, Gold, Platinum	N	Standard, to which first-, second-, or third-party certification is possible	Draft standard being developed; no certification program available yet.
Green Seal	Green Seal	Wide range	n/a	Y	Third-party certification	Uses various ASTM standards depending on product type.
EcoLogo/ Environmental Choice	TerraChoice Environmental Marketing	Wide range of products	n/a	Y	Third-party certification	Backed by the Canadian government.

Figure 11.9 Certification programs available for materials selection for US high-performance buildings. Many of these certification programs are referenced by US building assessment systems such as LEED. (BuildingGreen, Inc.)

*Y = referenced in LEED credit language, N = not referenced, ID = referenced, in Innovation in Design options

TABLE 11.4

BuildingGreen's Top 10 Green Products for 2011

Product	Manufacturer	Description
Carpet tiles with PFC-free carpet fibers	InterfaceFLOR	Eliminates perfluorinated (PFC) compounds from carpet tiles
PVC-free resilient flooring	Lifeline	Tough, built-in wear layer with no PVC
Knight Wall CI-Girt rainscreen system	Knight Wall Systems	Designed to allow continuous insulation, although it also contains an interchangeable cladding
Waterborne ceramic coating	EonCoat LLC	Waterborne ceramic coating is made out of phosphoric acid and milk of magnesia for highly corrosive industrial applications
Aqua2use graywater system	Water Wise Group, Inc.	Collects and purify the water that goes down the drain from sinks and washing machines for outdoor irrigation
Analog-to-digital wireless thermostat	Cypress Envirosystems	Seamlessly replaces an analog pneumatic thermostat with wireless digital controls
Ritter XL solar thermal system	Regasol	Combines evacuated tubes, compound parabolic reflectors, and water, a more efficient heat transfer fluid than glycol, and can produce very hot water even in very cold climates
Ductless heat pumps and variable refrigerant flow systems with tenant submetering	Misubishi Electric	Can be used in multifamily and hotel applications, where custom set points and even submetering may be desirable, work well even at very low temperatures
AllSun Trackers	Allearth Renewables	Use Global Positioning System (GPS) to track the sun's path across the sky from dawn to dark
EnduraLEDs	Phillips	A 12-watt LED replacement for 60-watt incandescent lights

green building marketplace. A second set of recent green building products cited by *EBN* as the top 10 products of 2011 is shown in Table 11.4.[17]

The following sections address the current issues and status of major classes of construction materials. A comprehensive discussion of the wide range of green building materials and products, both existing and emerging, is beyond the scope of this book. Therefore, the materials discussed here are those considered most important because of the scale of their application in construction: wood and wood products, concrete and concrete products, metals, and plastics.

WOOD AND WOOD PRODUCTS

Wood and products made of wood are very important construction materials, made all the more important because of their renewability. Enormous areas of the United States are considered to be covered with trees, some 747 million acres (302 million hectares), or about one-third of the US landmass. Of this, 504 million acres (204 million hectares) are classified as timberland, that is, productive forest capable of growing at least 20 cubic feet (0.6 cubic meter) of commercial wood per acre per year. Approximately 67 million acres (27 million hectares) are owned by the forest products industry, 291 million (118 million hectares) are held by 10 million individual private landowners, and another 49 million acres (20 million hectares) contained within the National Forest System are available for forest management.[18]

A wide variety of wood products are used in construction, including dimensional lumber, engineered wood products, plywood, oriented strand board, and composite materials with wood fiber content. Wood products used in high-performance green buildings should originate in sustainably managed

TABLE 11.5

FSC Principles for Management of Forests (2002 Edition)

Principle 1: Compliance with Laws and FSC Principles. Forest management shall respect all applicable laws of the country in which they occur, and international treaties and agreements to which the country is a signatory, and comply with all FSC Principles and Criteria.

Principle 2: Tenure and Use Rights and Responsibilities. Long-term tenure and use rights to the land and forest resources shall be clearly defined, documented, and legally established.

Principle 3: Indigenous Peoples' Rights. The legal and customary rights of indigenous peoples to own, use, and manage their lands, territories, and resources shall be recognized and respected.

Principle 4: Community Relations and Workers' Rights. Forest management operations shall maintain or enhance the long-term social and economic well-being of forest workers and local communities.

Principle 5: Benefits from the Forest. Forest management operations shall encourage the efficient use of the forest's multiple products and services to ensure economic viability and a wide range of environmental and social benefits.

Principle 6: Environmental Impact. Forest management shall conserve biological diversity and its associated values, water resources, soils, and unique and fragile ecosystems and landscapes, and, by so doing, maintain the ecological functions and the integrity of the forest.

Principle 7: Management Plan. A management plan—appropriate to the scale and intensity of the operations—shall be written, implemented, and kept up to date. The long-term objectives of management, and the means of achieving them, shall be clearly stated.

Principle 8: Monitoring and Assessment. Monitoring shall be conducted—appropriate to the scale and intensity of forest management— to assess the condition of the forest, yields of forest products, chain of custody, management activities, and their social and environmental impacts.

Principle 9: Maintenance of High Conservation Value Forests. Management activities in high conservation value forests shall maintain or enhance the attributes that define such forests. Decisions regarding high conservation value forests shall always be considered in the context of a precautionary approach.

Principle 10: Plantations. Plantations shall be planned and managed in accordance with Principles and Criteria 1 through 9, and Principle 10 and its criteria. While plantations can provide an array of social and economic benefits, and can contribute to satisfying the world's needs for forest products, they should complement the management of, reduce pressures on, and promote the restoration and conservation of natural forests.

forests and should bear labels certifying this fact. The major organization governing sustainable forestry internationally is the FSC, whose US-based certifiers are the SmartWood Program and Scientific Certification Systems.[19] A third organization that collaborates with both the FSC and the USGBC to foster the use of certified wood products is the Certified Wood and Paper Association (CWPA).[20] The FSC program is based on a set of 10 principles used as the basis for the criteria to qualify forests for certification (see Table 11.5). And the USGBC LEED building assessment standard provides a point related to certified wood products and for rapidly renewable resources, that is, wood products grown in plantation forests. FSC principle 10 addresses the forestry practices required to earn certification for these types of forests.

The Sustainable Forestry Initiative (SFI) program is a comprehensive system of principles, objectives, and performance measures developed by professional foresters, conservationists, and scientists that combines the perpetual growing and harvesting of trees with the long-term protection of wildlife, plants, soil, and water quality. On January 1, 2007, the SFI program became a fully independent forest certification program. The multistakeholder board of directors of the Sustainable Forestry Initiative is now the sole governing body over the SFI Standard (SFIS) and all aspects of the program. The diversity of the board members reflects the variety of interests in the forestry community.

The SFIS spells out the requirements of compliance with the program. The SFIS is based on nine principles that address economic, environmental, cultural, and legal issues, and a commitment to continuously improve sustainable forest management (see Table 11.6).

TABLE 11.6

SFIS Principles

1. **Sustainable Forestry**
   To practice sustainable forestry to meet the needs of the present without compromising the ability of future generations to meet their own needs by practicing a land stewardship ethic that integrates reforestation and the managing, growing, nurturing, and harvesting of trees for useful products with the conservation of soil, air, and water quality, biological diversity, wildlife and aquatic habitat, recreation, and aesthetics.

2. **Responsible Practices**
   To use and to promote among other forest landowners sustainable forestry practices that are both scientifically credible and economically, environmentally, and socially responsible.

3. **Reforestation and Productive Capacity**
   To provide for regeneration after harvest and maintain the productive capacity of the forest-land base.

4. **Forest Health and Productivity**
   To protect forests from uncharacteristic and economically or environmentally undesirable wildfire, pests, diseases, and other damaging agents, and thus maintain and improve long-term forest health and productivity.

5. **Long-Term Forest and Soil Productivity**
   To protect and maintain long-term forest and soil productivity.

6. **Protection of Water Resources**
   To protect water bodies and riparian zones.

7. **Protection of Special Sites and Biological Diversity**
   To manage forests and lands of special significance (biologically, geologically, historically, or culturally important) in a manner that takes into account their unique qualities and to promote a diversity of wildlife habitats, forest types, and ecological or natural community types.

8. **Legal Compliance**
   To comply with applicable federal, provincial, state, and local forestry and related environmental laws, statutes, and regulations.

9. **Continual Improvement**
   To continually improve the practice of forest management and also to monitor, measure, and report performance in achieving the commitment to sustainable forestry.

Only companies and organizations that have successfully completed an audit by an independent and accredited certification body can claim certification to the SFIS. SFI certification audits are rigorous, on-the-ground assessments, conducted by highly qualified and objective individuals.

Of the leading certification schemes in operation in the United States, only the SFI program has a strict separation between standard setting and accreditation of certifying bodies. Recognized international protocols (ISO) for auditing explicitly require that these functions be separate. To date, over 127 million acres have been independently certified to the SFIS.[21]

It should be noted that there are several other third-party certification systems for sustainably harvested wood, including the American Tree Farm System (ATFS) and CSA Sustainable Forest Management (SFM). The Green Globes building assessment system takes all of these into account in awarding points, while the USGBC relies solely on the FSC certification system.

CONCRETE AND CONCRETE PRODUCTS

As one of the mainstays of construction, and one of its oldest and best-known materials, concrete has an enormous and increasing number of roles in construction. Concrete is normally composed of coarse aggregate (rock), fine aggregate (sand), cement, water, and various additives. With respect to high-performance buildings, concrete has many positive qualities: high strength, thermal mass, durability, and high reflectance; is generally locally available; can be used without interior and/or exterior finishes; does not off-gas and affect indoor air quality; is readily cleanable; and is impervious to insect damage and fire. Concrete can be designed to be pervious or cast into open-web pavers, thus allowing water to infiltrate directly into the ground to reduce the need for stormwater systems.

The key issue with concrete is the carbon dioxide emitted in the cement manufacturing process. Cement, which comprises 9 to 14 percent of most concrete mixes, is second only to coal-fired utilities in carbon dioxide emissions. For each ton of powder cement produced, up to an equal mass of carbon dioxide is generated. However, during the life cycle of a concrete element, the cement reabsorbs about 20 percent of the carbon dioxide generated in the manufacturing process, at least partially mitigating this effect. Minimizing the quantity of cement in a concrete mix is a strategy that has a number of potential benefits. Fly ash and ground blast furnace slag, both of which have cementitious properties, can be at least partially substituted for cement and result in increased concrete performance. Fly ash can be readily substituted for over 30 percent of the cement volume, blast furnace slag for more than 35 percent. These substitutions have the advantage of making beneficial use of otherwise industrial waste while simultaneously reducing the quantity of carbon dioxide associated with concrete production. Fly ash and blast furnace slag can also be blended with cement in the cement manufacturing process, resulting in reduced carbon dioxide emissions, reduced energy consumption, and expanded production capacity.

The recycling properties of concrete are generally satisfactory. Crushed concrete can be used as subbase for roads, sidewalks, and parking lots. In the Netherlands, recycled concrete aggregate can substitute for one-third of the virgin aggregate in concrete mixes. In general, recycled concrete aggregate is in high demand and has relatively high value.

METALS: STEEL AND ALUMINUM

Metals in general have high potential for recycling, and most metal products used in typical building applications have significant recycled content. The performance of metal products in building applications can be outstanding, providing high strength and durability with relatively light weight. Additionally, metals are readily recycled, and their dissipation into the environment during the recycling process is benign. Although the LCA and embodied energy impacts associated with metals may appear to be higher than those of alternatives, the inherent recyclability of metals, their durability, and their low maintenance make them competitive for high-performance building applications.

Steel production today incorporates used steel products in the manufacture of new steel in the two production processes still being used. The basic oxygen furnace (BOF) uses 25 to 35 percent scrap steel for products that require drawability—for example, automobile fenders and cans—while the electric arc furnace (EAF) uses almost 100 percent scrap steel for products whose main requirement is strength—for example, structural steel and concrete reinforcement. Steel made from the BOF process generally has a total recycled content of 32 percent, which is composed of 22.6 percent postconsumer content and 8.4 percent postindustrial content. EAF-produced steel generally has a recycled content of about 96 percent, with a postconsumer content of 59 percent and postindustrial content of 37 percent.[22] Recycled steel consumes a fraction of the resources and energy of steel produced from iron ore. Each ton of recycled steel saves 2500 pounds (1134 kilograms) of iron ore, 1400 pounds (635 kilograms) of coal, and 120 pounds (54 kilograms) of limestone. Only one-fifth of the energy needed to produce steel from iron ore is required to recycle scrap steel.

Steel recycling systems in the United States are well established, so much so that recycling is dictated less by environmental concerns than by economics.

Aluminum recycling also has marked environmental benefits. Recycled aluminum requires only 5 percent of the energy needed to produce aluminum from bauxite ore, thus eliminating 95 percent of the greenhouse gases that would be generated by manufacturing aluminum from bauxite. Approximately 55 percent of the world's aluminum production is powered by hydropower, which, although controversial because of its environmental impacts, is a renewable resource. Recycling 1 pound (0.45 kilogram) of aluminum saves 8 pounds (3.6 kilograms) of bauxite and 6.4 kWh of electricity. Aluminum recycling in the United States is highly successful and well established, with about 65 percent of aluminum being recycled. The recycled content of the average aluminum can is about 40 percent, and improved engineering means that, today, 1 pound (0.45 kilogram) of aluminum produces 29 cans versus 22 cans in 1972. Although there has been controversy over the value of recycling aluminum cans, the industry claims that they can be profitably recycled by individuals and groups. Recycling rates for building applications range from 60 to 90 percent in most countries.[23] Aluminum panels used in buildings are corrosion-resistant, lightweight, and virtually maintenance-free; aluminum also has high reflectivity, making it extremely useful as a roofing material. Aluminum is also used extensively in electrical wiring applications, as a casing for appliances, and in moldings and extrusions for windows.

PLASTICS

Along with wood and metals, plastics, which are composed of chains of molecules known as *polymers*, are a major constituent of building products, both as virgin materials and as recycled content. Plastics have a high potential for recycling, and the industry has developed a systematic method for designating and labeling the seven major classes of plastics. The Society of the Plastics Industry, Inc. (SPI), introduced this system in 1988 to facilitate recycling of the growing quantity of plastics entering the marketplace and the waste stream. Large quantities of postconsumer plastics, particularly HDPE and polyethylene terephthalate (PET), are being recycled into a range of building products, such as plastic lumber. Construction products are the second highest user of plastics in the United States, exceeded only by packaging.

At present, however, there is little, if any, recycling of plastic building products into other end uses, which is a serious problem. Closed-loop behavior is, of course, desirable. But there are some success stories in plastic recycling. One is the development of processes that recycle HDPE into high-quality plastic lumber, a product with very high durability that is impervious to rot, insects, and saltwater damage, and with a lifetime measured in hundreds of years. The holy grail of any recycling effort is to develop technologies that can recycle products back into their original use. The United Resource Recovery Corporation (URRC) technology, recently developed in Germany, can recycle PET plastics back into very high quality flakes, which can then be used to produce the ubiquitous clear plastic of soft-drink bottles. Recycling rates for HDPE and PET in this country are in the 20 percent range, the highest for the common classes of plastics used in consumer products.

Manufacturers of plastics derived from chlorine or that employ chlorine in their production are under severe pressure from environmental groups such as Greenpeace because of the various impacts associated with their manufacture and disposal. PVC, a ubiquitous product in construction (it appears in piping, siding, flooring, and wiring, to name a few), is the main focus of these struggles. To date, recycling rates for PVC are among the lowest for the seven major classes of plastics covered by the SPI—less than 1 percent. And in the United States, PVC is being defended by its industry based on its technical and economic merits, meaning that fundamental changes to the product or its manufacture are not anticipated in the near future. In contrast, the European PVC industry is exploring how to make fundamental changes in the production and disposal of its products, positioning PVC to be regarded as an environmentally responsible product. A green paper on PVC was released by the European Commission in 2000, indicating that the major problems with PVC are the use of certain additives (lead, cadmium, and phthalates) and the disposal of PVC waste.[24] According to the green paper, only 3 percent of PVC waste is recycled; 17 percent is incinerated and the remaining 80 percent is landfilled, with the total waste stream amounting to 3.6 million tons per year. The risks associated with landfilling PVC, especially the loss of phthalate from soft PVC, were highlighted, along with the problems caused by incineration, namely, the generation of dioxins, which are very hazardous chemicals. Unquestionably, PVC recycling must be improved, and reformulation of the basic product must be considered in order to remove the barriers to its recycling. PVC product recycling faces many of the same problems associated with other plastics, namely, the use of additives such as plasticizers, stabilizers, fillers, flame retardants, lubricants, and colorants, which are used to provide specific properties.

A relatively new development in the plastics industry is the production of biobased polymers such as PLA. In 2002, Cargill Dow Polymers (CDP) opened a large facility in Blair, Nebraska, to manufacture a plastic product from PLA, the first of its kind, thus marking the introduction of a polymer technology based on a renewable resource, rather than oil, a nonrenewable resource. The product is known as NatureWorks PLA, which CDP says can be produced from other agricultural products such as sugar beets and cassava. Not to be outdone, Dow Chemical introduced a product called BIOBALANCE polymers, which are advanced polyurethane polymers designed to be used as commercial carpet backing. One of the polyurethane components, polyol, is derived from renewable resources. Another Dow Chemical product, WOODSTALK, is manufactured from formaldehyde-free polyurethane resin and harvested wheat straw fiber, a renewable resource. It is a boardlike material that can be used as an alternative to medium-density fiberboard (MDF) for millwork, cabinetry, and shelving. BioBase 501 is a relatively new, low-density, open-cell polyurethane foam insulation partially made from soybeans. The polyol component of Bio-Base 501 is made of SoyOl, the soy-based component that is also used in carpet backing. And in Stockholm, Sweden, a new process developed by the Royal Institute of Technology uses wood to create polymers known as *hemicellulose-based hydrogels*, as announced in late 2003. In addition to being produced from renewable resources such as agricultural products and wood, biobased polymers hold the promise of being recyclable via natural processes.

Design for Deconstruction and Disassembly

It is undeniable that the current state of construction is wasteful and will be difficult to change. As noted at the start of this chapter, closing materials loops in construction remains the most challenging of all green building efforts. More specifically, choosing building materials and products is by far the most daunting challenge.

Criteria for materials and products for the built environment should be similar to those for industrial products in general. Many materials used in

TABLE 11.7

Factors That Increase the Difficulty of Closing Materials Loops for the Built Environment

1. Buildings are custom-designed and custom-built by a large group of participants.
2. No single "manufacturer" is associated with the end product.
3. Aggregate, for use in subbase and concrete, brick, clay block, fill, and other products derived from rock and earth, are commonly used in building projects.
4. The connections of building components are defined by building codes to meet specific objectives (e.g., wind load, seismic requirements), not for ease of disassembly.
5. Historically, building products have not been designed for disassembly and recycling.
6. Buildings can have very long lifetimes exceeding those of other industrial products; consequently, materials have a long "residence" period.
7. Building systems are updated or replaced at intervals during the building's lifetime (e.g., finishes at 5-year intervals; lighting at 10-year intervals; HVAC systems at 20-year intervals).

buildings, most notably metals, are the same as those used in other industries. But buildings have a distinct character compared to other industrial products. The major factors that make closing materials loops in this segment of the economy particularly difficult are delineated in Table 11.7. The vision of a closed-loop system for the construction industry is, by necessity, one that is integrated with other industries to the maximum extent possible. Many materials—again, metals—can flow back and forth for various uses, whereas others, such as aggregates and gypsum drywall, are unique to construction, so their reuse or recycling would stay within construction. Closing materials loops for the built environment will be much more difficult due to the factors that make its materials cycles differ significantly from those of other industries.

To move from wasteful materials practices to closed-loop materials behavior will require that the green building movement embrace the concepts of deconstruction and design for disassembly (DfDs). Deconstruction is the whole or partial disassembly of buildings to facilitate component reuse and materials recycling; DfDs is the deliberate effort during design to maximize the potential for disassembly, as opposed to demolishing the building totally or partially, to allow the recovery of components for reuse and materials for recycling and to reduce long-term waste generation. To be effective, DfDs (a notion that emerged in the early 1990s) must be considered at the design stage.

Experiments in DfDs conducted at Robert Gordon University in Aberdeen, Scotland, included a wide range of approaches that can facilitate a greatly improved materials cycle: handling, materials identification, simplicity of construction techniques, exposure of mechanical connections, independence of structure and partitioning, and making short-life-cycle components readily accessible. Research indicates that DfD must be implemented at three levels of the entire materials system in buildings in order to produce sound product design and construction strategies: the systems or building level, product level, and materials level. A number of examples exist to test various DfD ideas. One, a multistory residential housing project in Osaka, Japan, employs a reinforced concrete frame to support independently constructed dwellings that can be replaced on 15-year cycles without removing the supporting frame. Ultimately, closing construction materials loops will necessitate the inclusion of product design and deconstruction together in a process that might be labeled *design for deconstruction and disassembly* (DfDDs).

TABLE 11.8

Principles of DfDs as Applied to Buildings

1. Use recycled and recyclable materials.
2. Minimize the number of types of materials.
3. Avoid toxic and hazardous materials.
4. Avoid composite materials and make inseparable products from the same material.
5. Avoid secondary finishes to materials.
6. Provide standard and permanent identification of material types.
7. Minimize the number of different types of components.
8. Use mechanical rather than chemical connections.
9. Use an open building system with interchangeable parts.
10. Use modular design.
11. Use assembly technologies compatible with standard building practice.
12. Separate the structure from the cladding.
13. Provide access to all building components.
14. Design components sized to suit handling at all stages.
15. Provide for handling components during assembly and disassembly.
16. Provide adequate tolerance to allow for disassembly.
17. Minimize the number of fasteners and connectors.
18. Minimize the types of connectors.
19. Design joints and connectors to withstand repeated assembly and disassembly.
20. Allow for parallel disassembly.
21. Provide permanent identification for each component.
22. Use a standard structural grid.
23. Use prefabricated subassemblies.
24. Use lightweight materials and components.
25. Identify the point of disassembly permanently.
26. Provide spare parts and storage for them.
27. Retain information on the building and its assembly process.

Philip Crowther of Queensland Technical University in Brisbane, Australia, suggests 27 principles for building DfDs that are enumerated in Table 11.8.[25] This comprehensive list covers a wide range of thinking about materials selection, product design, and deconstruction.

Crowther's work serves as an excellent starting point in the discussion of a comprehensive approach to developing a seamless framework for closing construction materials loops. Importantly, these principles perhaps generate as many questions as they answer. An example is principle 4, which calls for avoiding composite materials. In the context of materials, "composite" can have many meanings—for example, mixed materials (concrete, steel) or homogeneous layered materials (PVC pipe, laminated wood products). Composites may be very acceptable under certain conditions, where recycling the composite mixture is feasible or where the ability to readily disassemble the layers has been designed into the product. The question is how to develop a systematic approach for determining the acceptability of composites as building materials within the context of attempting to increase reuse and recycling.

Deconstruction offers an alternative to demolition that has two positive outcomes: first, it is an improved environmental choice; second, it can serve to create new businesses, to dismantle buildings, transport recovered components and materials, remanufacture or reprocess components, and resell used components and materials. Existing buildings, although not designed to be taken

Figure 11.10   One of the innovations in the design of Rinker Hall at the University of Florida in Gainesville was to include DfD as a design criterion. One of the design features is the use of bolted, exposed steel connections to permit their ready removal. (Photograph courtesy of M. R. Moretti)

apart, are, in fact, being disassembled to recover materials. There are distinct benefits to be gained from increasing the recycling rates of materials from buildings from the 20 percent range to in excess of 70 percent, because waste from demolition and renovation activities can comprise up to 50 percent of national waste streams. Economic and noneconomic policy instruments can assist in the shift from demolition to deconstruction by providing financial incentives and aiding in allotting the time needed for deconstruction. In developing countries, building deconstruction practices offer a source of high-quality materials to assist in improving the quality of life and the potential for new businesses, which may provide economic opportunity for their citizens.

In spite of its many benefits, designing buildings for deconstruction has rarely occurred in the United States. Rinker Hall at the University of Florida in Gainesville is likely the only LEED-certified building, out of thousands that have been certified, that was designed to be disassembled, receiving an innovation credit from the USGBC for its deconstructability (see Figure 11.10).

Case Study: Project XX Office Building, Delft, Netherlands

According to the architect Jouke Post, office buildings typically have a life span of just 20 years due to inevitable changes in technology and corporate management. Demolition produces an enormous amount of waste from materials that have not reached their useful life expectancy. The XX Office Building in Delft, Netherlands, explored a solution to this waste problem by planning for a shorter building life and by planning for deconstruction and materials reuse in the initial design (see Figure 11.11A–F). The semipermanent design concept challenges designers to think in terms of reality: a 20-year building life rather than the ideal 100-year life of a typical framed structure. Once its practical use has ended, the XX Office Building can be deconstructed and the materials can be reused or recycled.

Figure 11.11   (A) The XX Office Building located in Delftech Park in Delft, Netherlands, has a ceiling-to-floor rectangular glass façade. Standard-sized glazing will be reused after the building is deconstructed. (J. M. Post, XX Architecten)

Figure 11.11   (B) The columns and beams, shown during construction. (J. M. Post, XX Architecten)

Figure 11.11   (C) The columns and beams in a completed office, are exposed and connected by stand-off steel rod lower chords and bolts to promote ease of construction and deconstruction. (J. M. Post, XX Architecten)

Figure 11.11   (D) Ceiling-to-floor window screens control the amount of daylight entering the building. (J. M. Post, XX Architecten)

Figure 11.11 (E) Cardboard ductwork is inexpensive, resourceful, and recyclable and will be close to the end of its life expectancy by the time the XX Office Building is ready for deconstruction. (D. Stephany)

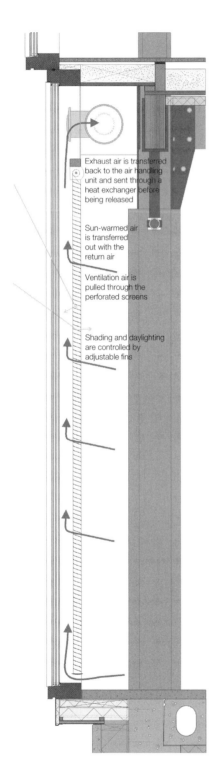

Exhaust air is transferred back to the air handling unit and sent through a heat exchanger before being released

Sun-warmed air is transferred out with the return air

Ventilation air is pulled through the perforated screens

Shading and daylighting are controlled by adjustable fins

Figure 11.11 (F) An air inlet between the screen and window creates a thermal buffer zone, resulting in energy savings and improved climate control. (J. M. Post, XX Architecten)

The 19,200-square-foot (2000-square-meter), two-story building was constructed in 1998 and is a simple, open, and unified rectangular plan. The structure is primarily laminated wood, which was chosen after an analysis of steel, aluminum, concrete, stone, synthetic material, and cardboard for their durability, strength, cost, and future recyclability. The exposed columns and beams are connected by steel rod chords and bolts to provide ease of construction and deconstruction.

The ground floor consists of a concrete slab with 20 percent recycled aggregate. Between levels, sandwiched panels (600 cm × 500 cm) filled with sand are used to improve the acoustical separation. The roof is made of fibrous concrete and recyclable bitumen roof covering. Originally, the roof was held down by weights in the pattern of two Xs representing Roman numerals, hence the building's name. The façade consists of wooden frames attached to the main structure by brackets for ease of deconstruction. These frames have standardized triple-paned windows (approximately 2 m × 5 m) fastened to them. Each frame segment has its own ceiling-to-floor window screen controlling the amount of daylight entering the building. The screens are perforated and help keep heat from entering the building by creating a double-façade system, or "Mercator climate façade." The return air ductwork is composed of cardboard tubes that run along the perimeter of the building. The design uses the energy generated by its 80 occupants and their electric office equipment in place of a heating system.

THOUGHT PIECE: CLOSING MATERIALS LOOPS

The notion of closing materials loops is central to sustainable construction, but it is likely the most difficult and challenging of all the concepts emerging in the shift to a much more environmentally responsible built environment. Buildings are simply not generally made of materials that are recyclable or reusable; the materials are simply optimized, at lowest cost, for their function. As a result, future high-performance buildings are likely to be composed of materials and systems that have a much greater closed-loop potential than those being utilized today. In this thought piece, Brad Guy, an international expert on the subject of deconstruction, discusses the practicality of more sustainable materials practices in the near term.

Closing Materials Loops

Bradley Guy, School of Architecture and Planning, Catholic University of America, Washington, DC

Closing materials loops is a necessary paradigm for any attempts to minimize the impacts of the built environment on the human and ecological environment now and for the survival prospects of future generations. The resource flows in the United States and globally that are dedicated to the built environment materials have consequences across the entire spectrum of resources and in every environmental impact category. The single activity of cement production alone is responsible for between 5 and 7 percent of global greenhouse gases (European Commission, 2011). While harvesting of lumber is a relatively low energy activity, the effects of deforestation and changes in land use because of building impacts have a globally significant impact on the use of land and timber resources for building activities. The list goes on for the upstream impacts of the provision of materials resources into the built environment. This ecological rucksack can be very high in relative mass and environmental impacts in proportion to the final material. Alan Durning (1992) proposed that the average consumer product requires 16 times more resources than will end up in the final product. This suggests that, for every kilogram of building material avoided, reduced, reused, remanufactured, or recycled, another 16 times its mass of materials have been conserved.

It is not remotely sufficient to strive for more benign new building materials or high-performance new buildings when the even higher-than-expected levels of CO_2 reported by the Intergovernmental Panel on Climate Change (IPCC), resulting from current greenhouse gas (GHG) emissions pose the long-term threats for which operationally efficient buildings will come too late when providing their full benefits over 10 or 20 years of building life. While predictions of operational energy efficiencies are typically based on models and subject to the whims of users and often underappreciated maintenance—the one certainty in the life of a building is the materials of which it is made and the systems used to construct them. The other certainty is that the materials-use impacts are in real time and the environmental degradation and emissions caused by their extraction, manufacture, transport, and construction can be reduced at the moment of a building project's conception through the choices that are made by the owner, designer, engineer, and builder who have the expertise and stake in the outcome at building sign-off. In the cause of consideration for future generations, I would like to also posit that it is a professional and ethical responsibility for any architect or builder to consider the end-of-life consequences of his or her design and materials choices. To not do so would be the equivalent of standing in a crowded city and shooting a bullet into the air. It will come down somewhere.

For every kilogram of material avoided through effective design, and reuse of buildings and building parts, a kilogram of raw resources will be preserved. Some suggest that this will not occur because of the Jevons paradox, whereby efficiencies of use of a resource will tend to increase its consumption. Even if this were true in areas of personal finance, at least for the built environment, the relatively static and long-lived nature of buildings ensure that materials investments will remain for some period. The consequences of the reuse of a building, the reuse of materials, and designing for the adaptation of buildings to extend their invested structure and designing for disassembly at end of life of nonstructural and structural building assemblies are physical manifestations in a physical realm.

Design for recovery, reuse, and recycling is a fundamental precept of cradle to cradle, zero waste, extended producer responsibility, and so forth. All matter degrades, and there is no perpetual-motion machine of materials flow. In order to design for recycling, there must be a recycler, and for the recycler to function, there must be a materials flow allowed by the materials producer and the architect and builder. The constant refrain of the catch-22 that without waste there will be no reuse and recycling infrastructure and without infrastructure we cannot propose design for recovery of materials does not aid the way forward. Many green building systems, for example, have developed design for adaptation and/or disassembly within their systems, as a means to use green building standards as an aid to the market of these design practices. One is the Australia Green Star, which has a credit for use of design for disassembly and for "dematerialization," to use less steel for a structure of equivalent performance. The current version of LEED for Healthcare awards points for "design with flexibility," and the proposed next version of LEED for Commercial Interiors has proposed adding design for flexibility. The state of California has recently, as of 2011, put into effect legislation requiring all carpet sold in that state to have an extended producer responsibility system by the manufacturer. This legislative policy and the voluntary market transformation that LEED has proven to be possible are essential ingredients in extending the life of materials and their maintenance in the social and economic system of materials flow. Once extracted no building material should leave the economic loop until it has reached the true end of its utilitarian or energetic value, and materials of high order should never be substituted for materials of lower order.

The design paradigm for closing materials loop is slowly changing and will continue to progress as the realization of resource constraints become more severe. In some cases this will be political. Local resources and the reuse of blighted vacant land, and existing buildings and existing salvaged materials cannot be exported and hence provide the most basic element of resource conservation.

Summary and Conclusions

Many new materials and products are being developed to serve the high-performance green building movement. But in the face of the rapid changes taking place in this area, there is no clear philosophy that precisely articulates the criteria for this new class of products and materials. One proposal is that LCA should determine what constitutes greenness in the context of building materials and products. But LCA, too, has limitations in that it does not adequately address closed-loop materials behavior, which is how nature behaves. Nor does LCA address whether a product or building can be disassembled and recycled, or the recyclability of products and materials. A material or product could conceivably appear to be very beneficial according to the LCA data but may not be recyclable and subject to disposal after use. LCA does, however, provide an excellent account of the resources and environmental impacts of a given decision and allows side-by-side comparisons of various approaches—for example, a steel versus a concrete structural system. Combined with other criteria, LCA offers a good way of evaluating the appropriateness of labeling a product or material green. At this point in the evolution of high-performance green buildings, considering both the production and the fate of materials and products should be a high priority. And as pointed out in the Cardinal Rules for a Closed-Loop Building Materials Strategy, the products and materials must be harmless in use and in recycling before they can be considered truly green.

Notes

1. *Green Building Materials: A Guide to Product Selection and Specification* was written by Ross Spiegel and Dru Meadows.
2. See "Navigating the Maze of Environmentally Preferable Products" in *Environmental Buildings News* (2003) for a wide-ranging discussion of EPPs.
3. The Blue Angel ecolabel website is www.blauer-engel.de.
4. The Athena Environmental Impact Estimator, version 3.0, is available for purchase from the Athena Sustainable Materials Institute online at www.athenasmi.ca. Athena uses a location-specific database of materials to provide LCA information about whole-building systems—for example, wall or roof sections. A demonstration version of Athena is available for download from the website.
5. BEES is a product of the Building and Fire Research Laboratory (BFRL) of the National Institute of Standards and Technology (NIST). BEES measures the environmental performance of building products using the LCA approach specified in the ISO 14000 standards. It is available from www.bfrl.nist.gov/oae/software/bees.html.
6. For a fuller discussion on using LCA as the primary tool in making building materials decisions, see Trusty and Horst (2003).
7. The website of the Natural Step's US branch is www.naturalstep.org.
8. Adapted from a working paper by the Oregon Natural Step Construction Industry Group, "Using the Natural Step as a Framework Toward the Construction and Operation of Fully Sustainable Buildings" (2004).
9. Ibid.
10. Excerpted from "Building Materials: What Makes a Product Green?" in *Environmental Building News* (2008). This article provides a detailed description of each of the various attributes in the five major categories of green building products.
11. Excerpted from Trusty and Horst (2003).
12. An overview of the International Organization for Standardization (ISO) can be found at its website, www.iso.ch/iso/en/ISOOnline.frontpage. ISO 14000 is one of many standards promulgated by this organization. The ISO 14040 series is the member of the ISO 14000 family of standards that addresses LCA.
13. Excerpted and adapted from Trusty (2003).
14. BEES Online is located at www.nist.gov/el/economics/BEESSoftware.cfm.
15. Product life-cycle information is from InterfaceFLOR Commercial, Inc., environmental product declaration for GlasBac dated January 2011.
16. The figure is adapted from "Behind the Logos: Understanding Green Product Certification" in *Environmental Building News* (2008).
17. Detailed information on the BuildingGreen Top Ten Products for 2012 is available at www.buildinggreen.com.
18. Forestry statistics are from the American Forest and Paper Association website, www.afandpa.org.
19. The website of the US branch of the FSC is www.fscus.org. The two US certifiers maintain websites at these addresses: the SmartWood Program, www.rainforest-alliance.org/forestry/certification, and Scientific Certification Systems, www.scscertified.com/.
20. The Certified Wood and Paper Association's website is www.scscertified.org.
21. The SFI Standard can be found at the AFPA website, www.afandpa.org.

22. Excerpted from "2002: The Inherent Recycled Content of Today's Steel," available on the Steel Recycling Institute's website, www.recycle-steel.org.
23. Data on aluminum are from the International Aluminum Institute at www.world-aluminum.org.
24. The European Community's "Green Paper: Environmental Issues of PVC" (2000) can be found at http://ec.europa.eu/environment/waste/pvc/pdf/en.pdf.
25. Philip Crowther's list of DfDs principles was first published in his doctoral dissertation at Queensland University of Technology with the title "Design for Disassembly: An Architectural Strategy for Sustainability" (2002).

References

"Behind the Logos: Understanding Green Product Certification." 2008. *Environmental Building News* 17 (1): 1–14.

"Building Materials: What Makes a Product Green?" 2000. *Environmental Building News* 9 (1): 1, 10–14.

Crowther, Philip. 2002. "Design for Disassembly: An Architectural Strategy for Sustainability." Doctoral dissertation, School of Design and Built Environment, Queensland University of Technology, Brisbane, Australia.

During, Alan T. 1992. *How Much Is Enough? The Consumer Society and the Future of the Earth.* Washington, DC: The Worldwatch Institute.

European Commission, 2011. "Innovative Ways to Reduce CO_2 Emissions from Cement Manufacturing," DG Environment News Service, October 11. Available at http://ec.europa.eu/environment/integration/research/newsalert/pdf/258na1.pdf.

Demkin, Joseph, ed. 1996. *Environmental Resource Guide.* Hoboken, NJ: John Wiley & Sons.

"Navigating the Maze of Environmentally Preferable Products." 2003. *Environmental Building News* 12 (11): 1–15.

Oregon Natural Step Construction Industry Group. 2004. "Using the Natural Step as a Framework toward the Construction and Operation of Fully Sustainable Buildings." Available at www.ortns.org/documents/THSConstructionPaper.pdf.

Spiegel, Ross, and Dru Meadows. 2012. *Green Building Materials: A Guide to Product Selection and Specification*, 3rd ed. Hoboken, NJ: John Wiley & Sons.

Trusty, Wayne B. 2003. "Understanding the Green Building Toolkit: Picking the Right Tool for the Job." Proceedings of the U.S. Green Building Council Annual Conference, Pittsburgh, PA. Available at www.athenasmi.ca.

Trusty, Wayne B., and Scot Horst. 2003. "Integrating LCA Tools in Green Building Rating Systems." Proceedings of the U.S. Green Building Council Annual Conference, Pittsburgh, PA. Available at www.athenasmi.ca.

Chapter 12

Indoor Environmental Quality

P roviding excellent indoor environmental quality (IEQ) has emerged as one of the key goals in the design of high-performance green buildings, on a par with energy efficiency and ecological system restoration. The US Centers for Disease Control and Prevention (CDC) defines IEQ as the quality of the air in an office or other building environment. Although the quality of indoor air is indeed very important, the high-performance green building movement considers a much wider range of health, safety, and comfort factors. In addition to indoor air quality (IAQ), other aspects of IEQ that are routinely considered include lighting quality, daylighting and exterior views, acoustics, noise and vibration control, thermal comfort and control, odors, electromagnetic radiation, potable water monitoring, and ergonomics. In this chapter, we first discuss the problems that have stimulated such enormous interest in IEQ in general and IAQ in particular. These include sick building syndrome, building-related illness, and evidence that poor lighting quality, noise and vibration, and other factors are impacting the health and quality of life of the people using or living in buildings. Then we cover the best practices being used to address these issues and the integration of these solutions into the design of high-performance buildings. The specific issues of ventilation and emissions from materials are addressed, followed by a discussion of the potential financial benefits of providing excellent IEQ in buildings.

Indoor Environmental Quality: The Issues

Most prominent of all the issues addressed by IEQ is air quality. According to the National Safety Council and the US Environmental Protection Agency (EPA), air quality in buildings can be up to 100 times worse than the quality of outside air. This is especially important because Americans spend a large fraction of their day indoors, about 90 percent of the total time with 65 percent of our time spent in our homes. Chemical contaminants like volatile organic compounds (VOCs) and radon plus biological pollutants like mold, pet dander, and plant pollen produce a toxic environment in homes and buildings. However, as noted above, air quality is not the only factor affecting the health and performance of workers in office buildings; students and teachers in schools; the workforce in factories; and people using fitness centers, theaters, and retail outlets. A wide range of other factors that affect people's health are being considered and becoming part of the integrated design process. In the following section, we discuss some of these contributions to poor IEQ and the dangers they pose.[1]

INDOOR ENVIRONMENTAL FACTORS

The indoor environment of a building has a complex makeup. Table 12.1 provides a list of building elements that are thought to affect the indoor

TABLE 12.1

Building Elements Affecting IEQ

Operation and Maintenance of the Building	Ventilation and performance standards
	Ventilation system operational routines and schedules
	Housekeeping and cleaning
	Equipment maintenance, operator training
Occupants of the Building and Their Activities	Occupant activities: occupational, educational, recreational, domestic
	Metabolism: activity and body characteristic dependent
	Personal hygiene: bathing, dental care, toilet use
	Occupant health status
Building Contents	Equipment: heating, ventilation, and air conditioning (HVAC), elevators
	Materials: emissions from building products and the materials used to clean, maintain, and resurface them
	Furnishings
	Appliances
Outdoor Environment	Climate, moisture
	Ambient air quality: particles and gases from combustion, industrial processes, plant metabolism (pollen, fungal spores, bacteria), human activities
	Soil: dust particles, pesticides, bacteria, radon
	Water: organic chemicals including solvents, pesticides, by-products of treatment process chemical reactions
Building Fabric	Envelope: material emissions, infiltration, water intrusion
	Structure
	Floors and partitions

environment. The factors that comprise IEQ can be classified as chemical, physical, and biological.[2] The sensory systems of the inhabitants interact directly with some factors, such as sound level, light, odor, temperature, humidity, touch, electrostatic charges, and irritants.[3] Hundreds of other substances can also be harmful to inhabitants yet go undetected by the sensory systems. Some of these can actually be more dangerous than those that are detected, as their presence can only be determined through testing. Inhabitants may be exposed to high concentrations of these substances for long periods of time without even knowing it—among them radioactive substances, many toxic substances, carcinogens, and pathogenic microorganisms.

Physical indoor environmental problems are traceable primarily to the electrical and mechanical infrastructure of a building. They include sound/noise transmission, lighting quality, thermal conditions, and odors. Physical factors are generally nontoxic but are at least a nuisance to building occupants and can lead to health problems after exposure for extended periods.

A wide variety of chemicals can contaminate the indoor environment. Chemicals may be introduced into the indoor environment by painting, installation of carpets, or cleaning products. Chemical factors are classified according to the form they take at room temperature: vapor, gas, liquid, or particulate. Particulates include inorganic fibers, respirable particulates such as dust and dirt, metals, and a variety of organic materials. Because small particulates can penetrate deep into the lungs, they are a serious concern. The size and density of particulates determine how deeply they can penetrate the respiratory system. Radon, a naturally occurring radioactive gas that has been connected to health problems, is a problem in many regions of the United States; thus, taking measures to mitigate it is important for ensuring a good indoor environment.

Biological contaminants include bacteria, fungi, viruses, algae, insect parts, and dust, which may result in allergenic or pathogenic reactions. There are many sources for these pollutants: pollens from outdoors, viruses and bacteria from humans, and hair and skin flakes from household pets, to name but a few.

Sound/Noise Transmission

Control of sound and noise transmission in buildings is a major problem. Noise from air-handling systems, lights, transformers, and other sources can cause discomfort and even health problems for building occupants. Building designers and engineers are often intimidated by the challenge of dealing with sound and noise transmission because it is a somewhat intangible concept in a world of mostly tangibles: steel, size, color, and so on.

The basic premise in creating an acoustically acceptable indoor environment is to ensure that sound levels in particular areas of a building are at or below an acceptable range for the specific application. For instance, it would be a mistake to locate a helicopter pad just outside of a library. It is clear that sections of a building where low noise levels are required must be separated and insulated from noise-generating areas. When it comes to acoustics, designers can easily prevent obvious problems—for example, taking care not to locate a conference room next to a chiller plant. But more subtle problems may be overlooked, such as neglecting to insulate a wall that separates a restroom from a private office.

A less obvious requirement for ensuring good indoor sound quality is to eliminate as much as possible the subtle background noises that, although not necessarily apparent to building occupants, can be irritating and may, over time, lower morale and decrease productivity. Building systems can generate a wide variety of annoying sounds. Fluorescent light ballasts often buzz when they are not in perfect order, and ventilation systems produce a host of grating yet seemingly untraceable noises. Fan vibrations, too, are a nuisance inherent in ventilating systems that, when isolated, can be dealt with effectively and cheaply. Duct air noises are more problematic and much more difficult to fix. High-speed air in a duct can create whistling sounds and vibrations that are difficult to eliminate. The solution is to reduce the air velocity. To maintain the same quantity of air at a lower velocity, a duct with a larger cross-sectional area must be used. But this solution itself can pose problems when the ductwork is installed in a tight ceiling space or an HVAC chase. The best answer to this problem is to address it before it happens by including an acoustic specialist in the design of the HVAC system.

As noted, high noise levels in commercial buildings can lead to morale problems and loss of productivity when occupants become irritated and annoyed and thus distracted from their work. The other major noise-related problem for building occupants is caused by exposure to unhealthy noise levels generated by air handlers, transformers, lighting, elevators, machinery, and motors.

Lighting Quality

Problems associated with lighting quality are similar to those associated with noise in that the cause is a poorly understood building support system. As a requirement for a high-quality indoor environment, lighting is probably better understood than sound, but it is nevertheless often overlooked in building design.

TABLE 12.2

General Color Characteristics of Typical Building Lighting Systems

Type of Light	Color Characteristics
Incandescent (argon-surrounded filament)	White with yellow tint
Incandescent (halogen-surrounded filament)	White
Fluorescent	White with blue tint
LED	White or white with blue tint
Mercury vapor	White with blue tint
Metal halide	White with blue-green tint
Sodium vapor (high-pressure)	Amber white
Sodium vapor (low-pressure)	Yellow

It is widely acknowledged that natural sunlight is the best light source for the eye. Unfortunately, these days most people spend an inordinate amount of time indoors and away from natural sunlight. Thus, the ideal healthy indoor light environment is one that allows natural light indoors or whose lighting system replicates natural light as closely as possible. Natural sunlight has an equal spectral distribution of the visible light frequencies combined to appear as white light. In contrast, artificial light sources are bound by the laws of physics, and hence they are limited in the frequencies of visible light that they emit. A list of common artificial light sources and their general color characteristics is shown in Table 12.2.

Incandescent lights, particularly the halogen type, give the best color rendition of natural light. Fluorescent and mercury vapor lights emit white light with a distinct preponderance of blue frequencies. Fluorescents can be made to offer more in the warm color range, but the color of fluorescent lighting is not natural and typically tends to produce a too-bright, sterile atmosphere. Sodium-based lights produce a yellowish light and are commonly used for outdoor applications.

In commercial buildings, the primary sources of artificial light are incandescent and fluorescent lighting fixtures. Mercury vapor and metal halide sources are also used in large rooms or high-bay areas. In a typical building, general lighting in office areas is almost entirely fluorescent. Incandescent lights are used for more direct applications where a fluorescent tube is not applicable—for example, accent lighting. Incandescent lights are also used predominantly in dimming applications such as recessed lighting in lecture halls and meeting rooms. Dimming fluorescent fixtures are available, but they have not yet replaced incandescent lights as the choice for dimming applications.

The glow of fluorescent lighting in office settings can often be irritating to occupants. The obvious complaints caused by too much fluorescent light are sore eyes and headaches, lowered morale, and decreased productivity. Poor lighting also has more subtle effects on mood. The eye, it is now known, is most comfortable with natural sunlight, which changes in intensity and color throughout the day. Because indoor artificial light is basically unchanging in color and intensity, there may be adverse effects on the health and well-being of those subjected to it. This is an important new field of study in the area of IEQ, so it is not entirely understood.

Flickering lights can also cause irritation and health problems. Ballasted lights—for example, fluorescent, mercury vapor, metal halide, and sodium lights—are subject to flickering when the ballast malfunctions. This can easily lead to sore eyes and headaches and, ultimately, lower productivity. Glare is also a problem; however, unlike the others described here, it is not a consequence of artificial light but rather involves the light source and reflector positioning. Windows, desktops, and computer screens, even shiny paper, are all reflectors that can cause uncomfortable glare. Glare, depending on the intensity of the light, can quickly lead to discomfort and headaches, especially when reading, typing, or looking at a computer screen.

Thermal Conditions

The climatic setting in which a person is working has a profound impact on how he or she behaves and how well he or she works. But because everyone is different, what is perfectly comfortable to one person in an office may be profoundly uncomfortable to his or her neighbor. In general, the indoor comfort range is considered to be located in the center of the psychometric chart. Generally accepted ranges for comfort are as follows: in winter, temperatures between 68 and 75°F (20 and 24°C) and relative humidity between 30 and 60 percent; in summer, temperatures between 72 and 80°F (22 and 27°C) and relative humidity between 30 and 60 percent. Relative humidity below 30 percent in any season is considered too dry and will lead to discomfort. Typically, lower humidity levels can be tolerated in the winter and higher humidity levels can be tolerated in the summer, but relative humidity levels outside the 30 to 60 percent range are generally uncomfortable in all seasons.

Air velocity, mentioned briefly above, is another variable in the indoor climate that is not a fundamental property of the air. Air velocity varies greatly, depending on where one is in relation to vents, doors, windows, and fans. It is an integral aspect of air conditioning (heating and cooling) that indoor air be circulated; hence, it must have a certain velocity. The goal of HVAC designers is to introduce the highest-velocity air where it has little or no effect on the building occupants, usually along ceilings or walls, so that by the time it comes in contact with people, it has slowed to an undetectable rate. High-velocity air is more likely to cause discomfort in cool indoor climates and, conversely, be welcome in warm indoor climates.

Odors

Odors are one of the most common and annoying indoor environmental problems. Solving these problems is not easy, because the human olfactory system is highly complex and not well understood; moreover, the chemical sources that create many of these odors also are poorly understood. Even simple odors in office settings are complex, consisting of many substances. Typical sources of odors in the indoor environment include tobacco smoke, human body odor, and cleaning and personal grooming products. Off-gassing of building materials is another common source of smells. Complicating this issue is the pronounced difference in individual sensitivity to odors. Visitors to an office are generally far more sensitive to odors than its long-time occupants, for example. Because human reactions to odors are so varied, it is nearly impossible to predict how any one person or group of people will react.

Volatile Organic Compounds

VOCs are carbon-containing compounds that readily evaporate at room temperature and are found in many housekeeping, maintenance, and building products made with organic (carbon-based) chemicals. Paints, glues, paint strippers, solvents, wood preservatives, aerosol sprays, cleansers and disinfectants, air fresheners, stored fuels, automotive products, and even dry-cleaned clothing and perfume are all sources of VOCs. In any indoor environment, there can be up to 100 different VOCs in varying concentrations. Carbon filters can be used to adsorb VOCs, but they must be replaced regularly, as the odors deplete the carbon.

There are six major classes of VOCs: aldehydes (formaldehyde), alcohols (ethanol, methanol), aliphatic hydrocarbons (propane, butane, hexane), aromatic hydrocarbons (benzene, toluene, xylene), ketones (acetone), and halogenated hydrocarbons (methyl chloroform, methylene chloride).

Formaldehyde is highly reactive and may be found in all three states of matter. It is highly soluble in water and can irritate body surfaces normally containing moisture—for example, the eyes and the upper respiratory tract. Formaldehyde gas is pungent and easily detectable by its odor at concentrations well below 1 part per million (ppm). It is perhaps the most commonly occurring VOC in construction, found in many common products such as paints, wood products, and floor finishes. When combined with other chemicals, it can be used as glues and binders in numerous products. Urea formaldehyde foam insulation, particleboard, interior-grade plywood, wallboard, some paper products, fertilizers, chemicals, glass, and packaging materials contain significant amounts of formaldehyde.

Radon

Radon, a colorless and odorless gas, is the product of the decay of the radium isotope that results from the disintegration of uranium-238. An inert gas, radon itself is fairly harmless, but as it decays, the resulting materials, known as *radon daughters*, are not. Radon daughters are not chemically inert, and they form compounds that bind to dust particulates in the atmosphere. When inhaled, these particles can lodge in the respiratory system and cause damage due to the alpha particle radiation they emit. The half-life of the daughters is relatively short: they disintegrate in 1 hour or less. Despite this rapid disintegration, radon is a major concern because it may take 10 to 20 years for the first signs of exposure to develop, and it has serious consequences. The inhalation of radon is the second leading cause of lung cancer in America, is suspected in the deaths of 2000 to 20,000 individuals a year, and is considered one of the most deadly indoor air pollution problems.

Anthony Nero of the Indoor Environment Radon Group at Lawrence Berkeley National Laboratory noted that the average indoor level of radon represents a radiation dose about three times larger than the dose most people get from X-rays and other medical procedures in the course of their lifetime.[4,5] Hundreds of thousands of Americans living in houses with high radon levels are exposed yearly to as much radiation as people who were living in the vicinity of the Chernobyl nuclear power plant in 1986, when one of its reactors exploded. According to the EPA, an acceptable maximum level of radon is 4 picocuries per liter (pCi/L) of air. In Canada, the Atomic Energy Control Board (AECB) also set a level of 4 pCi/L for the general public in homes and other nonoccupational settings. If this level is exceeded, action must be taken to reduce it.

In buildings, radon occurs primarily through diffusion from the underlying subsoil into the building structure. Radon gas can enter a building through cracks or openings such as sewer pipe openings, cracks in concrete, wall-floor joints, hollow masonry walls, and other similar pathways. If the foundation of the building is tight, very little or no radon will enter. Because of the ground-up infiltration process of radon, a multistory building will have lower radon concentrations than a single-story building with an identical foundation. Indoor radon concentrations also relate directly to ventilation and fresh air intake of buildings. Due to energy conservation techniques and the resultant tighter buildings, new buildings may actually encourage the infiltration of radon gas by negative pressurization.

Asbestos

Asbestos is another potentially deadly IAQ problem. Unlike radon, however, the health implications of asbestos have been documented in detail, for it has been a major environmental problem for many years. When it was discovered that the threadlike particles in asbestos could lodge in human lungs, its use began to be phased out. Exposure to asbestos has been definitively linked to stomach and lung cancers.

The term *asbestos* refers to a group of silica-based minerals in fibrous bundles. Introduced in the 1930s and widely used in the United States from 1940 to 1973, asbestos comprises a large number of naturally occurring materials that are processed to produce a manageable form for use in construction, insulation, and fire retardation materials. Indoor building materials containing asbestos include thermal insulation on ceilings and walls; insulating materials used on pipes, ducts, boilers, and tanks; and finishing materials such as ceiling and floor tiles and wall boards. The materials that pose the greatest threat are those that can be easily crumbled or powdered by hand pressure.

High-quantity release of asbestos into the airstream usually occurs during maintenance, renovation, and other construction activities, when it becomes dangerous. There is very little danger to human health if the material is left undisturbed; asbestos becomes a health hazard only when its fibers are released into the air. Most experts agree that if asbestos surfaces are not deteriorating or being abraded, thus releasing asbestos fibers, they are best left alone. Removal of asbestos is very costly and can be done safely only by professionals. An unsafe removal process can do more harm than good by releasing more particles into the air, where they can continue to contaminate a building for years.

Combustion By-Products

Combustion by-products are created under conditions of incomplete combustion. The primary sources of combustion by-products that contribute to the contamination of indoor air are gas, wood, and coal stoves; unvented kerosene space heaters; fireplaces under downdraft conditions; and tobacco smoke. The major by-products include carbon dioxide, carbon monoxide, nitrogen dioxide, sulfur dioxide, and particulates. Their health effects can vary, depending on the type of by-product produced.[5] Each of these will now be described more fully.

Carbon dioxide. Carbon dioxide is a colorless, odorless, and tasteless gas. Although it is a by-product of combustion, it is relatively harmless; it is, after all, also a natural product of respiration. That said, and despite the fact that it is nontoxic, if the concentration of carbon dioxide is too high, the result can be unpleasant and perhaps unhealthy for a building's inhabitants. And since it is a natural product of respiration, it can also be an indicator of the quality of ventilation and IAQ.

Carbon monoxide. Carbon monoxide is another colorless, odorless, and tasteless gas, but it must not be confused with carbon dioxide. The effects of high-level carbon monoxide exposure can range from nausea and vomiting to headaches and dizziness to coma and death. The health effects of low-level carbon monoxide exposure are not clearly defined, but its toxicity is unquestionable. The symptoms of carbon monoxide poisoning, which include nausea, dizziness, confusion, and weakness, may be confused with those of the flu. People with anemia or a history of heart disease can be especially sensitive to carbon monoxide exposure.

Nitrogen dioxide. Concentrated nitrogen dioxide is a dark-brown gas with a strong odor. Exposure can cause irritation of the skin and eyes and other mucous membranes. Controlled human exposure studies and epidemiological studies in homes with gas stoves illustrate that, depending on the level of exposure, nitrogen dioxide can alter lung function and cause acute respiratory symptoms. Because of its ability to oxidize, nitrogen dioxide has been shown to damage the lungs directly. Symptoms of exposure may include shortness of breath, chest pains, and a burning sensation or irritation in the chest. People with chronic respiratory illnesses, such as asthma and emphysema, may be especially sensitive to nitrogen dioxide.

Sulfur dioxide. Sulfur dioxide is a colorless gas with a suffocating odor. It is highly soluble in water and is thus readily absorbed by the mucous membranes. Once it is inhaled, sulfur dioxide is dissolved and forms sulfuric acid, sulfurous acid, and bisulfate ions. During normal nasal respiration, sulfur dioxide is absorbed primarily by the nasal tissues; only 1 to 5 percent reaches the lower respiratory tract. However, when a person breathes through the mouth, for example, during heavy exercise, significant quantities of sulfur dioxide can penetrate the lower respiratory tract even at low concentrations. The primary physical effect of sulfur dioxide exposure is bronchoconstriction. This begins at considerably lower levels for asthmatics than for healthy individuals. The constriction will develop almost immediately upon exposure, but it will also subside just as quickly when exposure ends. The intensity of the constriction is directly related to the amount of sulfur dioxide per unit of time that reaches the lower respiratory tract, not necessarily the level of the exposure. Also, the effect of sulfur dioxide does not increase with time.

Combustion particulates. Particulates produced by combustion can directly affect lung function. The smaller the particulates, the more deeply they penetrate the lungs and thus the more dangerous they become. The particles can serve as carriers for other contaminants or as mechanical irritants that interact with chemical contaminants.

Mold and Mildew

Humidity and airflow rates significantly affect the concentrations of biological contaminants. Moisture can act as a breeding ground for molds, bacteria, and mites. Mites are the most prominent cause of house dust allergies. They are found in beds and pillows, especially when humidity levels are high. An indoor moisture level of 30 to 50 percent relative humidity is recommended to maintain good health as well as comfort. Relative humidity is the amount of moisture in the air relative to the amount of moisture the air can hold when it is completely saturated. Biological contaminants may also multiply in standing water, in cooling towers, in water-damaged ceilings, and on surfaces where moisture in the air condenses on cold walls. Additionally, damp organic materials like leather, cotton, furniture stuffing, and carpets can be contaminated with fungi. Airflow rates also have an important effect on the concentrations of airborne biological pollutants. Reduced flow rates tend to provide a favorable medium for molds, dust, and fungi. The HVAC equipment in a building plays a very important role in maintaining proper airflow rates.

Sick Building Syndrome and Building-Related Illness

Of the wide variety of issues associated with IEQ, in the recent past two in particular stand out: sick building syndrome (SBS) and building-related illness (BRI). Though both refer to health problems associated with IAQ, there is a very important difference between them. SBS describes an assortment of symptoms experienced by a majority of building occupants for which no specific cause can be identified. Typically, SBS is diagnosed when the affected employees' symptoms disappear almost immediately on leaving the building. In contrast, BRI refers to symptoms of a diagnosable illness that can be attributed directly to a defined IAQ problem.

SBS, also known as *tight building syndrome*, is the "condition in which at least 20 percent of the building occupants display symptoms of illness for more than two weeks, and the source of these illnesses cannot be positively identified." Most of the structures that fall victim to SBS are modern office buildings, the majority of which have been constructed over the past two decades and are tightly sealed, mechanically ventilated, and have few or no operable windows. Symptoms of SBS may include headache; fatigue and drowsiness; irritation of the eyes, nose, and throat; sinus congestion; and dry, itchy skin. These symptoms can occur alone or in combination. The most common complaints include flulike symptoms or respiratory tract infections. Some occupants relate SBS to stresslike headaches, coughs, and the inability to concentrate, while others experience dry skin or rashes.[6]

The economic impact of SBS can be tremendous, making it a building owner's worst nightmare. The EPA has estimated that the United States spends over $140 billion in direct medical costs attributable to IAQ problems.[7] SBS is also believed to be responsible for marked decreases in productivity coupled with increases in absenteeism. Vacant buildings and nonrenewed building leases may also be a direct result of SBS. An example of the high costs associated with SBS is the Polk County Court House in Florida. Located in Lakeland, a community in central Florida, the court house was constructed for $37 million and opened in the summer of 1987. Due to a severe case of SBS, it was closed in 1992; its occupants, including prison inmates, had to be evacuated and temporarily relocated. It took three years and $26 million to literally rebuild the facility to correct the original toxic mold problems that were attributed to design and construction problems.

The wide range of conditions associated with both SBS and BRI, some chemical and some biological—including multiple chemical sensitivity, legionellosis, and allergic reactions—are described in the following sections.

Multiple Chemical Sensitivity

Multiple chemical sensitivity (MCS), a relatively recently identified condition related to IAQ, is marked by sensitivity to a number of chemicals, all at very low concentrations.[8] MCS is characterized by severe reactions to a variety of VOCs and other organic compounds that are released by building materials and many consumer products. These reactions may occur following one sensitizing exposure or a sequence of exposures. It should be noted, however, that there is currently a great deal of debate over the legitimacy of the condition. Some contend that it is a physical illness, while many others believe the cause to be psychosomatic.

Legionellosis

Legionellosis refers to two important bacterial diseases: Legionnaire's disease and Pontiac fever, caused by the bacterium *Legionella pneumophila*. The diseases are not spread via person-to-person contact, but rather through the soil-air and water-air links both indoors and outdoors. The bacteria can survive in water for up to a year under certain conditions. *Legionella* prefers stagnant water, which is found in the drain pans of HVAC units and cooling towers. Fans then can transfer the bacteria, to be inhaled by unsuspecting victims. Sources of *Legionella* in residences and other buildings may also include hot tubs, vaporizers, humidifiers, and contaminated forced-air heating systems. Algae and other aquatic life forms can promote the growth of *Legionella* by providing the bacteria with food.

Pontiac Fever

In July 1968, 95 out of 100 people employed in—ironically—a public health building in Pontiac, Michigan, became ill with a flulike ailment. In fact, if the number of cases had not comprised such a high proportion of the employees, the disease probably would have been diagnosed as the flu. The employees all claimed to suffer from headaches, fevers, and muscle aches and pains. Called *Pontiac fever*, the disease was eventually traced back to a faulty HVAC system. However, it was not until the discovery of Legionnaire's disease, nearly 10 years later, that the bacterium that caused Pontiac fever was finally identified.

Pontiac fever is a mild form of legionellosis. It is characterized by a high attack rate (90 percent) and a short incubation period of 2 to 3 days. The disease lasts for only 3 to 5 days and requires no hospitalization. Symptoms include those exhibited by the employees in 1968, as well as chills, sore throat, coughing, nausea, diarrhea, and chest pain. Many people may never suspect that they have Pontiac fever, as only an estimated 5 to 10 percent of those seeking medical care have lab tests done.

Legionnaire's Disease

Legionnaire's disease is a type of pneumonia caused by *Legionella*. Both the disease and the bacterium were discovered following an outbreak traced to a 1976 American Legion convention in Philadelphia, Pennsylvania. This disease develops within 2 to 10 days after exposure to *Legionella*, and early symptoms may include loss of energy, headache, nausea, aching muscles, high fever [often exceeding 104°F (40°C)], and chest pains. Later, many bodily systems, as well as the mind, may become affected. The disease eventually causes death if high fever and antibodies cannot defeat it. Victims who survive may suffer permanent physical or mental impairment. The CDC has estimated that the disease infects 10,000 to 15,000 persons annually in the United States; others have estimated as many as 100,000 annual US cases.

Legionnaire's disease is a severe multisystem illness that can affect the lungs, gastrointestinal tract, central nervous system, and kidneys. It is characterized by a low attack rate (2 to 3 percent), a long incubation period (2 to 10 days), and severe pneumonia. Unlike Pontiac fever, hospitalization is required. Most victims are men in their 50s and 60s who are smokers and/or have underlying respiratory problems. Alcohol consumption, diabetes, and recent surgery can also be contributing factors.

Allergic Reactions

Allergies are reactions to a form of indoor air pollution that occur when the body responds to nontoxic substances, like pollen, as threats. The body will mimic the effects of a real illness by stimulating the production of white blood cells to combat the allergen. An individual usually does not experience an allergic reaction until after the second exposure to a specific allergen. The first exposure results in the manifestation of the allergy. Allergens that cause an allergic response include viable and nonviable agents. Viable agents include bacteria, fungi, and algae. Common nonviable agents include house dust, insect and arachnid body parts, animal dander, mite fecal pellets, remains of molds and their spores, pollens, and dried animal excretions.

Preventing encounters with offending allergens is easier said than done. They constitute a new variation on the IAQ problem in that the reactions of a building's inhabitants to an allergen can vary more than with other environmental factors. What may send one person gasping to the emergency room may have absolutely no effect on another. Regular cleaning to remove dust, the use of high-efficiency filters, and regular filter changing can help to reduce or eliminate biological contaminants.[9]

Integrated IEQ Design

Clearly, the complex range of IEQ issues warrants an integrated approach to the design of buildings to maximize the quality of human occupied spaces. The *Whole Building Design Guide* provides a good overview of integrated IEQ design and suggests that the following measures for attaining good IEQ in buildings:[10]

- Facilitate quality IEQ through good design, construction, and operating and maintenance practices.
- Value aesthetic decisions, such as the importance of views and the integration of natural and man-made elements.
- Provide thermal comfort with a maximum degree of personal control over temperature and airflow.
- Supply adequate levels of ventilation and outside air for acceptable indoor air quality.
- Prevent airborne bacteria, mold, and other fungi through building envelope design that properly manages moisture sources from outside and inside the building, and with heating, ventilating, air-conditioning (HVAC) system designs that are effective at controlling indoor humidity.
- Use materials that do not emit pollutants or are low-emitting.
- Assure acoustic privacy and comfort through the use of sound absorbing material and equipment isolation.
- Control disturbing odors through contaminant isolation and removal, and by careful selection of cleaning products.
- Create a high-performance luminous environment through the careful integration of natural and artificial light sources.
- Provide quality water.

These important recommendations are covered in more detail in the following sections.

FACILITATE QUALITY IEQ THROUGH GOOD DESIGN, CONSTRUCTION, AND OPERATING AND MAINTENANCE PRACTICES

The project design team can make major contributions to the quality of the project's IEQ through the specification of products and materials, as well as the design of lighting, daylighting, air-conditioning, ventilating, and other systems that have a direct bearing on the environmental quality of the building. Specifying materials that contain zero or low VOCs and entryway systems that remove chemicals and dust particles from people entering the building are examples of materials and product specifications that can contribute to good IEQ. Designing an integrated daylighting/lighting system often involves computer simulation and the selection of appropriate types of windows that can both facilitate good daylighting and minimize solar thermal heat gains in the building. The construction phase of the project is also very important in ensuring a high-quality indoor environment because best practices can eliminate possible future causes of indoor environmental problems. An example is the potential contamination of air handlers, ductwork, diffusers, and grilles that carry air throughout the building by dust and debris generated during the construction process. Good construction practices can eliminate this possible threat to air quality. Clearly, the operations and maintenance phase is key to good long-term IEQ and facilities managers must be aware of best practices in retaining the environmental quality provided by the project team.

VALUE AESTHETIC DECISIONS

Designers have a responsibility to ensure a high degree of aesthetic quality in buildings, which not only contributes to the cultural value of the facility over the long term but also promotes indoor environmental quality. For example, operable windows, which have long been considered problematic for buildings due to the issue of coordinating their use with the operation of the mechanical systems, are making a comeback in the current era, because of their ability to provide natural ventilation. Buildings are also now being designed to connect people to nature, and the provision of good views for building occupants is often a goal of the project team.

PROVIDE THERMAL COMFORT

Thermal comfort for building occupants is a major objective of most high-performance building projects and involves the interplay of several parameters: air speed, temperature, humidity, and radiant temperature. The first three parameters—air speed, temperature, and humidity—are provided by the designers of the building's HVAC system, while radiant temperature, which is the result of direct solar radiation on the skin, is controlled by the selection of windows, shading devices, and other approaches that can affect solar radiation through windows. ASHRAE 55-2010, Thermal Environmental Conditions for Human Occupancy, is the basis for thermal comfort in high-performance buildings in the United States. Provision of control over thermal comfort by building occupants is also an important consideration for high-performance buildings because the health and productivity of the occupants is, at least in part, a function of their ability to adjust their surroundings to make them comfortable and, hence, more productive.

SUPPLY ADEQUATE LEVELS OF VENTILATION AND OUTSIDE AIR

Contamination levels inside buildings deteriorate over time depending on the number of occupants, the activities in the building, the materials and products of construction, and, most importantly, the ventilation in the building. ASHRAE 62.1-2010, Ventilation for Acceptable Indoor Air Quality, provides the framework for designing effective insulation systems for buildings. Carbon dioxide levels in buildings are an important surrogate for the overall air pollution levels in the buildings; that is, as carbon dioxide levels rise, so too do the levels of other contaminants such as VOC fine particulates in a wide range of other chemicals. More recently, designers have learned how to optimize building ventilation systems by monitoring carbon dioxide levels and then using this as feedback for the control of the ventilation system, such that, as carbon dioxide levels rise and fall, the ventilation rates are adjusted accordingly by the automatic control system. This strategy has the benefit of both providing precise ventilation and minimizing the energy required to condition outside air being brought into the building for conditioning purposes. System designers have also learned how to separate the ventilation air system from the recirculating air system in order to provide even more precise control over ventilation rates.

PREVENT AIRBORNE BACTERIA, MOLD, AND OTHER FUNGI

Mold has become a major issue of IEQ systems design. It is important that the building envelope be designed to prevent intrusion of water into the building through careful detailing and the incorporation of moisture barriers into the exterior wall system of the building. Controlling humidity with the HVAC system is also essential to preventing the growth of mold in the building, especially at extreme load conditions, when outside humidity is either very high or very low. Mold is measured by the number of spores per cubic meter of air, and it is important to ensure that the level of mold in the indoor air is less than that of the outside air, and in no case more than 700 spores per cubic meter of air.

USE MATERIALS THAT DO NOT EMIT POLLUTANTS OR ARE LOW EMITTING

VOCs are complex chemicals that are both synthetic and naturally occurring. Many, like formaldehyde, are incorporated into building materials to enhance their properties, for example, making paint more durable and more rapidly drying. Toluene, xylene, and benzene are other examples of synthetic VOCs that are toxic and harmful to human and ecosystem health. In spite of the benefits they may provide, VOCs pose a threat to the health of building occupants and are being eliminated in high-performance green buildings. Chemicals that are used in the building, for example, in cleaning supplies and copy machines, should be specially stored in spaces that prevent their migration into the surrounding building environment. Radon control should also be considered in areas where it has been identified as present in the local soils. For renovation projects, the removal of asbestos and lead-based paint should be accomplished in a manner that prevents exposure to workers and future exposure to building occupants.

ASSURE ACOUSTIC PRIVACY AND COMFORT

Transmission of noise and sounds through buildings can affect both the health and comfort of building occupants, and significant effort should be made to minimize noise generation and transmission by the use of sound-absorbing materials; sound and noise attenuating walls, floors, and ceilings; isolating air handlers and other rotating machinery from the building; and the design of HVAC systems that are quiet and that do not transmit conversations between spaces.

CONTROL DISTURBING ODORS THROUGH CONTAMINANT ISOLATION AND PRODUCT SELECTION

Some building spaces such as copying rooms, janitors closets, storage rooms, and designated smoking areas should be negatively pressurized and isolated from the other spaces in the building, and they should be exhausted directly to the exterior of the building. This strategy prevents the migration of chemicals and odors that are typical of these spaces to the occupied areas of the building.

CREATE A HIGH-PERFORMANCE LUMINOUS ENVIRONMENT

Daylighting has enormous benefits because it contributes directly to human health and also can provide significant energy savings when part of a well-designed, integrated lighting system. A wide range of high-performance lighting systems are now available that can provide high-quality, high-efficiency light.

PROVIDE QUALITY WATER

The building water system should be designed to provide the appropriate quality of water for all purposes. Potable water is needed for drinking, kitchen sinks, water fountains, lavatories, and dishwashers, and its quality should be monitored to ensure that it does not contain inappropriate levels of various metals and bacteria. Good-quality water, which does not meet potable water standards, can also be incorporated into the building for uses such as flushing toilets and urinals and for landscape irrigation. As is the case with all aspects of high-performance building, periodic maintenance helps ensure that the water systems of the building provide the quality of water required for the activities in the building.

Addressing the Main Components of Integrated IEQ Design

In the following sections, we will address the design of the major subsystems that affect IEQ, including integrated lighting, daylighting, and views; thermal comfort and comfort control; acoustic comfort; electromagnetic radiation; and the design of building HVAC systems.

INTEGRATED LIGHTING, DAYLIGHTING, AND VIEWS

Building lighting systems are complex, and their design should consider optimizing a balance between contributing to human health and reducing energy consumption. Lighting systems in buildings consume about 30 percent of total US energy, a significant cost and a significant contribution to climate change. Yet good lighting design is important to human health, and providing an inadequate lighting system would be counterproductive because it would result in increased illness and absenteeism and decreased productivity. Daylighting has the dual benefit of both contributing to human health and significantly reducing building lighting energy. An important consideration in the design of an integrated lighting system for buildings is the provision of views to the outside for the building occupants. The health and productivity of building users is directly affected by their ability to see the outside world, especially nature, during their normal workdays or school days. This concept was first articulated Edward O. Wilson in his book, *Biophilia*, when he suggested that humans crave connection with nature and that improving the ability of people inside buildings to connect with the outside world provides positive benefit for their psyche and health.[11] The idea is that humans evolved deeply enmeshed with the intricacies of nature and that we still have an affinity with nature ingrained in our genes. Not all daylighting systems provide views. Buildings that rely on rooftop clerestory windows and skylights for daylighting provide views of the sky but not of nature at the ground level. Vision windows that extend from near the floor to the ceiling provide this type of visual access to nature (see Figure 12.1).

The degree of visual comfort in a building is a function of both daylight and artificial lighting levels. Generally, these two forms of lighting can be evaluated separately, since artificial lighting must be provided for those situations where there is no or insufficient daylight available, for example, in the evenings or on cloudy days. However, there is a transition point in modern green buildings where artificial lighting and daylighting are traded off, depending on the

Figure 12.1 The outside is brought in at Hillside Middle School in Salt Lake City, Utah. Floor-to-ceiling windows throughout the design provide not only transparency and daylighting but also excellent views from common areas (A) as well as from the library (B), classrooms, and office clusters. (GSBS Architects and Benjamin Lowry Photographer)

availability of daylighting. Modern lighting control systems have the capability of throttling artificial light levels in response to the availability of daylighting, thus optimizing the use of electrical energy in the building for lighting purposes. Note that the energy benefits of daylighting are covered in Chapter 9.

Daylighting, of course, has long been important to architecture for obvious reasons. In the era prior to electricity, building illumination was provided largely by openings, windows, and glazing, and significant design effort was invested in using natural light to the maximum extent possible. A wide variety of design features are available to architects for the purposes of enhancing the daylighting system (see Table 12.3 and Figure 12.2A–D).

According to the *Whole Building Design Guide*, a daylighting design consists of systems, technologies, and architecture. The following list indicates some of the components of a typical daylighting system design, although all of them may not be present at the same time:

TABLE 12.3

Design Features Available to Architects to Maximize Daylighting in Buildings

Atrium	Open area that interconnects a number of floor spaces within a building
Sawtooth roof	Comprised of a number of triangular-shaped parallel sections
Roof monitor	A raised section of roof that includes a vertically (or near vertically) glazed aperture for the purpose of illumination
Skylight	A relatively horizontal glazed roof aperture for the admission of daylight
Light court	A large shaft sometimes using the walls of its surroundings to reflect light
Clerestory windows	Vertical glazing high on a wall
Light shelf	A reflective horizontal surface that can be installed on both the exterior and interior of a building
Heliostat	Mirror that tracks the sun to reflect light
Synthetic wall window	Wall glazing located at ground level to provide natural light to below-grade areas
Deadlight	Fixed glass segment embedded into cast iron stair or sidewalk frames to facilitate natural light to subsurface areas

- Daylight-optimized building footprint
- Climate-responsive window-to-wall-area ratio
- High-performance glazing
- Daylight-optimized fenestration design
- Skylights (passive or active)
- Tubular daylight devices
- Daylight redirection devices
- Solar shading devices
- Daylight-responsive electric lighting controls
- Daylight-optimized interior design (such as furniture design, space planning, and room service finishes)

As is the case with most aspects of passive design, the design of the daylighting system begins with the building footprint. In general, for good daylighting, buildings should be oriented on an east–west axis that maximizes north and south exposures. It is important that the width of the building footprint in the north–south direction be minimized and in no case be less than 60 feet wide. German regulations stipulate that, in an office building, the occupants must be within 15 meters of an outside window to support both daylighting and views.

Lighting controls integrated with daylighting are important in the design of lighting systems because they help provide a constant level of illumination by using the artificial lighting to compensate for changing levels of daylight. These types of lighting controls consists of photocells that control either continuous dimming or stepped ballasts in the light fixtures. Nowadays, occupancy sensors, which turn the lights on and off in response to the presence of people in the space, are also integrated with the daylight-responsive lighting controls.

Design of the daylighting system must consider glare control, which deals with direct sunlight entering a space. Clearly, maximizing daylighting is

Figure 12.2   (A) A clerestory conducts diffuse light into a lobby below at St. Johns River State College in St. Augustine, Florida. (D. Stephany)

Figure 12.2   (B) A sawtooth roof design at Manassas Park Elementary School in Manassas Park, Virginia, a suburb of Washington, DC, is oriented to allow diffuse light to enter the building. This building also uses solar tubes that illuminate separate interior spaces on sunny days. (© Prakash Patel for VMDO Architects)

important but not at the expense of creating unpleasant working conditions in space. This is particularly important in daylighting design because the south side of the building will generally provide the bulk of the natural light for the building, and care must be taken to ensure that direct-beam sunlight is controlled by internal and external shading devices, horizontal and vertical louvers, and light shelves.

The following are some of the design considerations in designing daylighting systems:

Figure 12.2   (C) A light shelf allows sunlight in while protecting users from glare that is often associated with tall windows. (Decorating with Fabric)

Figure 12.2   (D) An atrium at EDS corporate headquarters in Plano, Texas, connects several floors through the use of daylighting. (Source: National Institute of Building Sciences)

- Increase the number of perimeter daylight zones.
- Promote daylight penetration high in the space by locating windows high on the wall or by providing roof monitors and clerestories.
- Reflect daylight by using light colors to increase room brightness.
- Slope ceilings to direct more light into space. Avoid direct-beam sunlight on critical visual tasks.

- Filter daylight with vegetation, curtains, and louvers to help distribute the light.

- Be aware that different building orientations require different daylighting strategies. For example, light shelves, while effective on the south side of a building, would not be effective on the east and west sides.

In general, it is a good idea to use either a computer model or a physical model to assist in the design of an integrated lighting and daylighting system. Computer software such as Radiance and Ecotect is available to perform a detailed design of the lighting system. Similarly, a physical model of the building can be constructed and used to test different daylight strategies, such as glazing, orientation, and trade-offs between daylighting and energy savings.

THERMAL COMFORT AND COMFORT CONTROL

ASHRAE 55-2010, Thermal Environmental Conditions for Human Occupancy, defines thermal comfort as the state of mind in humans that expresses satisfaction with the surrounding environment. It describes a person's psychological state of mind and is usually referred to in terms of whether someone is feeling too hot or too cold. More accurately, it describes the combination of environmental factors that can provide good thermal comfort. For example, according to the ASHRAE standard, office spaces have a suggested summer temperature between 74.3 and 77.9°F (23.5 and 25.5°C) and an airflow velocity of 0.59 ft/s (0.18 m/s). In the winter, the recommended temperature is between 70 and 73°F (21.0 and 23.0°C) with an airflow velocity of 0.49 ft/s (0.15 m/s).

In the United States, maintaining constant thermal conditions in offices is important, and even a minor deviation from comfort may be stressful and affect performance and safety. Workers already under stress are less tolerant of uncomfortable conditions. In other countries, such as Germany, where there is enormous emphasis on low-energy buildings, the comfort zone is not as rigid and there is more acceptance of a wider range of comfort conditions.

Providing thermal comfort in building spaces is a complex undertaking because it is a function of four environmental and two personal factors. The environmental factors are temperature, thermal radiation, humidity, and air speed, and the personal factors are clothing and metabolism. Three of the four environmental factors (temperature, humidity, and air speed) are familiar. Thermal radiation is the affect of direct solar or other radiation on the skin, and it can affect people in building spaces where there is direct sunlight, without shading or glare protection, into the space. Thermal comfort can be achieved through a wide variety of combinations of these factors. For example, it is well known that air speed can compensate for higher temperatures and thus the use of ceiling fans in rooms to provide airflow that makes the higher temperatures more tolerable. Similarly, lower humidity can make higher temperatures more acceptable. In addition to the four environmental factors noted, other factors such as clothing, activity levels, and personal factors such as individual health affect thermal comfort. Thermal comfort control is the ability of the occupant(s) to adjust at least one of the four environmental factors to their liking. Giving building users at least some degree of control over thermal comfort is generally recognized as contributing to the health and productivity of the occupants and is considered to be a significant measure in the design of high-performance green buildings.

Thermal comfort is based on research conducted on the four environmental factors by Ole Fanger and others at Kansas State University in the 1970s. Perceived comfort was found to be a complex interaction of the four factors. It was found that the majority of individuals would be satisfied by an ideal set of values. As the range of values deviated progressively from the ideal, fewer and fewer people were satisfied. This observation could be expressed statistically as the percentage of individuals who expressed satisfaction by comfort conditions and the predicted mean vote (PMV). The PMV index predicts the mean response of a larger group of people who vote according to the ASHRAE thermal sensation scale where:

+3 hot

+2 warm

+1 slightly warm

0 neutral

−1 slightly cool

−2 cool

−3 cold

In general, the rule is that if 80 percent of a population agrees that the thermal conditions are comfortable, then the combination of environmental factors is providing comfortable conditions. This is the basis for the recommendations in ASHRAE 55-2010 for acceptable combinations of environmental factors.

Inclusion of clothing and metabolic rates in the determination of thermal comfort is also included in ASHRAE 55-2010. The Clothing Level (CLO) is a numerical value describing thermal insulation provided by clothing and ranges from 0.5 to 1.5. The CLO valuation assumes that a person is standing. If an individual spends most of the day sitting, the CLO value may need to be increased, depending on the type of chair. CLO values are determined for the average occupant for each season on a space-by-space basis.

The Metabolic Rate (MET) estimates the typical level of activity of the occupants within a given space. MET is expressed on a decimal scale and ranges from 0.7 to 8.7. The 0.7 level represents sleeping or resting, while above 1.0 is light activity; greater than 2.0 represents moderate activity and perspiration. When values rise above 1.0, evaporation of perspiration becomes a factor in an individual's level of comfort. An estimate of the average metabolic rate of the occupants in a given space is determined as an input to assessing thermal comfort.

The 2010 version of ASHRAE 55 addresses the use of various new technologies to deliver thermal comfort using lower-energy approaches (see Figure 12.3). Air movement, in general, is becoming a more popular strategy for cooling occupants as opposed to lower operational temperature because the energy requirements are lower. ASHRAE 55-2010 includes a new method for determining the cooling effect of air movement above 30 feet per minute (fpm) (1.64 meters/second). This allows ceiling fans or other means of elevating air speed to provide comfort at higher summer temperatures than were previously permissible. New provisions based on field-study research allow elevated air speed to broadly offset the need to cool air in warm conditions. ASHRAE 55-2010 allows modest increases in operative temperature beyond the predicted PMV limits as a function of air speed and air turbulence, both of which increase the cooling sensation by using convection to remove heat from the skin.

ASHRAE STANDARD

Thermal Environmental Conditions for Human Occupancy

See Appendix I for approval dates by the ASHRAE Standards Committee, the ASHRAE Board of Directors, and the American National Standards Institute.

This standard is under continuous maintenance by a Standing Standard Project Committee (SSPC) for which the Standards Committee has established a documented program for regular publication of addenda or revisions, including procedures for timely, documented, consensus action on requests for change to any part of the standard. The change submittal form, instructions, and deadlines may be obtained in electronic form from the ASHRAE Web site (www.ashrae.org) or in paper form from the Manager of Standards. The latest edition of an ASHRAE Standard may be purchased from the ASHRAE Web site (www.ashrae.org) or from ASHRAE Customer Service, 1791 Tullie Circle, NE, Atlanta, GA 30329-2305. E-mail: orders@ashrae.org. Fax: 404-321-5478. Telephone: 404-636-8400 (worldwide), or toll free 1-800-527-4723 (for orders in US and Canada). For reprint permission, go to www.ashrae.org/permissions.

ISSN 1041-2336

American Society of Heating, Refrigerating and Air-Conditioning Engineers, Inc.
1791 Tullie Circle NE, Atlanta, GA 30329
www.ashrae.org

Figure 12.3 The 2010 version of ASHRAE 55 created several other approaches to achieving thermal comfort, among them the use of increased air speeds.

ACOUSTIC COMFORT

Consideration of acoustic comfort in buildings generally falls far down the list of priorities, both in sustainable construction and in conventional building design. Acoustics are, in fact, very important to the health, well-being, and productivity of people in offices, schools, and virtually every other type of facility. Providing a good acoustical environment for building occupants helps increase their performance and reduces the incidence of illness and lost workdays. Acoustic comfort is part of a bigger picture of overall space comfort and includes not only acoustics but also other issues such as thermal comfort, lighting quality, the availability of daylight, and other similar factors. Noise is prevalent in buildings and can come from outside traffic noise, voices within the building, mechanical equipment in adjacent spaces, copiers, phones, and numerous other sources. In order to produce a good acoustical environment, there are several problems that must be addressed: (1) noise outside the building, (2) noise from adjacent spaces,

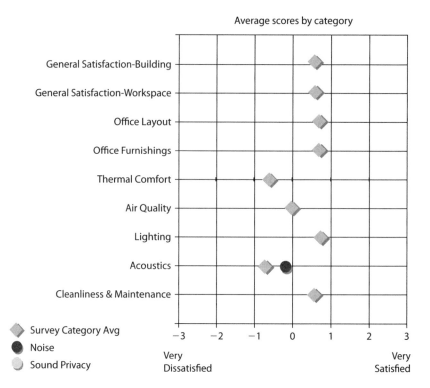

Figure 12.4 A survey conducted by the Center for the Built Environment (CBE) at the University of California at Berkeley indicated that of the various factors that constitute work space comfort, occupants were most dissatisfied with the acoustics of the spaces. (*Source:* National Institute of Building Sciences)

and (3) lack of self-control in the spaces of the building. Although the noise in the spaces may not be harmful to hearing, the presence of distracting noise reduces concentration on work or study and decreases the productivity of individuals. The Center for the Built Environment (CBE) at the University of California at Berkeley conducted postoccupancy evaluations of 15 buildings through a survey of 4096 respondents and found that over 60 percent of the occupants of office cubicles think that acoustics interfere with their productivity (see Figure 12.4).[12] Clearly, acoustics are a major concern of people and the workforce, and poor acoustics design can result in the building performing in a manner that compromises, rather than contributes, to human health.

Starting Point: Sound and Noise Control Terminology

In the United States, noise reduction is measured by the *Sound Transmission Class* (STC), which is a number that represents the noise reduction, in decibels (dB), of a building element such as a wall or window. Note that the decibel scale is logarithmic; thus, a 10 dB reduction in sound between two spaces corresponds to about a 50 percent reduction in the sound volume. For example, a 40 dB sound is about half as loud as a 50 dB sound. With respect to STC ratings, a wall with an STC rating of 20 provides a 20 dB sound reduction. A wall with an STC rating of 20 and with a 60 dB sound level on one side would reduce the noise level to 40 dB on the other side. A typical home interior wall constructed of one sheet of ½-inch drywall on either side of a wood frame has an STC of about 33. The scale most commonly used to measure decibels in a space is referred to as the dBA scale. Typical STC values and their effects on sound levels are indicated in Table 12.4.[13] Outside the United States, the *Sound Reduction Index* (SRI) is used instead of STC ratings.

The *Noise Criterion* (NC) is a rating for interior noise and noise from a variety of sources, including air-conditioning equipment. The lower the required NC rating for a space, the quieter the space will be. Table 12.5 shows the

TABLE 12.4

Effects of STC Rating on Sound Transmission through a Building Element

STC	Sound Level
25	Normal speech can be understood quite easily and distinctly through the wall.
30	Loud speech can be understood fairly well, normal speech heard but not understood.
35	Loud speech audible but not intelligible.
40	Onset of "privacy."
42	Loud speech audible as a murmur.
45	Loud speech not audible; 90% of statistical population not annoyed.
50	Very loud sounds such as musical instruments or stereo can be faintly heard; 99% of population not annoyed.
60+	Superior soundproofing; most sounds inaudible.

TABLE 12.5

Recommended NC and Equivalent Sound Levels (dBA) for Various Typical Building Spaces

Type of Space	Recommended NC Level	Equivalent Sound Level (dBA)
Assembly halls	25–30	35–40
Churches	30–35	40–45
Factories	40–65	50–75
Private offices	30–35	40–45
Conference rooms	25–30	35–40
Classrooms	25–30	35–40
Libraries	35–40	40–50
Homes	25–35	35–45
Restaurants	40–45	50–55
Concert halls	15–20	25–30
Motel rooms	25–35	35–45

recommended range of NC and dBA values for various typical building spaces.[14]

Reverberation time is an important description of the acoustic environment of a space. It is the time, in seconds, that it takes a sound to decay 60 dB below its original level. Spaces with longer reverberation times, above 2 seconds, are characterized by hard surfaces and the ability to hear conversations or lectures is impaired because of the presence of past sounds, but they make sense as spaces for concerts. A space with a long reverberation time is referred to as a "live" environment. When sound dies out quickly within a space, it is referred to as being an acoustically "dead" environment. An optimum reverberation time depends highly on the use of the space. For example, speech is best understood within a "dead" environment. Music can be enhanced within a "live" environment as the notes blend together. Different styles of music will also require different reverberation times (see Table 12.6).[15]

Reverberation time is affected by the size of the space and the amount of reflective or absorptive surfaces within the space. A space with highly absorptive surfaces will absorb the sound and stop it from reflecting back into the space.

TABLE 12.6

Recommended Maximum Reverberation Times for Speech and Music

	Reverberation Time Range (seconds) and Acceptability			
Type of Sound	0.8–1.3	1.4–2.0	2.1–3.0	Optimum Reverberation Time (sec)
Speech	Good	Fair–Poor	Unacceptable*	0.8–1.1
Contemporary Music	Fair–Good	Fair	Poor	1.2–1.4
Choral Music	Poor–Fair	Fair–Good	Good–Fair	1.8–2.0+

*With an adequately designed and installed sound system, speech intelligibility concerns can be mitigated. The optimum reverberation time can be somewhat subjective and can shift based on numerous variables.

This would yield a space with a short reverberation time. Reflective surfaces will reflect sound and will increase the reverberation time within a space. In general, larger spaces have longer reverberation times than smaller spaces. Therefore, a large space will require more absorption to achieve the same reverberation time as a smaller space. Notre Dame Cathedral in Paris, France, has a reverberation time of over 8 seconds and is a good space to hear pipe organ music, but a speech would be virtually unintelligible.

Reverberation time can also be adjusted within an existing space. Tests can be performed in a space to determine the existing reverberation time. Absorptive materials can then be added to or removed from a space to achieve the desired reverberation time. Whenever possible, it is highly advisable to consider reverberation time and other aspects of acoustics at the design stage. Making revisions to a space after the fact can be more costly and compromise aesthetics.

The Noise Reduction Coefficient (NRC) is a single-number index determined in a lab test and used for rating how absorptive a particular material is. This industry standard ranges from 0 (perfectly reflective) to 1 (perfectly absorptive). Acoustical ceiling tiles are normally specified to have an NRC of at least 0.75. Although they sound similar, NRC and STC have very different meanings. STC is the sound attenuation, in dBA, of a building element such as a wall, while NRC is the fraction of the sound that is absorbed by a material.

Exterior Noise Issues and Control

Producing a good indoor acoustical environment requires good planning and site selection to handle potential problems with high external noise levels. In general, sites that are in high noise areas such as near industrial areas and highways, should be avoided, and site selection should include locations that are suitable for the given purpose. For example, it is a good idea to site a school in a relatively quiet area so that the ambient external noise levels are relatively low, and extreme measures are not required to reduce the noise transmission into the building. If there is noise from a nearby highway, for example, the building can be designed such that storage areas, restrooms, janitors closets, and mechanical rooms are on the side facing the source of the noise. More sensitive areas such as classrooms can be located on the quiet side of the building. Earth berms or other structural solutions such as concrete barriers may be required if there is more than one direction from which significant noise is generated. Selection of building components is important in providing good acoustical protection from exterior noise sources. Windows, for example, are an important consideration because although they allow daylight and control heat and glare, they are vulnerable to noise transmission and must be selected with special consideration under acoustical characteristics. Double- and triple-pane glass with inert gas infill may be the best solution for situations where there is significant exterior noise yet maximum daylighting is desirable for health and energy reasons.

Interior Space Acoustic Requirements

Each type of interior space has different considerations and requirements, depending on the types of activities occurring in the space. Private offices, for example, require a space where private conversations can occur without being heard in adjacent spaces and where the acoustic conditions support the health and productivity of the worker. These types of spaces generally have problems of noise transmission through partitions, excessive noise levels in the room, and noises from the air-handling system of the building. Some of the solutions recommended by the *Whole Building Design Guide* are to extend walls from floor to structural deck above, insulate partitions to achieve the required STC value to reduce noise transmission from adjoining spaces, and locate offices and conference rooms so that they are not adjacent to mechanical equipment rooms.

Classrooms are spaces designed for learning, and modern classrooms generally will have multimedia communications environments. Good acoustics are needed for effective verbal communication, which means that there must be relatively low noise levels and vertical reverberation. Some of the types of noises that interfere with the learning process are noises from outside the school such as the nuclear traffic and aircraft flying over, hallway noise, other adjacent classrooms, mechanical equipment and ductwork, and noises within the classroom itself. The recognition of the need to have a high-quality acoustical learning environment in classrooms resulted in the publication of ANSI/ASA S 12.60, American National Standard Acoustical Performance Criteria, Design Requirements, and Guidelines for Schools. ANSI/ASA S 12.60 provides acoustical performance criteria, design requirements, and design guidelines for new school classrooms and other learning spaces. It requires both maximum background noise levels and maximum reverberation times for core learning spaces such as classrooms.

- ANSI/ASA S 12.60 requirements for background noise set the tone for acoustic comfort in core learning spaces in schools. Background noise is composed of noise from building systems, exterior sound transmission, and sound transmission from adjacent spaces. Excessive background noise can seriously degrade the ability to communicate.

- For core learning spaces with internal volumes of 20,000 cubic feet or less, one-hour steady-state background noise levels should not exceed 35 dBA.

- For core learning spaces with internal volumes of 20,000 cubic feet or more, one-hour steady-state background noise levels should not exceed 40 dBA.

- If the noisiest one-hour period during which learning activities take place is dominated by transportation noise, the maximum noise limits are increased by 5 dB.

Controlling the background noise levels within a space involves careful consideration of several building systems. Noise from the HVAC system, electrical fixtures, light fixtures, and plumbing system should all be considered in the noise control design. According to this standard, it is the architect's or designer's responsibility to specify systems and installation methods in order to meet the background noise levels required in the standard. The implementation of the noise control design is the responsibility of the contractor.

The following are the key reverberation time requirements for core learning spaces:

- The maximum reverberation time for core learning spaces with internal volumes greater than 10,000 cubic feet should not exceed 0.6 second.
- For core learning spaces with internal volumes of more than 10,000 but less than 20,000 cubic feet, the maximum reverberation time is 0.7 second.
- Reverberation time for spaces with more than 20,000 cubic feet of internal volume is not specified; however, guidelines are given in Annex C of the standard.

Sound Masking

Sound masking is the introduction of unobtrusive background sounds in the office environment to reduce interference from distracting office sounds and render speech from nearby workers virtually unintelligible. Sound masking means that the stable background noise of the office is raised controllably to minimize the intelligibility of nearby speech without creating a new source of distraction. A sound-masking level of 40 to 45 dBA would be typically recommended for office use. It is often used in open and closed offices where the ambient sound level is too low and, as a result, privacy is compromised. Sound masking works by electronically producing sounds similar to softly blowing air and projecting it through speakers installed above tiles in the ceiling. The sound is evenly distributed throughout the area being masked and can be adjusted to the individual privacy requirements in any given area. In an open-plan office without a suspended ceiling, speakers can be set on the systems furniture or even under the raised floor. Appropriate sound masking can be used to achieve acceptable speech privacy between two neighboring workstations. Optimum sound masking is smooth and unnoticeable and similar to ventilation system noise. The sound pressure level and spectrum need to be considered to obtain a balance between acoustic comfort and efficient masking performance. In many cases, ventilation creates an appropriate masking. In large and high open offices, constant occupant activities and babble can create an appropriate masking. But, in many cases, the creation of optimum masking requires an electronic audio system (see Figure 12.5).

The use of electronic masking has not become common practice although the importance of masking is emphasized in acoustic design guidelines worldwide. One reason may be that very few research reports have been published in this area and the human health impacts have not been established. However, in general, due to the relatively low noise levels of these systems, they tend to be within the range of normal office noise and are not considered harmful to the building occupants.

ELECTROMAGNETIC RADIATION

Exposure to electromagnetic radiation is fairly commonplace. Natural electromagnetic radiation occurs in the form of light and heat and, aside from direct sunlight power, naturally occurring radiation levels are rather low. However, through advances in technology, additional radio radiation sources are having an impact on humans. Figure 12.6 shows frequently occurring radiation sources, arranged according to their frequency ranges and their effect on humans.[16] Electromagnetic radiation from technology is often referred to as *electrosmog*, which can further be defined as the invisible electromagnetic radiation resulting from the use of both wireless technology and electricity in the power system of buildings. The most common sources of wireless electrosmog are cordless

Figure 12.5 A networked sound-masking system manufactured by Lencore Acoustics Corporation includes digital signal processors, electronic noise generators, amplifiers, wiring, loudspeakers, controls, and other components to generate, amplify, distribute, and reproduce digitally synthesized and stabilized background sound masking to create speech privacy. (Photograph courtesy of Lencore Acoustics Corp. www.lencore.com)

phones, cordless baby alarms, mobile/cellular phone masts/towers/transmitters, mobile/cellular phones, and wireless networks. Table 12.7 shows the contribution of various common communications devices on human performance.[17] High-frequency radiation, like ultraviolet (UV) light and X-rays, has an ionizing effect that has been proven to harm body cells. Other frequency ranges have proven heat and irritation impacts on humans. These include electromagnetic fields caused by, for example, communications systems such as telephones and computer systems. This leads to tissue warming and, depending on intensity and duration, high blood pressure. At present, the short- and long-term impacts are as yet unknown. However, it is well known that high levels of electromagnetic radiation in the frequency range of communications can have a negative impact on sleeping patterns, brain performance, the immune system, and nervous and cellular systems. With the rapid rise in touch communications, electromagnetic loads on humans have also increased. Until current long-range and short-term studies have been scientifically interpreted, buildings should be designed with the precautionary principle in mind; that is, recommendations of international expert panels ought to be adhered to and therefore there should be a detailed analysis of particular critical areas with high radiation loads.

The aspects of electromagnetic radiation that should be considered are frequency range, field intensity, distance to the emitter, and the length of exposure. Radiation intensity is measured in watts per square meter, and the intensity of the radiation decreases with the square of the distance from the emitter. This means that a high-capacity emitter that is farther away, such as a cell phone tower, may be less harmful than a small emitter in the vicinity of a body, such as a cell phone. The radiation load from a cell phone at the ear is 100 times more than when it is 3.1 feet (1 meter) from the body. For high-performance green buildings, reducing electromagnetic radiation loads should be considered, and work tools such as telephone systems and cell phones need to be taken into account as well as computers and other electronic devices. Table 12.8 shows the critical values for electromagnetic radiation in different countries.[18] It is noteworthy that the countries and regions listed have allowable electromagnetic radiation levels that are up to 1000 times lower than international recommendations. In each case,

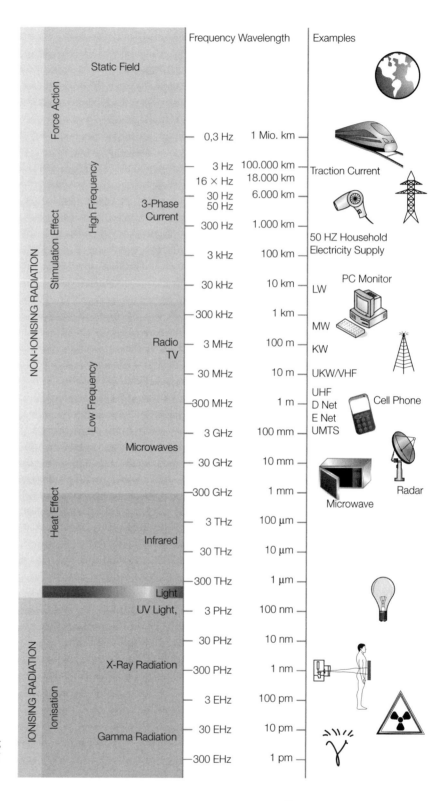

Figure 12.6 Overview of different radiation sources with their corresponding frequency ranges. (Illustration courtesy of Drees & Sommer)

these are significantly lower than the $10-50$ w/m^2 recommended by the International Commission on Non-Ionizing Radiation Protection (ICNIRP). Russia, which is rarely thought of as having high human or environmental standards, has one of the most stringent electromagnetic limits in the world at 0.02 w/m^2. Equally important is that this problem is not yet recognized as being important by the green building community in the United States. In contrast, in Germany the problem of electrosmog is taken far more seriously. Figure 12.7 shows a German

TABLE 12.7

Relative Effects of Electromagnetic Radiation from Common Office Communications and Computer Equipment

Office Equipment	Low	Medium	High	Extreme
Computer monitor		X		
Flat screen	X			
Normal keyboard and mouse		X		
Radio/infrared keyboard and mouse			X	
Bluetooth and wireless local area network (WLAN)				X
Printer	X			
Fax	X			
Copier	X			
Lights		X		
Laptop		X		
Normal telephone			X	
Portable telephone				X
Personal computer		X		
Desk lighting			X	
Ceiling lighting		X		
Beamers			X	
Office Furnishings				
Chairs	X			
Tables with metal frames		X		
Shelving	X			

TABLE 12.8

Critical Values for Electromagnetic Radiation in Different Countries

Country/Region	Critical Value for Electromagnetic Radiation (watts per square meter)
Germany	2–9
Australia/New Zealand	2
Italy	0.1
Poland	0.1
Czech Republic	0.24
Russia	0.02
Salzburg, Austria	0.001
Switzerland	1/10 of ICNIRP critical values*

*ICNIRP is the International Commission on Non-Ionizing Radiation Protection. Critical values set by the ICNIRP are 10 W/m² for general public exposure and 50 W/m² for occupational exposure for the range between 10 and 300 gigahertz (GHz).

office worker with instrumentation to determine the effects of electromagnetic radiation from office equipment on brain activity and the effects of efforts to neutralize this radiation. German research indicates that frequency modulators that neutralize the extraneous electromagnetic radiation can be effective in countering this radiation. Figure 12.8 indicates how frequency modulators can neutralize electromagnetic radiation superimposed on a power line. Figure 12.9 shows how this approach was adapted by the Institut für Physikalische Raumenstörung (IPR) in Berlin, Germany, to neutralize similar radiation affecting brain wave activity.

Figure 12.7 An office worker in Germany equipped with a portable electroencephalogram device to determine the effects of neutralizing electromagnetic radiation from office equipment. (Photograph courtesy of Drees & Sommer)

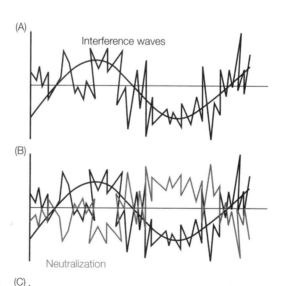

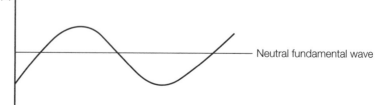

Figure 12.8 (A) Extraneous electromagnetic radiation is superimposed on an alternating current electrical wave. (B) A neutralizing wave that is the opposite of the extraneous electromagnetic radiation in magnitude and polarity is introduced. (C) The result is a clean wave from which the extraneous electromagnetic radiation has been removed. (Illustration courtesy of Institut für Physikalische Raumentstörung, Berlin, Germany)

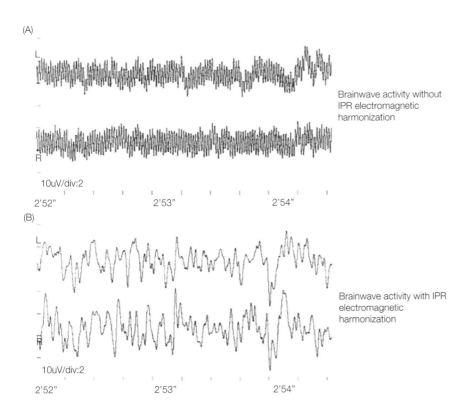

(A)

Brainwave activity without IPR electromagnetic harmonization

10uV/div:2

2'52"   2'53"   2'54"

(B)

Brainwave activity with IPR electromagnetic harmonization

10uV/div:2

2'52"   2'53"   2'54"

Figure 12.9   (A) The effects of electromagnetic radiation on brain wave activity and (B) normal brain wave activity after neutralization by frequency modulators. (Illustration courtesy of Institut für Physikalische Raumentstörung, Berlin, Germany)

HVAC SYSTEMS AND IEQ

Proper design of a building's HVAC system is perhaps the most important approach for providing a healthy indoor environment. Conversely, a poorly designed HVAC system can be a harbinger of trouble. The HVAC system provides a means for moving, exchanging, filtering, and conditioning all of the air in a building. Because the HVAC system plays such an important role in IEQ, it is imperative that it be understood and maintained properly. This section describes the advantages offered by an effective HVAC system and the problems caused by a poorly designed system.

HVAC System Design

HVAC systems differ greatly from building to building, from simple facilities with a simple forced-air furnace to hospitals with state-of-the-art, computer-controlled, automated systems. In all cases, however, the HVAC system affects IEQ because it moves and conditions air. A typical office building HVAC system is a complex arrangement of equipment, with sources of chilled water and hot water coupled to air handlers. The chilled and hot water can be either generated in the building via chillers and boilers or obtained from a central plant serving a group of buildings. The air handlers are composed of fans, cooling coils, heating coils, filters, and other components arranged in a large container that condition and circulate air through the building. Conditioning means that the air is heated or cooled, cleaned, and humidified, if needed, to ensure that the desired temperature and humidity conditions in the various building spaces and zones are provided. The total HVAC system consists of one or more air handlers (depending on building size), each of which is responsible for conditioning a specific zone of the building. The HVAC system is responsible for ensuring that the proper quantity of outside ventilation air is provided for the building occupants. The outside ventilation air is probably the most important

ANSI/ASHRAE Standard 62.1-2010
(Supersedes ANSI/ASHRAE Standard 62.1-2007)
Includes ANSI/ASHRAE addenda listed in Appendix J

ASHRAE STANDARD

Ventilation
for Acceptable
Indoor Air Quality

See Appendix J for approval dates by the ASHRAE Standards Committee the ASHRAE Board of Directors, and the American National Standards Institute.

This standard is under continuous maintenance by a Standing Standard Project Committee (SSPC) for which the Standards Committee has established a documented program for regular publication of addenda or revisions, including procedures for timely, documented, consensus action on requests for change to any part of the standard. The change submittal form, instructions, and deadlines may be obtained in electronic form from the ASHRAE Web site (www.ashrae.org) or in paper form from the Manager of Standards. The latest edition of an ASHRAE Standard may be purchased from the ASHRAE Web site (www.ashrae.org) or from ASHRAE Customer Service, 1791 Tullie Circle, NE, Atlanta, GA 30329-2305. E-mail: orders@ashrae.org. Fax: 404-321-5478. Telephone: 404-636-8400 (worldwide), or toll free 1-800-527-4723 (for orders in US and Canada). For reprint permission, go to www.ashrae.org/permissions.

ISSN 1041-2336

**American Society of Heating, Refrigerating
and Air-Conditioning Engineers, Inc.**
1791 Tullie Circle NE, Atlanta, GA 30329
www.ashrae.org

Figure 12.10 ASHRAE 62.1-2010 is the current version of the US standard that governs the design of building ventilation systems and ventilation rates.

contribution of the HVAC system to a quality indoor environment. In terms of IAQ, the higher the ventilation rate, the better is the air quality in the building. In the United States, ASHRAE 62.1-2010 governs the design of building ventilation systems (see Figure 12.10). A 2000 study by Wargocki, Wyon, and Fanger showed that a so-called productivity index based on tasks performed by office workers increased as ventilation rates were increased, resulting in a decrease in pollution loads (see Figure 12.11).[19]

Climate Control

Climate control is the general objective of the HVAC system; most likely, it is the reason the system was installed in the first place. Surprisingly, though, keeping in mind that the system is designed primarily for climate control, there is often a tremendous amount of dissatisfaction in this area. Before the advent of

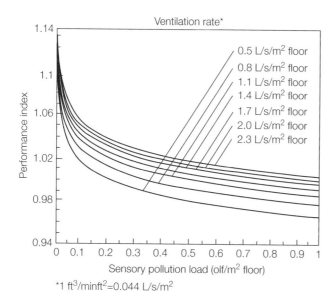

0.5 L/s/m² floor
0.8 L/s/m² floor
1.1 L/s/m² floor
1.4 L/s/m² floor
1.7 L/s/m² floor
2.0 L/s/m² floor
2.3 L/s/m² floor

Performance index

Sensory pollution load (olf/m² floor)

*1 ft³/minft²=0.044 L/s/m²

Figure 12.11   The performance of office workers increases (y-axis) as ventilation rates increase (series of curves) for specific pollution levels (x-axis). The ventilation rate is given in liters per second per square meter (L/s/m²) of floor area. (Courtesy SenseAir)

air conditioning, when opening windows was the only way to help cool a space, building occupants accepted any discomfort as unavoidable.

The state of the air in a space is defined by its psychrometric properties: temperature, relative humidity, enthalpy, and moisture content. Any two properties uniquely define the state of the air. The two most common properties that are controlled by an HVAC system are the temperature and the relative humidity (or moisture content, as the two are different manifestations of the same property). The proper balance of temperature and humidity must be provided by the HVAC system in order to maintain a comfortable indoor environment. Temperatures in the range of 65 to 78°F (18 to 26°C) and relative humidity levels between 30 and 60 percent are considered the comfort range for the majority of the population.

The HVAC system must be capable of controlling the supply air in spite of changing conditions in the return of outside sources. For example, if a summer thunderstorm saturates the outside air and suddenly increases the humidity level of the air flowing into the HVAC system, the system must be able to adapt and maintain the proper humidity level for the outgoing supply air. Moisture control is a critical yet difficult-to-achieve purpose of the HVAC system. When air is too dry, discomfort is a problem; when air is too moist, discomfort and contaminant generation become problems.

Contaminant Generation and Circulation

The HVAC system can often be the source of several types of airborne contaminants. It is a potential breeding ground for many types of biological contaminants, including molds, spores, and fungi. Certain components of the HVAC system can be more easily contaminated than others, particularly porous ductwork linings that are used for insulation and sound control. The HVAC system, because it is responsible for humidification and dehumidification, can also be the cause of uncomfortable humidity levels in the air. High humidity levels help to accelerate the production of biological contaminants. Excess water buildup in the HVAC system, particularly in locations near the evaporator coils or humidifiers, is a major breeding ground for biological contaminants. Water buildup inside components of the HVAC system is sometimes difficult to detect and often expensive to fix. If a well-designed HVAC system is running properly, there should be no excess water buildup.

When it becomes a circulator of airborne contaminants generated both inside and outside the building, the HVAC system is negatively affecting IAQ. ASHRAE 62.1-2010 defines the requirements for the quantities of outside air ventilation required to remove excess carbon dioxide generated by people. Unfortunately, ventilation sometimes has the effect of introducing new pollutants from the outside. The ASHRAE standard is based on the National Ambient Air Quality Standards (NAAQS) and provides guidelines to follow if the supply of outside air does not meet the standard.

Care must be taken to ensure that the outside air source for a building is not unnecessarily or inadvertently contaminated by an isolated pollutant source. IAQ problems frequently arise when air intakes are positioned near loading docks or other possible sources of pollutants. Air intakes should be located away from exhaust sources, both from automobiles and from other buildings, and they should be high enough above the ground to avoid bringing in ground source contaminants like radon and pesticides.

Interior source contaminant circulation is another major problem in buildings. The most probable cause of unnecessary circulation of internal contaminants is improper zoning. Most buildings have areas designated for specific purposes, and the HVAC system must meet the needs of each area. Some areas, such as laboratories and machine shops, have a greater need for ventilation than others—for example, office space. If the HVAC system is not properly zoned, contaminants from one area may affect the air quality of another area. For example, if an area of a building is turned into a metal shop where welding is performed, but the HVAC system continues to function as if it were an office space, contaminants will be spread to other parts of the building.

TABLE 12.9

Building Materials of Particular Concern Because of Their IAQ Impacts

Site Preparation and Foundation
Soil treatment pesticides
Foundation waterproofing
Mechanical Systems
Duct sealants
External duct insulation
Internal duct lining
Building Envelope
Wood preservatives
Curing agents
Glazing compounds
Thermal insulation
Fireproofing materials
Interior Finishes
Subfloor or underlayment
Carpet backing or pad
Wall coverings
Paints, stains
Partitions
Ceiling tiles

Emissions from Building Materials

All materials have emissions, some more than others, and they all may contribute to deterioration of the air quality. Many health complaints have been linked to new materials installed during the construction or renovation of buildings. "Engineering out" materials known to have an adverse effect on IEQ is perhaps the easiest means of ensuring excellent IAQ. Proper materials selection offers a type of quality control that can save millions of dollars in remediation and lessen legal liability. Table 12.9 lists materials of particular concern that warrant careful selection because of their potential adverse effects on IAQ.[20]

The primary concern with respect to building materials is the types of contaminants they emit. But of additional concern is that some materials act as "sinks" for emissions for other materials or for contaminants that enter the building from other sources. For example, many building materials readily absorb VOCs and rerelease them into the air. In fact, the majority of harmful building material constituents are VOCs, which are typically components of the manufacturing and installation processes. Usually, however, the emission rate will be reduced in proportion to the time the contaminant is exposed to the air.

Due to the increasing awareness of issues related to IAQ, both public agencies and private industry are promoting the use of low-emission building materials. Communicating IEQ requirements to subcontractors and suppliers is an important step in the process of creating a healthy building, but there is still

TABLE 12.10

Chronic RELs for Selected Organic Chemicals Associated with IAQ

Chemical Name	Chemical Abstracts Service (CAS) Number	REL (ppb*)	REL (μg/m³†)
Benzene	71-43-2	20	60
Chloroform	67-66-3	50	300
Ethylene glycol	75-00-3	200	400
Formaldehyde	50-00-0	2	3
Naphthalene	91-20-3	2	9
Phenol	108-95-2	50	200
Styrene	100-42-5	200	900
Toluene	100-88-3	70	300
Trichloroethylene	79-01-6	100	600
Xylenes	Several	200	700

*ppb = parts per billion.
†μg/m³ = micrograms per cubic meter.

debate about how to include materials emissions requirements in specifications. The MasterFormat form of specifications developed by the Construction Specifications Institute (CSI) is generally employed to describe the methods and materials of construction. MasterFormat had 16 divisions until 2005, when it was expanded to 50 divisions. Each division covers major aspects or systems of the building; further, each division is divided into sections that cover subsystems. Each section has three parts: Part 1—General, Part 2—Products, and Part 3—Execution. One suggestion for addressing the issue of how to include the required environmental attributes is to expand this three-part format to four parts for each section and to include information on materials emissions requirements and other environmental attributes in the new Part 4—Environmental Attributes.

Another option is to simply introduce an entire section into the general division (Division 1) of the CSI MasterFormat that addresses all the environmental requirements of the project, including materials emissions. This approach is being implemented in California with the creation of Section 01350—Special Environmental Requirements, which includes emissions requirements for materials.[21] Section 01350 covers product selection guidelines, emissions testing protocols, and nontoxic performance standards for cleaning materials. It requires that material safety data sheets (MSDS) be submitted for each material and that these materials be tested by an acceptable testing laboratory in accordance with American Society for Testing and Materials (ASTM) D5116-97, Standard Guide for Small-Scale Environmental Chamber Determinations of Organic Emissions from Indoor Materials/Products. Section 01350 also provides information about so-called chemicals of concern, which are carcinogens, reproductive toxicants, and chemicals with an established Chronic Reference Exposure Level (REL). A Chronic REL is an airborne concentration level that would pose no significant health risk for individuals indefinitely exposed to that level. Chronic RELs have been developed for 80 hazardous substances; another 60 chemicals are under review. The modeling of total concentrations of airborne emissions must show that the maximum indoor air concentration of any of the chemicals of concern must not exceed half of the REL. Table 12.10 lists some of the RELs for common VOCs present in building materials.

ADHESIVES, SEALANTS, AND FINISHES

Adhesives, sealants, caulks, coatings, and finishes are placed in the building when wet and are expected to dry, or cure, on the premises. The release of VOCs is an inherent part of this process. The solvents used in formulating these materials are the source of most VOCs emitted during drying and later during building occupation.

Adhesives and Sealants

Adhesives are materials or substances that bind one surface to another. They affect a wide range of construction materials; adhesives can be applied with floorings and wall coverings, or they may be a component of a material like plywood, particleboard, movable wall panels, and office workstations. Adhesives are applied in a liquid or viscous state, then cure to a solid or more solid state to achieve bonding. The majority of adhesives release VOCs and pose the greatest threat during their application and curing. When applied, adhesives should be used in areas with increased ventilation (at normal room temperature) for 48 to 72 hours to avoid accumulation of VOCs. The packaging label or other installation information for adhesives should always be consulted for additional product-specific precautions.

One method of characterizing adhesives in terms of their influence on IAQ is to identify the resin used in the base. Resins can be natural or synthetic. Natural resins usually have low emission potential, but in synthetic resins this potential can vary dramatically. Currently, advances are being made in the development of adhesives with no or low emissions.

Sealants are applied to joints, gaps, or cavities to eliminate penetration of liquids, air, and gases. (Note: Although the construction industry differentiates between indoor and outdoor sealants, the former being referred to as *caulks* and the latter as *sealants*, this discussion does not make this distinction.) Sealants are usually selected on the basis of their flexibility and resin base. Like adhesives, sealants can be hazardous during installation and curing. Their emission potential is directly related to the percentage of base resins and solids. Sealants, which definitely raise a concern with regard to their VOC emission potential, fortunately are used indoors in small quantities. Alternate water-based sealants manufactured using nontoxic components are now available. One such product for interior use is a vinyl adhesive sealant. An acrylic latex exterior sealant for building joints is also on the market.

The USGBC LEED and Green Globes building assessment systems provide credit for the use of low-emission adhesives and sealants. To earn this credit, adhesives, sealants, and sealant primers must meet the VOC content limits of the South Coast Air Quality Management District (SCAQMD) Rule 1168 (see Table 12.11).[22] Aerosol adhesives must meet the requirements of Green Seal GS-36, Standard for Commercial Adhesives.

Finishes

Finishes encompass a wide range of products, including paints, varnishes, stains, and sealers. Finishes are a major component of building materials and furnishings whose primary purpose is to provide protection against corrosion, weathering, and damage. Secondarily, they may also add aesthetic value to building materials. All finishes have similar characteristics. They require resins and oils to form a film and to aid adhesion by promoting penetration into the substrate. All coatings require carriers (water or organic solvents) that provide viscosity for application. Carriers also improve adhesion through evaporation.

TABLE 12.11

Sample VOC Limits on Adhesives Established by South Coast Air Quality Management District Rule 1168*

Architectural Applications	VOC Limit (grams per liter less water)
Indoor carpet adhesives	50
Carpet pad adhesives	50
Wood flooring adhesives	100
Rubber floor adhesives	60
Subfloor adhesives	50
Specialty Applications	**VOC Limit (grams per liter less water)**
PVC (polyvinyl chloride) welding	510
CPVC (chlorinated polyvinyl chloride) welding	490
ABS (acrylonitrile-butadiene-styrene) welding	325
Plastic cement welding	250
Adhesive primer for plastic	550
Substrate-Specific Applications	**VOC Limit (grams per liter less water)**
Metal-to-metal	30
Plastic foams	50
Porous material (except wood)	50
Wood	30
Fiberglass	80
Sealants	**VOC Limit (grams per liter less water)**
Architectural	250
Nonmembrane roof	300
Roadway	250
Single-ply roof membrane	450

*Through January 7, 2005, amendments.

TABLE 12.12

Hazardous Chemicals in Pigments

Antimony oxide	Titanium dioxide	Rutile titanium oxide
Cadmium lithopone	Chrome yellow	Molybdate orange
Strontium chromate	Zinc chromate	Phthalocyanine blue
Chrome green	Chromium oxide	Phthalocyanine green
Hydrated chromium oxide	Copper powders	Cuprous oxide

Paints and stains require solids, including pigments, to provide various colors. The amount of solids is a good indicator of the VOC emission potential of the finish. Table 12.12[23] lists the hazardous chemicals associated with particular pigments used in paints. The sanding or burning of finishes generates potential IEQ hazards such as dust from talc, silica, mica, and especially lead.

Water-based finishes are typically low-emitting; however, organic solvent-based finishes are more likely to be high-emitting. The current trend is to replace conventional finishes with water-based alternatives, although it is primarily paints that have been targeted in this effort. Very few stains, sealers, and varnishes have been successfully adapted for low VOC emissions, because, to date, alternative finishes generally do not perform as well as their traditional counterparts. The new products often require more applications to achieve results similar to those of traditional products. The color selection of alternative paints is limited as well. And although hypoallergenic, preservative-free paints are available, their shelf life and color selection are limited.

It is also important to point out that water-based products may have low VOCs but contain other hazardous materials. Unlike organic solvent-based paints, water-based paints require preservatives and fungicides such as arsenic disulfide, phenol, copper, and formaldehyde. These additives are considered chemical hazards by the National Institute for Occupational Safety and Health (NIOSH).

The USGBC LEED and Green Globes building assessment standards provide credit for the use of low-emission paints and coatings if their VOC emissions do not exceed the VOC and chemical component limits of Green Seal's GS-11 requirements. This standard specifies VOC limits of 150 grams/liter for nonflat interior paints and 50 grams/liter for flat interior paints.[24] There is growing interest in the use of paints with recycled content, and Green Seal issued GS-43, Environmental Standard for Recycled Content Latex Paint, in 2006, setting VOC limits of 250 grams/liter. Although having the environmental attribute of recycled content, paints just meeting this standard would not qualify as low-emission paints under GS-11.[25]

PARTICLEBOARD AND PLYWOOD

Adhesives containing urea formaldehyde (UF) are an integral part of the composition of particleboard and plywood. These materials emit the UF after they have been manufactured and installed in construction. The rate of emission of UF is affected by the temperature and humidity of the installation location.

Particleboard

Particleboard is a composite material made from wood chips or residues, bonded together with adhesives under heat and pressure. Particleboard is relatively inexpensive and is available in sheets that measure 4 by 8 feet (1.2 by 2 meters). The major IAQ concern with particleboard is the off-gassing of formaldehyde. Most particleboard (about 98 percent) contains UF. The remaining 2 percent contains phenol formaldehyde (PF). Particleboard containing PF emits far less formaldehyde than board made with UF. PF is used in particleboard where a high-moisture environment is anticipated, specifically restrooms and kitchens.

The most common construction application of particleboard is as a core material for doors, cabinets, and a wide variety of furnishings, such as tables and prefabricated wall systems. Particleboard is also used in wood-framed housing, primarily for nonstructural floor underlayment. Usually, a finished floor is installed over particleboard. Particleboard is also used as a backing for paneling. Once the board is covered, the VOC content is inconsequential because the formaldehyde emissions are delayed for as long as it remains covered.

Particleboard, although it can now be manufactured with lower formaldehyde emissions, is still of great concern because of the possibility of large exposed surface areas in relation to the volume of a given space. Emissions of trace amounts of formaldehyde can continue for several months or even years. These emissions do decrease over time, but rates increase as temperature and/or humidity rises. It is estimated that emission rates double with every increase of 12°F (7°C).

Plywood

Plywood is composed of several thin wood layers oriented at alternating 90° angles that are permanently bonded by an adhesive. The exterior plies are referred to as *faces*, and the interior plies are known as the *core*. Plywood is generally classified as hardwood or softwood. Approximately 80 percent of all softwood plywood is used as wall and roof sheathing, siding, concrete framework, roof decking, and subflooring. Hardwood plywood is used for building furniture, cabinets, shelving, and interior paneling.

The type of adhesive used to bond the plies plays a major role in assessing the effects of the plywood on IAQ. The surface area of the plywood in relation to the volume of the space is another determining factor for proper IAQ. Interior-grade plywood is generally bonded with UF resins. The off-gassing of UF in plywood can be compounded by finishes or sealants used in conjunction with the plywood. Size, temperature, humidity of the space, surface area, and finish of the plywood all can affect the concentration of formaldehyde emissions.

FLOOR AND WALL COVERINGS

Carpet, resilient flooring, and wall coverings may have VOC-emitting components and may use adhesives that emit VOCs as part of their installation process. New products with zero or low emissions are, fortunately, entering the marketplace to serve the green building industry. As competition and demand increase, the quality of the products will also improve; at the same time, the price will decrease, making these new products very competitive with conventional products.

Carpet

Of all building materials, carpet has generated the most debate, which is ironic considering that emissions from carpet systems are relatively low compared with emissions from other building materials. The majority of the emissions associated with carpeting are actually due to the adhesives used to secure it. Thus, when selecting a carpet, the entire system and the emissions of each constituent must be evaluated. The components of a carpet system are the carpet fiber, carpet backing, adhesive, and carpet pad (generally used in residential applications only).

Carpet backing is used to hold the fibers in place. Often two backings are used: one keeps the fibers in place, and the other adds strength and stability. The secondary backing is made from fabric, jute, or polypropylene bonded with either styrene-butadiene rubber (SBR) latex or a polymer coating such as synthetic latex. SBR latex contains the chemicals styrene and butadiene, which are known irritants to mucous membranes and skin. SBR latex adhesives are found in primary and secondary backings and emit low but steady amounts of the by-product 4-phenylcyclohexene (4-PC), the chemical that is responsible for the "new carpet" smell and is suspected of being a possible source of building occupant illness complaints.

Adhesives may be used twice in common carpet systems: to glue the backing to the fiber and/or to glue the carpet system to the substrate.

Carpet pads are an optional part of the carpet system. They generally do not contribute to IAQ problems. There are five basic types of pads: bonded urethane, prime polyurethane, sponge rubber, synthetic fiber, and rubberized jute.

There are five basic carpet fiber materials. Wool is the only natural fiber, and it accounts for less than 1 percent of the carpet market. The remaining four—nylon, olefin, polyester, and polyethylene terephthalate—are synthetic fibers. Derived from petrochemicals, synthetic fibers are stronger, more durable, and usually less expensive; they are also less likely than wool to release small fibers into the air.

Both the USGBC LEED-NC and Green Globes building assessment standards provide credit for using low-emission carpeting systems if the system meets or exceeds the requirements of the Carpet and Rug Institute's Green Label Plus program.[26]

Resilient Flooring

Resilient flooring is a pliable or flexible flooring. Tile and sheet are the two basic forms, both of which are attached to a substrate using adhesives. Resilient flooring can be composed of vinyl, rubber, or linoleum. Vinyl flooring is made primarily of PVC resins, with plasticizers, to provide flexibility; fillers; and pigments for color. Rubber flooring comes in two basic forms—smooth-surface or molded—and is made from a combination of synthetic rubber (styrene butadiene), nonfading organic pigments, extenders, oil plasticizers, and mineral fillers. Linoleum is a natural, organic, and biodegradable product. Its main components are linseed oil, pine rosin, wood flour, cork powder, pigments, driers, and natural mildew inhibitors. Linoleum tiles are durable, greaseproof, and water- and fire-resistant. They are also easily maintained and long-lasting.

Typically, no individual compound in resilient flooring has high VOC emissions. The plasticizers are the main source of emissions. Using a more rigid, less plastic tile is recommended to avoid potential hazards. Note, however, that low-emitting tiles may be glued with high-emitting adhesives.

Wall Coverings

Wall coverings are a popular alternative to paints. The majority of available coverings pose little or no threat to IAQ. The three basic types of wall coverings are paper, fabric, and vinyl. Paper itself has no impact on IAQ, but the adhesives used to apply it may contain formaldehyde. However, the majority of paper adhesives are purchased as a powder and mixed with water; thus, they emit little or no VOCs.

Fabric wall coverings may contain formaldehyde, which is sometimes used to keep the material from fading and to improve resistance to water. Fabric coverings can also act as a sink by absorbing extraneous VOCs in a building and reemitting them into a space. Two major concerns with vinyl wall coverings are the environmental conditions of the project location and the construction of the walls receiving the finish. In temperate climates, when moisture may not be readily evaporated, vinyl-covered walls can become moldy.

INSULATION AND CEILING TILES

Insulation and acoustical ceiling tiles can contribute VOC and particulate contaminants from a variety of sources. Depending on their composition, these materials may incorporate a variety of adhesives and fibrous materials that can combine to complicate the IAQ issue.

Insulation

Most insulation is made of fiberglass, mineral wool, and cellulose (made from recycled wood). Asbestos was also used frequently until the late 1970s. Fiberglass and mineral wool have raised IAQ concerns because of the small fibers that are produced when the material is disturbed. Fiberglass is listed by the International Agency for Research on Cancer as a possible carcinogen. Cellulose insulation is generally spray-applied and is considered a nontoxic material. In this materials category, foam insulation has received most of the attention because of its impacts on the environment rather than on IAQ. That said, VOCs are emitted from synthetic foam during manufacturing or while spray foam is used.

Acoustical Ceiling Tile

The suspended acoustical ceiling is one of the most common structures found in commercial buildings today. Most acoustical ceiling tile (ACT) is made from mineral or wood fibers, which are wetted and compressed to the desired thickness, size, and pattern. They are usually coated with a latex paint at the factory. The primary concern regarding the effects of ACTs on IAQ is the occurrence of microbial growth on either mineral fiber or fiberglass tile exposed to moisture. Another concern is that porous tiles can absorb VOCs and reemit them.

Economic Benefits of Good Indoor Environmental Quality

The key emerging economic benefits of high-performance green buildings appear to be their health and productivity benefits, with paybacks that may be as much as 10 times higher than their energy savings. More and more hard evidence of the effects of good indoor air quality (IAQ) is emerging, supporting design and construction efforts that provide excellent building air quality. More recently, the range of health problems connected to buildings has shifted from air quality alone to include a far wider range of human health effects associated with lighting quality, noise, temperature, humidity, odors, and vibration. This broader range of impacts is referred to as *indoor environmental quality* (IEQ) and includes the subject of IAQ.

The impact of buildings on human health is substantial and results from a combination of building design, construction practices, and the activities of the occupants. A study by Fisk and Rosenfeld in 1998, updated in 2002, placed the annual cost of IAQ problems at $100 billion.[27,28] Table 12.13, which is adapted from this study, shows estimated productivity gains from improvements made to indoor environments. In the United States, people spend about 90 percent of their time indoors—in their homes, workplaces, schools, shopping malls, fitness centers, or numerous other types of structures. Air quality in some of these buildings is often cited as being far worse than that of the outside air. This poor air quality can be attributed to a number of factors: tight buildings, materials that off-gas pollutants into the indoor environment, poor ventilation, and poor moisture control, to name a few. In addition, poor construction practices can contribute to significant IEQ problems. For example, ductwork that has been stored and handled without being covered and sealed can be contaminated with particulates that are blown into occupied spaces during building operation, potentially affecting the health of the people in the building.

The high-performance green building movement has been highly successful in integrating indoor environmental issues into the criteria for green buildings, in essence taking ownership of IEQ when it comes to new buildings, so it is now expected that a high-performance green building will have excellent IEQ. In

TABLE 12.13

Estimated Potential Productivity Gains from Improvements Made to Indoor Environments

Source of Productivity Gain	Strength of Evidence	US Annual Savings or Productivity Gain
Respiratory disease	Strong	$6–$14 billion
Allergies and asthma	Moderate to strong	$1–$4 billion
Sick building syndrome	Moderate to strong	$10–$100 billion
Worker performance	Moderate to strong	$20–$200 billion
Total range		$37–$318 billion

particular, the USGBC's LEED suite of standards addresses IEQ and provides points for incorporating at least some of the major IEQ components, generally those concerned with air quality and individual control of temperature and humidity. Green Globes, an emerging competitor to LEED, addresses the same issues as LEED but also includes other important IEQ matters such as acoustic comfort for building occupants and neighbors. Green Globes addresses noise from air-conditioning systems, plumbing, preventing noise generated in the building from affecting neighbors, protecting building occupants from outside noise, and noise attenuation for the structural system.

The actual savings attributed to a high-quality interior building environment are substantial and are thought to be greater than even the energy savings. A study by Greg Kats of Capital E indicated 20-year life health and productivity savings of $36.89 per square foot (square meter) for LEED-certified silver buildings and $55.33 per square foot (square meter) for LEED-certified gold and platinum buildings.[29] The productivity and health benefits of high-performance green buildings, a result of designing a high-quality indoor environment, dominate the discussion of benefits. For gold and platinum buildings, the claim is that the health and productivity benefits are almost 10 times greater than the energy savings, which amount to $5.79 per square foot (square meter). These results are not only impressive but startling as well. However, the basis for these claims is rarely scientific; thus, using these results in life-cycle costing (LCC) or in economic analyses should be done only with extreme caution to avoid compromising the justification of an otherwise sound approach.

Summary and Conclusions

IEQ is perhaps the most important human-related issue of green building, as it directly affects the health of the building occupants. Although IEQ covers a wide range of effects, LEED focuses on IAQ, with far less emphasis on noise and lighting quality. Green Globes does address a wider range of IEQ issues such as acoustic comfort and lighting quality, a definite step forward in the evolution of green building rating tools. As a consequence of relatively recent efforts to address building health, a number of new products have emerged, among them paints, carpets, adhesives, furniture, and wood products for millwork and cabinetry, which have zero or low emissions. Furthermore, greater attention is being paid to the proper sizing of HVAC equipment and control of humidity in spaces. The important issue of moisture infiltration and the consequent problems caused by mold and mildew growth are also being addressed by appropriate architectural detailing and the proper design of the building's air distribution system. Daylighting is receiving increased emphasis because of its demonstrated health benefits and its contribution to reductions in energy consumption. Providing exterior views to the building occupants to enhance their well-being is also a component of IEQ, which both the LEED and Green Globes rating systems acknowledge by allocating points for providing exceptional views. Future versions of these building assessment systems should consider increasing the importance of quiet, relatively noise-free building systems as an aspect of an important health issue. And to cover the full array of IEQ issues, lighting quality should receive additional focus and consideration.

Notes

1. As stated on the US EPA Region 1 website at www.epa.gov/region1/communities/indoorair.html.

2. Excerpted from Levin (1999).
3. As described in "IAQ Guidelines for Occupied Buildings Under Construction" (1995).
4. From Nero (1988).
5. Summarized from Meckler (1991).
6. See Bass (1993).
7. These are estimated productivity losses quoted by Mary Beth Smuts, a toxicologist with the US EPA, in Zabarsky (2002).
8. Excerpted from "Building Air Quality" (1991).
9. See Hays, Gobbell, and Ganick (1995) and Bass (1993).
10. From the online Whole Building Design Guide at www.wbdg.org.
11. From Wilson (1988).
12. From Jensen and Arens (2005).
13. From NAIMA (1997).
14. Adapted from "Comparing Noise Criteria," at The Engineering Toolbox website, www.engineeringtoolbox.com/noise-criteria-d_726.html.
15. Adapted from the ReveberationTime.com website at www.reverberationtime.com.
16. From Bauer et al. (2010).
17. From Bauer et al. (2010) and Gustavs (2008).
18. From Bauer et al. (2010).
19. From Wargocki et al. (2000)
20. Adapted from Hansen (1991).
21. The latest version of Section 01350 can be found on the California Integrated Waste Management Board (CIWMB) website, www.ciwmb.ca.gov/GreenBuilding/Specs/Section01350.
22. The latest version of the SCAQMD Rule 1168 can be found at www.arb.ca.gov/DRDB/SC/CURHTML/R1168.PDF.
23. See Hays, Gobbell, and Ganick (1995).
24. The Green Seal GS- 11 Standard can be found at the Green Seal website, www.greenseal.org/Portals/0/Documents/Standards/GS-11/GS-11_Paints_and_Coatings_Standard.pdf.
25. The Green Seal Environmental Standard for Recycled Content Latex Paint (August 2006) can be found at www.greenseal.org/Portals/0/Documents/Standards/GS-43/GS-43_Recycled_Content_Latex_Paint_Standard.pdf.
26. The criteria for the Green Label Carpet Testing Program can be found at www.carpet-rug.org/documents/glp/120101_GLP_Carpet_Criteria.pdf.
27. From Fisk and Rosenfeld (1998).
28. The updated information is contained in Fisk (May 2002).
29. Cited in Kats (2003), Executive Summary.

References

Bass, Ed. 1993. *Indoor Air Quality in the Building Environment.* Troy, MI: Business News Publishing.

Bauer, Michael, Peter Mösle, Michael Schwarz. 2010. *Green Building: Guidebook for Sustainable Architecture.* Berlin: Springer Verlag.

"Building Air Quality: A Guide for Building Owners and Facility Managers." 1991. EPA/400/1-91/033, US Environmental Protection Agency, Washington, DC.

CBE. 2007. *Acoustical Analysis in Office Environments Using POE Survey.* Berkeley, CA: Center for the Built Environment. Summary available at http://www.cbe.berkeley.edu/research/acoustic_poe.htm.

Fisk, W. J. 2002. "How IEQ Affects Health, Productivity." *ASHRAE Journal* 44 (5): 56, 58–60.

Fisk, W. J., and A. H. Rosenfeld. 1998. "Potential Nationwide Improvements in Productivity and Health from Better Indoor Environments." *Proceedings of ACEEE Summer Study '98* 8: 85–97.

Gustavs, Katharina. 2008. "Options to Minimize Non-Ionizing Radiation (EMF/RF/Static Fields) in Office Environments," University of Victoria. Available at www.buildingbiology.ca/pdf/2008_low_emr_office_environments.pdf.

Hansen, Shirley. 1991. *Managing Indoor Air Quality.* Liliburn, GA: Fairmont Press.

Hays, S. M., R. V. Gobbell, and N. R. Ganick. 1995. *Indoor Air Quality Solutions and Strategies.* New York: McGraw-Hill.

Hennessey, John F., III. 1992. "How to Solve Indoor Air Quality Problems." *Building Operating Management* 39 (7): 24−28.

Jensen, K., and E. Arens. 2005. "Acoustic Quality in Office Workstations, as Assessed by Occupant Surveys," *Proceedings of Indoor Air 2005*, Beijing, China, Sept. 4-9.

Kats, Gregory H. 2003. "The Costs and Financial Benefits of Green Buildings." A report developed for California's Sustainable Building Task Force. Available at the Capital E website, www.cap-e.com.

Levin, Hal. 1999. "Commercial Building Indoor Air Quality." A report prepared for the Northeast Energy Efficiency Partnerships, Inc. Available at www.buildingecology .com/articles/commercial-building-indoor-air-quality-introduction-to-the-problem/.

Meckler, M., ed. 1991. *Indoor Air Quality Design Guidebook.* Lilburn, GA: Fairmont Press.

NAIMA. 1997. "Sound Control for Commercial and Residential Buildings," North American Insulation Manufacturer's Association. Available at www.guardianbp .com/litlib/Naima_BI405.pdf.

Nero, A. V., Jr. 1988. "Controlling Indoor Air Pollution." *Scientific American* 258 (5): 42−48.

SMACNA. 1998. *Indoor Air Quality—A Systems Approach.* Chantilly, VA: Sheet Metal and Air Conditioning Contractors' National Association (SMACNA).

———. 2000. *Duct Cleanliness for New Construction.* Chantilly, VA: Sheet Metal and Air Conditioning Contractors' National Association (SMACNA).

———. 2009. *IAQ Guidelines for Occupied Buildings under Construction.* Chantilly, VA: Sheet Metal and Air Conditioning Contractors' National Association (SMACNA).

Wargocki, P., D. P. Wyon, and P. O. Fanger. 2000. "Productivity Is Affected by the Air Quality in Offices." *Healthy Buildings* 1: 635−640.

Wilson, Edmund O. 1984. *Biophilia.* Cambridge, MA: Harvard University Press.

Zabarsky, Marsha. 2002. "Sick-Building Syndrome Gains a Growing Level of National Awareness." *Boston Business Journal.* Available at www.bizjournals.com/boston/ stories/2002/08/19/focus9.html.

Part IV

Green Building Implementation

Part III provided an overview of the major systems of a green high-performance building: land and landscape, energy, water, materials, and indoor environmental quality (IEQ). Proper design of these systems is the starting point for green building. But without careful execution of the construction phase of the project and thorough commissioning of the finished building, a green building project is incomplete. Part IV of this book addresses these two important aspects of a project and how they fit into the overall green building process. In addition, this part covers the economics of green building and offers an overview of the possible life justifications for green buildings, including energy savings, water and wastewater savings, the benefits of commissioning, operations and maintenance savings, and other approaches to addressing the economics of green buildings. This part concludes with an overview of the future of green building and the variety of directions in which this movement may evolve and includes the following chapters:

Chapter 13: Construction Operations and Commissioning

Chapter 14: Green Building Economics

Chapter 15: The Cutting Edge of Sustainable Construction

Chapter 13 elaborates on two major aspects of green building that are not covered separately in LEED or Green Globes but that warrant additional consideration. The construction managers or general contractors who actually execute the design must be made clearly aware of their responsibilities. Therefore, the importance of developing a site protection plan, a health and safety plan, and a construction and demolition waste management plan is addressed in Chapter 13. Each plan is an extension, or elaboration, of current building assessment system requirements. The site protection plan includes the erosion and sedimentation control plan requirements found in the Sustainable Sites category of LEED and the Site category of Green Globes, as well as other measures designed to protect the biological and physical integrity of the site. The health and safety plan elaborates on issues during the construction phase and indoor air quality (IAQ) requirements, and includes additional measures designed to protect the workforce and the building's future occupants. The construction and demolition waste management plan is addressed in the

Materials and Resources category of LEED-NC, which was described in Chapter 5. Building commissioning, which has emerged as a key step in the third-party certification of high-performance buildings, is also thoroughly explored in Chapter 13. The building commissioning process continues to evolve, from its original role of testing and balancing heating, ventilation, and air conditioning (HVAC) systems to a more complete check of all building systems, including, for example, building finishes, ensuring that the owner receives the exact building called for in the design. Commissioning is becoming a service that occurs throughout the entire project, from the onset of design, rather than only at the completion of construction. Initial economic analyses of high-performance green buildings indicate that the savings due to building commissioning are truly staggering, even outstripping the financial benefits of energy savings. This is a remarkable outcome, and if future analyses were to confirm this result, these findings would transform a number of fundamental assumptions about buildings. For example, if the savings from commissioning were so marked at the onset of building operation, ongoing commissioning would also have notable benefits.

Economic analysis of green buildings is addressed in Chapter 14. Life-cycle costing (LCC) is the key tool for justifying the decisions to create a high-performance building. Initial studies indicate that the added costs for a LEED-NC new building are about 2 percent for a silver or gold certification and that total 20-year savings, using conservative financial assumptions, are on the order of $50 to almost $70 per square foot ($500 to $700 per square meter) for an initial additional investment of about $4 per square foot ($40 per square meter) for a $140-per-square-foot ($1400-per-square-meter) base building construction cost. Some studies report a one-year simple payback for a green building when all savings are included—energy, water, emissions, and health/productivity benefits.

The future of green building is covered in Chapter 15, the final chapter of this book. LEED, as might be expected, pushes green building in a given direction because the point system for achieving the various levels of certification, although generally performance based, tends to result in a fairly limited range of outcomes. At present, only a few attempts are being made to define the "ultimate" green buildings, those that will emerge in 20 years or more. The Living Building Challenge described in Chapter 4 is perhaps pushing the envelope the farthest of any of the building assessment systems. The purpose of this chapter is to attempt to remedy this oversight. To that end, three potential future strategies are described: one based on technology, a second on vernacular architecture, and a third on biomimetic models. No one of these is likely to provide the long-range solution; instead, most likely, a synthesis of the key ideas in these three strategies will be the outcome. Future versions of LEED, such as the proposed LEED v4, will ideally pave the way for green building and raise the bar for everyone engaged in this movement, from owners to materials suppliers, designers, and builders.

Chapter 13

Construction Operations and Commissioning

The role of the construction team in executing a green building project and making it a reality is extremely important and should not be underestimated. A general contractor or construction management company (GC/CM) that orients its employees and its subcontractors to the purposes of the project can make an enormous difference in the overall outcome. Several types of construction activities are specifically identified in the Leadership in Energy and Environmental Design (LEED) and Green Globes building assessment systems as potentially providing credit for certification, including construction waste management, erosion and sedimentation control, limiting the footprint of construction operations, and construction indoor air quality (IAQ). In addition to these aspects of the high-performance green building project, the construction team may make other contributions that are not specifically covered by building assessment systems. Examples include improving materials handling and storage; reusing site materials such as topsoil, lime rock, asphalt, and concrete; metering site electrical and water usage; and reducing pollution generation activities. It is important for the GC/CM to administer construction operations in a fashion that clearly communicates the unique aspects and requirements of high-performance green buildings to all the subcontractors and suppliers involved in the construction process. This chapter focuses on identifying how construction operations for high-performance green buildings may differ from conventional construction practices. Specific areas of focus in this chapter are site protection planning, materials handling and installation, construction and demolition waste management, managing IAQ during construction, and building commissioning.

Site Protection Planning

A *site protection plan* is used to ensure that disturbances to the site ecology and soils are minimized during construction operations. The potential impacts that can result from construction activities must be understood by the GC/CM in order to effectively establish and implement a site protection plan. Currently, neither LEED nor Green Globes has specific requirements for the components of a site protection plan; however, there are many construction activities that clearly have the potential to negatively impact site ecology and soils. Addressing these activities in the site protection plan will enhance the high-performance green building project by involving contractors and subcontractors in the process. A site protection plan includes erosion and sedimentation control, pollution control, reduced site disturbance, and on-site construction management operations. These topics are discussed in more detail below.

EROSION AND SEDIMENTATION CONTROL

Erosion and sedimentation control measures are important for reducing soil loss and the pollution of nearby water bodies. Erosion and sedimentation are caused by soil particles from the site being carried by wind or water to other locations. The result may be clogged sewer drains, contaminated adjoining sites and water bodies, and possibly costly site rework and cleaning in order to restore the site and surrounding areas to the required condition. Projects located on a site larger than 1 acre must meet the National Pollutant Discharge Elimination System (NPDES) requirements of the US Environmental Protection Agency (EPA) by implementing a Stormwater Pollution Prevention Plan (SWPPP). Projects seeking LEED certification must establish and implement an erosion and sedimentation control plan. Erosion-prone areas are identified by design professionals and construction managers so that a plan can be designed that controls water flow in the event of precipitation. Silt fences, storm drain inlet protection, and sediment traps are temporary solutions that must be continuously monitored due to the potential damage from construction activity. If these types of control devices are implemented on a project, a log containing daily and weekly walk-through inspections is required, along with photos and corrective actions taken if the control devices have been damaged. Figure 13.1 is an example of a temporary sedimentation control device that prevents soil-carrying water from entering the stormwater system and clogging it. More permanent water control devices may include infiltration trenches, vegetated swales, and bioretention cells. Information on these devices can be found in Chapter 10. Grading can control not only the direction of water flow but also the velocity through strategies such as lengthened flow paths, reduced gradients, and sheet flow.[1] Sheet flow is a strategy that causes water to flow at a low depth across a wide area to increase surface friction and minimize erosion. Seeding can also be used to help stabilize soil conditions and reduce water flow. Depending on the construction operations, seeding can be either a temporary or permanent means to control water flow.

Figure 13.1 Storm drain inlet protection implementation on a newly constructed site. This type of device must be continuously monitored during construction operations to ensure it is functioning properly. (Don Thieman, CPESC, ASP Enterprises)

POLLUTION PREVENTION

Controlling pollution is a daily responsibility of the GC/CM, and it is an activity that protects both workers and areas adjacent to the site. Pollution can be anything that is harmful, whether it is a substance or an effect introduced into the environment as a by-product of another activity. Noise, dust, air pollution, and light are a few types of pollution that can result from construction activities and that must be mitigated by corrective measures. Neither LEED nor Green Globes requires an overall construction pollution prevention plan; however, it is important to identify the short- and long-term effects of construction activities and the appropriate measures to reduce their impact. These measures can be either reactive, meaning that the construction activity assumes that pollution problems are going to happen, or proactive whereby pollution problems are entirely prevented. Established approaches for reducing pollution at its source can virtually eliminate the problem for those directly and indirectly involved with the project. Table 13.1 lists generic pollution sources together with reactive and proactive measures for handling construction site pollution.[2] These types of activities should be included in the site protection plan.

REDUCED SITE DISTURBANCE

The very act of constructing a building and the supporting infrastructure that supplies power, water, communications, sidewalks, and roads causes tremendous changes to the existing site. It is often said that "the greenest building is the one that has never been constructed." From an ecological system point of view, it is important to preserve as much of the site's existing biological systems and ecological functions as possible. Procedures for reducing the physical footprint of the construction process must be managed by the GC/CM. One way to approach constructing in an environmentally friendly manner is to first determine whether there are any endangered or threatened species located on or near the project site. By definition, an *endangered species* is an animal or plant listed by regulation as being in danger of extinction. A *threatened species* is any animal or plant that is likely to become endangered within the foreseeable future. Determining if endangered or threatened species exist on or near the site can be accomplished by contacting the local US Fish and Wildlife Service, the National Marine Fisheries Service, state agencies, or tribal heritage centers or by researching online for information on locations of endangered or threatened species. If there is a possibility that an endangered or threatened species is located in the area, it is important to conduct visual inspections, formal biological surveys, and an environmental assessment as required by the National Environmental Policy Act (NEPA). These contacts and research will indicate whether there may be a potential problem and if the Endangered Species Act (ESA) Requirements for Construction Activities should be implemented. Although this may seem like a difficult task, addressing this before construction begins will prevent potential delays in the project.

There are many possibilities for reducing site disturbance during construction. Examples include reducing the number of on-site parking spaces, specifying additional areas to be kept traffic-free, staging equipment and materials off-site, allowing only one accessible lane of traffic around the perimeter of the project, and having an active and aggressive pollution control policy. Adequate fencing and signage must be used to clearly communicate construction goals and avoid damage from construction equipment and activities. The establishment of contractual penalties can be used to minimize site disturbance, prevent damage to trees, and protect ecological systems.

TABLE 13.1

Examples of Reactive and Proactive Measures for Handling Construction Site Pollution

Pollutant	Source(s)	Reactive (Mitigation) Measures	Proactive (Prevention) Measures
Light	Night operations Welding or cutting operations Temporary lights left on at night	Shielding or redirecting light fixtures to focus only on work site Turning off temporary lights at end of workday	Revising construction schedules to avoid night operations Using smaller lights focused directly on task areas
Noise and vibration	Equipment operation	Arranging work shifts to allow worker breaks Perimeter fencing for noise barrier Personal protective equipment (PPE) for workers	Revising construction schedule to avoid operations during sensitive times Choosing equipment with lower noise production
Dust and airborne particles	Equipment operation Wind erosion of exposed soils	PPE for workers Surface treatment of exposed soils with water or dust suppression chemicals	Limiting site disturbance Covering exposed soil with temporary or permanent seeding Leaving existing vegetation intact
Airborne chemical emissions	Volatile organic compounds (VOCs) from the off-gassing of new synthetic materials	Increasing ventilation rates during product installation PPE for workers	Using low- or no-VOC products Designing for exposed surfaces Using prefinished materials
Soil and groundwater pollution	Engine drippings Refueling Accidental spills Improper disposal	Spill cleanup plans/equipment Providing contained storage for chemicals and hazardous materials Spill countermeasures such as berms, absorbent mats, and barriers	Centralized refueling Spill prevention training for employees Proper equipment maintenance Using nonhazardous materials where possible
Surface water pollution (heat and contaminants)	Engine drippings Accidental spills Exposed soil without erosion control measures Paved surfaces	Spill countermeasures Perimeter silt fences Spill cleanup plans/equipment Providing contained storage Stormwater detention basins	Proper equipment maintenance Pervious or high-albedo surfaces Seeding exposed soil Limiting construction disturbance Infiltration basins
Tracked soil on neighboring streets	Vehicle wheels	Vehicle wash stations	Limiting construction disturbance Off-site materials staging Just-in-time delivery

Identifying responsibilities and clearly communicating site-specific requirements to the entire team will greatly improve efforts to minimize site disturbance. Preserving habitat biodiversity is important, especially for greenfield sites. Reducing site disturbance also makes it easier to restore the site when the project is complete.

CONDUCTING ENVIRONMENTALLY FRIENDLY CONSTRUCTION OPERATIONS

There are numerous opportunities to enhance the conduct of construction operations from an environmental standpoint. For instance, the incorporation of a recycling facility for paper, commingled plastics, and other types of recyclable waste can be made available to the workforce. Additionally, containers can be made available for the collection of rechargeable batteries, compostable food waste, or other types of waste. The GC/CM can further reduce waste by sourcing reclaimed materials such as office furniture, cabinets, and tables for the construction trailer. Paper waste can be reduced through the use of a printer that is defaulted to double-sided printing. Strategies that avoid direct material ownership by the GC/CM, such as renting temporary construction barriers and fencing, fosters preservation and reuse of the materials used to facilitate the construction process. Identifying sources of material waste and implementing procedures that redirect those materials from entering landfills will reduce tipping fees.

Material efficiency is not the only practice that can be improved. Other opportunities include reducing the consumption of fuel and water and using energy-efficient equipment. Practices that increase efficiency and reduce waste should be included in the site protection plan so that they can be clearly communicated and enforced. Some examples of these types of practices include the following:

- Using conference calls and webinars to reduce transportation time and fuel costs for scheduled meetings. In situations where progress meetings are held on a regular basis, it may be advantageous to host the meeting at an appropriate location with strategically placed webcams to indicate the progress of construction operations.

- Incentivizing a carpool system to reduce site disturbance and fuel costs. This is particularly useful for subcontractors in order to reduce the number of vehicles brought to the site, reduce on-site congestion, and increase construction site flexibility.

- Using alternatively fueled vehicle for errands in order to reduce fuel costs.

- Monitoring energy and water consumption to help identify potential areas of excessive consumption. Identifying these problem areas will result in cost savings that will directly benefit the GC/CM by increasing profit margins. An example of improving energy efficiency is the use of light-emitting diode (LED) lighting technology, as shown in the Figure 13.2. Depending on the amount of construction lighting used on a site, LEDs may be an option because of their ability to reduce energy consumption by more than 67 percent compared to conventional incandescent and metal halide fixtures.

Figure 13.2   These modular, water-resistant LED fixtures from Clear-Vu Lighting mount on low-voltage wires powered by remote LED drivers, and provide dramatic energy and labor savings on job sites. (Clear-Vu Lighting LLC)

By implementing and executing a site protection plan, the builder will ensure that the existing ecosystems are protected and that the workforce and neighbors have all been considered in the construction process. Additionally, a site protection plan is a public sign that the construction firms managing the project are fully committed to the concept of high-performance green building.

Managing Indoor Air Quality during Construction

Perhaps the most important actors in a building construction project are the subcontractors. It is generally true in today's construction industry that general contractors are themselves performing less of the work involved in the actual erection of the building. Instead, the general contractor or construction manager organizes and orchestrates a diverse group of subcontractors to erect the building. For a green building project to meet its objectives, the subcontractors must be made aware of how the building project differs from a conventional construction project. Green building projects demand the utmost attention to worker and future occupant safety and health. Chronic exposure to occupational hazards can cause serious long-term health effects for the subcontractor workforce. These hazards include noise, dust, chemicals, and vibrations. Immediate job hazards, such as moving equipment, unstable earthwork, and working at heights, can also result in injury or death.

One significant area where the overall safety of the workforce can be improved is IAQ during construction. It is always good practice for the GC/CM to generate and implement a *construction IAQ management plan* for use both during construction activities and before occupancy. A construction IAQ management plan aids in communicating the specific plan to protect air quality and establishes the process for accomplishing this. Typical steps for developing and executing a good construction IAQ plan are shown in Table 13.2.[3] Proper management of an IAQ plan will also aid in earning credit toward green building certification under both LEED and Green Globes. Table 13.3 indicates the measures that builders can take to ensure good IAQ in the occupied building and which should be included in an IAQ plan.

In the development of the IAQ management plan, it is critical to include tangible measures to improve working conditions. The Sheet Metal and Air Conditioning Contractors' National Association (SMACNA) has produced several guidelines that can be used to assist the process of ensuring good air quality during and after construction. The SMACNA publication, *IAQ Guidelines for Occupied Buildings under Construction* (2007), provides a comprehensive approach to be applied during construction, demolition, or renovation of occupied spaces.[4] Chapter 3 of this standard focuses on control measures and guidelines to be used during construction. These areas of concern include (1) heating, ventilation, and air conditioning (HVAC) system protection before and after installation; (2) source control; (3) pathway interruption; (4) housekeeping; and (5) scheduling.

HVAC PROTECTION

Careless installation of the HVAC system components during construction can pose a health hazard to both the construction workforce and the future occupants of the facility. Dust, VOCs, and emissions from equipment can infiltrate the building and be circulated by the air-handling units. It is therefore important

TABLE 13.2

Steps for Managing IAQ during Construction

1. **Identify potential threats to IAQ.** This is typically associated with the type of construction task required to complete the job. Identify the risks associated when installing specific products, materials, and systems and evaluate the solutions in terms of cost and benefit for the overall project.
2. **Incorporate IAQ goals into the bid and construction documents.** These goals will help reduce risks that would conventionally be present.
3. **Ensure that all members of the project team are knowledgeable about IAQ issues.** Ensure they have defined responsibilities for implementation of good IAQ practices.
4. **Require the development and use of an IAQ management plan.** The purpose of the management plan is to prevent residual problems with IAQ in the completed building and to protect workers on the site from undue health risks during construction. The plan should identify specific measures to address:
    a. Problem substances, including construction dust, chemical fumes, off-gassing materials, and moisture. The plan will make sure that these problems are not introduced during construction or, if they must be, that they will be eliminated or their impact reduced.
    b. Areas of planning, including product substitutions and materials storage, safe installation, proper sequencing, regular monitoring, and safe, thorough cleanup.
5. **Conduct regular inspection and maintenance of IAQ measures.** These include ventilation system protection and ventilation rate.
6. **Conduct safety meetings, develop signage, and establish subcontractor agreements that communicate the goals of the construction IAQ plan.** The IAQ construction plan is also a good place to proscribe behaviors unacceptable to the owner that represent a potentially negative impact on long-term IAQ, such as smoking, using chewing tobacco, or wearing contaminated work clothes.
7. **Require contractors to provide information on product substitutions.** This information should be sufficient to allow operations and maintenance (O&M) staff to properly maintain and repair low-emitting or otherwise healthy materials in place.

TABLE 13.3

Measures for Builders to Implement to Ensure Good IAQ for Building Occupants

Keep building materials dry. Building materials, especially those like wood, porous insulation, paper, and fabric, should be kept dry to prevent the growth of mold and bacteria.

Dry water-damaged materials quickly. Water-damaged materials should be dried within 24 hours. Due to the possibility of mold and bacteria growth, materials that are damp or wet for more than 72 hours may need to be discarded.

Clean spills immediately. If solvents, cleaners, gasoline, or other odorous or potentially toxic liquids are spilled onto the floor, they should be cleaned up immediately.

Seal unnecessary openings. Seal all unnecessary openings in walls, floors, and ceilings that separate conditioned space (heated or cooled) from unconditioned space.

Ventilate when needed. Some construction activities can release large amounts of gases into a facility, and if the building is enclosed with walls, windows, and doors, outdoor air can no longer easily flow through the structure and remove the gases. During certain construction activities, temporary ventilation systems should be installed to quickly remove the gases.

Provide supplemental ventilation. During installation of carpet, paints, furnishings, and other VOC-emitting products, provide supplemental (spot) ventilation for at least 72 hours after work is completed.

Require VOC-safe masks for workers installing VOC-emitting products (interior and exterior).

Reduce construction dust. Minimize the amount of dust in the air and on surfaces. Examples include the use of vacuum-assisted drywall sanding equipment and the use of vacuums instead of brooms to clean construction dust from floors.

Use wet sanding for gypsum wallboard assemblies.

Avoid use of combustion equipment indoors.

to store and protect all HVAC equipment including ductwork, air handlers, and other air movement components, from dust, moisture, and odors during construction. This protection is accomplished by requiring that the equipment be wrapped with protection film as it is delivered on-site, as shown in Figure 13.3A. Once installed, the HVAC system must be sealed, as shown in Figure 13.3B, to prevent the introduction of moisture and contaminants. For ventilation purposes, the HVAC system must have installed filters with a Minimum Efficiency Reporting Value (MERV) of 8 on either all return air registers or the negative-pressure side of the system. These filters must be replaced whenever dirty and once again before occupancy with filters with a MERV of 13.

Figure 13.3 (A) Ductwork should be protected during storage and prior to installation. (B) Openings should be sealed during the installation process to prevent contamination. (Photographs courtesy of DPR Construction, Inc.)

CONTAMINATION SOURCE CONTROL

Improving IAQ can be accomplished by mitigating contamination levels at their source. One way to do this is to establish and monitor an IAQ baseline as described in the *EPA Protocol of Environmental Requirements, Baseline IAQ and Materials, for Research Triangle Park Campus*, Section 01445. Establishing and monitoring an IAQ baseline will help increase awareness of air quality during the project and help reduce airborne pollutant and emission discharges. For example, using low- or zero-emission materials wherever possible helps reduce exposure to toxic chemicals, such as VOCs. In situations where some level of formaldehyde or other VOC may be present, proper control measures such as space isolation and ventilation is essential. At a minimum, supplying workers with personal protective equipment (PPE) is an important consideration when needed. Workers may be tempted to avoid PPE if they believe the projects pose no hazards. Proper training and work policies are essential to ensure that construction materials and products are safely installed. Dust collection systems for all equipment used for cutting or sanding should be utilized to protect both workers and building IAQ.

PATHWAY INTERRUPTION

In order to keep dust down, construction activities should be physically isolated from clean or occupied areas. This can be accomplished with temporary barriers, such as plastic sheeting, tape, and entrance control measures such as sticky mats, as shown in Figure 13.4. When used, temporary barriers must be regularly inspected to identify actual or potential leaks or tears that need to be repaired. Clean, completed areas must be positively pressurized, with the construction areas negatively pressurized and exhausted directly to the outside. The use of a high-efficiency vacuum to frequently clean up construction dust will reduce the spread of potential contaminants.

HOUSEKEEPING

Proper maintenance and cleaning should be regularly undertaken on any construction project. Construction site cleaning consists of more than just picking up

Figure 13.4   Sticky mats and walk-off mats are entrance control measures to help reduce contaminants entering into a clean area. (Photograph courtesy of D. Stephany)

scrap materials or sweeping the floor. It also includes cleaning and storing porous materials that tend to absorb liquids and gases that are commonly present on a construction site. Porous materials include drywall, insulation, and ceiling tiles, to name a few. Materials that are porous act as a sink, absorbing contaminants such as formaldehyde during construction and slowly releasing them over time. Contaminant gases are absorbed from other materials that off-gas, such as furniture, adhesives, mastics, varnishes, paints, or carpeting, or as combustion by-products, fuel fumes, and particulates from engines, motors, compressors, or welders. If possible, porous materials should be staged in an area isolated from off-gassing materials and be routinely checked for excessive levels of moisture prior to installation. Porous materials can also affect IAQ if they become wet and moldy. In the event that these materials must be cleaned, it is best to either wipe them with a dry cloth or use a high-efficiency vacuum system.

SCHEDULING

Sequencing of construction activities can be used to minimize exposure to dust, mold, emissions, and debris from contaminating previously installed materials. For example, "wet" construction procedures such as painting and sealing should occur before storing or installing "dry," porous materials. Additionally, increasing the outside air and ventilation exchange rates will decrease indoor air contamination levels. This process is known as a *building flush-out* and is conducted after construction has been completed. For LEED projects, the requirement is a building flush-out with a minimum of 14,000 cubic feet of outdoor air for every square foot of building floor area prior to occupancy. Air supplied to these internal spaces must be at least 60°F with no more than 60 percent relative humidity; otherwise, problems may occur such as mold or damage to electrical equipment. A typical building flush-out requires about 2 weeks, depending on the HVAC capacity and indoor air conditions. In the event that occupancy is desired prior to completion of a flush-out, LEED requires a minimum of 3500 cubic feet of outdoor air for every square foot before occupancy. Once occupied, a minimum ventilation rate of 0.30 cubic feet per minute per square foot is needed at least 3 hours prior to occupancy and must be continued during occupancy until the required 14,000 cubic feet of outdoor air is provided. The schedule should also include a reminder to replace all filtration media prior to occupancy.

Poor job-site construction practices can undermine even the best building design by allowing moisture and other contaminants to become potential long-term problems. Preventive job-site practices can preclude residual IAQ problems in the completed building and reduce undue health risks for workers.

Construction Materials Management

Effective materials management improves project sustainability, with the potential to reduce project costs. Working with vendors on product procurement and delivery practices can reduce solid waste. Appropriate storage helps prevent damage to products and also saves the cost of replacement and the disposal of damaged products. Finding alternative uses for excess materials reduces disposal costs and may also offer benefits such as tax credits.

PRODUCT PROCUREMENT AND DELIVERY

Product procurement involves identifying and selecting a source for products. It also involves communicating product requirements and delivery expectation to that source. It is followed up with ensuring that the delivered products meet these requirements. Additionally, it involves working with the vendor or supplier to correct any problems. Procuring green products may require using different vendors and suppliers than is customary for a company. New relationships, accounts, and lines of communication may need to be established. The use of some products may include some risk due to unfamiliar product lead times as well as subcontractor training. Additional effort is needed in order to address these types of issues. Continual familiarity with the green products selected will reduce risk and improve the sustainability of a building.

An important part of green delivery is the means and methods of the transportation chosen to deliver the product to the site. Typically, construction materials are delivered by either flatbeds or dump trucks. Distance and delivery times and routes must all be analyzed to ensure that the materials arrive on time without excessive fuel consumption. Another consideration is the packaging that is used to transport and protect the product during delivery. Incorporating packaging that is harmless to the environment is desirable. Some manufacturers can supply their product with returnable or reusable packaging, resulting in less packaging materials being landfilled. For instance, delivery of small quantities of sand and aggregate can be arranged using returnable heavy-duty bulk bags that are removed from the delivery truck by crane, as shown in Figure 13.5. These bags allow multiple types of materials to be delivered at once. This approach has lower transportation impacts than bringing loose material in a truck bed. It also keeps these materials from being contaminated on-site or spreading to unwanted areas, thus minimizing cleanup activities. Shipping peanuts and sheet polystyrene can also be reused as long as there is a

Figure 13.5 A crane moving bulk materials in large, reusable heavy-duty bulk bags from a truck for use in construction. (Photograph courtesy of Custom Packaging Products)

local shipping outlet that accepts the material. Other types of packaging may be compostable or biodegradable. For instance, many types of plastic are being made from plant products, such as corn and soybean starches, and may be compostable. Other packaging materials such as wood and corrugated cardboard can be recycled. Pallets that are no longer usable can be chipped and used as mulch.

PRODUCT STORAGE AND STAGING

Prior to installation, it is important to have adequate space for product storage and staging to ensure their protection. There are several possible issues when materials are stored on-site. These can include damage due to environmental conditions, such as moisture or temperature changes, or damages due to material handling, such as crushing or puncturing. Other problems may occur from chemical spills or absorption of contaminants from the surrounding environment.

Protecting products from moisture is clearly important for materials that are water-absorptive. Examples include drywall, carpets, acoustic ceiling tiles, and insulation. Exposure to moisture results in mold growth, swelling, and damage to adhesives. Damage can be prevented by covering materials as well as stacking them loosely to allow for good air circulation (see Figure 13.6). Manufacturers provide materials storage and handling instructions that should be followed.

Another potential source of damage is exposure to ultraviolet (UV) radiation. Products containing plastics must be protected from UV exposure, or they may photodegrade. When exposed to UV radiation, rigid plastics become

Figure 13.6 Construction products and materials such as the gypsum board shown here should be kept dry, covered, and off the ground to prevent future mold and IAQ problems. (Photograph courtesy of DPR Construction, Inc.)

TABLE 13.4

Potential Risks in Storing Materials and Mitigation Measures

Threats to Material Integrity	Mitigation Measures
Moisture	Moisture-proof indoor product storage
Exposure to precipitation	Placement to allow ventilation
Excess humidity	Preventing ground contact
Absorption from ground contact	Adequate covering
	Active ventilation/heating
Photodegradation	Indoor product storage
Exposure to UV radiation	Adequate covering
	Organized laydown yard
Material security	Indoor product storage
	Protected/locked storage
Temperature fluctuation	Indoor product storage
	Active ventilation/heating
Physical damage	Indoor product storage
By equipment during handling	Adequate support
By equipment while stored	Following manufacturer
Improper orientation/support	stacking/protection recommendations
Contamination	Adequate covering
Exposure to spills	Active ventilation
Exposure to dust	Sealed openings
Absorption of contaminants from	Clean before installation
surrounding materials	Separate storage of absorptive items from
	potential contaminants

brittle, softer plastics become chalky and lose their integrity, and color-critical products may start to fade. It is important to shade or cover these materials to prevent damage. Table 13.4 outlines problems that may occur in storing materials and the measures that should be taken to reduce risk.[5]

One way to minimize both the need to store material and the risk of damage to the product is *just-in-time delivery*. Just-in-time delivery is fairly common for projects involving large components for which on-site storage would be difficult.[6] For these projects, considerable planning and coordination are required because the components are likely to be custom-fabricated. Just-in-time delivery can also be used when scheduling constraints are looser or for situations where materials are more readily available. Certain commodities, such as drywall, ceiling tiles, carpet, carpet pad, and insulation, are best delivered as close to the time of installation as possible.

Preventing damage is not the only action necessary for materials storage. Materials themselves may require attention prior to installation. For instance, materials containing synthetic components or adhesives may need to be unwrapped and allowed to off-gas before installation. This prevents potential contamination of indoor air from fabrics, foam, composite wood, adhesives, and finish materials that may need to off-gas.

Construction and Demolition Waste Management

Construction and demolition (C&D) waste management takes advantage of opportunities for source reduction, materials reuse, and waste recycling. Source reduction is most relevant to new construction and large renovation projects, as it involves reduced waste factors in materials ordering, tighter contract language assigning waste management responsibilities to trade contractors, and value engineering of building design and components. During renovation and demolition, building components that still have functional value can be reemployed on the current project, stored for use on a future project, or sold on the ever-growing salvage market (see Figure 13.7). Recycling of building materials can be accomplished whenever sufficient quantities can be collected and markets are readily available. The difference in each opportunity must be understood in order to redirect materials from entering landfills. In doing that, the first step is identifying areas in which construction activities generate C&D waste.

WASTE GENERATION AND OPTIONS FOR DIVERSION AND REUSE

According to the last EPA study on the subject, C&D waste totaled more than 135 million tons (122.5 million metric tons) in the United States in 1998, about 77 million tons (70 million metric tons) of which resulted from commercial work alone.[7] Based on a typical developed country C&D rate of about 0.5 ton per capita annually, the current total C&D waste generation in the United States is likely to be about 170 million tons (154 million metric tons). Per unit area waste generation ranges from about 4 pounds (19.5 kilograms per square meter) for new construction and renovation to about 155 pounds (757 kilograms per square meter) for building demolition. On many construction projects, recyclable materials such as wood, concrete and masonry, metals, and drywall make

Figure 13.7 Example of proper waste separation to enhance the potential for materials reuse. (WasteCap Resource Solutions, Milwaukee, WI)

up as much as 75 percent of the total waste stream, presenting opportunities for significant waste diversion. As more C&D landfills reach capacity, new ones become increasingly difficult to site, and as more municipal waste landfills exclude C&D waste, tipping fees will continue to rise. Construction waste—and costs—can be managed just like any other part of the construction process, with positive environmental impacts on land and water resources. Many opportunities exist for reducing C&D waste. Construction managers are responsible for managing waste throughout the entire project. Handling such activities in a sustainable manner will further reduce the environmental impact of the building.

On-site fabrication of building components creates a large amount of construction scrap that is wasted. The likelihood of reusing scrap materials is much less on a job site than in either an off-site shop or centralized facility where similar products are regularly made. The option of sourcing building components or modules from off-site also permits the materials to be delivered when they are needed instead of being staged on-site where they pose an obstacle for construction work. Off-site prefabrication of building components or modules may include walls, kitchen equipment, stairways, ductwork assemblies, precast concrete, shelving and cabinetry, entire rooms that can be craned into place, and other specialized assemblies. If possible, materials should be ordered already cut to size to reduce construction time and on-site waste generation.

Purchasing materials in bulk can often avoid significant packaging waste as well as unit costs. Ideally, leftover materials should decrease if proper storage and staging have been executed; however, there are times when inefficient procurement results in excess products and materials on-site. Sustainable construction projects should prevent pollution by not ordering more material than necessary to complete the job. Careful attention to materials use can result in the ability to more precisely order materials for future projects. In some cases, manufacturers will buy back construction materials from a job site and restock them in their warehouse as long as those excess materials are not customized for the project and have been protected from damage.

Proper coordination with the various subcontractors is important to identify the scope of work and proper materials handling, staging, and waste separation. Clearly communicating with subcontractors will help in preventing potential rework as well as physical damage to installed systems. This is especially important when finishes are incorporated into the project and either have high exposure to foot traffic or are located in tight-fit areas. Rework not only generates waste but also increases project cost and extends the completion schedule.

Renovation projects require the removal of existing materials before new construction can begin. This removal can occur through either demolition or deconstruction. Demolition is the complete destruction of an existing building, structure, or space, leaving a mixture of materials that is difficult to separate. Deconstruction is construction in reverse in which the building and its components are dismantled for the purpose of reusing them or enhancing recycling. Demolition is not necessarily more cost effective than deconstruction if valuable materials and components are recovered that more than offset the additional time needed for deconstruction. Before either deconstruction or demolition occurs, a materials audit should be conducted to create an inventory of materials that may have value and that should be salvaged. Windows, doors, and brick are examples of building components that may have value and that should be removed in a manner that preserves their integrity. The materials audit may benefit from the opinion of a building materials reuse provider. Reuse providers may be found locally around the United States and Canada through the Building Materials Reuse Association (BMRA). In other cases, usable leftover materials can be donated to charities, such as Habitat for Humanity. Local universities, colleges, or trade schools may also be interested in using leftover materials as part of education and training.

If materials cannot be readily deconstructed, there still may be value in recycling the demolished materials. Recycling requires establishing areas on the construction site for scrap storage, cutting areas, recycling, and disposal. This includes developing procedures for separating hazardous waste by-products of construction (e.g., paints, solvents, oils, and lubricants) and for disposing of these wastes in accordance with federal, state, and local regulations. Establishing this type of area not only improves the potential for diverting C&D materials from the landfill but also establishes a visible first impression of how green the project may be to those watching the construction process.

Another alternative is to process specific demolition debris and use it as on-site fill. A variety of materials can be used, including concrete, brick, concrete masonry units (CMUs), and biodegradable and compostable materials. Concrete, brick, and CMUs can be crushed and used as a subbase or for a drainage field for water management purposes.

Commissioning

One of the major contributions of the high-performance green building delivery system is to require building commissioning as a standard practice. This has come about because at least a basic level of commissioning is required for certification under the US Green Building Council (USGBC) LEED-NC building assessment system and is highly recommended by Green Globes. Building commissioning provides the owner with an unprecedented level of assurance that the building will function as designed, with resultant high reliability and reduced operating costs. The success of building commissioning has culminated in the formation of specialist building commissioning companies and in the development of building commissioning departments in engineering firms whose purpose is to service the green building market. Building commissioning services can be executed in two ways: installation inspection and performance testing. Installation inspection identifies how specified components are installed on-site before equipment start-up. The inspection uses a checklist to verify compliance with construction drawings, specifications, and manufacturers' requirements. Areas of nonconformity can be documented with photos and written descriptions to facilitate resolution. Figure 13.8 shows an example of a defect found by a commissioning authority (CxA) during an installation inspection. Performance testing, or functional testing, happens when all components of a system have been installed. The purpose is to verify that the system, as a whole, is operating properly under full and partial load conditions. Sequence-of-operation testing is used to imitate all expected modes of building operation, including start-up, shutdown, capacity modulation, and emergency operations. Alarms are checked to ensure they are functioning properly, and piping and electrical connections to other equipment are inspected for proper installation and function.

Studies of the effects of building commissioning indicate that it may reduce building operating costs by a larger margin than energy conservation measures. A report by Greg Kats of Capital E put the 20-year savings in operations and maintenance due to building commissioning at $8.47 per square foot, compared with energy savings of $5.79 per square foot.[8] Unquestionably, then, building commissioning is a powerful tool for ensuring that the design intent—to reduce resource consumption and environmental impacts—is indeed carried out in the construction process. Building commissioning is, however, an additional

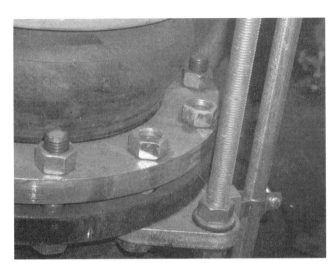

Figure 13.8 Commissioning installation inspection identified improper fastening of a pump flange. This equipment is exposed to significant vibration, which can loosen bolts that are not properly connected and torqued. (Photograph courtesy of John Chyz, Cross Creek Initiative, Inc.)

Figure 13.9 Logos of the (A) AABC Commissioning Group and (B) the Building Commissioning Association. [(A) Logo courtesy of the AABC Commissioning Group (ACG). Reprinted with permission. (B) Courtesy of BCA]

service, meaning that it adds to the first, or construction, cost of the building project. Furthermore, and unfortunately, during the cost reduction exercises that are now common practice during design, these additional fees are subject to being cut, regardless of their benefits.

Two organizations heavily engaged in and committed to improving building commissioning are the AABC Commissioning Group (ACG) and the Building Commissioning Association (BCA) (see Figure 13.9). According to the BCA, "The basic purpose of building commissioning is to provide documented confirmation that building systems function in compliance with criteria set forth in the Project Documents to satisfy the owner's operational needs. Commissioning of existing systems may require the development of new functional criteria in order to address the owner's current systems performance requirements."[9] The ACG has established a commissioning guideline and a certification program for commissioning agencies. The process defined in the ACG commissioning guideline can apply to any building system, and the same steps of planning, organizing, systems verification, functional performance testing, and documenting the tasks of the commissioning process that apply to building mechanical systems can also be applied to building electrical systems, control systems, telecommunications systems, and others.

ESSENTIALS OF BUILDING COMMISSIONING

Federal and state governments are increasingly requiring commissioning of their facilities; in fact, several government organizations also publish building commissioning guidelines. For example, the US General Services Administration's Public Buildings Service published *The Building Commissioning Guide*, and the Federal Energy Management Program produced *The Continuous Commissioning Guidebook for Federal Managers*.[10,11] The first provides an overall framework and process for building commissioning from project planning through tenant occupancy, while the latter establishes building commissioning as an ongoing process for use in resolving operating problems in buildings. Another publication, *New Construction Commissioning Handbook for Facility Managers*, was prepared for the Oregon Office of Energy by Portland Energy Conservation, Inc. (PECI), as part of a regional program involving four northwestern states. Its aim is to make building commissioning standard practice.[12]

ESSENTIALS OF BUILDING COMMISSIONING

According to the BCA, the building commissioning process is controlled and coordinated by a CxA. The following are the essential elements of building commissioning as carried out by the CxA:[13]

1. The CxA is in charge of commissioning process on behalf of the owner, is an advocate for the owner's interests, and makes recommendations to the owner about the performance of the commissioned systems.

2. The CxA must have adequate experience to perform the commissioning tasks and must have recent hands-on experience in building systems commissioning; building systems performance and interaction; operations and maintenance procedures; and building design and construction processes.

3. The scope of commissioning must be clearly defined in the commissioning contract and commissioning plan.

4. The roles and scope of all building team members in the commissioning process should be clearly defined in the design and engineering consultants' contracts; [in] the construction contract; [in] the General Conditions of the Specifications; in the divisions of the specifications covering work to be commissioned; and in the specifications for each system or component for which a supplier's support is required.

5. A commissioning plan must be produced to describe how the commissioning process will be carried out, and should identify the systems to be commissioned; the scope of the commissioning process; the roles and lines of communications for each team member; and the estimated commissioning schedule. The commissioning plan is a single document that reflects specified criteria identified from the contracts and contract documents.

6. For new construction, the CxA should review systems installation for commissioning issues throughout construction.

7. Commissioning activities and findings are documented exactly as they occur, distributed immediately, and included in the final report.

8. A functional testing program, composed of written, repeatable test procedures, is carried out, indicating expected and actual results. The installation inspection program should be carried out in a similar manner.

9. The CxA should provide constructive input for the resolution of system deficiencies.

10. A commissioning report is produced that evaluates the operating condition of each system; deficiencies that were discovered and measures taken to correct them; uncorrected operational deficiencies accepted by the owner; functional test procedures and results; documentation of all commissioning activities; and a description and estimated schedule for deferred testing.

MAXIMIZING THE VALUE OF BUILDING COMMISSIONING

As noted previously, building commissioning has tremendous potential for generating savings for the building owner. To ensure that the maximum value is obtained for building commissioning, the BCA recommends that the scope of building commissioning also include the following:

1. Prior to design, seek assistance in evaluating the owner's requirements such as energy conservation, indoor environmental quality (IEQ), training, operations, and maintenance.

2. During each design phase, review construction documents for compliance with design criteria, commissioning requirements bidding issues, construction coordination and installation concerns, performance, and facilitation of operations and maintenance.

3. Review equipment submittals for compliance with commissioning issues.

4. Review and verify schedules and procedures for system start-up.

5. Ensure that training of operating staff is conducted in accordance with project documents.

6. Ensure that operations and maintenance manuals comply with contract documents.

7. Assist the owner in assessing system performance prior to expiration of the construction contract warranty.

HVAC SYSTEM COMMISSIONING

Today's process of building commissioning has its roots in the science of testing, adjusting, and balancing (TAB). For more than 40 years, the standards developed by the Associated Air Balance Council (AABC) have been traditionally used to verify that a building's HVAC systems are operating as designed. The TAB agency is an independent organization hired to check that air handlers, fans, pumps, dampers, energy recovery systems, hot water heating units, and other components are functioning properly; that the flow rates of hot and chilled water are as designed; and that airflows are properly adjusted so that the quantities of supply air, return air, and ventilation air in each space are also as designed. With the advent of the high-performance green building movement, AABC expanded its activities and nomenclature to also cover building commissioning through the development of ACG. Although commissioning of HVAC systems remains the most common commissioning activity (so that the initial evolution from TAB to commissioning was fairly straightforward), commissioning providers today are increasingly called upon to perform "total building commissioning" and address a much broader range of building systems. ACG defines the key commissioning activities as those shown in Table 13.5.[14]

COMMISSIONING OF NONMECHANICAL SYSTEMS

Although the commissioning of mechanical systems is at the heart of building commissioning, the commissioning process should include all building systems: all electrical components; telecommunications and security systems; plumbing fixtures; rainwater harvesting systems; graywater systems; electronic water controls; items such as finishes, doors, door hardware, windows, millwork, and

TABLE 13.5

Key HVAC System Commissioning Activities as a Function of Project Phase

Phase	Key Commissioning Activities
Predesign	Establish commissioning as an integral part of the project.
	Owner selects the CxA.
	Develop the scope of commissioning.
	CxA reviews the design intent.
Design	Review of design to ensure [that] it accommodates commissioning.
	Write commissioning specifications defining contractor responsibilities.
	CxA produces the commissioning plan.
	The project schedule is established.
Construction	CxA reviews contractor submittals.
	CxA updates commissioning process.
	Continued coordination of commissioning process.
	Carry out and document system verification checks.
	Carry out and document equipment and system startup.
	Carry out and document TAB activities.
Acceptance	Carry out functional performance tests for all HVAC systems.
	Train O&M [operations and maintenance] staff for effective ongoing operations and maintenance of all systems.
	Provide full documentation of HVAC systems.
Postacceptance	Correct any deficiencies and carry out any required testing.
	Carry out any required "off season" tests.
	Update documentation as required.

ceiling tiles; and any other component in the building's drawings and specifications. The commissioning tasks for nonmechanical systems are to:[15]

- Ensure appropriate product selection during design and the design intent review.

- Ensure that product specifications are clear by conducting a specification review.

- Ensure [that] the construction manager or subcontractor selects an acceptable product during the submittal review.

- Ensure [that] the construction manager or subcontractor properly installs the product.

- Ensure [that] adequate operations and maintenance (O&M) documentation is provided so that facility staff can properly maintain the item through an O&M documentation review.

- Ensure [that] facility staff receive adequate training to operate and maintain the item through a training verification.

- Ensure [that] the O&M plan addresses all items through an O&M plan review.

- Ensure [that] the building's indoor environmental quality (IEQ) meets the design objectives.

The more involved the commissioning process, the greater is the need for a diverse commissioning team whose members can handle the range of systems included in the building commissioning process. The members involved during the design process may be different from those who test the systems when construction is complete. For example, the commissioning team members engaged during design may be experts in multidisciplinary work and have an overall understanding of the process. This knowledge may include experience in selecting products and ensuring complete documentation to support a clear direction to all design decisions. Figure 13.10 shows an example of how detailed systems knowledge can help detect future operational problems as part of the commissioning process.

COSTS AND BENEFITS OF BUILDING COMMISSIONING

Building commissioning provides a wide range of benefits—and, it is important to note, the earlier in the building design and construction process that building commissioning is implemented, the greater will be the benefits. The ideal arrangement is for the CxA to be hired at the onset of the project, along with the design team and construction manager. The CxA provides another set of inputs to the building team and ensures that, throughout design and construction, issues related to commissioning are included in the construction documents. The following are some of the typical benefits that can be expected as a result of including a full scope of building commissioning services from an independent CxA:

Figure 13.10   The CxA identified boiler exhaust condensation and dripping onto an exterior outlet and proposed the relocation of the electrical outlet to prevent future corrosion as well as increase safety. (Photograph courtesy of John Chyz, Cross Creek Initiative, Inc.)

- Reduced operating costs due to an energy efficiency increase of 5 to 10 percent, attributed to building commissioning

- Increased employee productivity due to improved IEQ resulting from building commissioning

- Improved construction documents resulting from the participation of the CxA in the review process during each design phase and the potential for greatly reducing change orders

- Fewer errors in equipment ordering due to the continual review of equipment requirements by the CxA

- Fewer equipment installation errors because the CxA reviews equipment installation during the construction process

- Fewer equipment failures during building operation due to the testing, calibration, and reporting carried out by the CxA

- Complete documentation of systems provided to the owner

- A fully functioning building from the first day of operation

The Oregon Office of Energy provides information on the benefits of building commissioning for energy savings for several building types. These are listed in Table 13.6.[16]

TABLE 13.6

Energy Savings Attributable to Building Commissioning for Various Building Types

Building Type	Dollar Savings	Energy Savings
110,000 ft^2 office	$0.11/ft^2/yr ($12,276/yr)	279,000 kWh/yr
22,000 ft^2 office	$0.35/ft^2/yr ($7,630/yr)	130,800 kWh/yr
60,000 ft^2 high-tech manufacturing	$0.20/ft^2/yr ($12,000/yr)	336,000 kWh/yr

TABLE 13.7

Cost of Commissioning Services by an Independent Third-Party Service

Construction Cost	Total Commissioning Cost*	Note
<$5 million	2%–4%	Costs include moderate travel, but building complexity, number of site visits, and other factors may also affect the cost.
<$10 million	1%–3%	
<$50 million	0.8%–2.0%	
>$50 million	0.5%–1.0%	
Complex projects (labs)	Add 0.25%–1%	

*As a percentage of the construction cost.

TABLE 13.8

Costs of Commissioning during Design and Construction Phases for Typical Systems

Phase	Commissioned System	Total Commissioning Cost
Design	All	0.1% to 0.3%
	HVAC and controls	2.0% to 3.0% of total mechanical cost
Construction	Electrical system	1.0% to 2.0% of total electrical cost
	HVAC, controls, and electrical system	0.5% to 1.5% of total construction cost

The cost of commissioning is a function of the size of the project, its complexity, and the level of commissioning selected by the owner (see Table 13.7).[17] Buildings with simple conditioning systems, few zones, and simple control systems would be at the lower end of the commissioning cost range shown for various levels of construction cost, whereas buildings with complex conditioning systems and control systems would be at the higher end of the range. Similarly, the benefits of commissioning for more complex buildings are far greater than those for buildings with relatively simple systems.

It is important to point out that Table 13.7 does not separate the costs of fundamental commissioning, required by the LEED rating system, from those of enhanced commissioning, which is optional under LEED. Fundamental commissioning may be carried out by personnel from the design firms on the building team as long as they are not directly involved in the project, and these costs are sometimes rolled into the design fee to minimize costs.

The Oregon Office of Energy provides another viewpoint of commissioning costs, as shown in Table 13.8.[18]

It should be noted that the commissioning costs in the design phase include the costs for the CxA and the architect, with the allocation being approximately 75 percent for the CxA and 25 percent for the architect. Similarly, for the construction phase, additional costs are charged for the engineers to attend meetings, create checklists, and participate in testing. These costs are not listed in Table 13.8 and amount to 10 to 25 percent of the CxA's fee. There may also be additional costs for the architect's involvement in reviewing the commissioning plan and attending meetings, in the range of 5 to 10 percent of the CxA's fee.

Due to the nature of construction, virtually every building is a unique, one-off design, including the design of complex mechanical and electrical systems and their control systems. The consequence of this sophistication and complexity is that high-performance buildings need to be carefully tuned and calibrated to ensure their operation is as designed. The commissioning process has been shown to be invaluable in providing a high degree of quality assurance for buildings with sophisticated energy and conditioning systems and is now virtually standard practice for green building certification. As a result of its success, the commissioning process is being extended to other building systems such as the building envelope and even interior finishes. This thought piece by John Chyz of the Cross Creek Initiative addresses the important and evolving role of commissioning in the production of high-performance facilities.

The Role of Commissioning in High-Performance Green Buildings

John Chyz, Managing Director, Cross Creek Initiative, Inc., Gainesville, Florida

Peter J. Wilson argues in *The Domestication of the Human Species* that settling down into a built environment was the most radical and far-reaching innovation in human development, having a pivotal effect on human psychology and social relations. It is no wonder, then, that the Moore's law—like trajectory in the evolution of human intelligence has given rise to notions of sustainability, reduced carbon footprints, and the defining of a healthy relationship between human beings and their natural world. This trend in thought is coincidentally paralleled by recent strides to enhance the health of the human body—organic diets, alternative medicine, and exercise. Built environments, in addition to being fabricated essentially out of the very building blocks of nature, create visual barriers, define boundaries between outside and inside and between various socioeconomic groups, house user-defined activities, and overwhelmingly serve as physical representations for human interactivity.

Sustainable built environments specifically are realized through the successful implementation of the growing array of strategies that have evolved out of this environmentally conscious movement toward a greener planet. Typically, these efforts focus on energy and water conservation, fastidious materials selection, minimization of site disruption, and attention on healthy indoor environments, to name a few. Of growing importance for the successful delivery of high-performance green buildings is some measure of quality assurance and control during design and construction of new facilities in addition to the ongoing optimization of existing buildings. A green building may be carefully designed and engineered to deliver superior energy performance, but little of this sustainable feature will come to fruition if the mechanical systems have not been installed, programmed, and balanced correctly. One of the biggest challenges facing the construction industry today is the coordination of trades. The growing complexity and sophistication of building systems has driven a market response by way of highly specialized products and the trained professionals to design, install, troubleshoot, and maintain them. Despite the long-standing notion that the most current and widely implemented delivery protocols, including communication structures, for new construction and renovation projects are effective at translating the owner's vision and goals into a fully functional, optimized, and sustainable building, evidence from the field has demonstrated otherwise. The truth of the matter is that new buildings are erected by individual contractors installing their respective systems. At project turnover, rarely are these systems and subsystems tested as a whole living, breathing unit. This approach is analogous to car manufacturers designing, prototyping, and producing a new vehicle, then delivering it to you or me (the driver) without test driving it first. Using the same analogy, we all know that used vehicles require tune-ups and oil changes periodically and existing buildings are no different. In fact, it may surprise some to learn that in the absence of a retro-commissioning process, there are no other industry standard protocols for "tuning up" an existing building other than the execution of an energy audit. Energy auditing is essentially used to identify where a "vehicle may save on fuel costs" and how much it will cost to implement those measures that will achieve the desired mpg improvement.

We at the Cross Creek Initiative have commissioned well over 1 million square feet of commercial building space ranging from new university academic buildings to existing health-care facilities. A brief study of those projects has demonstrated an astonishingly wide range of deficiency issues encountered and subsequently rectified. The following list identifies those problems that have occurred with the greatest frequency:

- Envelope leaks/building pressurization
- Visual observations (incorrect installation, damage, drainage, missing equipment)
- Accessibility issues/housekeeping
- Engineering issues (over/undersized)
- Inadequate outdoor air delivery and CO_2 monitoring
- Sequence optimization/tuning/programming
- Variable-frequency drive (VFD) control/status/fault
- Runtime/overridden in hand position
- Incorrect labeling/documentation conflict
- Field/building management system (BMS) issue (incorrect wiring)

It becomes evident rather quickly that these kinds of issues collectively not only cost owners money in wasted energy each year but can also compromise occupant comfort, health, and overall safety.

The world's largest database of commissioning cost/benefit case studies was assembled by Evan Mills and his team at the Lawrence Berkeley National Laboratory in 2004 and updated in 2009. The results of the ensuing meta-analysis were eye-opening. Of data gathered for 643 buildings across 26 states, the median normalized cost to deliver commissioning was $0.30/ft^2 for existing buildings and $1.16/ft^2 for new construction projects. All told, according to Mills, this represented an average of 0.4 percent of the overall construction cost. Through the rectification of the 10,000+ deficiencies discovered, a median energy savings of 13 percent was realized for the new construction projects and 16 percent for the existing buildings, with payback times of 4.2 years and 1.1 years, respectively. Furthermore, project teams that elected to implement a comprehensive commissioning process enjoyed nearly twice the overall median energy savings. With regards to greenhouse gas emissions and using the same study, Mills maintains that the median cost of conserved carbon equaled −$25/metric ton for new construction projects and −$110/metric ton for existing buildings—figures that compare favorably with the current market prices of carbon offsets ($10–$30/metric ton).

Perhaps the most compelling figures derived from the study fall out of a simple extrapolation from the current stock of commercial buildings in the United States. Applying the median energy savings derived from the control group nationally results in a projected energy savings of $30 billion by 2030, the equivalent of approximately 340 megatons of CO_2 each year. Believe it or not, the current size of the commissioning industry serving existing buildings has only reached approximately $200 million per year. According to Mills, if each existing building in the United States were retro-commissioned every five years, the commissioning industry would quickly swell to $4 billion per year, requiring an additional 1500 to 25,000 full-time-equivalent employees.

Rather than simply acting as tool for the realization of energy savings, a well-executed building commissioning process may be more accurately described as a risk management strategy. It ensures that building owners have been delivered with a building that meets their expectations within the specified budget and provides insurance for policy managers that their initiatives accurately meet targeted goals. Furthermore, the building commissioning process serves to detect and rectify issues that would eventually prove far more costly to the owner in the future from the standpoint of operation, maintenance, safety, and unwanted litigation.

In response to the programmatic deficit that plagues the building design, construction, and maintenance industries and in light of the crucial importance of "getting it right" when it comes to the delivery of a high-performance green building, the building commissioning process has not only become an essential component of business as usual but also presents owners with the unique opportunity to save energy and reduce carbon emissions while simultaneously improving occupant health and comfort. If a quality commissioning process were to be embraced by the design, construction, and maintenance industries nationwide, the potential for job creation, environmental leadership, cost savings, and improvements in occupant health would be significant, to say the least.

Summary and Conclusions

The success of a high-performance green building project is at least in part dependent on the conduct of the construction phase. The construction manager must fulfill several specific responsibilities to ensure that the process embodies the intent of green building, namely, to be environmentally friendly and resource-efficient and to result in a healthy building. By protecting plants and other ecosystem components, keeping the footprint of construction operations as small as possible, and minimizing sedimentation and erosion, the builder can meet the first of these objectives—protecting the environment. The resource efficiency goal can be met by reducing C&D waste and by planning and executing a well-thought-out construction waste management plan. The quality of indoor air can be protected for future building occupants by protecting ductwork from fabrication through installation; by properly storing materials to avoid moisture penetration, mold, and mildew; and by appropriately ventilating and flushing out the building prior to occupancy. A thorough training program for subcontractors should be instituted, and requirements peculiar to the construction of a green building should be integrated with other standard training programs, such as construction safety. Finally, building commissioning is an important component of the delivery process for high-performance green buildings and has been shown to reap enormous benefits in the form of reduced O&M costs. A diverse range of firms provide building commissioning services in support of the high-performance green building movement. The economic returns for building commissioning are very high—greater, according to some accounts, than for energy savings.

For building commissioning to be truly effective, it must occur periodically throughout the building's life cycle because complex systems tend to drift out of specification and even fail. The high-performance green building system has brought the relatively new discipline of building commissioning to the forefront in terms of its value to the building project. The return on investment for building commissioning warrants consideration of an extensive building commissioning program for green building projects.

Notes

1. From NCCER (2011).
2. Ibid.
3. Adapted from "Maintaining Indoor Environmental Quality (IEQ) during Renovation and Construction" at the Center for Disease Control website www.cdc.gov/niosh/topics/indoorenv/ConstructionIEQ.html.

4. From SMACNA (2007).
5. From GPRO (2011) and NCCER (2008).
6. From NCCER (2011).
7. From Franklin Associates (1998).
8. From Kats (2003).
9. As described on the BCA website at https://netforum.avectra.com/eweb/StartPage .aspx?Site=BCA&WebCode=HomePage.
10. From GSA (2005).
11. From FEMP (2002).
12. From OEO (2000).
13. From the BCA website. See note 9.
14. From ACG (2005).
15. Ibid.
16. From OEO (1997).
17. Ibid.
18. Ibid.

References

ACG. 2005. "ACG Commissioning Guideline," available at www.commissioning.org/ commissioningguideline/ACGCommissioningGuideline.pdf.

FEMP. 2002. "The Continous Commissioning Guidebook for Federal Managers," prepared for the Federal Energy Management Program by the Energy Systems Laboratory, October. Available at www1.eere.energy.gov/femp/pdfs/ccg01_covers .pdf.

Franklin Associates. 1998. "Characterization of Building-Related Construction and Demolition Debris in the United States," prepared for the US Environmental Protection Agency by Franklin Associates, Report No. EPA530-R-98-010. Available at www.epa.gov/osw/hazard/generation/sqg/cd-rpt.pdf.

GPRO. 2011. *Construction Management: Green Professional Skills Training (GPRO)*. New York: Urban Green Council.

GSA. 2005. "The Building Commissioning Guide," April. Available at www.wbdg.org/ ccb/GSAMAN/buildingcommissioningguide.pdf.

Kats, Gregory H. 2003. "Green Building Costs and Financial Benefits," published for the Massachusetts Technology Collaborative. Available at the Capital E website, www .cap-e.com.

National Center for Construction Education and Research (NCCER). 2008. *Your Role in the Green Environment.* Upper Saddle River, NJ: Pearson/Prentice Hall.

National Center for Construction Education and Research (NCCER). 2011. *Sustainable Construction Supervisor Trainee Guide.* Upper Saddle River, NJ: Prentice Hall.

Oregon Energy Office (OEO). 1997. "Commissioning for Better Buildings in Oregon," available at www.oregon.gov/ENERGY/CONS/BUS/comm/docs/commintr.pdf?ga=t.

Oregon Energy Office (OEO). 2000. "New Construction Commissioining Handbook for Facility Managers," prepared by Portland Energy Conservation, Inc. (PECI) for the Oregon Energy Office. Available at www.oregon.gov/ENERGY/CONS/BUS/ comm/docs/newcx.pdf?ga=t.

SMACNA. 2007. "IAQ Measures for Occupied Buildings during Construction," 2nd Edition, ANSI/SMACNA 008–2008, Sheet Metal and Air Conditioning Contractors National Association.

Wilson, Peter J. 1991. *The Domestication of the Human Species.* New Haven, CT: Yale University Press.

Chapter 14

Green Building Economics

The market for green buildings in the United States continues to increase both in size and in market share. In *Green Outlook 2011*, McGraw-Hill Construction (MHC) reported that the market size of green construction, including both residential and nonresidential buildings, had jumped fourfold in just three years, from $10 billion in 2005 to $42 billion in 2008, and was expected to range between $55 billion and $71 billion in 2011. In 2010, it was estimated that new nonresidential green construction represented 28 to 35 percent of total construction volume, 50 percent higher than just two years earlier. By 2015, MHC forecasts that the scale of new nonresidential green construction could be in the $120 billion to $150 billion range, representing 40 to 48 percent of total nonresidential construction volume. Similar growth is occurring in building retrofits with MHC forecasting a market of $14 billion to $18 billion in 2015. What is clearly very remarkable, even startling, about this growth is that it is occurring in spite of the major downturn in construction resulting from the so-called great recession of 2008–2010. The three sectors with the greatest rate of market growth and penetration are education, health care, and office buildings. Green building data indicate that there are several major trends in the ongoing shift to green buildings.[1]

First, the bigger the building project, the more likely it is to be a high-performance building. Because health-care projects tend to be larger, the number of green health-care projects is growing very rapidly. Over 70 percent of projects at least $50 million in size are including the US Green Building Council (USGBC) Leadership in Energy and Environmental Design (LEED) building rating system in their specifications. Second, throughout the United States, schools at all levels from K–12 to university are high-performance green buildings, and green building activity in the educational sector was between $13 billion and $16 billion in 2010. This rapid growth rate is likely being propelled by a combination of state and local mandates that require schools to be certified as green buildings. Third, a significant number of federal, state, and local governments are requiring that publicly owned buildings be high-performance green buildings. At least 12 federal agencies, 33 states, and 384 local government programs have been enacted as of 2010.

In this chapter, we cover the business case for high-performance green buildings; the economics of green building, including how to quantify a wide variety of savings and benefits; and the management of the additional first, or capital, costs that may accompany a green building project. Finally, we discuss the topic of "tunneling through the cost barrier," which suggests that the synergies created in greening a building can be so significant that significant capital cost reductions can be achieved.

General Approach

Understanding building economics is important for any construction project, but it is especially important for high-performance green buildings because justifying this approach can involve somewhat more complex analysis than for conventional construction. High-performance buildings can produce benefits for their owners in a diverse range of categories: energy, water, wastewater, health and productivity, operations and maintenance (O&M), maintainability, and emissions, to name a few. To be able to address the scope of benefits, the building team must be able to either quantify the effects of their decisions by using simulation tools or rely on the best available research and evidence gathered from other projects.

This chapter addresses the economic and business arguments for high-performance buildings and approaches for quantifying the various benefits achievable by investing in environmentally beneficial buildings.

A report to the California Sustainable Building Task Force states that a 2 percent additional investment to produce a high-performance building would produce life-cycle savings that are 10 times greater than the incremental investment.[2] For example, an additional $100,000 investment in a $5 million building should produce at least $1 million in savings for a building with an assumed 20-year life cycle. This is a truly remarkable claim and, if verifiable, makes a virtually unshakable case for high-performance building.

Today, high-performance green buildings are thought to have a higher capital or construction cost than conventional buildings, on the order of 2 percent, or $2 to $5 per square foot.[3] The additional required capital is proportional, at least generally, to the level of the building's LEED-NC rating (see Table 14.1).[4]

An analysis of the financial benefits of high-performance green buildings concluded that significant benefits could be attributed to this type of delivery system and that there was a correlation between the LEED-NC rating and the financial return. Table 14.2 indicates that, for a typical high-performance building, the total net present value (TNPV) of the energy savings over a 20-year life cycle is $5.79 per square foot, with other notable per square foot savings from reduced emissions ($1.18), water ($0.51), and O&M savings resulting from building commissioning ($8.47).[5] Table 14.2 also shows productivity and health savings per square foot of $36.89 for LEED certified and silver buildings and $55.33 for LEED gold and platinum buildings. Clearly, the productivity and health benefits of high-performance green buildings dominate this discussion, and for gold and platinum buildings, the claim is that the savings are almost 10 times greater than the energy savings. It is important to point out, however, that although these claims are generally accepted by high-

TABLE 14.1

Cost Premiums Derived from 33 Buildings with a LEED-NC Rating

LEED-NC Rating	Sample Size	Cost Premium
Platinum	1	6.50%
Gold	6	1.82%
Silver	18	2.11%
Certified	8	0.66%
Average	—	1.84%

TABLE 14.2

Value of Various Categories of Savings for Buildings Certified by the USGBC

Category	20-Year Total Net Present Value (TNPV) per Square Foot*
Energy value	$5.79
Emissions value	$1.18
Water value	$0.51
Waste value—construction only, one year	$0.03
Commissioning O&M† value	$8.47
Productivity and health value (certified and silver)	$36.89
Productivity and health value (gold and platinum)	$55.33
Less green cost premium	($4.00)
Total 20-year NPV (certified and silver)	$48.87
Total 20-year NPV (gold and platinum)	$67.31

*Net present value (NPV) is the net savings for each year, taking into account the discount rate (time value of money). The 20-year total net present value (TNPV) is the sum of the NPVs for all 20 years and represents the total life-cycle savings.

†O&M commissioning ensures that the building is built and operated according to the design and results in substantially lower O&M costs.

TABLE 14.3

Comparison of Costs and Savings for NREL Prototype Buildings

Feature	Added Cost	Annual Savings
Energy efficiency measures	$38,000	$4,300
Commissioning	$4,200	$1,300
Natural landscaping, stormwater management	$5,600	$3,600
Raised floors, movable walls	0	$35,000
Waterless urinals	($590)	$330
Total	**$47,210**	**$44,530**

performance building practitioners, most of those made for productivity and health improvements are based on anecdotal information, not scientific research. The total 20-year net present value (NPV) is $48.87 for certified and silver buildings and $67.31 for gold and silver buildings. The magnitude of these benefits is very impressive when considering that, on average, the incremental construction cost ranges from about $1.50 per square foot for LEED-certified buildings to about $9.50 per square foot for LEED platinum buildings.

A side-by-side analysis of two prototype buildings by the US Department of Energy's Pacific Northwest National Laboratory (PNNL) and the National Renewable Energy Laboratory (NREL) compared the costs and benefits of investing in high-performance buildings. A base two-story, 20,000-square-foot (1858-square-meter) building with a cost of $2.4 million meeting the requirements of ASHRAE Standard 90.1-1999 was modeled using two energy simulation programs, DOE-2.1e and Energy-10, and compared to a high-performance building that added $47,210 in construction costs, or about 2 percent, for its energy-saving features. Table 14.3 summarizes the results of this study.[6] The features listed are those for which an additional investment was made to produce the high-performance version of the NREL prototype building:

- Building commissioning, as noted previously, can produce significant savings by ensuring that the mechanical systems are functioning as designed.
- Natural landscaping and stormwater management produce savings due to the elimination of infrastructure and the use of easily maintainable native plants.
- Raised floors and movable walls produce savings by improving the flexibility of a building, reducing renovation costs.

The results of this comparison are remarkable because they indicate that the annual savings produced by the high-performance version are about equal to the added construction cost, producing a simple payback in just over one year. It should be noted, however, that this study did not address this comparison as if the building were to undergo certification through the USGBC's LEED process.

The additional capital costs often associated with high-performance buildings are a function of several factors. First, these buildings often incorporate systems that are not typically present in conventional buildings, such as rainwater harvesting infrastructure, daylight-integrated lighting controls, and energy recovery ventilators. Second, green building certification (fees, compilation of information, preparation of documents, cost of consultants) can add markedly to the costs of a project. And, finally, many green building products cost more than their counterparts, often because they are new to the marketplace and demand is only in the process of developing. In this last category are many nontoxic materials such as paints, adhesives, floor coverings, linoleum, and pressed strawboard used in millwork, to name but a few of the many new green building products emerging to serve the high-performance building market. Conversely, cost reductions for some building systems are achievable in green buildings—for example, in heating, ventilation, and air conditioning (HVAC) systems—that can be downsized as a consequence of improved building envelope design. However, additional energy-saving components such as energy recovery ventilators (ERVs), premium high-efficiency motors, variable-frequency drives for variable air volume (VAV) systems, carbon dioxide sensors, and many others all add to the front-end capital cost.

As for every other type of project, understanding the economics of the situation and including them in the decision-making process is of crucial importance. As described earlier, the classical approach used in assessing high-performance building economics is life-cycle costing (LCC), which includes a consideration of both first cost (sometimes referred to as *construction cost* or *capital cost*) and operating costs (utilities and maintenance). These two major cost factors are combined in a cost model that takes into account the time value of money, the cost of borrowed money, inflation, and other financial factors. They are then combined into a single value, the total net present value (TNPV) of the annual costs, and the selection of alternatives is based on an evaluation of this quantity. In some cases, due to legislated requirements, only the capital cost is considered. For example, the state of Florida allows decisions on building procurement to be made solely on the basis of capital costs, whereas the US government requires that an LCC approach be used. Consequently, producing a high-performance public-sector building in Florida can be very challenging; therefore, finding creative mechanisms for investing in higher-quality construction is imperative. One potential mechanism is the creation of a revolving fund from which building owners or users can borrow and that can then be repaid through savings over time.

The Business Case for High-Performance Green Buildings

Making the case for high-performance buildings in the private sector must include a justification of why they make good business sense. In an attempt to address this issue, in 2003 the USGBC produced a brochure, "Making the Business Case for High Performance Green Buildings," that addresses the advantages to a business of selecting green buildings over conventional facilities.[7]

According to the USGBC, high-performance green buildings:

1. Recover higher first costs, if there are any. Using integrated design can reduce first costs, and higher costs for technology and controls reap rapid benefits.

2. Are designed for cost effectiveness. Owners are experiencing significant savings in energy costs, generally in the range of 20 to 50 percent, as well as savings in building maintenance, landscaping, water, and wastewater costs. The integrated design process, which is the hallmark of developing high-performance green buildings, contributes to these lower operational costs.

3. Boost employee productivity. Increased daylight, pleasant views, better sound control, and other soft features that improve the workplace can reduce absenteeism, improve health, and boost worker productivity.

4. Enhance health and well-being. Improved indoor environments can translate into better results in hiring and retaining employees.

5. Reduce liability. Focusing on the elimination of sick buildings and specific problems such as mold can reduce the incidence of claims and litigation.

6. Create value for tenants. Improved building performance can reduce employee turnover and maintenance and energy costs, thus contributing to better bottom-line performance. Additionally, the operating costs for building tenants will be substantially lower.

7. Increase property value. A key strategy of the LEED-NC building rating system is to differentiate green buildings in the marketplace, with the implicit assumption that lower operating costs and better indoor environmental quality will translate to higher value in the building marketplace. A building carrying a LEED-NC plaque will imply superior operational and health performance; hence, buyers will be willing to pay a premium for these features. This would, in turn, spur demand for more high-performance green buildings.

8. Take advantage of incentive programs. Many states, for example, Oregon, New York, Pennsylvania, and Massachusetts, have programs in place that provide financial and regulatory incentives for the development of green buildings. The number of these programs is likely to grow and may include, among other possibilities, shorter project approval times, lower permit fees, and lower property taxes.

9. Benefit your community. Green buildings emphasize infill development, recycling, bicycle use, brownfield rehabilitation, and other measures that reduce environmental impacts, improve the local economy, and foster stronger neighborhoods. Businesses opting for high-performance green buildings will be contributing to the overall quality of life in the community and earn a better reputation as a consequence of their efforts.

TABLE 14.4

Some of the Business Benefits of Green Building

	Green Retrofit and Renovation	New Green Buildings
Operating cost savings	—	13.6%
Over 1 year	8.5% for owners (10.5% tenants)	—
Over 10 years	16% for owners (15% tenants)	—
Building value increase	6.8%	10.9%
Return on investment (ROI) improvement	19.2%	9.9%
Occupancy increase	2.5%	6.4%
Higher rent	1%	6.1%

10. Achieve more predictable results. The green building delivery system includes improved decision-making processes, integrated design, computer modeling of energy and lighting, and life-cycle costing, and ensures that the owner will receive a final product that is of a predictable high quality. The best practices beginning to emerge in this era of high-performance buildings will also enable more accurate results forecasting.

In addition to these 10 factors, a number of other benefits can be claimed for high-performance buildings, many of them societal. For example, high-performance buildings can help address other problematic issues, among them:[8]

- High electric power costs
- Worsening power grid problems such as power quality and availability
- Possible water shortages and waste disposal issues
- State and federal pressure to reduce criteria pollutants
- Global warming
- Rising incidence of allergies and asthma, especially in children
- The health and productivity of workers
- The effect of school environments on children's ability to learn
- Increasing O&M costs for state facilities

There is also a range of benefits specifically for owners of commercial properties. McGraw-Hill Construction surveyed commercial building office tenants in 2006 and found that, on average, they would be willing to pay a 16 percent premium for green office space. Some additional business benefits were cited by McGraw-Hill in surveys conducted in 2008 and 2009, and these are shown in Table 14.4.[9]

The Economics of Green Building

There are two schools of thought with respect to the economics of green buildings. One school maintains that the construction cost of these buildings should be the same as or lower than that of conventional buildings. The argument for this line of thinking is that through integrated design and reducing the size of mechanical systems needed to heat and cool an energy-efficient building, the costs of high-performance building construction can be kept in line with those of conventional buildings. The ING Bank building, south of Amsterdam in the Netherlands, completed in 1987, is an example of a high-performance facility that cost about $1500 per square meter ($150 per square foot), including the land, the building, and its furnishings. At that time, this cost was comparable to or less than that of other bank buildings in the Netherlands.[10] This impressive feat was accomplished for an architecturally complex 50,000-square-meter (500,000-square-foot) building, featuring slanting brick walls and an irregular S-shaped footprint, with gardens and courtyards and a 30,000-square-meter (300,000-square-foot) underground parking lot. It is set in a high-density, mixed-use area, with retail, office, and residential buildings surrounding it. If all high-performance buildings could be produced at this high level of architectural quality and at the same or lower cost as conventional buildings, the case for these advanced buildings would be made.

In contrast, the second school of thought is that high-performance green buildings will inevitably have higher capital costs, and that by assessing total building costs on a life-cycle basis, the advantages of high-performance building will be achieved. The additional capital costs occur because high-performance buildings incorporate technologies and systems that are simply not present in conventional buildings, some of them complex and expensive. When attempting to assess the LCC of the many alternatives that can produce a high-performance green building, two distinctly different cost categories can be identified—hard costs and soft costs—defined as follows:

- *Hard costs* are those that are easily documented because the owner receives periodic billing for them—for example, electricity, natural gas, water, wastewater, and solid waste.
- *Soft costs* are those that are less easy to document and for which assumptions must be made for their quantification. Examples of soft costs are maintenance, employee comfort/health/productivity attributable to a building, improved indoor environmental quality (IEQ), and reduced emissions.

An LCC analysis using only hard costs is generally acceptable as justification for alternative strategies that include a trade-off of operational costs versus capital costs. Including soft costs in an LCC analysis is far more difficult to justify because the data cannot be verified with the same degree of rigor as for hard costs. If the results of an analysis of alternatives for a high-performance building are to be subjected to a strict review by financial decision makers, then verifiable hard costs should dominate the analysis. If there is greater latitude in the decision-making process, justifiable soft costs can be employed in the analysis.

The following four key points need to be considered when attempting to develop a case for high-performance buildings based on economic issues:

1. The primary life-cycle savings for a high-performance building will be a result of superior energy performance. For some types of buildings, HVAC or mechanical plants may indeed be downsized mechanically as a result of reducing external loads through the employment of superior passive design strategies and the design of a highly thermal-resistant building envelope. A significant reduction in HVAC plant size may also translate to a reduction in the size and cost of the electrical plant. However, for buildings that are dominated by interior loads (people and equipment), the HVAC plant may be unchanged in size compared to a

conventional building. A daylit building will certainly require far lower levels of electrically derived light during the day but will still require a full lighting system during the evening. As a result, although it will produce significant operational savings, the daylighting system will not lower the requirements for artificial lighting and, in some cases, may actually complicate the design of the conventional lighting system.

2. Life-cycle savings can also be easily demonstrated for water and wastewater conservation measures because these utilities, like energy, are well known. As water and wastewater costs rise, especially in water-short areas, their life-cycle savings may, in some cases, approach the scale of energy savings.

3. Savings due to good IEQ can potentially exceed all other savings. For example, for a typical office building, maximum energy savings may be $1 per square foot ($10 per square meter) annually, whereas the worth of a 1 percent improvement in employee productivity translates to $1.40 to $3.00 per square foot ($14 to $30 per square meter). Although these savings are far greater than those of any other category, it is difficult to justify their inclusion in an LCC unless the building owner is especially motivated to include this information in the analysis.

4. Savings due to materials factors are very difficult to demonstrate. In many cases, green or environmentally friendly materials may, in fact, cost more—sometimes far more—than the alternatives. For example, compressed wheatboard used for cabinetry currently costs as much as 10 times more than the alternative, plywood.

Quantifying Green Building Benefits

An LCC for a green building project can address both hard and soft cost issues, either individually or in a comprehensive LCC that includes all cost factors. The following are general benefits that can be included in the LCC and the range of benefits that can be expected (hard costs) or justified (soft costs).

QUANTIFYING ENERGY SAVINGS

Green buildings use substantially less energy than conventional buildings and generate some of their power on-site from renewable or alternative energy sources. In a Capital E survey of 60 LEED-rated buildings conducted by Greg Kats in 2003, these buildings consumed an average of 28 percent less energy than their conventional counterparts and generated an average of 2 percent of their energy on-site from photovoltaics, thus reducing total fossil fuel–based energy consumption by about 30 percent. Reducing energy consumption provides a second benefit: a reduction in the emissions of global warming gases, which can also be assigned a cost benefit.

Analyzing the energy advantages of a high-performance green building requires the use of an energy simulation tool such as the aforementioned DOE-2.2 and Energy-10. A series of alternatives can be tried out and tested to determine the best combination of measures for the particular building and its location. An LCC analysis is also generated at the same time to provide cost and payback information, which is used in tandem with the energy-savings data to optimize energy performance. Using this approach, first costs and operational costs are combined to provide a comprehensive picture of the building's energy performance over an assumed lifetime.

TABLE 14.5

Comparison of Energy Performance for a Building Meeting ASHRAE Standard 90.1-1999 with a High-Performance Green Building

	Base Case Building Annual Energy Cost	High-Performance Building Annual Energy Cost	Percent Reduction
Lighting	$6,100	$3,190	47.7
Cooling	$1,800	$1,310	27.1
Heating	$1,800	$1,280	28.9
Other	$2,130	$1,700	20.1
Total	$11,800	$7,490	36.7

TABLE 14.6

Costs, Economic Metrics, and Energy Use: Base Case Compared to High-Performance Green Buildings

	Base Case	High-Performance Case
First cost of building	$2,400,000	$2,440,000
Annual energy cost	$11,800	$7,490
Energy reduction from base case	NA	36.7%
Economic Metrics		
Simple payback (years)	NA	8.65
Life-cycle cost	$2,590,000	$2,570,000
Reduction in life-cycle cost from base case	NA	0.85%
Savings-to-investment ratio	NA	1.47
Energy Consumption, Annual		
Million BTU	730	477
Reduction from base case	NA	34.6%

Estimating the energy savings for a particular project relies on using a base case that meets a minimum standard. The case of the two-story NREL prototype buildings was used as an illustration at the beginning of the chapter to discuss the costs and benefits of high-performance green buildings. The two-story, 20,000-square-foot (1858-square-meter) building with a base cost that meets the requirements of ASHRAE Standard 90.1-1999 was modeled using DOE-2.1e and Energy-10 to simulate various measures that would substantially improve its performance. The results of this comparison are shown in Tables 14.5 and 14.6.[11]

QUANTIFYING WATER AND WASTEWATER SAVINGS

Reductions in water consumption produce significant benefits with respect to water and wastewater. A sample, from Falcon Waterfree Technologies, LLC, of the financial impacts of reducing water consumption through the use of waterless fixtures is shown in Table 14.7. This example indicates that the per fixture savings for a waterless urinal are on the order of $161 to $192 per year. Although the cost of a waterless urinal, on the order of $300, is much higher than that of a flush urinal, the installation costs are much lower because connection to a source of water for flushing is unnecessary. Consequently, the savings noted in this table are for systems with very similar installation costs.

TABLE 14.7

Annual Savings Using Waterless Urinals Instead of Flush Urinals

Assumptions	75 Units	100 Units	200 Units
Total facility population	1,500	3,000	5,000
Percent of males	55%	50%	60%
Number of males	825	1,500	3,000
Number of urinals	75	100	200
Uses/day/person	3	3	3
Gallons/flush old urinals	3	3	3
Water cost/1,000 gallons	$2.50	$2.50	$2.50
Sewer cost/1,000 gallons	$2.50	$2.50	$2.50
Operating days/year	260	260	260
Annual Water Savings			
Savings in gallons	1,930,500 gal	3,510,000 gal	7,020,000 gal
Savings in dollars	$4,826	$8,775	$17,550
Annual Sewer Savings			
Savings in gallons	1,930,500 gal	3,510,000 gal	7,020,000 gal
Savings in dollars	$4,826	$8,775	$17,550
Total Water and Sewer Savings	**$9,652**	**$17,550**	**$35,100**
Annual Operating Cost Comparison			
Flush urinal*	$5,625	$7,500	$15,000
Waterless urinal[†]	$3,217	$5,580	$11,700
Total Operating Cost Savings	**$2,408**	**$1,650**	**$3,300**
Total Annual Savings[‡]	**$12,060**	**$19,200**	**$38,400**
Annual Savings/Urinal	**$161**	**$192**	**$192**

*Total water savings (3 uses/day × 260 days × number of users × water cost).
[†]Total sewer savings (3 uses/day × 260 days × number of users × sewer rate).
[‡]Water/sewer savings plus operating cost savings.

TABLE 14.8

Projected Savings from Using Waterless Urinals Instead of Flush Urinals in Various Occupancies: Existing versus New Buildings

Building Type	No. of Males	No. of Urinals	Uses/Day	Gallons/Flush	Days/Year	Water Savings/Gallon	Water Savings/Liter
Small office	25	1	3	3.0	260	58,500	220,000
New office	25	1	3	1.0	260	19,500	73,800
Restaurant	150	3	1	3.0	360	54,000	204,000
New restaurant	150	3	1	1.0	360	18,000	68,100
School	300	10	2	3.0	185	33,300	126,000
New school	300	10	2	1.0	185	11,100	42,000

In fact, some studies report that the installation costs for waterless urinals are lower than those for flush urinals.

Another set of examples of waterless urinal savings is shown in Table 14.8 for various occupancies such as an office building, a restaurant, and a school, both existing buildings and new buildings.[12] The basic approach indicated in these examples can be extended to a range of other water alternatives, to include rainwater harvesting, graywater systems, ultra-low-flow fixtures, and composting toilets. That is, reductions in water and potentially wastewater costs can be used to develop an LCC analysis for assessing the financial performance of the alternatives versus conventional practice.

QUANTIFYING HEALTH AND PRODUCTIVITY BENEFITS

Factoring human benefits into LCC analyses must be done cautiously and conservatively. Although there is ample information about the health and productivity benefits of high-performance buildings, rarely has it been compiled scientifically; therefore, it cannot be said to have the same reliability as that for hard costs. Nevertheless, some of the major benefits that have been cited are impressive, for example:

- A paper by William J. Fisk of the Indoor Environment Department at Lawrence Berkeley National Laboratory suggests that enormous savings and productivity gains can be achieved through improved indoor air quality (IAQ) in the United States. He estimated $6 to $14 billion in savings from reduced respiratory disease; $1 to $4 billion from reduced allergies and asthma; $10 to $30 billion from reduced sick building syndrome (SBS)—related illnesses; and $20 to $160 billion from direct, non-health-related improvements in worker performance.[13]

- Daylighting benefits to human health and performance can potentially provide marked financial returns—if they can be quantified. A study of student performance in daylit schools indicates dramatic improvements in test scores and learning progress. One often-cited study of schools in Orange County, California, by the Heschong Mahone Group found that students in classrooms with daylighting improved their test scores 20 percent faster in math and 26 percent faster in reading than students in schools with the lowest levels of daylighting. The study also looked at students in Seattle, Washington, and Fort Collins, Colorado, where improvements in test scores were 7 to 18 percent.[14]

- Another study by the Heschong Mahone Group compared sales in stores with skylights versus nonskylit stores and found that the skylit stores had 40 percent higher sales.[15]

A reasonable approach to determining how to include productivity and health savings in green buildings was suggested in a report to California's Sustainable Building Task Force.[16] In this report, the authors recommend assigning a 1 percent productivity and health gain to buildings attaining a USGBC LEED-NC certified or silver level and a 1.5 percent gain for buildings achieving a gold or platinum level. These gains are derived in a conservative fashion from information about improvements in human performance (see Table 14.9). Savings are the equivalent of $600 to $700 per employee per year, or about $3 per square foot ($30 per square meter), for a 1 percent gain, and $1000 per employee per year, or $4 to $5 per square foot ($40 to $50 per square meter), for a 1.5 percent gain.

TABLE 14.9

Human Performance Improvements Associated with Green Building Attributes

Green Building Attribute	Productivity Benefits
Increased tenant control over ventilation	0.5%–34%
Increased tenant control over temperature and lighting	0.5%–34%
Control over lighting	7.1%
Ventilation control	1.8%
Thermal control	1.2%

QUANTIFYING THE BENEFITS OF REDUCING EMISSIONS AND SOLID WASTE

Emissions attributed to the operation of buildings are staggering in scope. High-performance buildings have the potential to dramatically lower these impacts. As a result of energy requirements, buildings in the United States are responsible for the creation of 48 percent of the nation's sulfur dioxide emissions, 20 percent of nitrous oxide, and 36 percent of carbon dioxide. Additionally, buildings produce 25 percent of solid waste, consume 24 percent of potable water, create 20 percent of all wastewater, and cover 15 percent of land area.[17] Construction and demolition waste in the United States amounts to about 150 million tons per year, or about 0.5 ton per capita annually. Converting avoided emissions to benefits attributable to high-performance buildings can be accomplished by calculating the societal costs of emissions. The societal impacts of these emissions can be quantified as follows:

- Sulfur dioxide: $91 to $6800 per ton ($100 to $7500 per metric ton)
- Nitrous oxide: $2090 to $10,000 per ton ($2300 to $11,000 per metric ton)
- Carbon dioxide: $5.50 to $10 per ton ($6 to $11 per metric ton)

For the NREL prototype building, Tables 14.10 and 14.11 provide a summary of benefits that can be claimed as a result of energy reductions and avoided emissions.

Including the maximum emissions reductions benefits has a significant impact on the payback time. The payback time due to energy savings is reduced

TABLE 14.10

Summary of Energy and Cost Savings for the NREL Prototype Building*

	Base Case	High-Performance Case
Area (square feet)	20,000	20,000
Total cost	$2,400,000	$2,440,000
Incremental cost	NA	$40,000
Annual energy use (BTU)	730 million	477 million
Annual energy cost	$11,800	$7,490
Reduction in energy use	NA	34.6%
Reduction in energy cost	NA	36.7%
Simple payback, energy	NA	8.7 years
Simple payback, energy and emissions	NA	6.0 years

*Base case and high-performance case, with simple payback for energy alone and for energy and emissions.

TABLE 14.11

Avoided Emissions and Annual Benefit for the NREL Prototype Building: High-Performance Case Compared to Base Case

Emission Type	Tons of Emissions Avoided per Year	Annual Benefit
Sulfur dioxide	0.16	$1090
Nitrous oxide	0.08	$800
Carbon dioxide	10.7	$107
Total	10.94	$1997

TABLE 14.12

Savings for Diverting Construction Waste from Landfill for the NREL Prototype Building*

Diversion Rate	$50/Ton Tipping Fee	$75/Ton Tipping Fee	$100/Ton Tipping Fee
0%	$0	$0	$0
50%	$1750	$2625	$3500
75%	$2625	$3938	$5250

*Assuming 7 pounds per square foot (32 kilograms per square meter) waste generation for various diversion rates and tipping fees.

from 8.7 years to 6.0 years when the societal costs of avoided emissions are included.

Savings from reduced solid waste generation can also be included in the life-cycle picture. For high-performance buildings, solid waste reductions are a result of three factors. First is construction and demolition waste reduction, which is addressed in high-performance building assessment systems such as the USGBC LEED-NC building rating system. For example, LEED-NC awards 1 point for diverting at least 50 percent of construction and demolition waste from landfilling and 2 points for diverting 75 percent or more of this waste stream. Second, high-performance buildings address the generation of solid waste by building occupants by calling for the allocation of building space for the collection and storage of recyclables. In fact, LEED-NC makes this allocation of space a prerequisite for achieving a rating, thereby making it a mandatory requirement. Third, high-performance buildings address the use of recycled content and reuse of building materials, thus creating incentives and demand for closing materials loops and reducing the landfilling of solid waste. LEED-NC provides 1 point for 5 percent resource reuse and 2 points for 10 percent resource reuse. For recycled content, 1 point is provided if 5 percent of materials have postconsumer recycled content or 2 points for 10 percent postconsumer recycled content. Alternatively, LEED-NC provides 1 point for a 10 percent total of postconsumer plus one-half of postindustrial content and 2 points for a 20 percent total of postconsumer plus one-half of postindustrial content.

The financial benefits of diverting construction and demolition waste from landfilling can be readily calculated. For a nominal US construction project, waste is generated at the rate of about 7 pounds per square foot (32 kilograms per square meter). The actual savings are a function of the diversion rate. Table 14.12 itemizes the savings from construction waste diversion as a function of diversion rate and tipping fees—that is, the cost of disposal.

QUANTIFYING THE BENEFITS/COSTS OF BUILDING COMMISSIONING

One of the hallmarks of high-performance buildings is that, upon completion of the building, all systems are carefully checked and validated through testing. As a consequence of this movement, building commissioning has become a new profession. Commissioning professionals are engaged in the project from the start, along with members of the design and construction professions. And although commissioning does add extra cost to a building, the value of this service is substantial, because it provides assurance that the building will perform as designed. Costs of commissioning for typical buildings are shown in Table 14.13.[18] The benefits of building commissioning are difficult to quantify, but current general practice is to attribute a 10 percent energy savings to

TABLE 14.13

Commissioning Costs for Typical New Construction

Scope of Commissioning	Cost
Whole building	0.5%–1.5% of construction cost
HVAC and control systems	1.5%–2.5% of mechanical system cost
Electrical systems	1.0%–1.5% of electrical system cost
Recommissioning existing buildings	$0.17 per square foot ($1.83 per square meter)

commissioning. In the case of the NREL prototype building used as an example in this chapter to quantify energy savings, the payback period for building commissioning is less than four years.

QUANTIFYING MAINTENANCE, REPAIR, AND MISCELLANEOUS BENEFITS/COSTS

In attempting to minimize LCC, high-performance buildings are specifically designed to lower maintenance costs, but they can also produce lower costs in other areas. The following are examples of design features that can provide these additional economic benefits for high-performance buildings:[19]

- Durable materials
  - Fluorescent lighting systems with long-life, 10,000-hour lights in place of short-life, 1000-hour incandescent lights. LED lights have huge potential with 50,000-hour lifetimes and rapidly decreasing manufacturing costs.
  - Fly ash and blast furnace slag concrete with higher durability compared to conventional concrete mix design
  - Low-emission paints with higher durability compared to conventional paints
  - Light-colored roofing materials that have longer life than conventional roofing materials
  - Polished concrete floors with very long lifetimes and low maintenance costs compared to carpeting and other floor finishes
- Repairability
  - Recycled-content carpet tiles that can be replaced in worn areas
  - Mechanical and electrical systems designed for ease of repair and replacement by virtue of space allocation and physical arrangement of equipment, piping, conduit, power and control panels, and other components
- Miscellaneous costs
  - Designing buildings with areas for recycling that reduce waste disposal costs
  - Sustainable landscape design that reduces the need for irrigation, fertilizer, herbicides, and pesticides
  - Stormwater management using constructed wetlands instead of sewers

Quantifying the financial benefits of improved maintenance and repair must, of course, be accomplished on a case-by-case basis and can be difficult to carry out because a database containing this type of information is not readily available. For the sustainable landscape design and stormwater management entries listed above under "Miscellaneous costs," an example of how to present

TABLE 14.14

Economic Comparison of Sustainable Stormwater Management and Landscape Practices for the NREL Prototype Buildings

	Incremental First Cost	Incremental First Cost/1000 Square Feet (100 square meters)	Total Incremental Cost	Annual Cost Savings/1000 Square Feet (100 square meters)	Total Cost Savings	Simple Payback (years)
Sustainable stormwater management	$3140	$157 ($169)	$3140	$28.30 ($30.45)	$566	5.6
Sustainable landscape design	$2449	$122 ($131)	$2440	$152.00 ($163.55)	$3040	0.8

the savings is provided in Table 14.14 for the NREL prototype buildings used to illustrate energy savings in this chapter.[20] The two site-related strategies for the NREL prototype buildings are as follows:

> *Sustainable landscape design.* A mixture of native warm-weather turf and wildflowers is used to create a natural "meadow" area. This strategy is compared with traditional turf landscaping of Kentucky bluegrass, which requires substantially more irrigation, maintenance, and chemical application.

> *Sustainable stormwater management.* An integrated stormwater management system combines a porous gravel parking area with a rainwater collection system, where rainwater is stored for supplemental irrigation of native landscaping. This porous, gravel-paved parking area is a heavy load-bearing structure filled with porous gravel, allowing stormwater to infiltrate the porous pavement (reducing runoff) and to be moved into an underground rainwater collection system. The water can be used to supplant freshwater from the public supply for uses that do not require potable water. This sustainable system is compared to a conventional asphalt parking area and a standard corrugated pipe stormwater management system without rainwater harvesting.

Although the particular sustainable stormwater system used for the prototype increases the total construction cost by a little over $3000 (about 0.1 percent of the total building construction cost), it saves over $500 annually in maintenance costs because less labor is required for patching potholes and performing other maintenance on an asphalt lot. The resulting payback period is less than six years. The sustainable landscaping approach shows even more favorable economics: the incremental first cost is nearly $2500, but this is repaid in less than one year with an annual O&M cost savings of $3040 in avoided maintenance, chemical, and irrigation costs.

Managing First Costs

For many organizations, especially state and local governments, the first, or capital, cost is the primary factor in making decisions about a project because legislation often dictates the maximum investment in a specific type of building. For example, in Florida, the new school construction cost per student station is limited to approximately $13,500 for elementary schools, $15,500 for middle schools, and $20,500 for high schools. For many other potential green building clients, a similar situation exists, with decision makers heavily constrained by construction cost limitations. Coping with these circumstances requires careful consideration of strategies for producing a high-performance building when LCC may be difficult to bring into the process. The following is a list of recommendations for managing first costs for high-performance building projects:[21]

1. Make sure that senior decision makers support the concept.

2. Set a clear goal early in the process. Ideally, the decision to go green should be made before soliciting design proposals so that contract language reflects the green goal, thus permitting more flexibility in decision making. Certain green measures that can save money (such as site planning) have to be done early.

3. Write contracts and requests for proposal (RFPs) that clearly describe your sustainability requirements. For example, specify whether the goal is a LEED silver rating or the equivalent.

4. Select a team that has experience with sustainable development. Hiring a mechanical, electrical, and plumbing (MEP) firm with green experience alone can save 10 percent of the MEP construction costs. Look for team members with a history of creative problem solving.

5. Encourage team members to get further training and develop sources of information on green materials, products, and components and technical/pricing information on advanced systems.

6. Use an integrated design process. Do not make the green components add-ons to the rest of the project. Integrate all the candidate green measures into the base budget. Establishing an integrated design can lead to capital savings. Investing 3 percent of total project costs during design can yield at least 10 percent savings in construction through design simplifications and fewer change orders.

7. Understand commissioning and energy modeling. To minimize up-front costs, use a sampling approach for building commissioning.

8. Look for rebates and incentives from states, counties, cities, and utilities.

9. Educate the decision makers without inundating them with technical information. Stay focused on their objectives. Respect their sense of risk aversion.

10. Manage your time carefully. Select one or two team members to oversee research on green products and systems. Set a specific deadline for research results and give the discovery manager the power to cut off research.

The following are some design and construction strategies that a team can use to reduce first costs:[22]

- Optimize site and orientation. One obvious strategy to reduce first costs is to apply appropriate siting and building orientation techniques to capture solar radiation for lighting and heating in winter, and shade the building using vegetation or other site features to reduce the summer cooling load. Fully exploiting natural heating and cooling techniques can lead to smaller HVAC systems and lower first costs.

- Reuse/renovate older buildings and use recycled materials. Reusing buildings, as well as using recycled materials and furnishings, saves virgin materials and reduces the energy required to produce new materials. Reusing buildings may also reduce the time (and therefore money) associated with site planning and permitting.

- Reduce project size. A design that is space-efficient yet adequate to meet the building objectives and requirements generally reduces the total costs, although the cost per unit area may be higher. Fully using indoor floor space and even moving certain required spaces to the exterior of the building can reduce first costs considerably.

- Eliminate unnecessary finishes and features. One example of eliminating unnecessary items is choosing to eliminate ornamental paneling, doors (when privacy is not critical), and dropped ceilings. In some cases, removing unnecessary items can create new opportunities for designers. For example, eliminating dropped ceilings might allow deeper daylight penetration and reduce floor-to-floor height (which can reduce overall building dimensions).

- Avoid structural overdesign and construction waste. Optimal value engineering and advanced framing techniques reduce material use without adversely affecting structural performance. Designing to minimize construction debris (e.g., using standard-sized or modular materials to avoid cutting pieces and thereby generating less construction waste) also minimizes labor costs for cutting materials and disposing of waste.

- Fully explore integrated design, including energy system optimization. As discussed previously, integrated design often allows HVAC equipment to be downsized. Models such as DOE-2 allow the energy performance of a prospective building to be studied and the sizing of mechanical systems to be optimized. Using daylighting and operable windows for natural ventilation can reduce the need for artificial lighting fixtures and mechanical cooling, thereby lowering first costs. Beyond energy-related systems, integrated design can also reduce construction costs and shorten the schedule. For example, by involving the general contractor in early planning sessions, the design team may identify multiple ways to streamline the construction process.

- Use construction waste management approaches. In some locations, waste disposal costs are very high because of declining availability of landfill capacity. For instance, in New York City, waste disposal costs exceed $75 per ton ($82 per metric ton). In such situations, using a firm to recycle construction waste can decrease construction costs because waste is recycled at no cost to the general contractor, thereby saving disposal costs.

- Decrease site infrastructure. Costs can be reduced if less ground needs to be disturbed and less infrastructure needs to be built. Site infrastructure can be decreased by carefully planning the site, using natural drainage rather than storm sewers, minimizing impervious concrete sidewalks, reducing the size of roads and parking lots (e.g., by locating near public transportation), using natural landscaping instead of traditional lawns, and reducing other man-made infrastructure on the site, when possible. For example, land development and infrastructure costs for the environmentally sensitive development on Dewees Island, off the coast of Charleston, South Carolina, were 60 percent below the local average because impervious roadway surfaces and conventional landscaping were not used.

An excellent study of construction costs for green buildings was conducted by Lisa Fay Matthiessen and Peter Morris of Davis Langdon, a cost consulting company.[23] Their report suggests that there is no statistical difference between high-performance green buildings that used LEED-NC for guidance and conventional buildings; that is, the cost per square foot falls into the same range of costs for both green and conventional buildings of a similar program type. The majority of LEED-NC certified buildings examined by the authors did not require additional funding, and where additional costs were incurred, they were due to certain extraordinary specific features such as photovoltaics. The factors that influence the cost of a green building are:

- *Demographic location.* The location of a project, rural versus urban, creates opportunities and problems in obtaining LEED-NC points. For example, points for transportation and urban development are readily available in urban settings, while stormwater management innovations are more likely in rural areas.

- *Bidding climate and culture.* In some states, such as California, contractors and subcontractors are far more familiar with LEED-NC and are less likely to perceive a project as risky, thus lowering costs.

- *Local and regional design standards, codes, and initiatives.* In states such as Oregon and Pennsylvania, where there has been significant government support of green building efforts, the costs are generally lower because green buildings are more likely to be considered the norm.

- *Intent and values of the project.* A clear statement that the owner is serious about the green building concept will motivate the project team members and ensure that green building features are incorporated from the onset of the project, thus lowering overall costs.

- *Climate.* The paybacks for energy-conserving features vary by location because the costs of energy also vary by geographic region. Additionally, some aspects of passive design may be difficult to achieve in very hot, humid, or very cold climates. As a result, more complex and costly active systems are needed to meet the operational requirements of the owner.

- *Timing and implementation.* Fully incorporating green features from the start of design and ensuring their detailed integration into the project will result in lower costs.

- *Size of the building.* Larger, more complex buildings will typically have higher costs for larger, more complex systems simply due to the scale of the project.

- *Synergies.* Selecting systems that have multiple benefits will produce lower costs. For example, a well-designed landscape can integrate stormwater management and building shading and can be designed to require no irrigation, saving infrastructure and lowering operational costs.

The Davis Langdon study also noted that a well-developed budget methodology could go a long way toward reducing construction cost impacts. The authors recommend that the following measures be followed at every step of design and construction to keep a green building construction within budget:

- Establish team goals, expectations, and expertise.
- Include specific goals in the program.
- Align the budget with the program.
- Stay on track during design and construction.

Integrating green building goals into the project, having appropriate expertise and commitment in the project team, and detailed planning are perhaps the key elements in keeping costs aligned with the budget. In this respect, green building projects are no different from any other well-organized and well-run building project except for the inclusion of team knowledge of the green building concept and requirements. Experience to date is that the learning curve for obtaining the requisite knowledge is not very steep and that training in and exposure to one green building project provide the foundation for successfully tackling other similar projects.

Tunneling through the Cost Barrier

The preferred design approach used to create a high-performance green building is sometimes referred to as *integrated design*, which is covered in detail in Chapter 7. The fundamental assumption of integrated design is that by bringing the various disciplines together and forcing them out of their silos, a wide variety of synergies is possible. One of the most commonly cited synergies is in the design of the building energy systems, where architects and mechanical engineers collaborate on the details of the building envelope, resulting in a smaller HVAC plant. The present approach to building design does not promote sustainability because the designers, architects, and engineers each optimize the systems they design, generally resulting in a suboptimal building. Additionally, the fee structures for design professionals are such that maximizing cost and complexity can result in higher fees, clearly the wrong motivation when it comes to creating superior buildings. Consequently, finding the synergies that will produce truly high-performance buildings is a struggle, requiring changes in both attitudes and design contracts.

Amory Lovins of the Rocky Mountain Institute (RMI) describes the effects of producing integrated design synergies as "tunneling through the cost barrier" because the result can be a dramatic reduction in first, or capital, costs. One example cited by Lovins was the design of an industrial process for the carpet maker Interface for a plant in Shanghai, China. The initial design for this process called for 95 horsepower of pumping power. When Jan Schilhan of Interface examined the design, he threw out the assumptions engineers normally use for sizing pipes, making the pipes larger in diameter, thus greatly reducing pipe friction because fluid velocity was greatly reduced. Because friction follows with the fifth power of the pipe diameter, doubling the pipe diameter results in a friction reduction of 86 percent, so that pumping power falls by the same amount. Also, contrary to common design practices, Schilhan laid out the pipes with minimal bends and with the pipe lengths as short as possible, because each bend and each foot of pipe causes additional friction losses. This redesign reduced pumping power from the original 92 horsepower to 7 horsepower, a 92 percent, or Factor 12, improvement. The result of these changes was not only a significant reduction in energy consumption but also a significant reduction in capital cost due to the far smaller pumps, reduced piping complexity, and a smaller electrical service, far offsetting the slightly higher cost of larger-diameter piping.[24]

Many of the tradition-rooted assumptions used by engineers and architects often result in poor design practices that persist for decades, even generations. Challenging these assumptions is important if superior buildings with lower capital costs are the desired outcome. The key, according to Lovins, is whole-system engineering, in which all the benefits of a technology are counted, not just, for example, the energy-savings benefits. High-efficiency electric motors have as many as 18 benefits, and superwindows have as many as 10 benefits, including better daylighting, radiant comfort, no condensation, and noise blocking, to name but a few. Buildings have ample opportunity for synergies and cost reductions, many of them as yet unexplored. One area ripe for exploration is the integration of buildings into local ecosystems and geological formations. Trees have enormous capacity for stormwater uptake and can selectively allow sunlight to fall on buildings, depending on the time of year, as their leaves can block and absorb solar radiation during the summer and, by dropping off the tree in the fall, allow penetration of the sun during winter days. Living roofs on buildings provide insulation, reduce the heat island effect, store stormwater, and replace the ecological footprint removed by the building. Greenery integrated into buildings contributes to a healthy experience for occupants, as suggested by the biophilia hypothesis (see Chapter 2). Coupling the building with the ground and groundwater can help provide heating and cooling while lowering energy consumption. Wetlands and constructed wetlands could also benefit the built environment via wastewater treatment and stormwater storage, leading to reduced capital costs.

Lovins suggests four principles as aiding the attempt to tunnel through the cost barrier:[25]

1. *Capture multiple benefits from single expenditures.* By dematerializing buildings, for example, it may be possible to provide more space at lower cost while proportionately reducing environmental impacts. High-efficiency lighting reduces electrical energy requirements and reduces the heat load to the space and can be coupled with occupancy and daylight sensors.

2. *Start downstream to turn compounding losses into savings.* Rather than focusing on the fan power required to push air through ductwork, more attention on reducing friction losses in ductwork through better layout, reducing the length of duct runs, eliminating unnecessary bends, and increasing the duct cross section results in far lower fan horsepower, smaller and less costly equipment, and quieter operations. Going one step further downstream, designing systems that heat and cool only the bottom 6 feet (1.8 meters) or so of vertical zones, where the occupants actually are, further reduces energy consumption. This is the strategy known as *displacement ventilation* (described in Chapter 9), and by delivering air from an underfloor plenum, it can help reduce floor-to-floor heights. Reducing this dimension results in lower overall building heights and lower material costs.

3. *Get the sequence right.* If the issue is health and productivity, thinking through how people will use the space and how they will have access to daylight, views, and, preferably, greenery, and maximizing the amount of natural light falling on their work spaces should be the first and foremost matter for consideration. The lighting systems should be designed only after the primary human factors are considered. The result: a better indoor environment and lower energy costs.

4. *Optimize the whole system and not just the parts.* This is the crux of whole-system engineering, a collaborative effort among architects and engineers to jointly and creatively design the building and its systems. In Germany, for example, the design disciplines have collaborated to create buildings that have superb passive design, totally eliminating the need for cooling systems, resulting in buildings using one-seventh of the primary energy of conventional US buildings. This represents the essence of integrated design and cost barrier tunneling.

Summary and Conclusions

High-performance buildings have enormous potential benefits: for their owners, for the environment, and for society in general. The ability to express and clearly justify these benefits in an economic analysis is an important factor in determining whether or not the project will be conventional or high-performance in its design and construction. Evidence is beginning to emerge that provides information and tools for the building team to use in developing a model that addresses both hard and soft costs. Hard-cost savings on energy, water, and wastewater are fairly straightforward to quantify and include in an economic analysis. Soft costs, such as human health and productivity savings, as well as savings due to building commissioning, are not so straightforward to justify; hence, care must be exercised when including them in a cost analysis. Hopefully, additional verifiable, peer-reviewed data will emerge in the coming years, and the decision to include these data in a green building project analysis will be far easier than it is at present.

Notes

1. The information on green building market size and share is from *Green Outlook 2011*.
2. Cited in the Executive Summary of "Green Building Costs and Financial Benefits" by Kats (2003b).
3. Derived from a survey of 33 green buildings conducted by Greg Kats of Capital E in 2003 for the state of California and the USGBC and reported in Kats (2003a).
4. From Kats (2003a).
5. Ibid.
6. From "The Business Case for Sustainable Design in Federal Facilities" (2003).
7. Available in the Members section of the USGBC website, www.usgbc.org.
8. Paraphrased from Kats (2003a).
9. The information in the table is as cited in *Green Outlook 2011* and attributed to two earlier MHC studies, *Green Building Retrofit and Renovation*, SmartMarket Report (2009); and *Commercial and Institutional Green Building*, SmartMarket Report (2008).
10. An excellent description of the ING Bank building can be found in von Weizsäcker, Lovins, and Lovins (1997). This book had great influence on high-performance buildings because it suggested that reducing resource consumption by 75 percent was necessary to achieve sustainability and that, furthermore, the technologies needed to support this reduction already existed. A follow-on concept, Factor 10, suggests that long-term sustainability would require a 90 percent reduction in resource consumption.
11. Both tables are excerpted from "The Business Case for Sustainable Design in Federal Facilities" (2003).
12. Excerpted from "Big Savings from Waterless Urinal," *Environmental Building News* (1998).
13. From Fisk (2000).
14. From "Daylighting in Schools" by the Heschong Mahone Group (1999). This company, which specializes in building energy efficiency, has published several landmark reports on the correlation between daylighting and student performance. Recent reports for the California Energy Commission are available from the company website, www.h-m-g.com/.
15. Reported in "Skylighting and Retail Sales" by the Heschong Mahone Group (1999).
16. Productivity and health gains are from "The Costs and Financial Benefits of Green Buildings" (Kats 2003a).
17. Adapted from *2002 Buildings Energy Databook* (2002).
18. Adapted from "What Can Commissioning Do for Your Building?" (1997).

19. Adapted from "The Business Case for Sustainable Design in Federal Facilities" (2003).
20. Ibid.
21. Adapted from Syphers et al. (2003).
22. From "The Business Case for Sustainable Design in Federal Facilities" (2003).
23. From Matthiessen and Morris (2004).
24. From Hawken, Lovins, and Lovins (1999), chapter 6.
25. Derived from Lovins (1997).

References

2002 Buildings Energy Databook. 2002. US Department of Energy. Available at http://buildingsdatabook.eren.doe.gov.

"Big Savings from Waterless Urinal." 1998. *Environmental Building News* 7 (2), 1–14.

"The Business Case for Sustainable Design in Federal Facilities." 2003. Resource Document, US Department of Energy. Available at www.eere.energy.gov/femp/pdfs/bcsddoc.pdf.

"Daylighting in Schools." 1999. Heschong Mahone Group. A report for the Pacific Gas and Electric Company. Available at www.coe.uga.edu/sdpl/research/daylighting study.pdf.

Fisk, William J. 2000. "Health and Productivity Gains from Better Indoor Environments and Their Relationship with Building Energy Efficiency." *Annual Review of Energy and the Environment* 25: 537–566.

Hawken, Paul, Amory B. Lovins, and L. Hunter Lovins. 1999. *Natural Capitalism.* New York: Little, Brown.

Kats, Gregory H. 2003a. "The Costs and Financial Benefits of Green Buildings." A report developed for California's Sustainable Building Task Force. Available at the Capital E website, www.cap-e.com.

———. 2003b. "Green Building Costs and Financial Benefits." A report written for the Massachusetts Technology Collaborative. Available at the Capital E website, www.cap-e.com.

Lovins, Amory B. 1997. "Tunneling through the Cost Barrier: Why Big Savings Often Cost Less Than Small Ones." *Rocky Mountain Institute Newsletter* 13 (2): 1–4.

"Making the Business Case for High Performance Green Buildings." 2003. US Green Building Council. Available at www.usgbc.org.

Matthiessen, Lisa Fay, and Peter Morris. 2004. "Costing Green: A Comprehensive Cost Database and Budgeting Methodology." Davis Langdon. Available at https://www.usgbc.org/Docs/Resources/Cost_of_Green_Full.pdf.

McGraw-Hill Construction. 2010. *Green Outlook 2011: Green Trends Driving Growth.* New York: McGraw-Hill Construction.

"Skylighting and Retail Sales." 1999. Heschong Mahone Group. A report for the Pacific Gas and Electric Company. Available at www.pge.com.

Syphers, Geof, Arnold Sowell, Jr., Ann Ludwig, and Amanda Eichel. 2003. "Managing the Cost of Green Buildings." Published in the "White Paper on Sustainability," a supplement to *Building Design and Construction.* Available at www.bdcmag.com.

von Weizsäcker, Ernst, Amory B. Lovins, and L. Hunter Lovins. 1997. *Factor Four: Doubling Wealth, Halving Resource Use.* London: Earthscan.

"What Can Commissioning Do for Your Building?" 1997. Portland, OR: Portland Energy Conservation, Inc. (PECI).

Chapter 15

The Cutting Edge of Sustainable Construction

The contemporary high-performance green building movement continues to gain momentum in the United States and other countries and is transforming the entire process of creating the built environment, from design through construction and operation. In the United States, green building is beginning to dominate the market for commercial and institutional buildings, with almost 50 percent of new buildings in this sector forecasted to be green by 2015. This movement is affecting not only new construction but also renovations to existing buildings, building products, design tools, and the education of built environment professionals.

In the United States, for all practical purposes, the US Green Building Council (USGBC) Leadership in Energy and Environmental Design (LEED) building assessment system defines what constitutes a high-performance green building. Although LEED has been an enormous success in the marketplace, two questions remain: What is the ultimate goal of building assessment standards such as LEED and Green Globes, and how will they evolve over time to improve the buildings currently being produced that are using them as guidance? Because of the success of the LEED building assessment system, the USGBC is focused almost exclusively on the implementation of the existing suite of LEED rating products for new construction and for existing buildings, and is working to generate and implement other LEED rating tools to cover areas of importance such as health care and retail. Consequently, a long-term vision of what constitutes the high-performance building of the next generation is lacking and, as a result, is hampering progress toward a truly sustainable built environment.

In this final chapter, the cutting edge and future high-performance green buildings are addressed for the purpose of stimulating thinking about the long-range goals of this movement. The first section addresses the emerging issue of passive survivability, a new building theme embraced by the green building community in the wake of Hurricane Katrina in 2005. Although not yet being incorporated into new buildings, it is on the cusp of consideration and fits nicely into the general philosophical approach underpinning high-performance green buildings. The second section contains several case studies of newer green buildings to illustrate the best practices being employed today. These buildings, of course, point the way to the future and what may possibly be the norm for the green buildings of the future. The future is uncertain, however, and many different outcomes can be hypothesized, and certainly not all can be covered in detail here. In the fourth section of this chapter titled *Challenges*, three main approaches to designing future green buildings are proposed, one based on history, another on technology, and a third on ecology. These represent the main attractors for strategies in this arena, although the likely outcome will be a hybrid of these widely differing but potentially equally successful approaches.

Passive Survivability

Recent severe weather events are causing a shift in thinking that will result in buildings having the capability of assisting human survival in the wake of natural or human-induced disasters. During the Chicago heat wave of 1995, the deaths of more than 700 people in their homes or apartments were attributed to high temperatures. In many apartments, temperatures remained in excess of 90°F (32°C), even at night. The death toll could have been far higher had Chicago lost power during the heat wave. Ten years after the Chicago heat wave, New Orleans was struck by Hurricane Katrina in August 2005, resulting in thousands of deaths, incredible suffering, enormous dislocation of residents, and severe economic impacts. Temperatures in the Louisiana Superdome rose to 105°F (42°C), creating dangerous conditions inside the very structure to which people were sent to survive the immediate aftermath of Katrina.

Passive survivability is a new term being used to describe how buildings should be designed and built to assist the survival of their human occupants in the wake of disasters. In an editorial in *Environmental Building News* (*EBN*) in November 2005, passive survivability was defined by Alex Wilson as ". . . the ability of a building to maintain critical life-support conditions if services such as power, heating fuel, or water are lost for an extended period."[1] The term *passive survivability* was first used by the military to describe measures taken to ensure that military vehicles are able to withstand attacks. It was included in a set of proposals called the *New Orleans Principles*, resulting from a reconstruction conference held in Atlanta in November 2005.[2] One of these proposals states, "Provide for passive survivability: Homes, schools, public buildings, and neighborhoods should be designed and built or rebuilt to serve as livable refuges in the event of crisis or breakdown of energy, water, and sewer systems."

The fact of climate change, and the probability of higher temperatures and more frequent and more violent hurricanes, should be sufficient to cause a shift to using passive survivability as a design criterion. Backup generators are unlikely to be able to provide the power needed for ventilation and air conditioning for extended periods of time; consequently, buildings need to have several key design features that help ensure passive survivability. Among these key green design features are cooling load avoidance, capability for natural ventilation, a high-efficiency thermal envelope, passive solar gain, and daylighting.

Most of the preliminary efforts at passive survivability have addressed the very real problem faced by regions prone to hurricane activity, thought to be on the increase due to climate change. The same basic principles apply to areas that may be subject to severe winter conditions such as blizzards and ice storms, the emphasis shifting to providing the capability for heating, either through passive solar design or the use of local energy resources such as wood. A 1998 ice storm in eastern Canada left 4 million people without power and forced 600,000 people from their homes, with 28 fatalities, indicating that persons living in colder climates also should consider passive survivability strategies for their built environment. Earthquakes have not yet been addressed in the preliminary literature on passive survivability, although, in principle, buildings that are designed to survive earthquakes may still have downed utilities and should have the added capability of passive survivability. It is clear, then, that different regions will have different approaches to passive survivability that will depend on the weather and the typical natural hazards in that region.

Passive survivability should also be extended to infrastructure. Cisterns can be located throughout a community and under streets for an emergency water supply and for fire protection. Key control and communications systems such as traffic signals and streetlights could have solar-charged power backups. Sewage infrastructure could also be planned to have normal and passive survivability functions.

Exactly how passive survivability can help mitigate the effects of terrorist attacks remains an open question. Clearly, any area of the country can be subject to the effects of terrorism. Attacks directed at utility infrastructure could be mitigated by a shift to passive survivability as a criterion for building. The impacts of biological or nuclear attacks could also be mitigated, at least for some period of time, by passive survivability, although it is likely that systems that would seal the building and protect the occupants from airborne biological agents or radioactivity would likely not be incorporated into typical construction. The wide variety of potential attacks makes designing buildings for all eventualities impossible. However, for attacks directed against infrastructure, buildings can certainly be provided with key features that assist the occupants in having a safe place, reasonable temperatures, ventilation, and potable water, the key elements of survival.

The list of measures that can be included in a strategy for passive survivability is remarkably similar to a list of typical green building measures (Table 15.1).[3] Indeed, an argument in support of incorporating passive survivability measures into buildings could also be considered an argument in favor of green building.

TABLE 15.1

Checklist for Designing Passive Survivability into Buildings

1. *Create storm-resilient buildings.* Design and construct buildings to withstand reasonably expected storm events and flooding.
2. *Limit building height.* Most tall buildings cannot be used during power outages due to their reliance on elevators and air conditioning, and a maximum height of six to eight stories is recommended.
3. *Create a high-performance envelope.* A well-insulated thermal envelope with high-performance glazings will assist in maintaining a reasonable interior temperature.
4. *Minimize cooling loads.* Proper building orientation, overhangs, shading, and high-performance glazing can minimize building heat loads.
5. *Provide for natural ventilation.* Provisions for natural ventilation, such as chimney effect air movement, even for buildings that would be normally air conditioned, would provide fresh air for the occupants.
6. *Incorporate passive solar heating.* In climates where heating may be the survivability issue, thermal mass and thermal storage walls can be used to help provide thermal energy for heating.
7. *Provide natural daylighting.* The same daylighting strategies used for green buildings also provide light in a passive survivability mode.
8. *Configure heating equipment to operate on PV power.* Gas- and oil-fired heating equipment is often dependent on electrical power for operation, and equipment may have to be configured to accept DC power from PV panels or have an inverter to provide AC power.
9. *Provide photovoltaic power.* PV can provide electrical energy during outages and, with battery storage, can also provide electricity at night. Note that PV panels need to be mounted and protected from high winds and flying debris.
10. *Provide solar water heating.* Solar thermal systems coupled with PV-powered pumps can provide hot water during power outages.
11. *Where appropriate, consider wood heat.* Especially in rural areas, low-pollution wood-burning stoves, masonry heaters, or pellet stoves can provide hearing.
12. *Store water on site; consider using rainwater to maintain a cistern.* Water storage for extended outages can be provided by a cistern. Storing water high in the building, for example, on the roof, can provide pressure with no need for pumps.
13. *Install composting toilets and waterless urinals.* Fixtures that do not rely on water for flushing have a distinct advantage in the aftermath of disasters.
14. *Provide for food production in the site plan.* Land can be set aside for fruit-bearing trees and shrubs as a source of food in passive survival mode.

Cutting Edge: Case Studies

Of all the high-performance buildings either registered or certified in the United States, several could be considered at the cutting edge of practice, among them the Federal Building in San Francisco, California, and the Forensic Science Center in Philadelphia, Pennsylvania. These projects, and the aspects that make them cutting-edge, high-performance buildings, are described below.

Case Study: The Federal Building, San Francisco, California

The new 18-story San Francisco Federal Building is referred to by its owner, the General Services Administration (GSA), as "a model of excellence" and rightfully so. Located on a 3-acre site in the South of Market Street neighborhood at the intersection of Seventh and Mission Streets, just a 10-minute walk from downtown, it is a long, slender, translucent tower, 60 feet (18 meters) wide and 234 feet (71 meters) high, providing 600,000 gross square feet (55,742 square meters) of usable space. It is a federal government complex serving the Social Security Administration, the Department of Labor, the Department of Health and Human Services, and the Department of Agriculture. The design was led by Thom Mayne of Morphosis Architects in a major collaboration with the Los Angeles office of Ove Arup for the integrated structural and mechanical design; with Horton Lees Brogden of Culver City, California, for lighting and daylighting design; and with the Building Technologies Department of the Lawrence Berkeley National Laboratory for modeling the natural ventilation system. The Smith Group of San Francisco served as executive architect and executed all interior space planning for the tenant agencies. The goal of the project was to provide a high-quality government work space within the project budget of $144 million. High quality in this context meant that the work space had to be efficient, secure, and flexible to allow change.

The San Francisco Federal Building actually consists of several components, the 18-story tower being the dominant feature (see Figure 15.1). A four-story, broader structure at the southwest base of the tower houses the Social Security Administration, an agency that generates substantial pedestrian traffic and is served by a separate entry for the public. In close collaboration with the ethnically diverse local community, a rich mix of Filipinos, Mexicans, Vietnamese, and other minority groups, the project team developed the building to provide a landscaped plaza that acts as a bridge to the local community, serving as a local asset and accommodating the substantial pedestrian activity in the area. The skin of the building unfolds to cover a day-care facility, and a freestanding cafeteria rounds out the facilities on the site (see Figure 15.2). The publicly accessible day-care center and cafeteria are used by both the employees of the building and the local community, providing an architectural solution with a socially responsible dimension. The design of the building responded to the local residents' desire not to have a massive building that would overshadow the two- and three-story light industrial, commercial, and residential structures (including artists' studios, senior housing, and single-room-occupancy units) that provide the eclectic character of the neighborhood.

EXCELLENCE IN DAYLIGHTING

Lighting for office buildings in the United States is the single largest energy consumer for this building type, accounting for up to 40 percent of the total energy. Consequently, minimizing artificial lighting can have significant economic and environmental

Figure 15.1 The Federal Building in San Francisco, California, designed by Morphosis Architects, is a breakthrough structure, with an outstanding passive design strategy coupled with active control systems. All building components are optimized in connecting the building to its surrounding environment for cooling, ventilation, and lighting. (Petros Raptis)

Figure 15.2   The folded, perforated metal skin covering portions of the San Francisco Federal Building assists in the flow of air through the structure and provides an interesting and appealing appearance for the structure, both at the ground level and at the upper façade. (Photograph courtesy of Jenna Hildebrand)

benefits. The narrow floor slab—just 65 feet (20 meters) wide—the use of floor-to-ceiling glazing, and a floor-to-floor height of 13 feet (4 meters) provide perfect conditions for substantial, deep-penetrating daylight. In contrast to normal practice, the perimeter of the building has open-plan offices, with 52-inch-high partitions that minimize the amount of light being blocked. The interior core contains meeting rooms and enclosed offices, all with clear glass panels to allow natural light to penetrate throughout the space. Fritted glass has been provided for these interior spaces for privacy when needed. The southeast face of the 18-story tower is covered with perforated panels that rotate to control light and provide unobstructed views across the city. The lighting system contains sensors that provide feedback to reduce artificial lighting as daylighting increases during the day and turn off lights when there are no occupants in a space. Task lights at workstations are on only when people are present in the spaces. The net result of the lighting strategies employed in the San Francisco Federal Building is a 26 percent reduction in lighting energy.

NATURAL VENTILATION STRATEGY

As was noted in Chapter 9, it is becoming standard practice in Germany to use natural ventilation as the strategy for cooling office buildings even during peak summer days. The result is that state-of-the-art German office buildings use less than 100 kWh/m^2 (30 kWh/ft^2) of annual primary energy, about 20 percent the consumption of code-compliant US buildings. Buildings employing passive cooling strategies based on natural ventilation are rare in the United States, especially large buildings. The San Francisco Federal Building embraces passive cooling and ventilation, taking advantage of the 49 to 65°F (9 to 18°C) air currents around the building and exploiting them via the design of building elements that allow and facilitate the deep penetration and circulation of outside air. A combination of computer-controlled air vents at floor level and occupant interaction with windows permits the use of these breezes to provide a comfortable and healthy interior environment. The air currents are admitted through openings on the northwest façade and vented through the southeast wall (see Figure 15.3). The open office

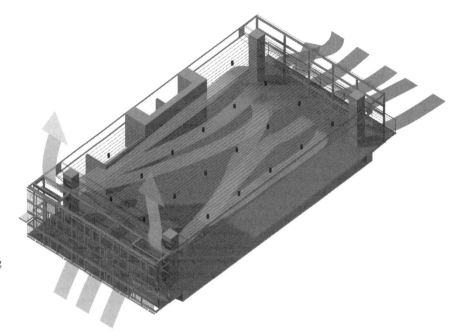

Figure 15.3 The San Francisco Federal Building is cooled and ventilated by using openings on either side of the building to direct outside air through the building. (Illustration courtesy of Morphosis Architects)

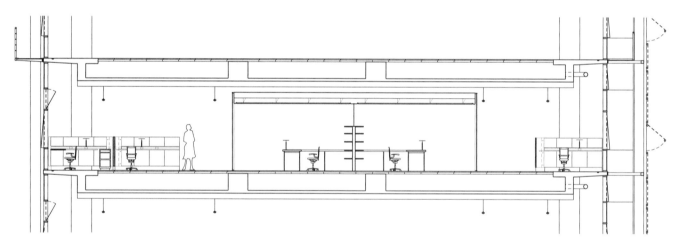

Figure 15.4 This section shows an interior conference room in the San Francisco Federal Building. Air flows from one side to the other via a pathway through and over the interior spaces. (Drawing courtesy of Morphosis Architects)

spaces are designed so as not to impede airflow across the floor, and even enclosed offices and meeting rooms have walls that stop short of the floor above, providing a pathway for air to cross the building (see Figure 15.4). In the evening, the air currents cool the concrete structure, providing a cool sink for the following day.

The southeast façade is covered with a perforated metal sunscreen that also helps induce airflow across the face of the building between the sunscreen and the façade, creating a pressure drop that induces warm airflow out of the building (see Figure 15.5). Solid narrow walls on the northeast and southwest sides contain the fire stairs and thus minimize heat gain on those sides of the building. Lower levels of the building require some mechanical cooling, and an innovative underfloor air distribution system combined with conventional heat pumps is used to meet the requirements of these zones.

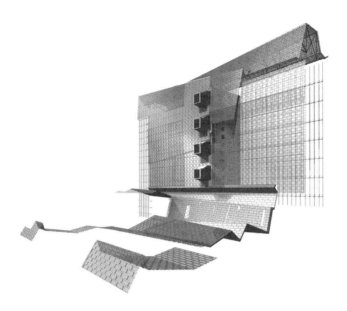

Figure 15.5 The perforated skin of the San Francisco Federal Building controls light and airflow through the building. As a result, as noted by architect Thom Mayne, the building "wears" the HVAC system. (Illustration courtesy of Morphosis Architects)

The natural ventilation strategy provides cooling for the building from mid-April through mid-October. November and March are swing months during which the building operates optimally with windows closed and no active heating. During the colder months of December through February, a hydronic heating system meets any heating demands; the heat is delivered through a finned-tube convector integrated into the exterior glazing along the entire length of the building. This scheme is estimated to save the federal government a substantial amount of money in annual energy costs, mostly through the reduction in size of mechanical systems. In the true spirit of sustainable construction, the savings realized from downsizing active, energy-consuming mechanical systems were shifted to an investment in intelligent façade design, allowing the employment of passive ventilation as a cooling strategy. As Thom Mayne of Morphosis described it, "The exterior envelope of the new building is a sophisticated metabolic skin, developed in direct response to light and climate conditions. In lieu of a conventional mechanical plant, the building actually 'wears' the air conditioning like a jacket."

The Federal Building is expected to require only 27,000 BTU/ft^2/yr (85 kWh/m^2/yr) in comparison to the GSA's national target of 55,000 BTU/ft^2/yr (173 kWh/m^2/yr) and in contrast to a typical consumption of 69,000 BTU/ft^2/yr (218 kWh/m^2/yr) for GSA buildings.

A FLEXIBLE AND INNOVATIVE INTERIOR STRATEGY

The San Francisco Federal Building also provides highly flexible spaces that can be changed as conditions and tenants change. A raised floor and an easily reconfigurable furniture system allow workstations to be arranged in grids or as single units. Each floor is modular, subdivided by circulation and support areas. The design of the building also promotes collaboration and teamwork through an innovative layout of the vertical transportation system. Starting at the third floor, the elevator stops only at every third floor, where there is a multistory lobby with stairs leading to the floor above and the floor below. A dedicated elevator bank serves the handicapped users of the building as well. The resulting circulation areas and waiting spaces bring people together in unexpected ways, facilitating the exchange of new ideas and information. A three-story interior sky garden, starting at the 11th floor, which is landscaped and has a variety of seating, provides a space for reflection and retreat, an inviting place with beautiful vistas (see Figure 15.6). It is a dramatic addition to a dramatic building.

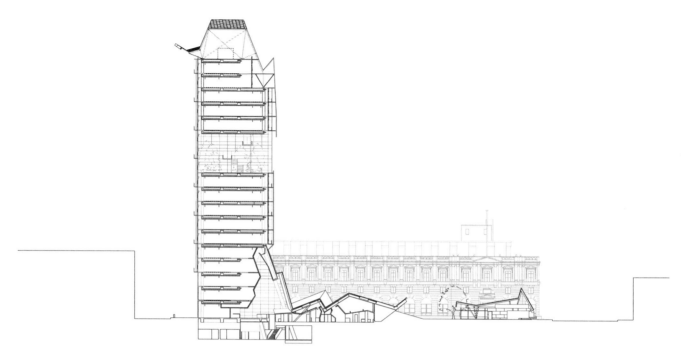

Figure 15.6   Section through the center of the San Francisco Federal Building showing the sky garden, which starts at the 11th floor. (Drawing courtesy of Morphosis Architects)

Articulating Performance Goals for Future Green Buildings

One of the major green building issues is to clarify the specific goals of high-performance green buildings. These goals can be expressed in a variety of suitable ways; this section describes four of them. One option is to apply the Factor 10 approach to buildings and focus on efforts that reduce the consumption of resources in the creation and operation of buildings to 1/10 of their present level, thereby aligning this movement with other sectors and institutions that are striving to behave sustainably.[4] A second option is to express the impact of a building in terms of its *ecological footprint*.[5] The unit of measurement for an ecological footprint is land area, which indicates the impacts by the peoples of different countries based on their lifestyles. The same concept could be applied to buildings, with impacts being stated in hectares or acres per unit area of building. Materials used in building could be measured in part by their *ecological rucksack*.[6] The ecological rucksack, a third approach, is the total mass of materials that must be processed to produce a unit mass of a specific metal or mineral. It is essentially a way to measure impact in terms of transformation of the surface of the planet—a serious matter because humans are now moving twice the amount of materials in natural systems. A fourth way to express the goals is through the routine use of life-cycle assessment (LCA), which describes the total inputs and outputs in the production of a given material. Comparisons could be made for different building solutions—for example, wall sections, to determine which approach consumes the least resources and has the fewest emissions.

For the high-performance building movement to make sense, establishing specific and reasonable goals is ultimately necessary to give the various players a direction for their activities. For the most part, the targets set in LEED are based on comparisons to a base building, that is, a building that just meets the requirements of the building code.

To project from an ideal future state to the present situation for the purpose of determining the steps that have to be accomplished to create the necessary change, a technique known as *backcasting* is used in the sustainable development arena. This immediately raises questions: What is the ideal future for high-performance green buildings? What do they look like? How do they differ from today's green buildings? Answering these challenging, even daunting, questions is critical if we are to make progress toward a future in which the buildings we construct come far closer to meeting the ultimate standard of high-performance building.

The Challenges

Chrissna du Plessis, a noted research architect and project leader on sustainable development at the Council for Scientific and Industrial Research (CSIR), the national building research institute of South Africa, located in Pretoria, has identified three major challenges we face in defining the future built environment:[7]

1. Taking the next technology leap
2. Reinventing the construction industry
3. Rethinking the products of construction

TAKING THE NEXT TECHNOLOGY LEAP

In the future, technology will undoubtedly play a powerful role in assisting and even accelerating change. In its simplest form, technology is nothing more than applied science, that is, using discoveries of basic science and mathematics for practical purposes, ideally for the benefit of people and natural systems. Technology is clearly a two-edged sword: along with its many benefits typically come a wide variety of impacts. Thus, the challenge is to foster technologies whose benefits are great and whose impacts are low. For the built environment, three general approaches are emerging:

1. Vernacular vision
2. High-technology approach
3. Biomimetic model

Each of these is accompanied by technological approaches. Even the vernacular vision, which focuses on relearning the lessons of history, is also about developing technologies that support today's implementation of those hard-learned lessons.

VERNACULAR VISION: RELEARNING THE PAST

Vernacular architecture embeds cultural wisdom and an intimate knowledge of place in the built environment. It comprises technology, or applied science, that has evolved by trial and error over many generations all over the planet as people designed and built the best possible habitat with the resources available to them. With respect to designing high-performance buildings, vernacular design comes closest to the ecological design capabilities available today.

Two contrasting examples of vernacular architecture are the traditional styles of the state of Florida and the southwest US Cracker architecture in Florida raises houses and buildings off the ground and creates flow paths for air around and through the structures, opening them to ventilation and conditioning by the prevailing winds. Originating in the early 1800s, the cracker house is well designed for the region's hot, humid climate. It emulates the chickee of the Seminole Indians, a covered structure with open sides, in which the floor, an elevated platform 3 feet (0.9 meter) above the often-wet ground, was used for both eating and sleeping. The galvanized metal roof of cracker buildings is durable and reflects Florida's daily intense solar radiation away from the structure. The structure is lightweight and sheds energy; rather than absorbing energy, it reflects it, thereby helping to maintain moderate interior temperatures.

Modern cracker architecture buildings, although they retain the appearance of their traditional predecessors, with metal roofs, cupolas, and porches, employ modern technology to meet the needs of contemporary businesses and homes. As is the case with much of today's vernacular architecture, some of the original features, such as the capability for passive ventilation, are, for all practical purposes, not useful due to year-round reliance on modern heating, ventilation, and air conditioning (HVAC) systems. Cracker architecture is generally limited to smaller buildings, as it is difficult to apply to large buildings, because the roof tends to become inordinately large, and for urban office buildings the porches lose their appeal (see Figure 15.7).

Adobe architecture, prevalent in the Southwest and Mexico, relies on local soils and a relatively massive structure made of adobe clay and straw brick. The large thermal mass of the structure enables the building to take advantage of the great diurnal temperature swings prevalent in high-desert areas for heating and cooling. During the day, the thermal mass absorbs solar radiation, storing it for later use, but also provides just enough thermal resistance to keep the interior temperature at a moderate level. As temperatures in the deserts and mountains plunge in the evening, the energy stored in the massive adobe structure is emitted by radiation and convection into the interior spaces (see Figure 15.8).

These two historical forms of vernacular architecture, in addition to taking advantage of experience with daily and seasonal weather patterns and the assets of the sites, made use of local materials—long-leaf pinewood in Florida and earth and straw in the Southwest. Incorporating local and regional materials is now a criterion in modern building assessment standards such as LEED. In this way, taking a vernacular approach promises an excellent start to incorporating passive energy design features into a building, because it implies using the site and structural design to assist heating and cooling. Fortunately, there are hundreds of examples of vernacular architecture worldwide that can be used as the basis for designing today's high-performance buildings. The challenge, of course, is to use the wisdom of the past to meet the requirements of modern buildings and current building codes while retaining the positive cultural, environmental, and resource aspects of vernacular design.

HIGH-TECHNOLOGY APPROACH

In contrast to the vernacular vision, which uses historical wisdom and cultural knowledge to design buildings, the high-technology approach generally follows the path of current trends in society. Contemporary society, especially in the developed world, has a love affair with technology. The prevalent attitude is that all our problems, including resource shortages and environmental dilemmas, can be solved simply by developing new technologies. For buildings, the high-technology approach centers on devising new energy technologies, such as photovoltaics and fuel cells, and on finding technical solutions to the question of

Figure 15.7 Vernacular architecture in northern Florida. Early cracker-style houses were lightweight, wood-framed structures with wooden siding and metal roofs. The passive aspects of these structures help them reflect solar radiation and facilitate cross-ventilation; they are raised off the ground for protection from flooding. Modern versions adapt the materials and energy strategies of early cracker architecture to produce hybrid structures that include high-technology windows, composited siding, and energy-efficient air conditioning. (A) The Geiger Residence, Micanopy, Florida (1906). (B) A small cracker vernacular office building in Gainesville, Florida (1996). (C) Interior of Summer House at Kanapaha Botanical Gardens, a larger, 10,000-square-foot (929-square-meter) cracker-style building near Gainesville, Florida (1998). [Photographs courtesy of (A) Ron Haase; (B) Jay Reeves; and (C) M. R. Moretti]

Figure 15.8 Examples of New Mexico adobe architecture. (A) As early as 350 AD, the Anasazi, the oldest-known inhabitants of New Mexico, began to build aboveground masonry structures, the foundations of which are visible here at the base of their cliff dwellings in Bandelier National Park. (B) Communities called *pueblos* flourished around 1250–1300 AD and contained intricate arrays of connected flat-roofed, multilevel adobe buildings. (C) A modern office building in Santa Fe, New Mexico, retains the appeal and function of traditional adobe architecture.

how to utilize renewable energy sources more effectively. Typical examples of this approach include windows with spectrally selective coatings and gas-filled panes, control systems and computer systems that respond to optimize energy use based on weather and interior conditions, energy recovery systems that incorporate desiccants to shift both heat and humidity, and materials incorporating postindustrial and postconsumer waste. Contemporary commercial and industrial buildings are equipped with a wide range of telecommunications and computer technologies that would challenge even the most advanced vernacular design approaches simply because of the need to remove the high levels of energy generated by today's workplace tools. Indeed, it could be argued that the technology of the building itself must be carefully matched to the technologies employed by the building occupants.

The high-technology approach to high-performance green building is, in short, an evolution of current practices. Over time, built environment professionals, backed up by experience, research, and the development of better systems and products, will be able to design buildings that are much more resource-efficient than today's green buildings and that will have far lower impacts in their construction and operation. Thus, the key characteristics of the ideal high-performance green building are based on making incremental—as opposed to radical—improvements in existing technology in these areas:

- *Energy.* The ultimate high-performance building consumes just 1/10 of the energy of current buildings and either uses only off-site-generated renewable energy or generates energy from renewable sources on-site for its entire needs. Passive design, assisted by extensive computer modeling, ensures the optimal use of natural ventilation, structural mass, orientation, building site, building envelope design, landscaping, and daylighting to minimize consumption of electricity and other energy sources so that the building can default to nature if it becomes disconnected from external energy sources.[8] Landscaping is carefully integrated into the project to assist in cooling and heating the structure.

- *Water.* The ideal high-performance building uses only 10 percent of the potable water of contemporary buildings and uses graywater, reclaimed water, or rainwater for nonpotable requirements. Wastewater is recycled for nonpotable building uses or is processed by constructed wetlands or Living Machines for discharge back into nature in as clean a state as it entered.

- *Materials.* All materials employed in the ultimate high-performance building are recyclable; building products can be disassembled and their constituent materials easily separated and recycled; buildings are deconstructable, capable of being disassembled and their components either reused or recycled. The cardinal rule for materials used in construction would be to eliminate those that are not recyclable, that are used in a one-off fashion and become waste after one use. An effective Factor 10 reduction in materials consumption would focus on reducing materials extraction by 90 percent, achievable by dramatically increasing the conservation of materials by deconstruction, materials recovery, and recycling and reuse. Increasing the durability and longevity of the built environment would also help achieve Factor 10 performance. However, this presumes that improvements in design would make buildings so much more valuable to society as cultural artifacts that their removal for economic reasons would be far less likely.

- *Natural systems interface.* The ultimate high-performance building is integrated with natural systems in a synergistic manner such that services and nutrients are exchanged in a mutually beneficial manner. Natural systems provide stormwater uptake and storage; assist cooling and heating; provide amenities; supply food; and break down waste from individual building scale to larger scales, up to the bioregional one. The building is carefully designed to take advantage of the natural assets of the site, the prevailing winds, and the microclimate at the building location.

- *Design.* Ideal high-performance buildings are designed using well-developed principles that are rooted in ecology. A robust version of ecological design is employed to ensure the integration of the building with its site and the natural assets. Architecture, landscape architecture, and engineering are carried out in a seamless, integrated process. The building professionals on the team work in a collaborative fashion, with fees based on the quality of design and construction and the building's performance. These same professionals work to minimize building complexity and maximize adaptability and flexibility.

- *Human health.* All aspects of indoor environmental quality (IEQ) in the ultimate high-performance building are carefully addressed, including air quality, noise, lighting quality, and temperature/humidity control. Ventilation rates are optimized to provide exactly the levels of fresh air that support health. Only zero-emission materials are permitted.

REINVENTING THE CONSTRUCTION INDUSTRY

The construction industry, referred to in its broadest sense to include design, construction, operation, renovation, and disposal of the built environment, has to change dramatically to meet the future challenges of building. Buildings have become commodities, with little to distinguish one from another in any serious manner, and with little effort to make them—as in the past—cultural artifacts of human existence. Low first cost is the normal order of business, so quality design receives minimal attention; materials and systems are employed that produce minimal performance; the construction process is carried out rapidly and at the lowest possible cost; and the norm is to demolish and landfill buildings at the end of their useful life. Scant attention is paid to the implications of this behavior, both for ecological systems and for human society. Owners focus on buildings that have minimal construction cost and that are designed just to accomplish their functions, with little or no attention given to their aesthetic features. Changing the mind-set of this cast of actors is an enormous challenge. To meet that challenge, these changes must take place:

- *Technology.* Technologies that minimize resource consumption and the environmental impact of the built environment need to be developed.

- *Policy.* As a general matter of policy, buildings need to be created based on life-cycle costs as well as first costs.

- *Incentives.* Government needs to develop financial incentives for high-performance construction, such as priority review by building departments, accelerated approval for projects of this type, and reductions in impact fees and/or property taxes for a specified period of time.

- *Education.* All the professionals in the industry need to be educated and trained in the need, process, and approaches for creating high-performance green buildings—owners, architects, engineers, landscape architects, interior designers, construction managers, subcontractors, materials and product manufacturers and suppliers, insurance and bonding companies, real estate agents, building commissioning consultants, and other professionals engaged in the process. This is also necessary for the workforce, the crafts workers, journeymen, and apprentices who work for the broad array of subcontractors that make buildings a physical reality.

- *Performance-based design fees.* Contracts for design and construction services need to be revised to offer incentives to the building team to meet and exceed project goals with respect to resource consumption and environmental impacts. These goals include targets for energy and water consumption, building health, construction waste, protection of the site's natural assets, and other objectives that contribute to the building's performance.

- *Construction process.* The physical process of construction needs to be changed to ensure that the activities involved in erecting the building have the lowest possible impact. Among these changes are reduce construction waste and recycle or reuse the residue, understand and implement effective soil and erosion control methods, protect flora and fauna on the site during the construction process, minimize soil compaction during construction, and store materials so that they are protected from wastage and are unlikely to cause IEQ problems.

RETHINKING THE PRODUCTS OF CONSTRUCTION

As this book has pointed out repeatedly, buildings consume enormous quantities of resources and can cause any number of negative impacts on their occupants. In addition to the resources required to build and operate individual buildings, a wide range of additional impacts are the consequence of decisions concerning how to distribute buildings across the landscape. For example, segregating buildings by type (residential, commercial, industrial, government, cultural, etc.) means that people are forced to use their automobiles to get from one type of building to another. The average American makes at least eight automobile trips per day, many of them for no reason other than to socialize. The concepts of new urbanism or traditional neighborhood development are seeking to reverse this trend by mixing building types and uses and by designing streets and neighborhoods for pedestrian movement. A general goal is that all daily needs must be available within a 10-minute walk from where the individual resides.

Other serious impacts result from the building stock itself. In the United States, buildings and houses are generally very large and consume large quantities of energy, water, and materials to both build and operate them. The extraction of resources to support the construction industry is profound. Some estimates state that 90 percent of all extracted resources in this country are used to create the built environment. Buildings consume two-thirds of all electricity and 35 to 40 percent of primary energy. Three important questions that need to be asked when a new building is proposed are:

- Is the building actually needed or is adequate space already available?
- Can the building be made smaller?
- Can an existing building be renovated for the new purpose?

Revamping Ecological Design

As noted in Chapter 3, contemporary ecological design has only very weak links to ecology. Although virtually any definition of high-performance green building makes reference to ecological design, to date there is little or no evidence of the application of ecology to design. To correct this situation, it is crucial that a new, comprehensive concept of ecological design be developed. In addition to considering ecology in far greater depth in building design, it is imperative to consider the potential for applying industrial ecology. Established as a new discipline in 1988, industrial ecology seeks to apply ecological theory to industrial production. Many of the issues and problems faced by an industrial system that builds automobiles and airplanes are also faced by those in the building design and construction professions. Consequently, the experience gained by applying industrial ecology to industrial production will be very useful in creating high-performance green buildings.

In a recent collaboration among architects, ecologists, and industrial ecologists, the possibility of applying current ecological theory to the creation of buildings was explored in great depth to determine which aspects of ecological theory and industrial ecology were applicable to buildings.[9] This collaboration offered a number of insights into how ecology and industrial ecology can better inform building design, construction, and operation. The results of this collaboration are summarized in the following lists:

GENERAL

1. Maximize second-law efficiency (effectiveness) and optimize first-law efficiency for energy and materials.[10]

2. As with natural systems, industry must obey the maximum power principle.[11]

3. Be aware that the ability to predict the effects of human activities on natural systems is limited.

4. Integrate industrial and construction activities with ecosystem functions so as to sustain or increase the resilience of society and nature.

5. Interface buildings with nature.

6. Match the intensity of design and materials with the rhythms of nature. In the built environment, move from the "weeds" stage to the "tree" stage for sites that are not frequently disturbed. "Weedy" structure (minimal built structure that is easily and cheaply replaced) may be much more adaptive to sites frequently disturbed by floods, storms, or fires.

7. Consider the life-cycle impacts of materials and buildings on natural systems.

8. Insist that industry take responsibility for the life-cycle effects of its products, to include take-back responsibility.

9. Address the consumption end of the built environment by integrating it with production functions.

10. Increase the diversity and adaptability of user functions in buildings through experiment and education.

11. Explore educational processes beyond academia that instruct through "learning by doing," by involving all stakeholders in processes that test different means by which the built environment is produced, sited, deconstructed, and resurrected.

12. Reduce information demands on producers and consumers by testing and improving the means by which materials, designs, and processes are certified as "green." This presupposes the development of a construction ecology based on nature and its laws.

13. Ensure that systems analyses look at system function, processes, and structure from different perspectives and at different scales of analysis.

14. Integrate ecological thinking into all decision-making processes.

15. Follow the precautionary principle to constrain and govern decision making.

MATERIALS

1. Keep materials in productive use, which also implies keeping buildings in productive use.[12]

2. Use only renewable, biodegradable materials or their equivalent, such as recyclable industrial materials.

3. Release materials created by the industrial system only within the assimilative capacity of the natural environment.

4. Eliminate materials that are toxic in use or release toxic components in their extraction, manufacturing, or disposal. Focus first on materials not well addressed by economics—the intermediate consumables (paints, lubricants, detergents, bleaches, acids, solvents) used to create wealth (buildings).

5. Eliminate materials that create "information" pollution—for example, estrogen mimics.

6. Minimize the use and complexity of composites and the numbers of different materials in a building.

7. Realize that not all synthetic materials are harmful and not all natural materials are harmless. Nature has many pollutants that are harmful; for example, natural fibers such as cotton are not necessarily superior to synthetic materials such as nylon.

8. Recognize that the impacts of natural materials extraction can be high—as is the case with agricultural products; or in forestry, in which pesticide use, transportation distances, processing energy, and chemical use are significant factors.

9. Standardize plastics and other synthetic materials based on recycling infrastructure and the potential for recycling and reuse.

10. Rather than for power generation, use fossil fuels to produce synthetic materials, and use renewable energy resources as the primary power source.

11. Acknowledge that it is not possible to rate or compare materials adequately based on a single parameter.

DESIGN

1. Model buildings based on nature.

2. Make structures part of the geological landscape.

3. Design buildings to be deconstructable, using components that are reusable and ultimately recyclable.

4. Design buildings and select materials based on intended use and then measure the outcomes of the design.

5. Incorporate adaptability into buildings by making them flexible for multiple uses.

6. Realize real savings by integrating the production, reuse, and disposal functions.

7. Focus on excellence of design and operation, with greenness as a critical component. Focusing exclusively on greenness trivializes it as a marginal movement.

8. Invest in design that improves building function while minimizing energy use and the number of materials. This will reduce the time and effort required to find and optimize new green materials.

9. Revise designs to take into account major global environmental effects such as global warming and ozone depletion. This is critical at this point in time.

10. Allow for experimentation in green building design to produce structures that, like nature, obey the maximum power principle.

11. Make sure that architects have a strong, fundamental education in ecology.

12. Use performance-based design contracts to develop greener buildings and better architects.

INDUSTRIAL ECOLOGY

1. Make changes needed to create an environmentally responsible industrial ecosystem intelligible to the members of the particular industry.

2. Focus on the clients and key stakeholders of the system. This is necessary due to limits on time, knowledge, and resources. Major stakeholders include the educational system and the insurance industry.

3. Make the new paradigm for industry the collaboration of actors versus the possession of technical expertise.

4. Reduce consumption. This is more important than increasing production efficiency as the change agent for industrial ecology.

5. Incorporate ecological engineering into industrial ecology.

CONSTRUCTION ECOLOGY

1. Ensure that construction ecology balances and synchronizes spatial and temporal scales to natural fluxes.

2. Recognize that the corporations leading the way in the production of new, green building materials are a "frontier species" that may be creating a new form of competition, which they are using to their advantage.

3. Be aware that green building probably can be implemented only incrementally because of resistance and potential disruptions from the existing production and regulatory systems.

OTHER ISSUES

1. Better educate government officials and code-writing bodies about ecology.

2. Establish performance standards for buildings and construction to replace existing prescriptive standards. The performance standards need to include provisions for using green building materials.

3. Regard the insurance industry as a major stakeholder in the built environment, as the threat of severe consequences from global warming will drive it to promote green building.

4. Rely on certification only as a starting point; do not rely on it entirely for information on products.

Today's Cutting Edge

In closing the last chapter of this book, it is useful to reflect on how far the high-performance green building movement has advanced and where the cutting edge of this field is at the present time. The following sections describe the areas in which high-performance green building has made significant progress over the past decade and where the cutting edge of change in high-performance green building can be found. These areas include:

- The development of green building standards
- The net zero built environment concept
- The Living Building Challenge building assessment system
- The emergence of environmental product declarations (EPDs)
- Carbon accounting for the built environment

HIGH-PERFORMANCE BUILDING STANDARDS

High-performance green buildings have evolved significantly since the development of the Building Research Establishment Environmental Assessment Method (BREEAM) in the United Kingdom in 1990 and the beta test version of LEED in 1998. One of the key challenges has been how to make the green building approach more readily available to all buildings, not just to the growing number of organizations that have been implementing green buildings. According to the USGBC, LEED addresses the top 25 percent of high-performance buildings. By developing a high-performance green building standard using an American National Standards Institute (ANSI) accredited process, the USGBC, in collaboration with the American Society of Heating, Refrigerating and Air-Conditioning Engineers (ASHRAE) and the Illuminating Engineering Society of North America (IESNA), is making it possible for green building requirements to be incorporated into building codes, thus addressing the other 75 percent of construction. The full name for the LEED-oriented standard is ASHRAE 189.1-2009, Standard for the Design of High-Performance Green Buildings Except Low-Rise Residential Buildings. If this standard were to become incorporated into building codes, it would free the USGBC and other green building organizations to set the bar for high-performance buildings even higher. It will also provide a baseline for sustainable design, construction, and building operation in order to drive green building into mainstream construction industry practices. ASHRAE 189.1 was rolled out in 2009 and applies to new commercial buildings and major renovations. It is modeled after the LEED building assessment system, including prescriptive measures drawn from the five main LEED categories: sites, water efficiency, energy and atmosphere, materials and resources, and indoor environmental quality. A second standard, ANSI/GBI 01-2010, Green Building Assessment Protocol for Commercial Buildings, was also developed using the ANSI standards development process and is based on the Green Globes building assessment system.

In addition to the two standards mentioned above, a full-fledged building code based on the LEED building assessment system has been developed and is gaining support. The *International Green Construction Code* (IgCC) effort was launched in 2009 and completed in 2012 as a final product. The purpose of this effort was to develop a model code focused on new and existing commercial buildings addressing green building design and performance. Jurisdictions that have already adopted the IgCC include Ft. Collins, Colorado; Richland, Washington; Kayenta Township, Arizona; and the state of Rhode Island. The

IgCC Public Version 2.0 offers a Zero Energy Performance Index (zEPI), requiring buildings to use no more than 51 percent of the energy allowable in the 2000 International Energy Conservation Code. Examples of provisions in Public Version 2.0 include:

- A 20 percent water savings beyond US federal standards for water closets in residential settings
- New requirements for identification and removal of materials containing asbestos
- Land use regulations, including new provisions addressing flood risk, development limitations related to "greenfields," use of turfgrass, and minimum landfill diversion requirements
- Clarification of responsibilities from the registered design professional to the owner to prevent potential conflicts with state and local requirements
- Greater consistency with industry standards for air-handling systems

The natural question to be asked when building assessment systems such as LEED are codified into standards is, Is there now a purpose for the organization that originated the assessment system? The building department in each jurisdiction would have the task of evaluating projects for their compliance with green building codes and standards. The likely answer is that organizations such as the USGBC will be developing the next-generation building assessment systems to push the envelope and maintain the momentum of the green building movement of the past three decades. Additionally, green building standards and codes do not provide an actual certification, which still may have value to a building owner, and only a building assessment system proponent such as the USGBC can provide this outcome.

THE NET ZERO BUILT ENVIRONMENT

A powerful movement centered on the concept of net zero is emerging and setting targets for resource consumption based on one of the core ideas of sustainability that suggests that humans should be surviving using the local resources provided by nature. In the case of energy, the main concept that has emerged is the design and construction of grid-tied buildings that are powered by photovoltaic electricity and that have been designed to generate at least the same amount of energy they consume over the course of a year. These buildings are commonly referred to as *net zero energy buildings* (NZEBs). The National Renewable Energy Laboratory (NREL) Research Support Facility described in Chapter 1 is an excellent example of an NZE office building that generates at least as much energy annually as it consumes. A second emerging net zero concept is known as *net zero water*, which requires that the building users depend on recycled water and water falling on the building site for 100 percent of their water needs. The US Army is a major proponent of this concept and defines a net zero water installation as ". . . one that limits the consumption of freshwater resources and returns water back to the same watershed so not to deplete the groundwater and surface water resources of that region in quantity and quality over the course of a year."[13] The Army is also engaged in a broader net zero initiative that addresses energy and waste as well as water. A recent European Union (EU) declaration requires "near" NZE buildings for all new construction and major renovations, with the deadline for public-sector compliance set for 2018. The California Energy Commission is recommending that the state require NZE for residential construction by 2020 and for commercial buildings by 2030. The Living Building Challenge requires both net zero energy and net zero water performance for buildings certified by its assessment system. This concept is also being extended to net zero emissions, net zero carbons, net zero land, and even net zero materials.

THE LIVING BUILDING CHALLENGE

The clear leader in building assessment systems in terms of degree of difficulty is the Living Building Challenge (LBC), a product of the Cascadia Green Building Council that joins American green building efforts in the Pacific Northwest with similar efforts in British Columbia, Canada. The LBC requires, among many other stringent measures, net zero energy, net zero water, and the processing of all sewage on-site. Additionally, unlike other assessment systems that provide several levels of certification, the building either is certified to LBC or is ineligible for certification because it fails to meet at least one of the imperatives spelled out in the requirements. One other outstanding feature of the LBC is that the building must actually demonstrate, via its actual operation, that it meets all the imperatives. Consequently, certification can be achieved only after one year of operation that demonstrates the building has met its objectives. The ambitious nature of the LBC and its stringent requirements make it the gold standard of building assessment systems and provide a truly remarkable and challenging approach that other assessment systems may want to emulate as they evolve over time. A far more detailed description of the LBC can be found in Chapter 4.

ENVIRONMENTAL PRODUCT DECLARATIONS

The emergence of environmental product declarations (EPDs) in which third-party entities provide an independent, public, and transparent environmental analysis of materials and products intended for construction is a significant step forward in the development of a greener built environment. By providing the equivalent of a nutrition label for building products, the stage is now set for competition among producers to develop the most environmentally friendly products. One of the early adopters of EPD for its products is InterfaceFLOR Corporation, a manufacturer of carpet tiles, that pledged to obtain third-party-validated EPDs on all their products globally by the end of 2012. The widespread use of EPDs also creates the potential for whole-building life-cycle assessment in which trade-offs of building materials for reduced energy consumption can be examined to find the optimal relationship. Companies such as InterfaceFLOR are viewing EPDs as an asset, not a liability, because it allows them to more fully communicate their corporate values to their customers, an especially important and useful outlook as the green building certification process moves toward an era where the majority of construction and major renovation projects are green buildings. More about EPDs and their significance can be found in Chapter 11.

CARBON ACCOUNTING FOR THE BUILT ENVIRONMENT

Of the 100 quadrillion BTU consumed annually in the United States, about 65 percent, or 65 quads (where 1 quad is equal to 1 quadrillion BTU), are used by the built environment, and this number, both as a percentage of total energy and in absolute terms, continues to rise. The energy system is powered largely by the combustion of fossil fuels, and as a consequence, any increase in energy consumption also tends to increase the carbon footprint of the activity. The potential consequences of climate change are so catastrophic that accounting for carbon is beginning to occur and project teams will soon be judged for merit based on how well the building minimizes the total carbon invested in its materials of construction and associated with its operational energy over its useful life. As the consequences of climate change become more apparent, stringent measures to control greenhouse gas emissions are likely, and due to the scale of its emissions, a likely target for increased and even draconian standards is the built environment. One interesting outcome of the NZE movement discussed earlier is that a building that uses renewable energy for all of its energy needs has, in effect, a net zero carbon footprint with respect to its operational energy. The reuse of existing buildings and materials extracted from buildings undergoing demolition are other measures that have been identified to reduce the carbon footprint of buildings because reuse has virtually no carbon associated with it. The more durable a building, the lower is its embodied carbon per unit time. Chapter 9 provides a more detailed description of carbon accounting practices for buildings and how it is affecting the design of contemporary high-performance green buildings.

THOUGHT PIECE: FRACTALS AND ARCHITECTURE

Ecological design is without a doubt the linchpin of sustainable construction and green building. At present, however, it is not well defined, which means that green building design has a shaky foundation. Kim Sorvig, a research professor at the School of Architecture and Planning at the University of New Mexico in Albuquerque and coauthor with Robert Thompson of *Sustainable Landscape Construction* (2000), has created the notion of fractal architecture as a bridge to ecological design. His reflections on the transition between the built and natural environments are excerpted here.

Processes, Geometries, and Principles: Design in a Sustainable Future

Kim Sorvig, Research Associate Professor, School of Architecture and Planning, University of New Mexico, Albuquerque

(© Kim Sorvig used by permission)

Sustainability is about integrating constructed and living systems. To integrate two different processes or entities, each must be clearly understood in its own right. I am convinced that deepening our understanding, not only of ecology but also of building, is an essential evolution of sustainable design.

Today's understanding of the inherent qualities of constructed systems is detailed and pragmatic, but we often fail to question important principles. Conversely, designers' understanding of ecological systems is generalized, frequently romanticized. In both areas, the relationships between processes (the use of a building, the life cycle of a watershed) and geometries remain unconsidered, with shapes and patterns designed by habit and without insight. There is a pressing need to understand, factually and concisely, the core qualities of constructed and natural systems and to use differences and similarities among these systems creatively.

The core qualities that pertain most to sustainable design can be called *processes, geometries,* and *principles*. The future of sustainable design may lie less in technical innovation than in whether designers can work out the conflicts between the core qualities of natural systems and those of human development.

Construction does not create its raw materials, but shapes existing substances into units and assembles those units into structures. This is so obvious that it is often overlooked, but it has a critical effect on the processes and geometry of construction.

The essential processes of construction are cutting and assembling, plus form casting. In natural systems, direct parallels to these form-making processes (especially cutting and assembly) are rare. Making bricks and building an arch are categorically different from erosion creating a stone arch or from a tree branch growing. Construction is a controlled system, dominated by a selected force (e.g., a saw blade) and excluding extraneous forces (jigs and clamps preventing unplanned movement).

The geometry of construction is based on its processes: cutting and assembling are most efficient when using regular, smooth, Euclidean forms. Such forms also lend themselves to easy measurement and calculation.

Form making in geological and biological systems is markedly different from construction processes. Nature's processes are growth, decay, deposition, and erosion, all radically different from the assembly processes of construction. Natural processes are part of an "open" system, with many forces interacting, no one dominating for long.

Mathematical understanding of the geometry of nature is recent, and designers are just beginning to appreciate it. Resulting from growth/decay processes, the forms characteristic of nature are called *fractals*. They result when multiple forces interact repeatedly over time. No matter what scale they are viewed at, or at what period in time, their forms remain self-similar (like endless variations on a constant theme). Two points are important here:

Fractals represent long-term dynamic stability among many forces in a system, with no single dominant force (virtually a definition of biodiversity and health).

Fractals are the optimal geometry for doing what natural systems do—collecting, transporting, and diffusing resources; filtering and recycling wastes; and so on.

Construction is ultimately about creating environments from which the forces of climatic and ecological change are excluded (temporarily). Every structure conflicts to some degree with both the processes and the forms of nature. Construction optimizes a few select functions; natural systems appear to optimize for diversity. Construction aims for structural permanence; natural systems self-organize stability through change. Both require resource efficiency, but achieve it differently.

Recognizing the core qualities of built and living systems can help generate new sustainability strategies and evaluate existing ones. One strategy is making built systems more fractal or naturalistic in form: biomimicry (to oversimplify). Cyclical buildings, stability-in-change, and literal integration of landscape with building are other strategies.

Sustainable geometries are, I believe, the next evolution for design. Energy-efficient, materials-efficient structures in the same old shapes and the same old locations are unlikely to be enough. Sustainable design's future is at the edge between buildings and landscapes, where the forms necessary for human structures interface with the forms essential to living systems. Designerly preoccupation with appearance—with buildings that stand out and that privilege machine-look over naturalism—is clearly not helping. We must apply our visual skills to understand how form and function interact, not just in architecture, not just in nature, but at the borders between the two.

Summary and Conclusions

Describing the qualities of the future high-performance green building is an essential and crucial step toward making real progress in this area. Three possible approaches have been described in this chapter: the vernacular vision, the high-technology approach, and the biomimetic model. Each, or a combination, may be able to answer some of the questions faced today by professionals involved in the high-performance green building movement, which, because it is relatively new, is heavily constrained by a narrow knowledge base, limited availability of appropriate technology, and the absence of a clear vision of the future. A robust theory of ecological design is sorely needed, as, fundamentally, that is what the design of high-performance green buildings is about: developing a human environment that functions in a mutually beneficial relationship with its natural surroundings and that exchanges matter and energy in a symbiotic manner.

Notes

1. The editorial by Alex Wilson in the December 2005 issue of *EBN* provided this definition for passive survivability. Wilson also noted that the requirements for passive survivability and the sustainable design features of many green buildings were remarkably similar.
2. The New Orleans Principles can be found at www.usgbc.org/ShowFile.aspx?DocumentID=4395.
3. The checklist can be found in Wilson (2006).
4. Factor 10, which is now part of EU policy, is influencing change to a sustainable system of production and consumption.
5. An ecological footprint is the land area, in hectares or acres, that a person or activity needs to function on a continuing basis. The term can also be applied to the built environment, for which a measurement such as ecological footprint per 1000 square feet (100 square meters) of building area is a potential metric for comparing building impacts. The term was popularized by Wackernagel and Rees (1996).
6. The ecological rucksack of a material is the total mass of materials that must be moved to extract a unit mass of the materials, expressed as a ratio. This term was coined by the Wuppertal Institute in Wuppertal, Germany, to draw attention to mass movements of materials that are changing the surface of the planet. Historically, attention has been paid to the impacts of toxic materials such as dichlorodiphenyltrichloroethane (DDT) and polychlorinated biphenyl (PCB), which are harmful in the microgram range. The ecological rucksack concept looks at the other end of the materials spectrum—the megaton-range movements of materials to extract resources. The bottom line is that both micrograms of toxic materials and megatons of less harmful materials should be accounted for with respect to their impacts.
7. As described in du Plessis (2003).
8. "Defaulting to nature" is an expression used by Randy Croxton of the Croxton Collaborative in New York City to describe the ability of a well-thought-out, passively designed building to provide heating, cooling, and lighting for its occupants, thus ensuring its operability in spite of being disconnected from external energy sources—for example, the electric power grid.
9. From "Conclusions" in Kibert, Sendzimir, and Guy (2002).
10. Natural systems match the energy source and its quality to energy use first (effectiveness) and then maximize the system's efficiency. Human-designed systems, in contrast, tend to focus on efficiency alone and neglect energy quality, thus often spending high-quality energy (e.g., electricity) on building needs that could be better served by low-quality energy (e.g., medium-temperature heat). Quality is a measure of the flexibility of applications for a particular energy source. Electricity can be used

to drive electric motors and generate power to move vehicles, while moderate-temperature heat, below that of the boiling point of water, has limited application and flexibility. The lower the temperature of the heat source is, the lower the quality of the energy. Using electricity for water heating has a very low level of effectiveness because it is using high-quality energy in an application that could use low-quality energy sources. Refer to Kibert, Sendzimir, and Guy (2002), chapter 3.

11. The maximum power principle was hypothesized by the eminent ecologist H. T. Odum, the founder of a branch of ecology known as *systems ecology*. In its simplest form, the principle states that the dominant natural systems are those that pump the most energy. Refer to Kibert, Sendzimir, and Guy (2002), chapter 2.

12. As is often stated in the green building area, there is no waste in nature; all materials are kept in productive use. Of course, this is very simplified and, strictly speaking, not even true.

13. The Army website for its net zero initiative is www.army.mil/asaiee.

References

Benyus, Janine M. 1997. *Biomimicry: Innovation Inspired by Nature.* New York: William Morrow.

du Plessis, Chrissna. 2003. "Boiling Frogs, Sinking Ships, Bursting Dykes and the End of the World as We Know It." *International Electronic Journal of Construction*, Special Issue on Sustainable Construction. Available at www.bcn.ufl.edu.

Kibert, Charles J., Jan Sendzimir, and G. Bradley Guy, eds. 2002. *Construction Ecology: Nature as the Basis for Green Building.* London: Spon Press.

Sorvig, Kim, and Robert Thompson. 2000. *Sustainable Landscape Construction.* Washington, DC: Island Press.

Wackernagel, Mathis, and William Rees. 1996. *Our Ecological Footprint.* Gabriola Island, BC: New Society Publishers.

Wilson, Alex. 2005. "Passive Survivability." *Environmental Building News* 14 (12): 2.

———. 2006. "Passive Survivability: A New Design Criterion for Buildings." *Environmental Building News* 15 (5): 1, 15–16.

Appendix A
Quick Reference for LEED 3.0

Quick Reference for LEED v3.0	Commercial Interiors	Core & Shell	Healthcare	New Construction	Retail: New Construction	Schools
Total Possible Points	**110**	**110**	**110**	**110**	**110**	**110**
Sustainable Sites (SS)	**21**	**28**	**18**	**26**	**26**	**24**
Construction Activity Pollution Prevention		Prereq	Prereq	Prereq	Prereq	Prereq
Environmental Site Assessment			Prereq			Prereq
Site Selection	5	1	1	1	1	1
Development Density and Community Connectivity	6	5	1	5	5	4
Brownfield Redevelopment		1	1	1	1	1
Alternative Transportation					10	
Alternative Transportation - Public Transportation Access	6	6	3	6		4
Alternative Transportation - Bicycle Storage and Changing Rooms	2	2	1	1		1
Alternative Transportation - Low-Emitting and Fuel-Efficient Vehicles		3	1	3		2
Alternative Transportation - Parking Capacity	2	2	1	2		2
Site Development - Protect or Restore Habitat		1	1	1	1	1
Site Development - Maximize Open Space		1	1	1	1	1
Stormwater Design - Quantity Control		1	1	1	1	1
Stormwater Design - Quality Control		1	1	1	1	1
Heat Island Effect - Nonroof		1	1	1	2	1
Heat Island Effect - Roof		1	1	1	1	1
Light Pollution Reduction		1	1	1	2	1
Site Master Plan						1
Joint Use of Facilities						1
Tenant Design and Construction Guidelines		1				
Connection to the Natural World - Places of Respite			1			
Connection to the Natural World - Direct Exterior Access for Patients			1			
Water Efficiency (WE)	**11**	**10**	**9**	**10**	**10**	**11**
Water Use Reduction - 20% Reduction	Prereq	Prereq	Prereq	Prereq	Prereq	Prereq
Minimize Potable Water Use for Medical Equipment Cooling			Prereq			
Water-Efficient Landscaping		4	1	4	4	4
Innovative Wastewater Technologies		2		2	2	2
Water Use Reduction	11	4	3	4	4	4
Water Use Reduction - Measurement and Verification			2			
Water Use Reduction - Building Equipment			1			

(Continued)

Quick Reference for LEED v3.0	Commercial Interiors	Core & Shell	Healthcare	New Construction	Retail: New Construction	Schools
Total Possible Points	**110**	**110**	**110**	**110**	**110**	**110**
Water Use Reduction - Cooling Towers			1			
Water Use Reduction - Food Waste Systems			1			
Process Water Use Reduction						1
Energy and Atmosphere (EA)	**37**	**37**	**39**	**35**	**35**	**33**
Fundamental Commissioning of Building Energy Systems	Prereq	Prereq	Prereq	Prereq	Prereq	Prereq
Minimum Energy Performance	Prereq	Prereq	Prereq	Prereq	Prereq	Prereq
Fundamental Refrigerant Management	Prereq	Prereq	Prereq	Prereq	Prereq	Prereq
Optimize Energy Performance		21	24	19	19	19
Optimize Energy Performance - Lighting Power	5					
Optimize Energy Performance - Lighting Controls	3					
Optimize Energy Performance - HVAC	10					
Optimize Energy Performance - Equipment and Appliances	4					
On-Site Renewable Energy		4	8	7	7	7
Enhanced Commissioning	5	2	2	2	2	2
Enhanced Refrigerant Management		2	1	2	2	1
Measurement and Verification	5		2	3	3	2
Measurement and Verification - Base Building		3				
Measurement and Verification - Tenant Sub-metering		3				
Green Power	5	2	1	2	2	2
Community Contaminant Prevention - Airborne Releases			1			
Materials and Resources (MR)	**14**	**13**	**16**	**14**	**14**	**13**
Storage and Collection of Recyclables	Prereq	Prereq	Prereq	Prereq	Prereq	Prereq
PBT Source Reduction - Mercury			Prereq			
Building Reuse	2					
Building Reuse - Maintain Existing Walls, Floors, and Roof		5	3	3	3	2
Building Reuse - Maintain 50% of Interior Nonstructural Elements			1	1	1	1
Construction Waste Management	2	2	2	2	2	2
Sustainably Sourced Materials and Products			4			
Materials Reuse	2	1		2	2	2
Materials - Furniture and Furnishings	1					
Recycled Content	2	2		2	2	2
Regional Materials	2	2		2	2	2
Rapidly Renewable Materials	1			1	1	1
Certified Wood	1	1		1	1	1
PBT Source Reduction - Mercury in Lamps			1			
PBT Source Reduction - Lead, Cadmium, and Copper			2			
Furniture and Medical Furnishings			2			
Resource Use - Design for Flexibility			1			
Tenant Space - Long-term Commitment	1					
Indoor Environmental Quality (EQ)	**17**	**12**	**18**	**15**	**15**	**19**
Minimum Indoor Air Quality Performance	Prereq	Prereq	Prereq	Prereq	Prereq	Prereq
Environmental Tobacco Smoke (ETS) Control	Prereq	Prereq	Prereq	Prereq	Prereq	Prereq

(*Continued*)

Quick Reference for LEED v3.0	Commercial Interiors	Core & Shell	Healthcare	New Construction	Retail: New Construction	Schools
Total Possible Points	**110**	**110**	**110**	**110**	**110**	**110**
Hazardous Material Removal or Encapsulation			Prereq			
Minimum Acoustical Performance						Prereq
Outdoor Air Delivery Monitoring	1	1	1	1	1	1
Increased Ventilation	1	1		1	1	1
Construction IAQ Management Plan - During Construction	1	1	1	1	1	1
Construction IAQ Management Plan - Before Occupancy	1		1	1	1	1
Low-Emitting Materials			4		5	4
Low-Emitting Materials - Adhesives and Sealants	1	1		1		
Low-Emitting Materials - Paints and Coatings	1	1		1		
Low-Emitting Materials - Flooring Systems	1	1		1		
Low-Emitting Materials - Composite Wood and Agrifiber Products	1	1		1		
Low-Emitting Materials - Systems Furniture and Seating	1					
Indoor Chemical and Pollutant Source Control	1	1	1	1	1	1
Controllability of Systems - Lighting and Thermal Comfort		1			1	
Controllability of Systems - Lighting	1		1	1		1
Controllability of Systems - Thermal Comfort	1		1	1		1
Thermal Comfort - Design and Verification			1			
Thermal Comfort - Design	1	1		1	1	1
Thermal Comfort - Verification	1			1	1	1
Daylight and Views - Daylight	2	1	2	1	1	3
Daylight and Views - Views	1	1	3	1	1	1
Enhanced Acoustical Performance						1
Mold Prevention						1
Acoustic Equipment			2			
Innovation and Design Process (ID)	**6**	**6**	**6**	**6**	**6**	**6**
Integrated Project Planning and Design			Prereq			
Innovation in Design	5	5	4	5	5	4
LEED Accredited Professional	1	1	1	1	1	1
Integrated Project Planning and Design			1			
The School as a Teaching Tool						1
Regional Priority (RP)	**4**	**4**	**4**	**4**	**4**	**4**
Regional Priority	4	4	4	4	4	4

Appendix B

The Sustainable Sites Initiative™ (SITES™) Guidelines and Performance Benchmarks 2009

1. **Site Selection (21 points)** Select locations to preserve existing resources and repair damaged systems
Prerequisite 1.1: Limit development of soils designated as prime farmland, unique farmland, and farmland of statewide importance
Prerequisite 1.2: Protect floodplain functions
Prerequisite 1.3: Preserve wetlands
Prerequisite 1.4: Preserve threatened or endangered species and their habitats
Credit 1.5: Select brownfields or greyfields for redevelopment (5–10 points)
Credit 1.6: Select sites within existing communities (6 points)
Credit 1.7: Select sites that encourage non-motorized transportation and use of public transit (5 points)

2. **Pre-Design Assessment and Planning (4 points)** Plan for sustainability from the onset of the project
Prerequisite 2.1: Conduct a pre-design site assessment and explore opportunities for site sustainability
Prerequisite 2.2: Use an integrated site development process
Credit 2.3: Engage users and other stakeholders in site design (4 points)

3. **Site Design—Water (44 points)** Protect/restore processes and systems associated with site hydrology
Prerequisite 3.1: Reduce potable water use for landscape irrigation by 50 percent from established baseline
Credit 3.2: Reduce potable water use for landscape irrigation by 75 percent or more from established baseline (2–5 points)
Credit 3.3: Protect and restore riparian, wetland, and shoreline buffers (3–8 points)
Credit 3.4: Rehabilitate lost streams, wetlands, and shorelines (2–5 points)
Credit 3.5: Manage stormwater on site (5–10 points)
Credit 3.6: Protect and enhance on-site water resources and receiving water quality (3–9 points)
Credit 3.7: Design rainwater/stormwater features to provide a landscape amenity (1–3 points)
Credit 3.8: Maintain water features to conserve water and other resources (1–4 points)

4. **Site Design—Soil and Vegetation (51 points)** Protect and restore processes and systems associated with a site's soil and vegetation
Prerequisite 4.1: Control and manage known invasive plants found on site

Prerequisite 4.2: Use appropriate, non-invasive plants

Prerequisite 4.3: Create a soil management plan

Credit 4.4: Minimize soil disturbance in design and construction (6 points)

Credit 4.5: Preserve all vegetation designated as special status (5 points)

Credit 4.6: Preserve or restore appropriate plant biomass on site (3–8 points)

Credit 4.7: Use native plants (1–4 points)

Credit 4.8: Preserve plant communities native to the ecoregion (2–6 points)

Credit 4.9: Restore plant communities native to the ecoregion (1–5 points)

Credit 4.10: Use vegetation to minimize building heating requirements (2–4 points)

Credit 4.11: Use vegetation to minimize building cooling requirements (2–5 points)

Credit 4.12: Reduce urban heat island effects (3–5 points)

Credit 4.13: Reduce the risk of catastrophic wildfire (3 points)

5. **Site Design—Materials Selection (36 points)** Reuse/recycle existing materials and support sustainable production practices

Prerequisite 5.1: Eliminate the use of wood from threatened tree species

Credit 5.2: Maintain on-site structures, hardscape, and landscape amenities (1–4 points)

Credit 5.3: Design for deconstruction and disassembly (1–3 points)

Credit 5.4: Reuse salvaged materials and plants (2–4 points)

Credit 5.5: Use recycled content materials (2–4 points)

Credit 5.6: Use certified wood (1–4 points)

Credit 5.7: Use regional materials (2–6 points)

Credit 5.8: Use adhesives, sealants, paints, and coatings with reduced VOC emissions (2 points)

Credit 5.9: Support sustainable practices in plant production (3 points)

Credit 5.10: Support sustainable practices in materials manufacturing (3–6 points)

6. **Site Design—Human Health and Well-Being (32 points)** Build strong communities and a sense of stewardship

Credit 6.1: Promote equitable site development (1–3 points)

Credit 6.2: Promote equitable site use (1–4 points)

Credit 6.3: Promote sustainability awareness and education (2–4 points)

Credit 6.4: Protect and maintain unique cultural and historical places (2–4 points)

Credit 6.5: Provide for optimum site accessibility, safety, and wayfinding (3 points)

Credit 6.6: Provide opportunities for outdoor physical activity (4–5 points)

Credit 6.7: Provide views of vegetation and quiet outdoor spaces for mental restoration (3–4 points)

Credit 6.8: Provide outdoor spaces for social interaction (3 points)

Credit 6.9: Reduce light pollution (2 points)

7. **Construction (21 points)** Minimize effects of construction-related activities

Prerequisite 7.1: Control and retain construction pollutants

Prerequisite 7.2: Restore soils disturbed during construction

Credit 7.3: Restore soils disturbed by previous development (2–8 points)

Credit 7.4: Divert construction and demolition materials from disposal (3–5 points)

Credit 7.5: Reuse or recycle vegetation, rocks, and soil generated during construction (3–5 points)

Credit 7.6: Minimize generation of greenhouse gas emissions and exposure to localized air pollutants during construction (1–3 points)

8. **Operations and Maintenance (23 points)** Maintain the site for long-term sustainability

Prerequisite 8.1: Plan for sustainable site maintenance

Prerequisite 8.2: Provide for storage and collection of recyclables

Credit 8.3: Recycle organic matter generated during site operations and maintenance (2–6 points)

Credit 8.4: Reduce outdoor energy consumption for all landscape and exterior operations (1–4 points)

Credit 8.5: Use renewable sources for landscape electricity needs (2–3 points)

Credit 8.6: Minimize exposure to environmental tobacco smoke (1–2 points)

Credit 8.7: Minimize generation of greenhouse gases and exposure to localized air pollutants during landscape maintenance activities (1–4 points)

Credit 8.8: Reduce emissions and promote the use of fuel-efficient vehicles (4 points)

9. **Monitoring and Innovation (18 points)** Reward exceptional performance and improve the body of knowledge on long-term sustainability

Credit 9.1: Monitor performance of sustainable design practices (10 points)

Credit 9.2: Innovation in site design (8 points)

Appendix C
Unit Conversions

Multiply	By	To Obtain	Multiply	By	To Obtain
Length					
Inch (in)	0.025	Meter (m)	Meter (m)	39.370	Inch (in)
Meter (m)	100.0	Centimeter (cm)	Centimeter (cm)	0.010	Meter (m)
Kilometer (km)	1000.0	Meter (m)	Meter (m)	0.001	Kilometer (km)
Mile (mi)	1.609	Kilometer (km)	Kilometer (km)	0.622	Mile (mi)
Inches (in)	2.540	Centimeter (cm)	Centimeter (cm)	0.394	Inches (in)
Yard (yd)	0.914	Meter (m)	Meter (m)	1.094	Yard (yd)
Feet (ft)	0.305	Meter (m)	Meter (m)	3.281	Feet (ft)
Centimeter (cm)	0.394	Inches (in)	Inches (in)	2.540	Centimeter (cm)
Feet (ft)	30.480	Centimeter (cm)	Centimeter (cm)	0.033	Feet (ft)
Meter (m)	3.281	Feet (ft)	Feet (ft)	0.305	Meter (m)
Area					
Hectares (ha)	10000.0	Sq. Meter (m^2)	Sq. Meter (m^2)	0.0001	Hectares (ha)
Acre (ac)	0.405	Hectares (ha)	Hectares (ha)	2.471	Acre (ac)
Acre (ac)	43560.0	Sq. Feet (ft^2)	Feet (ft^2)	0.000023	Acre (ac)
Sq. Yard (yd^2)	0.836	Sq. Meter (m^2)	Sq. Meter (m^2)	1.196	Sq. Yard (yd^2)
Sq. Mile (mi^2)	2.590	Sq. Kilometer (km^2)	Sq. Kilometer (km^2)	0.386	Sq. Mile (mi^2)
Sq. Yard (yd^2)	0.836	Sq. Meter (m^2)	Sq. Meter (m^2)	1.196	Sq. Yard (yd^2)
Sq. Feet (ft^2)	0.093	Sq. Meter (m^2)	Sq. Meter (m^2)	10.764	Sq. Feet (ft^2)
Sq. Inch (in^2)	645.150	Sq. Millimeter (mm^2)	Sq. Millimeter (mm^2)	0.0016	Sq. Inch (in^2)
Sq. Feet (ft^2)	929.000	Sq. Centimeter (cm^2)	Sq. Centimeter (cm^2)	0.0011	Sq. Feet (ft^2)
Sq. Inch (in^2)	6.452	Sq. Centimeter (cm^2)	Sq. Centimeter (cm^2)	0.155	Sq. Inch (in^2)
Volume					
Acre-feet (ac-ft)	1233.5	Cu. Meter (m^3)	Cu. Meter (m^3)	0.0008	Acre-feet (ac-ft)
Cu. Yard (yd^3)	0.765	Cu. Meter (m^3)	Cu. Meter (m^3)	1.308	Cu. Yard (yd^3)
Cu. Feet (ft^3)	0.028	Cu. Meter (m^3)	Cu. Meter (m^3)	35.315	Cu. Feet (ft^3)
Cu. Feet (ft^3)	28.317	Liter (L)	Liter (L)	0.035	Cu. Feet (ft^3)
Gallon (gal)	3.785	Liter (L)	Liter (L)	0.264	Gallon (gal)
Cu. Inch (in^3)	16.387	Cu. Millimeters (mm^3)	Cu. Millimeters (mm^3)	0.061	Cu. Inch (in^3)
Cu. Feet (ft^3)	7.481	Gallon (gal)	Gallon (gal)	0.134	Cu. Feet (ft^3)
Temperature					
Celsius (°C)	$(+17.78) \times 1.8$	Fahrenheit (°F)	Fahrenheit (°F)	$(-32) \times 0.556$	Celsius (°C)
Mass/Weight					
Kilogram (kg)	1000.0	Grams (g)	Grams (g)	0.001	Kilogram (kg)
Kilogram (kg)	2.205	Pounds (lbs)	Pounds (lbs)	0.454	Kilogram (kg)
Short Ton (US)	2000.0	Pounds (lbs)	Pounds (lbs)	0.001	Short Ton (US)
Metric Ton (mt)	2240.6	Pounds (lbs)	Pounds (lbs)	0.00045	Metric Ton (mt)

(Continued)

Multiply	By	To Obtain	Multiply	By	To Obtain
Short Ton (US)	0.907	Metric Ton (mt)	Metric Ton (mt)	1.102	Short Ton (US)
Short Ton (US)	0.893	Long Ton (UK)	Long Ton (UK)	1.120	Short Ton (US)
Metric Ton (mt)	0.984	Long Ton (UK)	Long Ton (UK)	1.016	Metric Ton (mt)
Gram (g)	0.0353	Ounce (oz)	Ounce (oz)	28.350	Gram (g)
Pressure/Force					
lbs/in^2 (psi)	6.895	Kilo Pascal (kpa)	Kilo Pascal (kpa)	0.145	lbs/in^2 (psi)
Pound-Force (lbf)	4.448	Newton (N)	Newton (N)	0.225	Pound-Force (lbf)
kg/cm^2	14.220	lbs/in^2 (psi)	lbs/in^2 (psi)	0.070	kg/cm^2
kg/m^2	0.205	lbs/ft^2 (psf)	lbs/ft^2 (psf)	4.883	kg/m^2
Energy					
Megawatt Hour (MWh)	1000.0	Kilowatt Hour (kWh)	Kilowatt Hour (kWh)	0.0010	Megawatt Hour (MWh)
Kilowatt Hour (kWh)	3415.0	British Thermal Unit (BTU)	British Thermal Unit (BTU)	0.00029	Kilowatt Hour (kWh)
Watt Hour (Wh)	3.415	British Thermal Unit (BTU)	British Thermal Unit (BTU)	0.293	Watt Hour (Wh)
Ton refrigeration	12000.0	British Thermal Unit (BTU)	British Thermal Unit (BTU)	0.000083	Ton refrigeration
Power					
Watt (W)	3.412	BTU/hour	BTU/hour	0.293	Watt (W)
Kilowatt (kW)	1000.0	Watt (W)	Watt (W)	0.001	Kilowatt (kW)
Ton of Refrigeration	12000.0	BTU/hour	BTU/hour	0.000083	Ton of Refrigeration
Horsepower (hp)	33000.0	Foot Pound-Force/Min.	Foot Pound-Force/Min.	0.00003	Horsepower (hp)
Horsepower (hp)	0.746	Kilowatt (kW)	Kilowatt (kW)	1.341	Horsepower (hp)
Speed/Flow Rate					
Cu. Feet/Minute (cfm)	0.472	Liters/Second (L/s)	Liters/Second (L/s)	2.119	Cu. Feet/Minute (cfm)
Feet/Minute (fpm)	0.508	Centimeters/Sec. (cm/s)	Centimeters/Sec. (cm/s)	1.969	Feet/Minute (fpm)
Feet/Minute (fpm)	0.305	Meters/Minute (mpm)	Meters/Minute (mpm)	3.281	Feet/Minute (fpm)
Gallon/Minute (gpm)	0.063	Liters/Second (L/s)	Liters/Second (L/s)	15.853	Gallon/Minute (gpm)
Feet/Second (ft/s)	0.305	Meters/Second (m/s)	Meters/Second (m/s)	3.281	Feet/Second (ft/s)
Miles/Hour (mph)	1.610	Kilometer/Hour (km/h)	Kilometer/Hour (km/h)	0.621	Miles/Hour (mph)

Abbreviations and Acronyms

4-PCH: 4-Phenylcyclohexene

AABC: Associated Air Balance Council

ACI: American Concrete Institute

ACT: Acoustical ceiling tile

ADA: Americans with Disabilities Act

AGC: Associated General Contractors of America

AGMBC: Application Guide for Multiple Buildings and On-Campus Building Projects

AIA: American Institute of Architects

ANSI: American National Standards Institute

ARI: Air-Conditioning and Refrigeration Institute

ASG: Aluminosilicate glass

ASHRAE: American Society of Heating, Refrigerating and Air-Conditioning Engineers

ASLA: American Society of Landscape Architects

ASTM: American Society for Testing and Materials

ATFS: American Tree Farm System

AWEA: American Wind Energy Association

BAS: Building automation system

BAU: Business as usual

BCA: Building Commissioning Association

BCVTB: Building controls virtual test bed

BD&C: Building design and construction

BEA: Building energy analysis

BEE: Building environmental efficiency

BEES: Building for environmental and economic sustainability

BIM: Building information modeling

BIPV: Building-integrated photovoltaic

BOD: Basis of Design

BOMA: Building Owners and Managers Association

BREEAM: Building Research Establishment Environmental Assessment Method (the United Kingdom's building assessment system)

BRI: Building-related illness

BRIC: Brazil, Russia, India, and China

BRMA: Building Materials Reuse Association

C&D: Construction and demolition

CARE: Carpet America Recovery Effort

CAS: Chemical Abstracts Service

CASBEE: Comprehensive Assessment System for Building Environmental Efficiency (the Japanese building assessment system)

CBE: Center for the Built Environment

CBECS: Commercial Buildings Energy Consumption Survey. Developed by the US Department of Energy's Energy Information Administration (EIA)

CCAEJ: Center for Community Action and Environmental Justice

CCS: Carbon capture and storage

CDC: US Centers for Disease Control and Prevention

CDD: Cooling degree day

CFC: Chlorofluorocarbon

CFD: Computational fluid dynamics

CIB: Conseil International du Bâtiment

CIR: Credit interpretation ruling

CLO: Clothing Level

CMP: Credential Maintenance Program (for USGBC LEED Green Associates and LEED-APs)

CO₂e: Carbon dioxide equivalent

CO: Carbon monoxide

COMNET: Commercial Energy Services Network

COP: Coefficient of performance

COTE: Committee on the Environment, AIA

CRI: Color Rendering Index

CRS: Center for Resource Solutions

CRT: Cathode ray tube

CSIR: Council for Scientific and Industrial Research

CWPA: Certified Wood and Paper Association

Cx: Commissioning

CxA: Commissioning authority

DCV: Demand-controlled ventilation

DES: Diethylstilbestrol

DfD: Design for disassembly

DfDD: Design for deconstruction and disassembly

DfE: Design for the environment

DGNB: Deutsche Gesellschaft für Nachhaltiges Bauen (the German building assessment system)

DOE: US Department of Energy

EA: Energy and Atmosphere, a LEED category

EBN: *Environmental Building News*

ECM: Energy conservation measures

EDC: Endocrine-disrupting chemicals

EDP: Environmental product declaration

EEA: European Environment Agency

EEM: Energy efficiency measure

EERE: US Office of Energy Efficiency and Renewable Energy

EIA: US Energy Information Administration

EIE: Environmental impact estimator

ELV: End-of-life vehicle

EMS: Environmental management system

EPA: US Environmental Protection Agency

EPD: Environmental product declaration

EPP: Environmentally preferable product

EQ: Environmental Quality, a LEED category

ERV: Energy recovery ventilator

ESA: Endangered Species Act

ESC: Erosion and sedimentation control

ESCO: Energy services company

EU: European Union

EUI: Energy Use Index

fc: Foot-candle

FEMA: Federal Emergency Management Agency

FEMP: Federal Energy Management Program

FSC: Forest Stewardship Council

FTE: Full-time equivalent

GBCA: Green Building Council of Australia

GBCI: Green Building Certification Institute

GBI: Green Building Initiative

gbXML: Green building XML

GC/CM: General contractor/construction manager

GDDC: Green design and delivery coordination

GDP: Gross domestic product

GGA: Green Globes Assessor

GGGC: Governor's Green Government Council

GGP: Green Globes Professional

GHG: Greenhouse gas

GMO: Genetically modified organism

GNP: Gross national product

GSHP: Ground source heat pump

GWP: Global warming potential

HCFC: Hydrochlorofluorocarbon

HDD: Heating degree day

HDPE: High-density polyethylene

HET: High-efficiency toilet

HEU: High-efficiency urinal

HFC: Hydrofluorocarbon

HVAC&R: Heating, ventilating, air conditioning, and refrigerating

IAI: Alliance for Interoperability

IAMAP: International Arctic Monitoring and Assessment Program

IAQ: Indoor air quality

IARC: International Agency for Research on Cancer

ICC: International Code Council

ICE: Inventory of Carbon and Energy

ID: Innovation and Design, a LEED category

IEEE: Institute of Electrical and Electronics Engineers

IEQ: Indoor environmental quality

IESNA: Illuminating Engineering Society of North America

IEWC: International Electric Wire and Cable

IFC: Industry Foundation Class

IgCC: International Green Construction Code

iiSBE: International Initiative for a Sustainable Built Environment

IPCC: Intergovernmental Panel on Climate Change

IPD: Integrated project delivery

IPMVP: International Performance Measurement and Verification Protocol

IPR: Institut für Physikalische Raumenstörung

ISO: International Organization for Standardization

LBNL: Lawrence Berkeley National Laboratory

LCA: Life-cycle assessment

LCC: Life-cycle costing

LCD: Liquid crystal display

LCGWP: Life-cycle global warming potential

LCODP: Life-cycle ozone depletion potential

LED: Light-emitting diode

LEED: Leadership in Energy and Environmental Design (the USGBC building assessment system)

LEED AP: LEED Accredited Professional

LEED-CI: LEED for Commercial Interiors

LEED-CS: LEED for Core and Shell

LEED-EB:O&M: LEED for Existing Building: Operations and Maintenance

LEED GA: LEED Green Associate

LEED-HC: LEED for Healthcare

LEED-ID&C: LEED for Interior Design and Construction

LEED-NC: LEED for New Construction

LEED-ND: LEED for Neighborhood Development

LEED-SCH: LEED for Schools

LID: Low-impact development

LPG: Liquid petroleum gas

LSG: Light-to-solar-gain ratio

M&V: Measurement and verification

MAK: Maximum Workplace Concentration

MCS: Multiple chemical sensitivity

MEP: Mechanical, electrical, plumbing

MERV: Minimum Efficiency Reporting Value

MET: Metabolic Rate

MIPS: Materials intensity per unit service

MOU: Memorandum of Understanding

MPC: Model predictive control

MPR: Minimum Program Requirement

MR: Materials and Resources

MSDS: Materials safety data sheet

MW: Megawatt

NAAQS: National Ambient Air Quality Standard

NAVFAC: Naval Facilities Engineering Command

NC: Noise Criterion

NCI: National Charrette Institute

NCR: Noise Reduction Coefficient

NEMA: National Electrical Manufacturers Association

NEPA: National Environmental Policy Act

NFIP: National Flood Insurance Program

NIBS: National Institute of Building Sciences

NIOSH: National Institute for Occupational Safety and Health

NIST: National Institute of Standards and Technology

NOAA: National Oceanic and Atmospheric Administration

NO$_x$: nitrogen oxide, produced by the burning of fossil fuels

NPDES: National Pollutant Discharge Elimination System

NPV: Net present value

NREL: National Renewable Energy Laboratory

NZE: Net-zero energy

ODP: Ozone depletion potential

OPR: Owner's Project Requirements

PAFC: Phosphoric acid fuel cell

PCB: Polychlorinated biphenyl

PCR: Carpet Reclamation Program

PET: Polyethylene terephthalate

PF: Phenol formaldehyde

PHA: Polyhydroxyalkanoate

PLA: Polylactic acid

PNNA: Pacific Northwest National Laboratory

PPA: Power purchase agreement

PPM: Parts per million

PV: Photovoltaic

PVC: Polyvinyl chloride

REL: Reference Exposure Level

RFP: Request for proposal

RFQ: Request for qualifications

RIBA: Royal Institute of British Architects

RMI: Rocky Mountain Institute

RP: Regional Priority, a LEED category

RSF: Research support facility

SBS: Sick building syndrome

SCAQMD: South Coast Air Quality Management District

SFHA: Special flood hazard area

SFI: Sustainable Forestry Initiative

SHGC: Solar heat gain coefficient

SITES: Sustainable Sites Initiative

SMACNA: Sheet Metal and Air Conditioning Contractors' National Association

SPI: Society of Plastics Industry, Inc.

SS: Sustainable Sites, a LEED category

STC: Sound Transmission Class

SWPPP: Stormwater pollution prevention plan

TAB: Testing and balancing

TNPV: Total net present value

UF: Urea-formaldehyde

UIA: International Union of Architects

UL: Underwriters Laboratories

UN: United Nations
URRC: United Resource Recovery Corporation
USDA: US Department of Agriculture
USGBC: US Green Building Council
VAV: Variable air volume
VFD: Variable-frequency drive
VOC: Volatile organic compound
VSD: Variable-speed drive
VT: Visible transmittance
WBCSD: World Business Council on Sustainable Development
WBDG: Whole Building Design Guide
WE: Water Efficiency
WMO: World Meteorological Organization
WWTP: Wastewater treatment plant

Glossary

Agrifiber building products are manufactured from agricultural fiber. Examples include particleboard, medium-density fiberboard (MDF), plywood, oriented strand board (OSB), wheatboard, and strawboard.

Air economizer is a system found in HVAC air-handling systems that takes advantage of favorable weather conditions to reduce mechanical cooling by introducing cooler outdoor air into the building.

Albedo, or solar reflectance, is a measure of the ability of a surface material to reflect sunlight on a scale of 0 to 1. Solar reflectance is also called *albedo*. Black paint has a solar reflectance of 0; white paint (titanium dioxide) has a solar reflectance of 1.

Baseline building energy performance is the annual energy cost for a building design intended for use as a baseline for rating above standard design, as defined in ANSI/ASHRAE/IENSA 90.1-2007, Appendix G.

Biobased product is a commercial or industrial product using at least 50 percent (by weight) biologically generated substances, including but not limited to cellulosic materials (e.g., wood, straw, natural fibers) and products derived from crops (e.g., soy-based, corn-based).

Biodiversity is the variety of life in all forms, levels, and combinations, including ecosystem diversity, species diversity, and genetic diversity.

Biomass is plant material from trees, grasses, or crops that can be converted to heat energy to produce electricity.

Biomimicry, sometimes called *biomimetic design*, is an emerging design discipline that looks to nature for sustainable design solutions.

Blackwater is wastewater from toilets and urinals. Wastewater from kitchen sinks (perhaps differentiated by the use of a garbage disposal), showers, or bathtubs is also considered blackwater under some state or local codes.

Building Research Establishment Environmental Assessment Method (BREEAM) is the primary building assessment system in the United Kingdom.

Brownfield is a property whose use may be complicated by the presence or possible presence of a hazardous substance, pollutant, or contaminant.

Building automation system (BAS) is a commonly accepted name for the sensors, controls, and computers that control building energy systems such as lighting, heating, ventilating, and air-conditioning systems with the objective of minimizing energy consumption.

Building information modeling (BIM) is the process of generating and managing building data during its life cycle. It also refers to software that generates a three-dimensional building representation and with the ability to accommodate plug-ins that can potentially perform energy modeling, daylight studies, and life-cycle assessment (LCA) of building systems.

Carbon accounting is the process of measuring the amount of carbon dioxide an entity is emitting and the equivalents that are being released into the atmosphere.

Carbon dioxide (CO_2) levels are an indicator of ventilation effectiveness inside buildings. CO_2 concentrations greater than 530 ppm above outdoor CO_2 conditions generally indicate inadequate ventilation.

Carbon dioxide equivalent (CO$_2$e) is a measure used to compare the impact of various greenhouse gases based on their global warming potential (GWP). CO$_2$e approximates the time-integrated warming effect of a unit of a given greenhouse gas, relative to that of carbon dioxide (CO$_2$). GWP is an index for estimating the relative global warming contribution of atmospheric emissions of a unit mass of a particular greenhouse gas compared to emission of a unit mass of CO$_2$. The following GWP values are used based on a 100-year time horizon: 1 for CO$_2$, 23 for methane (CH$_4$), and 294 for nitrous oxide (N$_2$O). *See also* Global warming potential.

Carbon footprinting is the total set of greenhouse gas (GHG) emissions caused by an organization, event, product, or person.

Comprehensive Assessment System for Building Environmental Efficiency (CASBEE) is the primary building assessment system used in Japan.

Chain of custody (COC) is a tracking procedure for a product from the point of harvest extraction to its end use, including all successive stages of processing, transformation, manufacturing, and distribution.

Charrette is a collaborative session in which a project team creates a solution to a design or project problem. The structure may vary, depending on the complexity of the problem or desired outcome and the individuals working in the group. Charrettes can take place over multiple sessions in which the group divides into subgroups. Each subgroup then presents its work to the full group as material for future dialogue. Charrettes can serve as a way of quickly generating solutions while integrating the aptitudes and interests of a diverse group of people.

Chlorofluorocarbons (CFCs) are hydrocarbons formerly used as refrigerants that cause depletion of the stratospheric ozone layer. CFCs were banned from use by international agreements such as the Montreal Protocol of 1987.

Climate zone, US, is any of the eight principal zones, roughly demarcated by lines of latitude, into which the United States is divided on the basis of climate for the purpose of energy calculations and selecting prescriptive energy conservation measures.

Combined heat and power (CHP), or cogeneration, generates both electrical power and thermal energy from a single fuel source.

Comfort criteria are specific design conditions that take into account indoor temperature, humidity, and air speed; and outdoor temperature, outdoor humidity, seasonal clothing, and expected activity.

Commissioning (Cx) is the process of verifying and documenting that a building and all of its systems and assemblies are planned, designed, installed, tested, operated, and maintained to meet the Owner's Project Requirements.

Commissioning authority (CxA) is the individual designated to organize, lead, and review the completion of commissioning process activities. The CxA facilitates communication among the owner, designer, and contractor to ensure that complex systems are installed and function in accordance with the owner's project requirements.

Composite wood consists of wood or plant particles or fibers bonded by a synthetic resin or binder. Examples include particleboard, medium-density fiberboard (MDF), plywood, oriented strand board (OSB), wheatboard, and strawboard.

Composting toilets (sometimes called *biological toilets*, *dry toilets*, and *waterless toilets*) contain and control the composting of excrement, toilet paper, carbon additive, and, optionally, food wastes.

Comprehensive Environmental Response, Compensation, and Liability Act, or CERCLA, is more commonly known as *Superfund.* Enacted in 1980, CERCLA addresses abandoned or historical waste sites and contamination by taxing the chemical and petroleum industries and providing federal authority to respond to releases of hazardous substances.

Constructed wetland is an engineered system designed to simulate natural wetland functions for water purification.

Construction and demolition debris includes waste and recyclables generated from construction and from the renovation, demolition, or deconstruction of preexisting structures. It does not include land-clearing debris such as soil, vegetation, and rocks.

Construction indoor air quality (IAQ) management plan outlines measures to minimize indoor air contamination in a building during construction and describes procedures to flush the building to remove contaminants prior to occupancy.

Cradle to cradle is a framework for designing manufacturing processes powered by renewable energy, in which materials flow in safe, regenerative, closed-loop cycles.

Daylighting is the controlled entry of natural light into a space, used to reduce or eliminate electric lighting.

Daylight-responsive lighting controls are photosensors used in conjunction with other switching and dimming devices to control the amount of artificial lighting relative to the amount and quality of natural daylight.

Demand-controlled ventilation is automatic ventilation control based on measured carbon dioxide levels.

Deutsche Gesellschaft für Nachhaltiges Bauen (DGNB) or German Association for Sustainable Building is both the primary building assessment system used in Germany and the name of the organization that is its proponent.

District cooling distributes chilled water to multiple buildings primarily for air conditioning. The chilled water is usually provided by a dedicated cooling plant.

District heating is the distribution of heat from one or more sources to multiple buildings.

Drip irrigation delivers landscape irrigation water at low pressure through buried mains and submains. From the submains, water is distributed to the soil through a network of perforated tubes or emitters. Drip irrigation is a high-efficiency type of microirrigation.

Ecological design is an approach to design that transforms matter and energy using processes that are compatible and synergistic with nature and that are modeled on natural systems.

Ecological sustainability is a school of sustainability that focuses on the capacity of ecosystems to maintain their essential functions and processes and retain their biodiversity in full measure over the long term.

Ecology is the study of the living conditions of organisms in interaction with each other and with the surroundings, organic as well as inorganic.

Economizer *See* **Air economizer.**

Ecosystem is a basic unit of nature that includes a community of organisms and their nonliving environment linked by biological, chemical, and physical processes.

Embodied energy is the energy associated with the entire life cycle of a product, including its manufacture, transportation, and disposal, as well as the inherent energy captured within the product itself.

Emissivity is the ratio of the radiation emitted by a surface to the radiation emitted by a black body at the same temperature.

Energy conservation measures are installations of, or modifications to, equipment or systems intended to reduce energy use and costs.

Energy simulation model, or energy model, is a computer-generated representation of the anticipated energy consumption of a building. It permits a comparison of energy performance, given proposed energy efficiency measures, with the baseline.

Energy Star Rating is a measure of a building's energy performance compared with that of similar buildings, as determined by the Energy Star Portfolio Manager. It has a scale of 1 to 100, with a score of 50 representing average building performance while a score of 75 or better represents very good performance.

Eutrophication is the increase in chemical nutrients, such as the nitrogen and phosphorus often found in fertilizers, in an ecosystem. The added nutrients stimulate excessive plant growth, promoting algal blooms or weeds. The enhanced plant growth reduces oxygen in the land and water, reducing water quality and fish to other animal populations.

Evapotranspiration (ET) rate is the amount of water lost from a vegetated surface in units of water depth. It is expressed in millimeters per unit of time.

Exhaust air is air that is removed from a space and discharged outside the building by mechanical or natural ventilation systems.

Fly ash is the solid residue derived from incineration processes. Fly ash can be used as a substitute for portland cement in concrete.

Foot-candle (fc) is the quantity of light falling on a 1-square-foot area from a 1-candela light source at a distance of 1 foot (or 1 lumen per square foot). Foot-candles can be measured both horizontally and vertically by a foot-candle meter or light meter. 1 fc = 10.764 lux. *See also* **Lux; Lumen.**

Formaldehyde is a naturally occurring VOC found in small amounts in animals and plants but is carcinogenic and an irritant to most people when present in high concentrations, causing headaches, dizziness, mental impairment, and other symptoms. When present in the air at levels above 0.1 ppm, it can cause watery eyes; burning sensations in the eyes, nose, and throat; nausea; coughing; chest tightness; wheezing; skin rashes; and asthmatic and allergic reactions.

Fuel-efficient vehicles have achieved a minimum green score of 40 according to the annual vehicle-rating guide of the American Council for an Energy-Efficient Economy.

Full-time equivalent (FTE) represents a regular building occupant who spends 40 hours per week in the project building. Part-time or overtime occupants have FTE values based on their hours per week divided by 40.

Fully shielded exterior light fixture, the lower edge of the shield is at or below the lowest edge of the lamp, such that all the light shines down.

Geothermal energy is hot water or steam from within the earth that is used to generate electricity.

Geothermal ground source heat pump (GSHP), or ground heat pump, is a central heating and/or cooling system that pumps heat to or from the ground. It uses the earth as a heat source (in the winter) or a heat sink (in the summer).

Glare is any excessively bright source of light within the visual field that creates discomfort or loss of visibility.

Global warming potential (GWP) is an index that describes the radiative characteristics of well-mixed greenhouse gases and that represents the combined effect of the differing times these gases remain in the atmosphere and their relative effectiveness in absorbing outgoing infrared radiation. This index approximates the time-integrated warming effect of a unit mass of a given greenhouse gas in today's atmosphere, relative to that of carbon dioxide, which has a GWP of 1.

Graywater   is untreated household wastewater that has not come into contact with toilet waste and has low organic content. Graywater includes used water from bathtubs, showers, bathroom wash basins, and water from clothes washers and laundry tubs. It must not include wastewater from kitchen sinks or dishwashers.

Green Associate   is a credential offered by the USGBC that designates a person as being knowledgeable about the fundamentals of green building. Passing the Green Associate examination is a prerequisite for an individual to become a LEED Accredited Professional (LEED AP).

Green building   is a facility designed using a holistic and collaborative process that addresses life-cycle resource consumption, environmental impacts, and the health of the occupants and local ecosystems.

Green Building Certification Institute (GBCI)   is a nonprofit, third-party organization that reviews the application for buildings applying for USGBC LEED certification and tests applicants for Green Associate or LEED AP credentials.

Green Building Initiative (GBI)   is a nonprofit organization whose mission is to accelerate the adoption of building practices that result in energy-efficient, healthier, and environmentally sustainable buildings by promoting credible and practical green building approaches for residential and commercial construction.

Green Globes   is a green building guidance and assessment program that offers an effective, practical, and affordable way to advance the overall environmental performance and sustainability of commercial buildings.

Green roof, or vegetated roof,   is a roof system that may include a water-proofing and root-repellant system, a drainage system, filter cloth, a light-weight growing medium, and plants. Vegetated roof systems can be modular, with drainage layers, filter cloth, growing media, and plants already prepared in movable, interlocking grids, or each component can be installed separately.

Green-e   is a program established by the Center of Resource Solutions to both promote green electricity products and provide consumers with a rigorous and nationally recognized method to identify those products.

Greenfield   is undeveloped lands such as fields, forests, farmland, and rangeland.

Greenhouse gases (GHGs)   absorb and emit radiation at specific wavelengths within the spectrum of thermal infrared radiation emitted by Earth's surface, clouds, and the atmosphere itself. Increased concentrations of greenhouse gases are a root cause of global climate change.

Halons   are substances used in fire suppression systems and fire extinguishers that deplete the stratospheric ozone layer.

Hard costs   are the costs of the land, materials, labor, and machinery used to construct a building and are sometimes referred to as *direct construction costs*.

Heat island effect   refers to absorption of heat by hardscapes, such as dark, nonreflective pavement and buildings, and its radiation to surrounding areas. Particularly in urban areas, other sources may include vehicle exhaust, air conditioners, and street equipment; reduced airflow from tall buildings and narrow streets exacerbates the effects.

High-performance green building   is the terminology used to more specifically define the intended outcome of a green building design and construction process.

HVAC systems   are equipment, distribution systems, and terminals that provide the processes heating, ventilating, or air conditioning.

Hydrochlorofluorocarbons (HCFCs) are refrigerants that cause significantly less depletion of the stratospheric ozone layer than chlorofluorocarbons.

Hydrofluorocarbons (HFCs) are refrigerants that do not deplete the stratospheric ozone layer but may have high global warming potential. HFCs are not considered environmentally benign.

Impervious surfaces have a perviousness of less that 50 percent and promote runoff of water instead of infiltration into the subsurface. Examples include parking lots, roads, sidewalks, and plazas.

Indoor air quality (IAQ) is the nature of air inside the space that affects the health and well-being of building occupants. It is considered acceptable when there are no known contaminants at harmful concentrations and a substantial majority (80 percent or more) of the occupants do not express dissatisfaction.

Indoor environmental quality (IEQ) is the nature of the overall quality of the environment inside a building resulting from attention to a broad range of effects, which includes air quality, lighting quality, daylighting, acoustics, noise, vibration, odors, thermal comfort, and electromagnetic radiation.

LEED, or Leadership in Energy and Environmental Design, is an internationally recognized green building certification system developed by the USGBC and administered by the Green Building Certification Institute.

LEED Accredited Professional, or LEED AP, is a credential earned by passing an examination administered by the Green Building Certification Institute (GBCI) that designates the holder as having specialized knowledge regarding the LEED building assessment system.

Life-cycle assessment (LCA) is an analysis of the environmental impacts and potential impacts associated with a product, process, or service.

Life-cycle costing (LCC) is an accounting methodology used to evaluate the economic performance of a product or system over its useful life. It considers operating costs, maintenance expenses, and other economic factors.

Light pollution is waste light from buildings and their sites that produces glare, is directed upward to the sky, or is directed off the site, wasting energy and creating navigation problems for some species, such as sea turtles.

Light trespass is obtrusive light that is unwanted because of quantitative, directional, or spectral attributes. Light trespass can cause annoyance, discomfort, distraction, or loss of visibility.

Lumen is a measure of the lighting power perceived by the human eye.

Lux is the SI unit of illuminance and luminous emittance measuring luminous power per area.

Minimum Efficiency Reporting Value (MERV) is a filter rating defined by ASHRAE Standard 52.2-1999. MERV ratings range from 1 (very low efficiency) to 16 (very high).

Mixed-mode ventilation combines mechanical and natural ventilation modes of ventilation system operation.

National Pollutant Discharge Elimination System (NPDES) is a permit program that controls water pollution by regulating point sources that discharge pollutants into the waters of the United States. Industrial, municipal, and other facilities must obtain permits if their discharges go directly to surface waters.

Native (or indigenous) plants are plants that live or grow naturally in a particular region.

Natural, or passive, ventilation is provided by thermal, wind, or diffusion effects openings in the building facade, roof, or other components for the purpose of creating low-energy air movement.

Net metering   is a metering arrangement that allows on-site generators to send excess electricity flows to the regional power grid. These electricity flows offset all or a portion of those drawn from the grid.

Noise Reduction Coefficient (NRC)   is the arithmetic average of sound absorption coefficients at 250, 500, 1000, and 2000 Hz for a material. The NRC is often published by manufacturers in product specifications, particularly for acoustical ceiling tiles and acoustical wall panels.

Off-gassing   is the emission of volatile organic compounds (VOCs) from synthetic and natural products.

Off-site renewable energy   is green power from an electrical utility or other source. There is no physical *renewable energy* system either on-site or specifically connected to the building.

On-site renewable energy   is energy derived from the sun, wind, water, earth's core, and biomass that is captured and used on the building site, using such technologies as wind turbines, photovoltaic solar panels, transpired solar collectors, solar thermal heaters, small-scale hydroelectric power plants, fuel cells, and ground source heat pumps.

Ozone (O_3)   is an oxygen molecule with three oxygen atoms that is both an air pollutant and that also comprises an atmospheric layer. It is not usually emitted directly into the air, but at ground level. It is the product of a chemical reaction between oxides of nitrogen (NO_x) and volatile organic compounds (VOCs) in the presence of sunlight. The ozone layer is an ultraviolet-absorbing layer in the atmosphere that was being destroyed by synthetic chlorine and bromine containing gases. Its protection was the main objective of the Montreal Protocol of 1987.

Ozone depletion potential (ODP)   is a number that refers to the amount of ozone depletion caused by a substance. The ODP is the ratio of the impact on ozone of a chemical compared to the impact of a similar mass of CFC-11. Thus, the ODP of CFC-11 is defined to be 1.0. Other CFCs and HCFCs have ODPs that range from 0.01 to 1.0. The halons have ODPs ranging up to 10. Carbon tetrachloride has an ODP of 1.2, and methyl chloroform's ODP is 0.11. HFCs have zero ODP because they do not contain chlorine.

Perviousness   is the percentage of the surface area of a paving system that is open and allows moisture to soak into the ground below.

Phenol formaldehyde   is used for exterior products, although many of these products and off-gases only at high temperatures.

Photovoltaic (PV) energy   is electricity from photovoltaic cells that convert sunlight into electricity.

Plug load   is an electrical load due to an appliance or other electrical device that is normally "plugged" into an electrical receptacle.

Postconsumer recycled content   is the percentage of material in a product that was consumer waste. The recycled material was generated by household, commercial, industrial, or institutional end users and can no longer be used for its intended purpose. It includes returns of materials from the distribution chain. Examples include construction and demolition debris, materials collected through recycling programs, discarded products (e.g., furniture, cabinetry, decking), and landscaping waste (e.g., leaves, grass clippings, tree trimmings).

Potable water.   *See* **Water, potable**.

Preconsumer recycled content   is the percentage of material in a product that is recycled from manufacturing waste. Examples include planer shavings, sawdust, bagasse, walnut shells, culls, trimmed materials, overissued publications, and obsolete inventories. Excluded are rework, regrind, or scrap materials capable of being reclaimed within the same process that generated them.

Process water is used for industrial processes and building systems such as cooling towers, boilers, and chillers. It can also refer to water used in operational processes, such as dishwashing, clothes washing, and ice making.

R-value indicates the thermal resistance of a material. The R-value of thermal insulation depends on the type of material, its thickness, and its density. The higher the R-value, the greater is the insulating effectiveness. In calculating the R-value of a multilayered installation, the R-values of the individual layers are added.

Rainwater harvesting is utilizing rainwater for potable, nonpotable, industrial, or irrigation applications.

Rapidly renewable materials are agricultural products, both fiber and animal, that take 10 years or less to grow or raise and can be harvested in a sustainable fashion.

Reclaimed water is wastewater that has been treated for reuse.

Recycled content is the proportion, by mass, of preconsumer or postconsumer recycled material in a product.

Regenerative design is a system of technologies and strategies, based on an understanding of the inner working of ecosystems, that generates designs to reinforce rather than deplete underlying life-support systems and resources.

Regionally extracted materials are raw materials mined or harvested within a 500-mile radius of the project site.

Regionally manufactured materials are assembled as finished products within a 500-mile radius of the project site. Assembly does not include on-site assembly, erection, or installation of finished components.

Relative humidity is the ratio of partial density of airborne water vapor to the saturation density of water vapor at the same temperature and total pressure.

Remediation is the process of cleaning up a contaminated site by physical, chemical, or biological means. Remediation processes are typically applied to contaminated soil and groundwater.

Renewable energy is energy from sources that are not depleted by consumption. Examples include energy, from the sun, wind, (low-head) hydropower, geothermal energy, and wave and tidal systems.

Renewable Energy Certificates (RECs) are tradable commodities representing proof that a unit of electricity was generated from a renewable energy resource. RECs are sold separately from electricity itself and thus allow the purchase of the attributes of green power for a green building project.

Resource Conservation and Recovery Act (RCRA) addresses active and future facilities and was enacted in 1976 to give the EPA authority to control hazardous wastes from cradle to grave, including generation, transportation, treatment, storage, and disposal. Some nonhazardous wastes are also covered under RCRA.

Restorative design is a design approach that combines the restoration of polluted, degraded or damaged sites back to a state of acceptable health through human intervention with biophilic designs that reconnect people to nature.

Reuse returns materials to active use in the same or a related capacity as their original use, thus extending the lifetime of materials that would otherwise be discarded. Examples of construction materials that can be reused include extra insulation, drywall, and paints.

Reverberation is an acoustical phenomenon that occurs when sound persists in an enclosed space because of its repeated reflection or scattering on the enclosing surfaces or objects within the space.

Reverberation time (RT) is a measure of the amount of reverberation in a space and is equal to the time required for the level of a steady sound to decay by 60 dB after the sound has stopped. The decay rate depends on the amount of sound absorption in a room, the room geometry, and the frequency of the sound. RT is expressed in seconds.

Salvaged materials or reused materials are construction materials recovered from existing buildings or construction sites and reused. Common salvaged materials include structural beams and posts, flooring, doors, cabinetry, brick, and decorative items.

Service life is the expected lifetime of a building.

Set points are the operating targets for building energy systems and for indoor air quality.

Site energy is the amount of heat and electricity consumed by a building, as reflected in utility bills.

Soft costs are expense items that are not considered direct construction costs. Examples include architectural, engineering, financing, and legal fees.

Solar reflectance. *See* **Albedo.**

Solar Reflectance Index (SRI) is a measure of a material's ability to reject solar heat, as shown by a small temperature rise. Standard black (reflectance 0.05, emittance 0.90) is 0, and standard white (reflectance 0.80, emittance 0.90) is 100. For example, a standard black surface has a temperature rise of 90°F (50°C) in full sun, and a standard white surface has a temperature rise of 14.6°F (8.1°C). Once the maximum temperature rise of a given material has been computed, the SRI can be calculated by interpolating between the values for white and black. Materials with the highest SRI values are the coolest choices for paving.

Solar thermal systems collect or absorb sunlight via solar collectors to heat water that is then circulated to the building's hot water system. Solar thermal systems can be used to heat water for residential and commercial use or for heating swimming pool water.

Sound absorption is the portion of sound energy striking a surface that is not returned as sound energy.

Sound Absorption Coefficient describes the ability of a material to absorb sound, expressed as a fraction of incident sound. The Sound Absorption Coefficient is frequency-specific and ranges from 0.00 to 1.00. For example, a material may have an absorption coefficient of 0.50 at 250 Hz, and 0.80 at 1000 Hz. This indicates that the material absorbs 50 percent of incident sound at 250 Hz, and 80 percent of incident sound at 1000 Hz. The arithmetic average of absorption coefficients at midfrequencies is the Noise Reduction Coefficient.

Sound Transmission Class (STC) is a single-number rating for the acoustic attenuation of airborne sound passing through a partition or other building element, such as a wall, roof, or door, as measured in an acoustical testing laboratory according to accepted industry practice. A higher STC rating provides more sound attenuation through a building component.

Source energy is the total amount of raw fuel energy required to operate a building; it incorporates all transmission, delivery, and production losses for a complete assessment of a building's energy use.

Submetering is used to determine the proportion of energy use within a building attributable to specific end uses or subsystems (i.e., lighting or HVAC systems).

Supply air is delivered by mechanical or natural ventilation to a space and is composed of a combination of outdoor air and recirculated air.

Sustainable development is development that meets the needs of the present without compromising the ability of future generations to meet their own needs.

Sustainable forestry   is the practice of managing forest resources to meet the long-term forest product needs of humans while maintaining the biodiversity of forested landscapes. The primary goal is to restore, enhance, and sustain a full range of forest values, including economic, social, and ecological considerations.

Systems thinking   is a framework for understanding interrelationships in a system rather than individual components, and for understanding patterns of change rather than static "snapshots." It addresses phenomena in terms of wholeness rather than in terms of parts.

Tertiary treatment   is the highest form of wastewater treatment and includes removal of organics, solids, and nutrients as well as biological or chemical polishing.

Thermal comfort   exists when occupants express satisfaction with the thermal environment.

Thermal efficiency   is a measure of the efficiency of converting a fuel to energy and useful work. Useful work and energy output is divided by the higher heating value of input fuel.

Tipping fees   are charged by a landfill for disposal of waste, typically quoted per ton.

Total suspended solids (TSS)   are particles that are too small or light to be removed from stormwater via gravity settling. Suspended solid concentrations are typically removed via filtration.

U-value (thermal transmittance)   is the rate of heat transmission per unit time per unit area for an element of construction and its boundary air films.

Urea formaldehyde   is a combination of urea and formaldehyde used in some glues that may emit formaldehyde at room temperature.

Variable air volume (VAV)   system is an HVAC system that provides temperature control by varying the supply of conditioned air in different zones of the building according to its heating and cooling needs. The air supply temperature may be constant or varied.

Vegetated roof.   *See* **Green roof**.

Ventilation   is the process of supplying air to or removing air from a space for the purpose of controlling air containment levels, humidity, or temperature within the space.

Visible light transmittance (VLT)   is the ratio of total transmitted light to total incident light (i.e., the amount of visible spectrum, 380−780 nanometers, of light passing through a glazing surface divided by the amount of light striking the glazing surface). The higher the VLT value, the more incident light passes through the glazing. VLT is also abbreviated as Tvis.

Vision glazing   is the portion of an exterior window between 30 and 90 inches above the floor that permits a view to the outside.

Volatile organic compounds (VOCs)   are any one of several organic compounds that are released to the atmosphere by plants or through vaporization of oil products, and which are chemically reactive and involved in the chemistry of tropospheric ozone production.

Waste diversion   is a management activity that disposes of waste other than by incineration or the use of landfills. Examples include reuse and recycling.

Wastewater   is the spent or used water from a home, community, farm, or industry that contains dissolved or suspended matter.

Water, potable,   is water that meets or exceeds the EPA's drinking water quality standards and is approved for human consumption by the state or local authorities having jurisdiction; it may be supplied from wells or municipal water systems.

Watergy refers to the relationship between water and energy and can have two distinct meanings. The first is the amount of energy required per unit of water to extract, treat, and distribute a given water source, for example, groundwater, reclaimed water, or rainwater. In this same context, it is the energy required per unit of water to move, treat, and dispose of wastewater. The unit of measurement is in kilowatt-hours/1000 gallons or kilowatt-hours/cubic meter of water or wastewater. For example, for a 150-foot-deep groundwater well, the watergy is typically around 1.5 to 2 kWh/1000 gal (0.4 to 0.5 kWh/m^3), and for conventional wastewater treatment and disposal, 2 to 4 kWh/1000 gal (0.5 to 1.0 kWh/m^3) are required. A second meaning of watergy is the water needed to produce a unit of energy from a specific energy source. One kilowatt-hour of energy would require 56 gallons (212 liters) of water for its production if the source was a high hydroelectric dam. For a coal-fired power plant, each kilowatt-hour would require 0.51 gallons (2 liters) of water for its production.

Waterless urinals are dry plumbing fixtures that use advanced hydraulic design and a buoyant fluid to maintain sanitary conditions.

Weighted decibel (dBA) is a sound pressure level measured with a conventional frequency weighting that roughly approximates how the human ear hears different frequency components of sounds at typical listening levels for speech.

Wetlands are natural or constructed areas that are inundated or saturated by surface water or groundwater at a frequency and duration sufficient to support, and that under normal circumstances do support, a prevalence of vegetation typically adapted for life in saturated soil conditions. Wetlands generally include swamps, marshes, bogs, and similar areas.

Xeriscaping is a landscaping method that makes routine irrigation unnecessary. It uses drought-adaptable and low-water plants as well as soil amendments such as compost and mulches to reduce evaporation.

Index

Lovins, Amory, 51
Lovins, L. Hunter, 51
Low-e glass. *See* Low-emissivity glass
Low-emissivity (low-e) glass, 357
Lyle, John, 87–88

Maintenance and repair, quantifying costs, 474–475
Materials:
　bio-, 107–108
　biobased, 107–108
　biological, 107–108
　certified wood, 163–164
　emissions, 422–423
　intensity per unit service (MIPS), 48
　low-emitting, 166–167
　rapidly renewable, 163
　recycled content, 163
　regional, 163
　reuse, 163
Materials and Resources (MR), LEED category, 161–164
McDonough, William, 89, 102
McHarg, Ian, 86
MCS. *See* Multiple chemical sensitivity
MDF. *See* Medium-density fiberboard
Meadows, Dru, 355
Measurement and verification, 161, 250
Mechanical systems, active, 268–273
Metal stocks, depletion of, 59–60
MIPS. *See* Materials intensity per unit service
Montreal Protocol, 55
Motors, electric, 280
MR. *See* Materials and Resources
Multiple chemical sensitivity (MCS), 397
Mumford, Lewis, 86

Natural Capitalism, 106
Natural Step, The, 50Net zero energy, 5, 18
Net Zero Energy Buildings, 18
NZE, *See* Net zero energy
nZEB, *See* Net Zero Energy Buildings
Neutra, Richard, 84–85
NREL prototype buildings, 468–469
NREL Research Support Facility (RSF), 18–20

Odum, H.T., 111, 113
Oil rollover point, 6
Orr, David, 95–96
Our Common Future, 43
OWP-11, 70–72
Ozone depleting chemicals, 289
Ozone depletion, 289
Ozone protection, 289

Parking capacity, 155
Particleboard, 426
Passive design:
　cooling, 268–269
　massing, 252–253
　orientation, 252–253

shape, 252–253
strategy, 251
ventilation, 268
Passive survivability, 484–485
Performance goals, 490–491
PET. *See* Polyethylene terephthalate
Peterson, Gary, 100
PF. *See* Phenol formaldehyde
Photovoltaics:
　building integrated (BIPV), 285–286
　description, 285–286
PLA. *See* Polylactic acid
Plumbing fixtures, 321
Plumbing fixture control, 321
Plastics, 377–378
Plywood, 426–427
Polyethylene (PE), 107
Polyethylene terephthalate (PET), 377–378
Polylactic acid (PLA), 107
Polyvinyl chloride (PVC), 377–378
Pontiac Fever, 398
Precautionary Principle, 37–38
Prime farmland:
　definition, 218–219
　loss of, 218–219
Polluter Pays Principle and Producer Responsibility, 39–40
Project XX Office Building, 381–384
Protecting nature, 42–43
Protecting the rights of the nonhuman world, 40–41
Protecting the vulnerable, 40
PVC. *See* Polyvinyl chloride

Radiant cooling, 281–282
Rainwater:
　definition, 315
　harvesting, 315, 324–326
Reclaimed water, 316
Recyclables, storage and collection of, 162
Recycled content, 163
Recycling:
　organic route, 362
　technical route, 362
　thermodynamic limits, 104
Reed, Bill, 108–109
Rees, William, 47
Reflectance of roofing materials, 157
Regenerative design, 108–111
Regional Priority, LEED category, 173
REL. *See* Reference Exposure Level
Renewable energy, 284–287
Renewable energy system, 284–287
Request for Proposals (RFP), 197–198
Request for Qualifications (RFQ), 197–198
Reuse:
　building, 162
　materials, 363
Reversibility Principle, 39
RFP. *See* Request for Proposals
RFQ. *See* Request for Qualifications